Probability and Mathematical Statistics (Continued)

MUIRHEAD • Aspects of Multiva
OCHI • Applied Probability and S
  Physical Sciences
PRESS • Bayesian Statistics: Princi
PURI and SEN • Nonparametric M
PURI and SEN • Nonparametric M
PURI, VILAPLANA, and WERTZ
  Applied Statistics
RANDLES and WOLFE • Introduction to the Theory of Nonparametric
  Statistics
RAO • Asymptotic Theory of Statistical Inference
RAO • Linear Statistical Inference and Its Applications, *Second Edition*
RAO • Real and Stochastic Analysis
RAO and SEDRANSK • W.G. Cochran's Impact on Statistics
ROBERTSON, WRIGHT and DYKSTRA • Order Restricted Statistical
  Inference
ROGERS and WILLIAMS • Diffusions, Markov Processes, and
  Martingales, Volume II: Îto Calculus
ROHATGI • An Introduction to Probability Theory and Mathematical
  Statistics
ROHATGI • Statistical Inference
ROSS • Stochastic Processes
RUBINSTEIN • Simulation and The Monte Carlo Method
RUZSA and SZEKELY • Algebraic Probability Theory
SCHEFFE • The Analysis of Variance
SEBER • Linear Regression Analysis
SEBER • Multivariate Observations
SEBER and WILD • Nonlinear Regression
SEN • Sequential Nonparametrics: Invariance Principles and Statistical
  Inference
SERFLING • Approximation Theorems of Mathematical Statistics
SHORACK and WELLNER • Empirical Processes with Applications to
  Statistics
STOYANOV • Counterexamples in Probability

Applied Probability and Statistics

ABRAHAM and LEDOLTER • Statistical Methods for Forecasting
AGRESTI • Analysis of Ordinal Categorical Data
AICKIN • Linear Statistical Analysis of Discrete Data
ANDERSON and LOYNES • The Teaching of Practical Statistics
ANDERSON, AUQUIER, HAUCK, OAKES, VANDAELE, and
  WEISBERG • Statistical Methods for Comparative Studies
ARTHANARI and DODGE • Mathematical Programming in Statistics
ASMUSSEN • Applied Probability and Queues
BAILEY • The Elements of Stochastic Processes with Applications to the
  Natural Sciences
BARNETT • Interpreting Multivariate Data
BARNETT and LEWIS • Outliers in Statistical Data, *Second Edition*
BARTHOLOMEW • Stochastic Models for Social Processes, *Third Edition*
BARTHOLOMEW and FORBES • Statistical Techniques for Manpower
  Planning
BATES and WATTS • Nonlinear Regression Analysis and Its Applications
BECK and ARNOLD • Parameter Estimation in Engineering and Science
BELSLEY, KUH, and WELSCH • Regression Diagnostics: Identifying
  Influential Data and Sources of Collinearity
BHAT • Elements of Applied Stochastic Processes, *Second Edition*
BLOOMFIELD • Fourier Analysis of Time Series: An Introduction
BOLLEN • Structural Equations with Latent Variables
BOX • R. A. Fisher, The Life of a Scientist
BOX and DRAPER • Empirical Model-Building and Response Surfaces
BOX and DRAPER • Evolutionary Operation: A Statistical Method for
  Process Improvement

Applied Probability and Statistics (Continued)

BOX, HUNTER, and HUNTER • Statistics for Experimenters: An Introduction to Design, Data Analysis, and Model Building

BROWN and HOLLANDER • Statistics: A Biomedical Introduction

BUNKE and BUNKE • Nonlinear Regression, Functional Relations and Robust Methods: Statistical Methods of Model Building

BUNKE and BUNKE • Statistical Inference in Linear Models, Volume I

CHAMBERS • Computational Methods for Data Analysis

CHATTERJEE and HADI • Sensitivity Analysis in Linear Regression

CHATTERJEE and PRICE • Regression Analysis by Example

CHOW • Econometric Analysis by Control Methods

CLARKE and DISNEY • Probability and Random Processes: A First Course with Applications, *Second Edition*

COCHRAN • Sampling Techniques, *Third Edition*

COCHRAN and COX • Experimental Designs, *Second Edition*

CONOVER • Practical Nonparametric Statistics, *Second Edition*

CONOVER and IMAN • Introduction to Modern Business Statistics

CORNELL • Experiments with Mixtures: Designs, Models and The Analysis of Mixture Data

COX • A Handbook of Introductory Statistical Methods

COX • Planning of Experiments

DANIEL • Applications of Statistics to Industrial Experimentation

DANIEL • Biostatistics: A Foundation for Analysis in the Health Sciences, *Fourth Edition*

DANIEL and WOOD • Fitting Equations to Data: Computer Analysis of Multifactor Data, *Second Edition*

DAVID • Order Statistics, *Second Edition*

DAVISON • Multidimensional Scaling

DEGROOT, FIENBERG and KADANE • Statistics and the Law

DEMING • Sample Design in Business Research

DILLON and GOLDSTEIN • Multivariate Analysis: Methods and Applications

DODGE • Analysis of Experiments with Missing Data

DODGE and ROMIG • Sampling Inspection Tables, *Second Edition*

DOWDY and WEARDEN • Statistics for Research

DRAPER and SMITH • Applied Regression Analysis, *Second Edition*

DUNN • Basic Statistics: A Primer for the Biomedical Sciences, *Second Edition*

DUNN and CLARK • Applied Statistics: Analysis of Variance and Regression, *Second Edition*

ELANDT-JOHNSON and JOHNSON • Survival Models and Data Analysis

FLEISS • The Design and Analysis of Clinical Experiments

FLEISS • Statistical Methods for Rates and Proportions, *Second Edition*

FLURY • Common Principal Components and Related Multivariate Models

FOX • Linear Statistical Models and Related Methods

FRANKEN, KÖNIG, ARNDT, and SCHMIDT • Queues and Point Processes

GALLANT • Nonlinear Statistical Models

GIBBONS, OLKIN, and SOBEL • Selecting and Ordering Populations: A New Statistical Methodology

GNANADESIKAN • Methods for Statistical Data Analysis of Multivariate Observations

GREENBERG and WEBSTER • Advanced Econometrics: A Bridge to the Literature

GROSS and HARRIS • Fundamentals of Queueing Theory, *Second Edition*

GROVES • Survey Errors and Survey Costs

GROVES, BIEMER, LYBERG, MASSEY, NICHOLLS, and WAKSBERG • Telephone Survey Methodology

GUPTA and PANCHAPAKESAN • Multiple Decision Procedures: Theory and Methodology of Selecting and Ranking Populations

GUTTMAN, WILKS, and HUNTER • Introductory Engineering Statistics, *Third Edition*

HAHN and SHAPIRO • Statistical Models in Engineering

Panel Surveys

Panel Surveys

Editors

Dr. DANIEL KASPRZYK
Bureau of the Census
Population Division
Washington, D.C.

Dr. GREG DUNCAN
Survey Research Center
Institute for Social Research
University of Michigan
Ann Arbor, Michigan

Dr. GRAHAM KALTON
Survey Research Center
Institute for Social Research
University of Michigan
Ann Arbor, Michigan

Dr. M. P. SINGH
Statistics Canada
Methodology Branch
Ottawa, Ontario

WILEY

A Wiley-Interscience Publication

JOHN WILEY & SONS

New York · Chichester · Brisbane · Toronto · Singapore

Library of Congress Cataloging in Publication Data

Panel surveys / editors, Daniel Kasprzyk...[et al.].
     p.   cm. – (Wiley series in probability and mathematical
statistics. Applied probability and statistics section)
    "A Wiley-Interscience publication."
    Includes bibliographies and index.
    ISBN 0-471-62592-2
    1. Sampling (Statistics)   I. Kasprzyk, Daniel.   II. Series.

QA276.6.P36   1989
001.4'222 – dc20   89-33697
                                         CIP

Printed in the United States

10 9 8 7 6 5 4 3 2 1

Contributors

Barbara A. Bailar
American Statistical Association
Alexandria, Virginia

Richard D. Burgess
Statistics Canada
Ottawa, Ontario, Canada

David Cantor
Westat, Inc.
Rockville, Maryland

Michael J. Colledge
Statistics Canada
Ottawa, Ontario, Canada

Larry S. Corder
Center for Demographic Studies
Duke University
Durham, North Carolina

Martin David
Department of Economics
University of Wisconsin
Madison, Wisconsin

Pat Doyle
Mathematica Policy Research, Inc.
Washington, D.C.

Greg J. Duncan
Survey Research Center
The University of Michigan
Ann Arbor, Michigan

Lawrence R. Ernst
U.S. Bureau of the Census
Washington, D.C.

Robert E. Fay
U.S. Bureau of the Census
Washington, D.C.

Stephen E. Fienberg
College of Humanities
    and Social Sciences
Carnegie-Mellon University
Pittsburgh, Pennsylvania

Ralph Folsom
Research Triangle Institute
Research Triangle Park, North Carolina

Martin R. Frankel
Bernard Baruch College
City University of New York
New York, New York

Wayne A. Fuller
Department of Statistics
Iowa State University
Ames, Iowa

James J. Heckman
Department of Economics
Yale University
New Haven, Connecticut

Jan M. Hoem
Institute for Social Research
University of Stockholm
Stockholm, Sweden

D. Holt
Department of Social Statistics
University of Southampton
Southampton, England

Daniel G. Horvitz
Research Triangle Institute
Research Triangle Park, North Carolina

Curtis A. Jacobs
U.S. Bureau of Labor Statistics
Washington, D.C.

Graham Kalton
Department of Biostatistics
School of Public Health
The University of Michigan
Ann Arbor, Michigan

Daniel Kasprzyk
U.S. Bureau of the Census
Washington, D.C.

Lisa LaVange
Research Triangle Institute
Research Triangle Park, North Carolina

James M. Lepkowski
Survey Research Center
The University of Michigan
Ann Arbor, Michigan

Denise Lievesley
International Statistical Institute
Voorburg, The Netherlands

Lee A. Lillard
The Rand Corporation
Santa Monica, California

Roderick J. A. Little
Department of Biomathematics
School of Medicine
University of California–Los Angeles
Los Angeles, California

David B. McMillen
U.S. Bureau of the Census
Washington, D.C.

Colm O'Muircheartaigh
Social and Community Planning
Research
London, England

Stanley Presser
Department of Sociology
University of Maryland
College Park, Maryland

B. Quenneville
Statistics Canada
Ottawa, Ontario, Canada

J. N. K. Rao
Department of Mathematics
and Statistics
Carleton University
Ottawa, Ontario, Canada

Richard Robb
Chicago Corporation
Chicago, Illinois

Willard L. Rodgers
Survey Research Center
The University of Michigan
Ann Arbor, Michigan

Fritz Scheuren
Internal Revenue Service
Washington, D.C.

Marita Servais
Survey Research Center
The University of Michigan
Ann Arbor, Michigan

Adriana R. Silberstein
U.S. Bureau of Labor Statistics
Washington, D.C.

M. P. Singh
Statistics Canada
Ottawa, Ontario, Canada

Peter Solenberger
Survey Research Center
The University of Michigan
Ann Arbor, Michigan

Gary Solon
Department of Economics
The University of Michigan
Ann Arbor, Michigan

K. P. Srinath
Statistics Canada
Ottawa, Ontario, Canada

Hong-Lin Su
Immunobiology Research Institute
Annandale, New Jersey

Jennifer Waterton
Post Office Headquarters
London, England

Rick L. Williams
Research Triangle Institute
Research Triangle Park, North Carolina

Halliman H. Winsborough
Center for Demography and Ecology
University of Wisconsin
Madison, Wisconsin

Contents

Preface   **xi**

Part One: Issues in the Design of Panel Surveys

Information Needs, Surveys, and Measurement Errors
Barbara A. Bailar   **1**

Part Two: Collection and Design Issues

Substantive Implications of Selected Operational
Longitudinal Design Features: The National Crime Survey
as a Case Study
David Cantor   **25**

Major Issues and Implications of Tracing Survey
Respondents
Richard D. Burgess   **52**

Collection and Design Issues: Discussion
Stanley Presser   **75**

Part Three: Statistical Design and Estimation

Coverage and Classification Maintenance Issues in
Economic Surveys
Michael J. Colledge   **80**

A Probability Sampling Perspective on Panel Data Analysis
Ralph Folsom, Lisa LaVange, and Rick L. Williams   **108**

Weighting Issues for Longitudinal Household and
Family Estimates
Lawrence R. Ernst   **139**

Statistical Design and Estimation: Discussion
Martin R. Frankel   **160**

Part Four: Data Base Management

Data Base Strategies for Panel Surveys
Pat Doyle   163

Data Base Management Approaches to Household
Panel Studies
Peter Solenberger, Marita Servais, and Greg J. Duncan   190

Managing Panel Data for Scientific Analysis: The Role of
Relational Data Base Management Systems
Martin David   226

Data Base Management: Discussion
Halliman H. Winsborough   242

Part Five: Sources of Nonsampling Error

Nonsampling Errors in Panel Surveys
Graham Kalton, Daniel Kasprzyk, and David B. McMillen 249

Sources of Nonsampling Error: Discussion
Colm O'Muircheartaigh   271

Part Six: Panel Conditioning

Symptoms of Repeated Interview Effects in the Consumer
Expenditure Interview Survey
Adriana R. Silberstein and Curtis A. Jacobs   289

Panel Effects in the National Medical Care Utilization and
Expenditure Survey
Larry S. Corder and Daniel G. Horvitz   304

Evidence of Conditioning Effects in the British Social
Attitudes Panel Survey
Jennifer Waterton and Denise Lievesley   319

Panel Conditioning: Discussion
D. Holt   340

Part Seven: Nonresponse Adjustments

Treatment of Wave Nonresponse in Panel Surveys
  James M. Lepkowski   348

Estimating Nonignorable Nonresponse in Longitudinal
Surveys through Causal Modeling
  Robert E. Fay   375

Item Nonresponse in Panel Surveys
  Roderick J. A. Little and Hong-Lin Su   400

Nonresponse Adjustments: Discussion
  Fritz Scheuren   426

Part Eight: Estimation of Cross-Sectional and Change Parameters

Comparisons of Alternative Approaches to the Estimation
of Simple Causal Models from Panel Data
  Willard L. Rodgers   432

Estimation of Level and Change Using Current
Preliminary Data
  J. N. K. Rao, K. P. Srinath, and B. Quenneville   457

Estimation of Cross-Sectional and Change Parameters:
Discussion
  Wayne A. Fuller   480

Part Nine: Modeling Considerations

The Value of Panel Data in Economic Research
  Gary Solon   486

Sample Dynamics: Some Behavioral Issues
  Lee A. Lillard   497

The Value of Longitudinal Data for Solving the Problem of Selection Bias in Evaluating the Impact of Treatments on Outcomes
James J. Heckman and Richard Robb   512

The Issue of Weights in Panel Surveys of Individual Behavior
Jan M. Hoem   539

Modeling Considerations: Discussion from a Modeling Perspective
Stephen E. Fienberg   566

Modeling Considerations: Discussion from a Survey Sampling Perspective
Graham Kalton   575

Index   587

Preface

Panel surveys—surveys in which similar measurements are made on the same sample at different points in time—attracted increasing attention in both the United States and Europe in the mid-1980s. In 1984, researchers from Australia, West Germany, Sweden, and the United States discussed uses and technical problems at the annual meeting of the American Statistical Association. Earlier in 1984, a conference in West Berlin reviewed specific panel surveys, with particular attention to their designs, implementation problems, and analytic uses. At approximately the same time, the United States Federal Committee on Statistical Methodology created a Subcommittee on Federal Longitudinal Surveys to review design, implementation, and analytic issues in longitudinal studies conducted by the United States government.

In 1985, the Social Science Research Council appointed a Working Group on the Comparative Evaluation of Longitudinal Surveys to develop criteria for evaluating the value of such surveys. At the same time, the Sloan Foundation awarded a grant to help promote cross-national research using panel data, encouraged by the fact that seven Western countries had launched large national sample surveys. In 1985, the Centenary Session of the International Statistical Institute featured a session of invited research papers focused on panel survey design and analysis.

At that time, Graham Kalton, then chair of the Section on Survey Research Methods of the American Statistical Association, was very interested in developing a series of single-topic symposia of interest to the section membership. These symposia would serve as complements to the diverse agenda of topics at the annual meetings of the American Statistical Association. Single-topic symposia offer the distinct advantage of bringing together the spectrum of current research on the topic. Central to the successful development of such symposia, aimed at providing a comprehensive treatment of a given subject, is the requirement that these state-of-the-art research results be made available subsequently to the research community in published form.

The high level of interest in panel surveys provided a good opportunity to see whether this idea was viable. Kalton asked Daniel Kasprzyk of the U.S. Bureau of the Census to present an outline of the topics to be covered in a symposium on panel surveys at the business meeting of the Section on Survey Research Methods in August 1985. The draft outline had two features: (1) it attempted to provide a comprehensive review of topics related to panel surveys, beginning with design issues, followed by issues in data collection, through to issues in the analysis of panel data, and (2) it acknowledged that the participants as well as the audience would be multidisciplinary—a mixture of survey statisticians, sociologists, demographers, mathematical statisticians, and economists. With the approval of the section, planning began in earnest for the International Symposium on Panel Surveys.

An organizing committee was formed consisting of Daniel Kasprzyk, U.S. Bureau of the Census, as Chair; Greg Duncan, The University of Michigan; and M. P. Singh, Statistics Canada. This committee, in consultation with Kalton, contacted researchers working with panel surveys and developed a program of invited papers for the symposium. The Section on Survey Research Methods of the American Statistical Association was the principal sponsor for the symposium. As such, staff from the American Statistical Association assisted the organizing committee in the program management and logistical arrangements. The International Association of Survey Statisticians (IASS) collaborated by drawing the symposium to the attention of its international membership.

Kasprzyk and Kalton contacted Beatrice Shube at John Wiley and Sons about publishing a book on issues related to panel surveys. She was both enthusiastic and supportive. Arrangements were made with Wiley to publish the research papers invited by the organizing committee.

As a result of these activities, the International Symposium on Panel Surveys was held on November 20–22, 1986, in Washington, D.C. Approximately 300 researchers attended. The program consisted of 25 papers invited by the organizing committee as well as 20 contributed papers.

This volume contains 22 invited papers presented at the symposium. All papers were reviewed by at least one editor and at least two independent referees and revised in light of the comments received. The volume follows the organization of the symposium, each section treating a general issue in the design and analysis of panel surveys:

Issues in the Design of Panel Surveys
Collection and Design Issues
Statistical Design and Estimation
Data Base Management
Sources of Nonsampling Error
Panel Conditioning
Nonresponse Adjustments
Estimation of Cross-Sectional and Change Parameters
Modeling Considerations

The chapters within each section are of three types—providing a general review of a topic related to panel surveys, presenting results related to methodological issues common to panel surveys, or presenting current research on panel survey problems. Formal discussions of the chapters in each section by invited discussants appear at the end of each section.

The symposium organizing committee is grateful for the efforts of many people. Ede Denenberg and Mary Esther Barnes of the American Statistical Association handled the logistical arrangements and publicity for the symposium with skill and dedication. Jill Stormer of the American Statistical Association very ably helped the organizing committee prepare the grant proposal to the National Science Foundation. Hazel Beaton of the U.S. Bureau of the Census answered numerous requests for information and provided excellent assistance with organizational matters. Our European colleagues—D. Holt (University of Southampton and IASS Scientific Secretary), Lars Lyberg (Statistics Sweden), Ute Hanefeld (Deutsches Institut fur Wirtschaftsforschung), and Martin Collins and Denise Lievesley (Social and Community Planning Research)—were very helpful in publicizing the symposium in Europe.

Financial support for the International Symposium on Panel Surveys and the preparation of this book came from the National Science Foundation, the U.S. Bureau of the Census, Statistics Canada, the National Agricultural Statistics Service, the National Center for Education Statistics, the National Center for Health Statistics, the U.S. Bureau of Labor Statistics, the National Institute on Aging, Westat, the National Opinion Research Center (NORC), the Research Triangle Institute, and the Survey Research Center of The University of Michigan. We gratefully appreciate their support, which was critical to the success of the venture.

The volume was prepared as camera-ready pages at The University of Michigan, using the TEXTEDIT program on the university's mainframe computing system. Katherine Metcalf undertook the massive task of entering the original drafts and handling initial revisions on TEXTEDIT. Jane Stanton handled subsequent revisions and expertly formatted and produced the final camera-ready pages. Sonya Kennedy organized and copyedited the material for the volume and supervised the production of the camera-ready pages. We thank all for their consummate professionalism and dedication throughout a complex and demanding process.

We are grateful to our colleagues who served as session chairs at the symposium: Gordon Brackstone (Statistics Canada), Dwight Brock (National Institute on Aging), Constance Citro (National Academy of Sciences), Charles Cowan (National Center for Educational Statistics), Cathryn Dippo (U.S. Bureau of Labor Statistics), Maria Gonzalez (Office of Management and Budget), D. Holt (University of Southampton), Daniel Horvitz (Research Triangle Institute), Graham Kalton (The

University of Michigan), Elizabeth Martin (U.S. Bureau of the Census), Edward Schillmoeller (A. C. Nielsen Company), Monroe Sirken (National Center for Health Statistics), and Robert Tortora (National Agricultural Statistics Service).

Finally, we gratefully acknowledge the contributions of the anonymous reviewers of each chapter. Their reviews improved the presentation and quality of the material.

Daniel Kasprzyk
Greg J. Duncan
Graham Kalton
M. P. Singh

Information Needs, Surveys, and Measurement Errors

Barbara A. Bailar

---

1. INTRODUCTION

By all measurable signs, the survey industry is growing. There seems to be a consensus that information needs can be satisfied by asking people questions. That consensus rests on the premises that we know which people to ask, that people will respond, and respond accurately, and that we have asked the correct questions. These premises are sometimes not examined very carefully.

Assuming that legitimate information needs exist and that a survey seems the best way to answer them, how do we decide what kind of a survey to do? Duncan and Kalton (1987) reviewed this question, listing a variety of types of estimates and surveys. Consider first the different kinds of information needs:

- Estimates of characteristics, activity, behavior, or attitudes for one point in time
- Estimates of net change between two or more time periods
- Estimates of gross change between two or more time periods
- Estimates of trends based on several time periods
- Estimates of durations, transitions, or frequency of occurrence for specific kinds of events for specific groups of people
- Estimates of characteristics based on cumulating data over time
- Estimates of rare events based on cumulating data over time
- Estimates of relationships among characteristics.

Given these needs, what kinds of surveys can satisfy them? First, I present terminology that will be used throughout this paper. There is a reason for trying to standardize our terminology. Longitudinal surveys have certain standard features, but beyond that contain many variations. The differences among survey types are related not only to design and population characteristics but also to efforts to estimate and understand measurement error.

Single-time surveys: These surveys are designed to produce estimates for a single point in time. However, because many of the questions are common to other surveys, they may be used to make estimates of net change. Such estimates confound real change in the population with change in survey methodology. An example of a single-time survey is

1

the Survey on Philanthropy, conducted by the Census Bureau in 1975 (Morgan et al., 1979).

Repeated surveys, no overlap of units: These surveys are often called periodic or recurring surveys. A survey organization repeats a survey on a given topic at regularly scheduled time points. There is nothing in the survey design that requires an overlap of sample units at different points in time. An example of a survey of this type is the National Health Interview Survey (National Center for Health Statistics, 1985).

Repeated surveys, partial overlap: These surveys are scheduled at regular intervals and include rotating panels; that is, people are inducted into the survey, surveyed a number of times, and then rotated out of the survey. The main purpose for the overlap is variance reduction. There is no attempt to follow people in sample units that move or to link records for identical units to make longitudinal estimates. An example of a survey of this type is the Current Population Survey (U.S. Bureau of the Census, 1978).

Longitudinal surveys, no rotation: These surveys are designed to follow a particular group of units over time, to create a longitudinal record for each observational unit, and to analyze the data based on the microrecords for units. An example of a survey of this type is High School and Beyond, a cohort study of 1980 high school sophomores and seniors (National Center for Education Statistics, 1981).

Longitudinal surveys, with rotation: These surveys are designed to follow a particular group of units for a specified period of time, to introduce new sample units at specified periods, to create longitudinal records for each observational unit, and to include longitudinal analysis. An example of a survey of this type is the Survey of Income and Program Participation (David, 1985; Nelson et al., 1985). Several other surveys that are now repeated surveys with partial overlap, such as the Current Population Survey or the National Crime Survey, would be in this category if the sample people were followed when they moved, if the microrecords were assembled, and if longitudinal analysis were included as part of the estimation plan.

Although we tend to think of these concepts mainly in terms of households and persons, they also apply to businesses. For example, the Monthly Retail Trade Survey is a repeated survey with partial overlap of units (U.S. Bureau of the Census, 1987). In fact, it is more complicated than that because it always includes the very largest businesses and rotates only the smaller businesses. The Survey of Manufacturers' Shipments, Inventories, and Orders is a longitudinal survey with no rotation (U.S. Bureau of the Census, 1986).

These types of surveys answer different information needs. Table 1 shows these five types of surveys by the kinds of estimates that they can produce. Notice that there are some things that each of the five survey types can produce: estimates for a single point in time; es-

TABLE 1. SURVEY TYPES BY KINDS OF ESTIMATES THAT CAN BE PRODUCED

Kind of Estimate	Type of Survey				
	Single Time	Repeated, No Overlap	Repeated, Partial Overlap	Longitudinal, No Rotation	Longitudinal, with Rotation
For one point in time	X	X	X	X	X
Durations, transitions, frequency of occurrence	X	X	X	X	X
Relationships among characteristics	X	X	X	X	X
Net change		X	X	X	X
Trends		X	X	X	X
Rare events— cumulated data		X	X		X
Gross change			X	X	X
Characteristics for longer time periods based on cumulated data				X	X

timates for durations, transitions, and frequency of occurrence; and estimates on relationships among characteristics.

Some people object to having durations, transitions, and so forth listed under single-time surveys, but these kinds of questions are frequently asked on single-time surveys. However, one hopes that a strength of repeated or longitudinal surveys is the ability to obtain more accurate data on these items through shorter recall periods.

In addition to producing estimates to satisfy different information needs, these types of surveys also provide different kinds of information on measurement error. Although there are many varieties of measurement error, I will concentrate on the following three: coverage and nonresponse, recall, and time-in-sample, defining these terms as follows:

Coverage and nonresponse error: The error that results from an incomplete accounting of the population of interest, either through a faulty frame due to lack of cooperation of the public or because of interviewer error.

Recall error: The error that results when people cannot remember whether or when events of a given type occurred, resulting in omitting events, incorrectly placing events in time, or reporting events that never took place.

Time-in-sample error: The error that results because the expected value of a person's response to a survey question is different depending on how many times he/she has responded to the survey.

Each of these is a serious problem and affects single-time surveys as well as repeated and longitudinal surveys. In the next sections of this paper, I will describe how each of these errors affects the various types of surveys, what we know about the problem, some techniques we use to cope with the problem, and what we need to know to design better surveys. I conclude with the need for a careful assessment about the trade-off in asking more questions and responding to new information needs, rather than responding to methodological needs.

2. COVERAGE AND NONRESPONSE

Coverage and nonresponse are serious considerations for all surveys. Coverage is related to the frames used to define the sample population. Nonresponse is related to failure to respond by persons selected into the sample. However, coverage and nonresponse are related when the frames used are compiled during the survey, and that happens in most household surveys. The sampling unit may well be a housing unit selected from a list of all housing units; yet the target population consists of all individuals within each housing unit. The roster of household members is frequently compiled during the interview at a household. Leaving someone off the roster is a form of nonresponse that is similar to item nonresponse but results in coverage error. This is the reason for reduced coverage in household surveys such as the Current Population Survey and the Survey of Income and Program Participation. Most undercoverage is related to within-household coverage problems rather than to missing whole housing units.

Nonresponse problems also occur quite separately from coverage problems. Housing units known to be in a sample never respond, or persons refuse to supply information for specific data items, although they may answer others. In this paper, the focus is on household or person nonresponse, rather than item nonresponse, because household or person nonresponse act together with undercoverage to cause problems for surveys, particularly longitudinal surveys.

Target populations must be defined for all surveys. In the National Longitudinal Study of the High School Class of 1972, the target population was all members of the senior class in U.S. schools, including private as well as public schools (Office of Management and Budget, 1986). In the Current Population Survey (CPS), the target population is all persons living in the United States, excluding institutions and individuals living in armed forces barracks. To the extent that we cannot provide a complete frame for the target population, we have coverage error. For example, in the case of high school seniors, a natural sampling unit is high schools with senior classes; however, if a good frame for all high schools does not exist, one often samples housing units and takes all or some of the people within. To the extent that an accurate list of housing units or an accurate list of those within housing units does not exist, coverage errors result.

The following data from the CPS, accumulated over the 12 months of 1985, provide an illustration of the coverage problem. Unit nonresponse is not a contributor to this coverage problem, since nonresponse adjustments are already incorporated in the survey data. Also, although CPS is a repeating survey, the coverage problem is not related to frequency of interview. Table 2 shows the ratio of the sample estimates of the population based on the CPS after nonresponse adjustment to independently derived estimates of the population brought forward from the 1980 census. The final estimates from the CPS reflect adjustments made for coverage bias, but, of course, assume that the characteristics of those missed are similar to the characteristics of those interviewed.

These data should give survey practitioners cause for concern. Population coverage for the black population is a serious problem. Contrary to popular belief, the problem exists not only for young black males but also for black males at ages 50–61. These data also show that, in general, coverage of males is worse than of females, and coverage of blacks worse than of whites. Coverage of other races is at least as good as whites, if not better.

Coverage errors affecting all surveys are exacerbated by nonresponse. And although nonresponse certainly exists in single-time surveys, the cumulative effect of nonresponse in repeated and longitudinal surveys is obviously greater when the same respondents are visited at intervals.

Unit nonresponse exacerbates the coverage problem since, in general, response rates are lower for blacks than for whites, and for males than for females, just as coverage rates were lower.

In longitudinal surveys and in repeated surveys with an overlap, nonresponse is usually a larger problem than in a single, cross-sectional survey, because one starts with a basic coverage problem, then has initial nonresponse, followed by additional nonresponse with each successive visit. Even though many longitudinal surveys have admirable success in attaining over 95% response at each reporting period, attrition does grow. It is possible to find longitudinal surveys that now survey around 50% of the original sample. (For more discussion of this topic, see the paper by Kalton et al. in this volume.)

In longitudinal surveys, nonresponse arises from the usual sources of people not being home or refusing, and then from the added source of

TABLE 2. RATIO OF SAMPLE ESTIMATES OF POPULATION FROM THE CURRENT POPULATION SURVEY TO INDEPENDENTLY ESTIMATED TOTALS: 1985

Age Groups	White		Black		Other Races	
	Male	Female	Male	Female	Male	Female
14–15	.9656	.9610	.9056	.9878	1.0556	.9239
16–17	.9651	.9496	.9187	.8939	.9467	.9876
18–19	.9243	.9167	.8575	.8389	.8924	.9772
20–21	.8961	.9208	.8025	.8503	.7761	.8704
22–24	.9144	.9340	.7350	.9059	.9067	.9404
25–29	.9149	.9316	.7990	.8917	.9882	1.0261
30–34	.9123	.9427	.7875	.9100	1.0071	1.0511
35–39	.9506	.9679	.8755	.9249	1.1036	1.0346
40–44	.9294	.9343	.8208	.8756	.9992	1.0062
45–49	.9208	.9456	.9079	.9080	.9964	.9895
50–54	.9406	.9727	.8071	.9228	.8846	.9211
55–59	.9639	.9428	.7917	.9022	.9880	.9380
60–61	.9445	.9846	.7600	.9493	1.0409	.7264
62–64	.9498	.9553	.9223	.9020	.7594	.9566
65–69	.9385	.9445	.9329	.8696	1.0173	.9102
70–74	.9518	.9821	1.0082	1.0189	.7056	.6554
75 and over	.9603	.9733	.9084	.9363	.6811	.7730
14 and over	.9347	.9492	.8408	.9092	.9576	.9698

Source: Current Population Survey, 1985.

not being able to locate the sample respondents in each wave of interviewing. The Survey of Income and Program Participation (SIPP) is a relatively new survey, conducted by the Census Bureau, and having eight interviews (except for the first panel, which had nine). Nonresponse rates for the survey are computed so that each of the sources of nonresponse can be examined. Type A rates refer to nonresponse caused by the inability to collect data at an eligible housing unit. Type D rates refer to nonresponse caused by an eligible household moving to an undetermined location or to a location more than 100 miles away from a location where an interviewer is available. Together, these rates represent the overall sample loss. They are not additive because of an adjustment made for unobserved growth in wave 1 Type A noninterviewed units. Table 3 shows the sample loss and its components for the 1984 and 1985 SIPP panels. Note that although the 1985 panel started out with somewhat higher rates of nonresponse, the final sample loss was slightly lower than the 1984 panel.

Many users of data would agree that these are fairly low attrition rates. The issue, however, is not whether these are similar to or better than other longitudinal surveys. Rather, the issue is whether attrition and the combined effect of attrition and undercoverage result in biased survey estimates.

TABLE 3. SAMPLE LOSS IN THE SURVEY OF INCOME AND
PROGRAM PARTICIPATION, 1984–1985 PANELS

Wave	Type A Rate		Type D Rate		Sample Loss	
	1984	1985	1984	1985	1984	1985
1	4.9	6.7	NA	NA	4.9	6.7
2	8.3	8.5	1.0	2.1	9.4	10.8
3	10.2	10.2	1.9	2.7	12.3	13.3
4	12.1	12.4	2.9	3.4	15.4	16.3
5	13.4	14.0	3.5	4.1	17.4	18.8
6	14.9	14.2	4.1	4.8	19.4	19.7
7	15.6	14.4	4.9	5.2	21.0	20.5
8	15.7	14.4	5.7	5.5	22.0	20.8
9	15.8	–	5.8	–	22.3	–

Sources: Nelson, D., Bowie, C. E., & Walker, A. (1987); and Bowie,
C. E. (1987).

How does nonresponse vary over the kinds of surveys? Single-time
and repeated surveys with no overlap probably do best, since they have
no attrition problem. Repeated surveys with partial overlap and longitudinal surveys with rotating panels with few visits probably are next
best. A longitudinal survey with no rotation usually has the largest
nonresponse problems. In the *New York Times* of November 16, 1986,
an editorial discussing the 1986 election contained a statement that
may apply equally well to longitudinal surveys: "Captivated by the
universe we can measure, we often slip into thinking it is the whole
universe."

These problems are not new. However, in longitudinal surveys, attrition adds a new dimension to the situation. Survey practitioners
work hard to develop techniques to mitigate the effect of these
problems. The solutions take the following form:

- Develop methods to improve coverage of whole housing units in
  sample
- Do occasional coverage checks for nonresponse classification er-
  rors
- Make intensive efforts to convert refusals
- Weight the data in various ways to reduce bias
- Make intensive efforts to follow individuals in longitudinal sur-
  veys.

Because so much time and effort go into these possible solutions,
evaluations of their effects are important to guide us to the best use of
resources. An examination of these evaluations gives fragmentary
results. It is important to improve the coverage of housing units, yet
the experience of the Census Bureau in the decennial census is that

over half of the coverage problem is with people who live in interviewed households. There seems to be a significant group of people who live in housing units that are visited and provide information but who are not reported as living there. Interviewers have said that they have seen these people, obviously very much at home, but reported as friends, neighbors, or relatives who live elsewhere. On a CPS interview, I once saw such a case where all the data were provided by a man who said he did not live at the household but was only there to look after the children. After completing the interview, the interviewer said that this same man had also provided the information for two previous interviews and also answered the telephone after dinner. It is obvious that some people do not want outsiders to know where they reside. It is also obvious that questions on "Who lives here?" do not take into account the ways in which many people live—for example, people with multiple residences. The participant-observer study by the Valentines (1971), sponsored by the Census Bureau in 1968–70, showed that many of the households that looked like female-headed households in a rather poor area of a large city actually had males associated with those units. Those males contributed to the economic sufficiency of the household and to the support of children, but they were not enumerated in the Census Bureau's surveys. Thus, efforts to improve whole household coverage, although useful, will not solve the problem. Efforts to improve within-household coverage are needed. This is particularly important for longitudinal surveys. People who are not enumerated in a longitudinal survey and consequently are not followed undoubtedly have different patterns of employment, income, illness, education, and expenditures from those interviewed.

Survey practitioners also do occasional coverage checks of housing units to be sure that noninterviews are classified correctly. Interviewers have the responsibility of classifying noninterview housing units as Types A, B, or C. Type A noninterviews consist of households occupied by persons eligible for interview. Type B noninterviews consist of units that are vacant or occupied by persons ineligible for interviews. Type C noninterviews are those units ineligible for the sample, such as demolished units, conversions to business, etc. Type A noninterviews are interviewer errors, and a high Type A rate for an interviewer results in remedial action. Type B and C noninterviews are not viewed as errors, and therefore interviewers may have some incentive to misclassify units.

In October 1966 and June 1967, an intensive coverage check was done in the CPS reinterview, with the emphasis totally on coverage. The 1.4% of missed persons found in October was more than three times the average of 0.4% for the preceding six months. For the part of the sample that was as area sample, the gains were usually whole households; for the list sample, where addresses were provided, the gains were mostly within households. The test also showed that the October reinterview had much higher detection of incorrect classification of noninterviews as Type B or Type C (U.S. Bureau of the Census, 1968). It is undoubtedly time to repeat this intensive reinterview to measure the potential effects on longitudinal surveys. Type B and C noninterviews are not necessarily visited at later interviews in a longitudinal survey, so a large number of misclassifications could lead to substantial coverage problems.

Because refusal rates are used as a measure of the quality of a survey, the field staff tries very hard to hold refusals to a minimum. When a refusal occurs, a more senior interviewer often tries to convert the refusal and is often successful. Nelson et al. (1987) report that a considerable effort is made in SIPP to convert refusals. About 30% of the refusal households in SIPP are converted. Often the respondents state that they became convinced of the benefits of the survey. However, there has as yet been no evaluation of the quality of responses from converted refusals.

There is some evidence, albeit anecdotal, that the data resulting from these conversions are of poor quality. In a recent meeting, the Senior Field Representatives, who work on many of the Census Bureau surveys and who counsel the interviewers, reported that these interviews are very fast, there is no probing, and the interviewers "tiptoe" through the interview. There was a general feeling that the quality of information collected was poor.

Other, more empirical evidence about the quality of converted refusal cases is contained in Sirken and Brown (1962). In a mail survey conducted in 1960, demographic information was collected for a sample of people who had recently died and a measure of the adequacy of the data reported was devised. The measure was based on the completeness and internal consistency of the information reported on the mail questionnaire. The study showed that the quality of data was not independent of the number of mailings it took to elicit the information. Informants who replied to the first mailing were most likely to report adequate data and those on the third mailing reported the least accurate data.

Statisticians weight data in various ways in an attempt to reduce coverage and nonresponse bias. There are many studies available to show that weighting can attenuate the effect of this bias (Chapman et al., 1986). The National Longitudinal Surveys of Labor Market Experience provide an interesting example. The original four cohorts were interviewed in 1966. A study done 15 years later concluded that those still being interviewed were not significantly different from those who had dropped out (Rhoton, 1986). Every five years, distributions of characteristics are compared to national estimates. Weights are adjusted accordingly, getting an adjustment that reflects noninterviews. Such weighting is done routinely in longitudinal surveys. For the characteristics used in the weighting procedure, the method has utility. For other characteristics, the effects are not known. This is a particularly important issue for longitudinal surveys, since a fundamental question exists concerning the use of weighting adjustments for wave nonresponse or the use of responses from previous reports (Kalton, 1986; see also Lepkowski's contribution to this volume).

Currently, much emphasis is placed on developing adjustments for nonresponse and coverage. Perhaps new approaches to increasing within-household coverage are necessary. Are survey practitioners really justified in assuming that people who are missed or who do not respond are like those who are respondents? More effort on reducing nonresponse misclassification errors will be useful since that will increase coverage. A serious question, which requires further research, is, "Are we spending too much time and effort on converting refusals?" Data are needed, particularly from longitudinal studies, to see whether the data from conversions are accurate. The problem is particularly acute for longitudinal surveys because it may be masking sample attrition. If the response is always that nothing has changed and the interviewer does not probe or if the quality of response is poor, resulting in extensive item nonresponse, a bias will result in the estimation of change, duration, or transitions.

Validation studies are needed, in which data are obtained from records or other sources for people who have left the sample. Do weighting adjustments reduce the bias in the variables of interest? It may be possible to weight the sample so that it reflects the demographic profile as given by recent population estimates, but that does not necessarily mean that these adjustments reduce bias for the characteristics being estimated. Finally, an important, but sensitive, issue: Do survey data support statements about minorities? Are the coverage and nonresponse problems so large that estimates of levels, trends, and comparisons are very weak and perhaps misleading? Is our ability to answer information needs that require repeated or longitudinal surveys compromised by undercoverage and nonresponse in various subpopulations? A careful review of data collected from converted refusals, estimates of undercoverage in certain sex–age race groups, and comparisons with records may provide insights.

3. RECALL AND RELATED ISSUES

Recall, the frequency of interviewing, the number of times to be interviewed, the reference period, and the use of bounding are all issues of serious concern to those who conduct repeated or longitudinal surveys. Because these surveys frequently offer alternatives to retrospective interviewing in cross-sectional surveys, recall problems are assumed to be diminished. This, of course, depends on the reference period and the length of recall. In a report issued by a government task force on federal longitudinal surveys, the advantages of longitudinal and rotating panel studies were discussed (Office of Management and Budget, 1986). One of these was:

> Longitudinal survey interviews usually have a shorter, bounded reference period that reduces recall bias in comparison to a retrospective interview with a long reference period. Rotating panels such as CPS and NCS also share this advantage. Longitudinal surveys with long intervals between interviews may lose this advantage.

The report made frequent reference to recall and related issues.

Survey designers obviously feel that longitudinal surveys have an advantage with respect to recall issues over single-time surveys or repeated surveys with no overlap. It is certainly true that shorter recall periods and reference periods are often used to improve reporting. Bounding is also frequently used. However, there is no general pattern that one sees over a wide range of surveys. Table 4 compares several surveys on whether bounding is used, length of recall, and other related items. Although different subject matter may call for different intervals between interviews or different recall periods, this table illustrates that there are no agreed-upon standards related to the use of bounding, the length of recall, and the frequency and number of interviews.

Bounding is used in some surveys, but not in all. Sometimes it is used only for prompting, and in some surveys the first interview data are thrown away. Reference periods vary widely, even within a survey. The frequency of interviews varies widely, as does the total number of interviews.

Some of these issues have stimulated a great deal of research. Consider first the length of recall period, which is also addressed elsewhere in this volume by Cantor as well as Kalton et al. There has been research in at least four surveys on this topic: the Survey of Residential Alterations and Repairs, the National Crime Survey, the San Jose Health Survey, and the Income Survey Development Program. A brief summary of the results of these studies follows.

Work on the Survey of Residential Alterations and Repairs (SORAR) was done in the 1960s by Neter and Waksberg (1965). This study used, among many other things, a 1-month and 3-month recall period for reporting expenditures for residential alterations and repairs. Lengthening the recall period from one to three months substantially reduced the number of repair or alteration jobs reported, but the loss was greater for small jobs than large jobs. The loss was about 40% for jobs under $10 but very small for jobs between $10 and $50. Therefore, the effect on expenditures was smaller than on the number of jobs reported.

In the National Crime Survey, experiments have been done comparing different recall periods. The current 6-month recall period was

TABLE 4. COMPARISON OF SELECTED SURVEYS
ON RECALL AND RELATED ITEMS

Survey	Key Variables	Bound-ing	Reference Period	Frequency of Interviews	Number of Interviews
CPS[a]	Labor force	No	Labor force—1 wk.; what done to look for work—4 wks.; supplements—vary, often annual	Monthly for 4 mos.; none for 8 mos.; monthly for 4 more mos.	8 times over 16 mos.
SIPP[b]	Income types and amounts; program participation	Yes	Core interviews—4 mos.; requires placement by mo.; supplements vary	Every 4 mos.	8 times over 2 yrs. and 8 mos.
NCS[a]	Crime incidents	Yes[d]	6 mos.; requires placement by mo.	Every 6 mos.	7 times over 3.5 yrs.
CES[a]	What people buy	Yes[d]	3 mos.	Every 3 mos.	5 times over 15 mos. (gives bounded data for 1 yr.)
NLSMYW[c]	Labor force; work experience; income, mobility	No	Labor force—1 wk.; other— varies up to 2 yrs.	1- or 2-yr. intervals	13 times since 1967 and 1968
SSE[c]	Movement into and out of scientific and engineering jobs; labor force	No	Education and training—2 yrs.; salary, wks. worked—1 yr.; employment—1 wk.	2 yrs., 1982, '84, '86; 3 yrs. to '89	4 times over 10 yrs.
NMCES[c]	Medical encounters and expenditures; access to care; medical insurance	Yes	1st interview—1 mo.; later interviews—3 mos.	About every 3 mos.	6 times over 18 mos.
LTCS[c]	Ability to perform everyday activities	No	12 mos.	2 yrs., 1982 & '84; 4 yrs., 1988	Not originally longitudinal; will be 3 times in 10 yrs.
PSID[c]	Income; labor force; program participation	No	12–18 mos.	Every yr.	18 times over 18 yrs.

Key: CPS = Current Population Survey; SIPP = Survey of Income and Program Participation; NCS = National Crime Survey; CES = Consumer Expenditure Survey; NLSMYW = National Longitudinal Surveys of Mature and Young Women; SSE = Survey of Scientists and Engineers; NMCES = National Medical Care Expenditure Survey; LTCS = Long-Term Care Survey; PSID = Panel Study of Income Dynamics.

[a]Repeated, partial overlap, with rotation; [b]longitudinal, with rotation; [c]longitudinal, no rotation; [d]first interview used only for bounding.

selected based on pretests carried out in 1970–71 comparing 12-month and 6-month recall in connection with a reverse record check. As reported by Bushery (1981), in 1978–79 an experiment was conducted in which about one-twelfth of the sample received an extra interview, so that these respondents had two consecutive interviews with a 3-month recall period. Another one-twelfth skipped an interview, so that their subsequent interview used a 12-month recall. The experiment was designed so that 3-month, 6-month, and 12-month reference periods covered the same time period, using respondents who had approximately the same average number of prior interviews.

The results showed that the 3-month recall period produced crime victimization rates on the order of 10%− 20% higher than the 6-month recall period, which in turn produced rates 10%−15% higher than the 12-month period. There was evidence that the difference between the 3-month and 6-month recall periods was higher for blacks than for non-blacks for some crimes, especially assault.

In the San Jose Health Survey, reported by Mooney (1962), respondents were asked about illnesses the day before the interview and then about illnesses the calendar month preceding the interview. Monthly incidence rates based on 1-day recall were more than four times the comparable rates for 1-month recall, for both acute and chronic illnesses.

Finally, in the Income Survey Development Program (ISDP) reported by Kasprzyk (1986), a 6-month recall period was contrasted with two consecutive 3-month recall periods and similar results were found.

Thus, the evidence points to loss of data with longer recall periods. Quantifying that loss for a large variety of characteristics is more difficult. As Neter and Waksberg (1965) point out, the loss is not the same for all kinds of items. Less expensive, smaller incidents were the most likely to be forgotten. Similarly, not all crime incidents, illnesses, or income amounts are forgotten at the same rate. Cost plays more of a part in determining recall periods than does the amount of loss of data. Survey practitioners need to develop quantifiable estimates of loss for one recall period compared with another to balance the cost figures that do quantify the cost of interviewing respondents more frequently. The cost argument often takes precedence because recall loss has not been quantified. Although specifying the recall loss for each item is not feasible, a generalized recall loss function similar to generalized variance functions may be possible.

The placement of responses by time of occurrence is closely related to recall period and length of reference period. In many surveys, respondents are asked to report events over the last several months and place them correctly by month. Extending the reference period usually leads to a decrease in the number of incidents reported in the immediately preceding month. Yet that month gives the highest estimate of all months reported on. Table 5 shows data from the National Crime Survey in which the reference period is six months. Respondents were asked to report whether they were subject to any criminal victimizations within the last six months and, if so, in what month did they occur.

The rates in the column for "one" indicate victimizations reported for the month prior to the interview. Although one would expect to see an equal number of victimizations by month, there is a rapid decrease in the number reported. This decrease could be due to forgetting, or to what has been called "internal telescoping," the tendency to shift the timing of events within the recall period. The results from the National Crime Survey would indicate that the direction of such internal telescoping is forward, bringing the events closer to the time of the interview. This same finding was reported by Neter and Waksberg (1965), indicating that internal shifting is a significant problem and that the results shown in Table 5 cannot be explained by forgetting alone.

Closely related to reference-period difficulties is a response error called the "seam" problem, observed in longitudinal surveys, where "seam" refers to the fact that two reference periods are matched together to produce a longitudinal record. The number of transitions observed between the last month of one reference period and the first month of another is far greater than the number of transitions observed between months within the same reference period. Since other studies have shown that the month closest to the interview has the most activity reported and that the month farthest away has the least, this is not an unexpected finding.

Burkhead and Coder (1985) report the results for SIPP. They studied changes occurring between waves of interviewing for the first 12 months of the 1984 panel. The authors report that changes from receiving to not receiving program benefits and vice versa are significantly higher for the months that span successive reference periods. This was true for all kinds of income and noncash benefits received and for both self- and proxy-response. They also showed that the larger proportion of gross change at the seam was generally followed by greatly reduced gross change the following month. The authors believe that questionnaire wording and recall error are the two most likely causes of this phenomenon. This phenomenon has also been observed in the Panel Study of Income Dynamics (Hill, 1987), and undoubtedly afflicts all longitudinal surveys. Obviously, either operational or statistical procedures are needed to reduce the bias in estimates of transitions, durations, and similar estimates that use data across waves of interviewing.

The use of bounding is a regular feature of many repeated and longitudinal surveys. Bounding is used to prevent external telescoping— the shifting of events from outside the recall period into the recall

TABLE 5. VICTIMIZATION RATES FOR EACH MONTH OF THE
RECALL PERIOD, JUNE 1973–SEPTEMBER 1974

Types of Crime	Number of Months Prior to the Interview					
	One	Two	Three	Four	Five	Six
Personal Crimes[a]						
Crimes of violence	72	47	42	37	31	26
Assault	59	36	33	27	23	18
Personal theft	189	137	121	102	94	74
Household Crimes[b]						
Burglary	185	135	128	112	101	81
Larceny	264	184	163	138	126	100
Motor vehicle theft	35	29	26	23	22	19

Source: National Crime Survey, 1973–74.

[a]Rates per 1,000 persons age 12 and over.

[b]Rates per 1,000 households.

period. Neter and Waksberg (1965) were among the first to study this and showed that the unbounded 1-month recall of jobs of $100 or more was about 55% higher than the corresponding bounded recall because of the forward shifting of events into the recall period.

Bounding is also used in the National Crime Survey (NCS). Indeed, the first NCS interview is not used in the estimation process; it only provides data to interviewers on incidents already reported. Since these cases are not used in estimation, the price to prevent external telescoping is high. Table 4 compares surveys along several dimensions, including the use of bounding. Only two, the NCS and the Consumer Expenditure Survey, actually throw away data from the first month's interviews. Why is unbounded recall used for some surveys and not for others? Why are data collected and used in estimation for some surveys but not for others? What research results have led to these decisions?

The final question I pose in this section relates more to analysis. What research determines how many visits will be made to respondents? The SIPP has eight interviews, but many analysts say that more would be desirable, while respondents and field staff say that fewer would be desirable. How was the decision made to revisit the high school class of 1972 in 1974, 1975, 1977, 1980, and 1984–85? What makes these intervals the desirable ones? Is there enough planning of the analysis prior to the start of the survey so that there is a rationale for the number of visits based on data needs, not just cost?

Survey practitioners use a variety of techniques to resolve these interesting nonsampling error problems. Longer recall periods give less accurate data, but because of insufficient research, quantifiable estimates of the loss incurred for a variety of characteristics by using longer recall periods are not available. Cost figures determine the recall period. It is clear from the design of many surveys that external telescoping is given more attention than internal telescoping. Sometimes, as in the National Crime Survey, data are collected but not used. Internal telescoping is recognized as a problem, but until the analysis of longitudinal data became important it received little attention. Since one of the primary purposes of these surveys was to measure net change, a simplified assumption was made that the amount of displacement of events within the reference period remained the same over all interviews; the assumption allows the assertion that estimates of change were not biased. Now, in estimating durations or transitions, these problems directly affect the analysis, and must be dealt with. Finally, the number of times we interview a respondent seems to be a function of three factors: analytic needs, the length of the interview, and the success of obtaining funding for additional interviews. For some characteristics, such as income, analysts have made a strong case for having enough interviews to provide annual data. The interaction of length of the panel with the quality of data provided has not been studied in any detail.

Learning more about recall, the reference period, and bounding is critical to all survey work, not just longitudinal surveys. The work by Cannell (1977) and his colleagues at The University of Michigan's Survey Research Center may prove very useful. Recent research on understanding how respondents process questions, reported on by Lessler and Sirken (1985) and Loftus (1986), provides insights. However, we need to know the extent of recall loss and displacement of events. We need to know the value of bounding interviews. And we need to understand interactions among the related variables of telescoping, recall, forgetting, and reference-period length.

4. TIME-IN-SAMPLE BIAS

Time-in-sample bias refers to the phenomenon that estimates from people reporting for the same time period but with different levels of exposure to a survey have different expected values. Some people call this "panel conditioning," but that attributes the entire reason for this phenomenon to respondents. There are many other factors at play as well.

The first reference to this bias is in a paper by Hansen et al. (1955) concerning estimates of unemployment. However, it has been shown to exist for surveys about expenditures, income, crime victimizations, and levels of illness.

Bias is observed when a significantly higher or lower level for a characteristic occurs in the first interview than in subsequent interviews, when one would expect the same level. This is the case for the number of employed and unemployed, the number of persons victimized by crime, the number of people with illnesses who visited a doctor, the number of people who worked more than 35 hours a week, the number of alterations and repairs on a housing unit, and the amount of money spent on consumer goods. Researchers disagree about whether estimates from the first or later interviews are more nearly correct. Time-in-sample biases are shown in Table 6 for the number of unemployed in the Current Population Survey for the time period January 1983 to October 1984.

These data are shown as index numbers, computed by dividing the total number of persons in a given rotation group having the characteristic of interest by the average number of persons having the characteristic of interest over all rotation groups and multiplied by 100. If an equal number of persons with the characteristic of interest were reported in each rotation group, the index for each would be 100. A rotation group in this survey is just an indication of how many times a person has been interviewed.

For the total population, the level of unemployment in month 1 is 7% higher than the average. The higher index in month 4 shows the effect of the "discouraged worker" questions. These questions are asked only of respondents in the sample for the fourth or eighth time. The respondents are asked about their intentions to look for work in the next 12 months. After asking those questions, the interviewers sometimes change responses to the employment questions, creating higher unemployment levels. The variances are high for these indexes, so it is difficult to make meaningful comparisons. However, it is obvious that the bias is present for men and women, although more pronounced for women, and for all population subgroups—whites, blacks, and Hispanics.

Another survey of interest is the Consumer Expenditure Survey Diary Sample, conducted in 1972–73. A diary was placed in a housing unit by an interviewer who collected background information at the time of placement and returned a week later to collect the completed diary and place a diary for the second week. A third visit was made to collect the diary at the end of the second week. If expenditure data were missing, the interviewers asked respondents for the data, and wrote the data in the diary. But in these cases respondents did not always provide daily expenditures, so the timing of the expenditures was unknown. Table 7 shows the results for expenditure data, by degree of respondent completion of the diary, again in terms of an index. Separate indexes are shown for total food expenditures for home consumption, for all other expenditures combined, and for a group of

- The first interview is not bounded, so it contains events from outside the reference period.
- Respondents change their behavior—buying behavior, looking for a job, voting habits, etc.
- Respondents become bored with the process and do not work as diligently or energetically at providing accurate responses in later interviews.

Probably no single factor is solely responsible for this bias. There are undoubtedly interactions. A few pieces of evidence rule out some factors. Data from the Consumer Expenditure Survey, in which the respondents filled out the diaries without help from any interviewers, showed the bias. Therefore, we know that the bias is not solely a function of the interviewers. In addition, Pearl (1979), in his evaluation of that survey, showed that a temporary alteration in buying habits did not seem to be the cause. He also believed that external telescoping was likely to be a factor, combined with poorer reporting behavior in later periods. This combination of factors is supported by the data since Pearl shows that the level of expenditures supported by the first few reporting days would far exceed expenditure data from independent sources. On the other hand, the data from later reporting days are far too low and would be very deficient compared with independent sources.

A somewhat similar effect was noted by Mooney (1962) in his evaluation of diaries kept by respondents on illnesses. For those respondents who had illnesses that were said to be medically attended, independent physicians' reports were also collected. A comparison was made of survey reports of medically attended illnesses with physician reports of illnesses for the respondents in both initial and repeat interviews. The reports fell into three categories: (1) the survey and physician reports agreed, (2) the survey report was not mentioned by a physician and was called an overreport, or (3) the physician reported an illness not mentioned in the survey, called an underreport. The results showed a high level of overreporting on initial interviews, 40% compared to 19% on repeat interviews. A high degree of underreporting also existed on initial interviews—46%—but an even higher degree on repeated interviews—72%. Mooney thought it was plausible that the same kind of pattern held for nonmedically attended illnesses as well. In these data, overreporting and underreporting almost balance out on the initial interview, but subsequent interviews are biased severely downward. Mooney's study is a landmark, since it is one of the few in which methodologists have tried to verify respondent reports to determine the accuracy of the initial and later interviews. At least for health incidence data, the initial interviews provided a better estimate.

The theory advanced by Williams and Mallows (1970) also deserves attention. Williams and Mallows show that differential response probabilities from one interviewing period to the next can cause changes in the measured unemployment rate. A condition for the observed unemployment rate not to change when the underlying "true" unemployment rate remains constant is that the ratio of response probabilities for certain groups remains constant. Let P_u and P_e be the probabilities of observing a respondent who is unemployed (P_u) or employed (P_e) at the time t_1. Then let P_{eu} be the probability of response for a person observed as employed at t_1 and unemployed at t_2. Similarly, P_{ue} is the response probability of a person observed as unemployed at t_1 and employed at t_2. Are the response probabilities of the employed the same as those for the unemployed? Many survey practitioners think that $P_u > P_e$. Data from the Current Population Survey show that it takes more visits to interview an employed person than an unemployed person. However, if an interviewer learns from the first visit when to schedule interviews, then nonresponse may be lower for the second time period and, consequently, response probabilities may change for the second interview. Only if $P_u/P_e = P_{eu}/P_{ue}$ will the estimated unemployment ratio remain unchanged when the "true" unemployment rate is unchanged. Williams and Mallows show how small changes in the response probabilities cause fairly large changes in the unemployment rate. For longitudinal surveys, it is important to keep response rates high. Undoubtedly, these ratios change, causing spurious changes in the estimates.

This type of bias is a serious problem in repeated and longitudinal surveys. It probably relates to a number of factors, including unbounded interviews, respondent boredom, and response probabilities. Obviously, the effect may be even more serious in longitudinal surveys having many contacts.

Some analysts have discounted the time-in-sample bias by comforting themselves with the notion that if the bias remains constant from one period to the next, it will affect only estimates of level rather than change. This should not comfort anyone, for at least three reasons. First, estimates of level are important in their own right. Second, this is true only under the additive model discussed by Bailar (1975). If one assumes a multiplicative model, suggested by Solon (1986), the bias affects both level and change. Finally, the bias does not remain constant. Precise measurements to determine the point at which it changes are not possible, but data show that it does change. The bias does not seem to increase as the characteristic of interest increases, but may be more influenced by the changing nature of who the respondents are, the pace at which respondents get bored, and the oversaturation of the public with polls and surveys. If the biases change from month to month, there is severe distortion in both estimates of level and change as shown by Bailar (1979).

Since much of the more interesting data from longitudinal surveys has to do with transitions from one time period to another, or trends over time, this bias should be of major concern. More research on its causes, validation of which data are more nearly correct, and statistical methods to reduce its effect are all needed.

5. SUMMARY

Many other topics challenge us that we do not fully understand—the use of additional survey modules on longitudinal surveys and how they affect data quality, the use of proxy respondents, and questionnaire wording, to name a few. A review of the literature shows a variety of studies, some of them excellent, but few building on previous work. Statistical solutions to survey methodology problems are being developed rather than methods that improve the survey process. More effort should be devoted to carrying out carefully controlled experiments to answer survey method questions related to longitudinal surveys.

Longitudinal surveys provide us unique opportunities, especially those in which new panels are introduced periodically. Evaluations and experiments that focus on recall, bounding, and time-in-sample bias should be developed as part of the ongoing survey operations. A better understanding of the data will ensure that the additional complexity of longitudinal surveys really does result in fulfilling more complex information needs.

We do, however, have to focus our attention on research. We need to make a commitment to move more of survey-taking onto a firm scientific foundation and away from the near-art-form status required when we get beyond sampling. Before we expend much more of our research in asking more and more questions, we ought to try to answer a few.

REFERENCES

Bailar, B. A. (1975), "The Effects of Rotation Group Bias on Estimates from Panel Surveys," *Journal of the American Statistical Association*, **70**, 23–30.

Bailar, B. A. (1979), "Rotation Sample Biases and Their Effects on Estimates of Change," *Bulletin of the International Statistical Institute*, **48**(2), 385–407.

Bowie, C. E. (1987, Nov. 25), *SIPP Operational Statistics* (Report No. 8), (Unpublished memorandum), U.S. Bureau of the Census, Washington, DC.

Burkhead, D., & Coder, J. (1985), "Gross Changes in Income Recipiency from the Survey of Income and Program Participation," *Proceedings of the Social Statistics Section*, American Statistical Association, 351–356.

Bushery, J. M. (1981), "Recall Biases for Different Reference Periods in the National Crime Survey," *Proceedings of the Section on Survey Research Methods*, American Statistical Association, 238–243.

Cannell, C. F. (1977), "A Summary of Research Studies of Interviewing Methodology, 1959–1970," in *Vital and Health Statistics: Series 2, Data Evaluation and Methods Research* (No. 69).

Chapman, D. W., Bailey, L., & Kasprzyk, D. (1986), "Nonresponse Adjustment Procedures at the Census Bureau: A Review," *Survey Methodology*, **12**(2), 161–180.

David, M. (1985), "Introduction: The Design and Development of SIPP," *Journal of Economic and Social Measurement*, **13**, 215–224.

Duncan, G., & Kalton, G. (1987), "Issues of Design and Analysis of Surveys across

Time," *International Statistical Review*, **55**(1), 97–117.

Hansen, M. H., Hurwitz, W. N., Nisselson, H., & Steinberg, J. (1955), "The Redesign of the Census Current Population Survey," *Journal of the American Statistical Association*, **50**, 701–719.

Hill, D. (1987), "Response Errors around the Seam: Analysis of Change in a Panel with Overlapping Reference Periods," *Proceedings of the Section on Survey Research Methods*, American Statistical Association.

Kalton, G. (1986), "Handling Wave Nonresponse in Panel Surveys," *Journal of Official Statistics*, **2**, 303–314.

Kasprzyk, D. (1986), "A Review of the Research and Design Issues in the Survey of Income and Program Participation," *Proceedings of the Section on Survey Research Methods*, American Statistical Association, 187–192.

Lessler, J. T., & Sirken, M. G. (1985), "Laboratory-Based Research on the Cognitive Aspects of Survey Methodology: The Goals and Methods of the National Center for Health Statistics Study," *Milbank Memorial Fund Quarterly: Health and Society*, **63**(3), 565–581.

Loftus, E. (1986), "Survey Remembering," *Proceedings of the U.S. Bureau of the Census Second Annual Research Conference*, 193–207.

Mooney, H. W. (1962), *Methodology in Two California Health Surveys* (Public Health Monograph No. 70).

Morgan, J., Dye, R., & Hybels, J. (1979), *Results from Two National Studies of Philanthropic Activity* (Research Report Series), Survey Research Center, Institute for Social Research, Ann Arbor, MI.

National Center for Education Statistics (1981), *High School and Beyond: National Longitudinal Study for the 1980s*.

National Center for Health Statistics (1985), "The National Health Interview Survey," in *Vital and Health Statistics, Series 1* (No. 18) (DHHS Pub. No. (PHS)85–1320, Public Health Service Report No. 8).

Nelson, D., Bowie, C. E., & Walker, A. (1987), "Survey of Income and Program Participation (SIPP) Sample Loss and Efforts to Reduce It," *Proceedings of the U.S. Bureau of the Census Third Annual Research Conference*, 629–643.

Nelson, D., McMillen, D. B., & Kasprzyk, D. (1985), *An Overview of the Survey of Income and Program Participation: Update 1* (SIPP Working Paper Series No. 8401), U.S. Bureau of the Census, Washington, DC.

Neter, J., & Waksberg, J. (1965), *Response Errors in Collection of Expenditures Data by Household Interviews: An Experimental Study* (Technical Paper No. 11), U.S. Bureau of the Census, Washington, DC.

New York Times (1986, Nov. 16), "Delusion of Democracy" (editorial).

Office of Management and Budget (1986), *Federal Longitudinal Surveys* (Statistical Policy Working Paper No. 13), National Technical Information Service, PB86–139730.

Pearl, R. B. (1979), *Reevaluation of the 1972–73 U.S. Consumer Expenditure Survey* (Technical Paper No. 46), U.S. Bureau of the Census, Washington, DC.

Rhoton, P. (1986), "Attrition and the National Longitudinal Surveys of Labor Force Behavior: Avoidance, Control, and Correction," *IASSIST Quarterly*, **10**(2).

Sirken, M. G., & Brown, M. L. (1962), "Quality of Data Elicited by Successive Mailings in Mail Surveys," *Proceedings of the Social Statistics Section*, American Statistical Association, 118–125.

Solon, G. S. (1986), "Effects of Rotation Group Bias on Estimation of Unemployment," *Journal of Business and Economic Statistics*, **4**(1), 105–109.

U.S. Bureau of the Census (1968), *The Current Population Survey Reinterview Program, January 1961 through December 1966* (Technical Report No. 19), Washington, DC.

U.S. Bureau of the Census (1978), *The Current Population Survey: Design and Methodology* (U.S. Bureau of the Census Technical Paper No. 40), U.S. Government Printing Office, Washington, DC.

U.S. Bureau of the Census (1986), *Manufacturers' Shipments, Inventories, and Orders: 1982–86* (Series M3–1(86)), Washington, DC.

U.S. Bureau of the Census (1987), *Monthly Retail Trade, Sales, and Inventories* (Current business report, monthly).

Valentine, C., & Valentine, B. (1971), *Missing Men: A Comparative Methodological*

Study of Underenumeration and Related Problems (Unpublished paper), U.S. Bureau of the Census, Washington, DC.

Williams, W. H., & Mallows, C. L. (1970), "Systematic Biases in Panel Surveys," *Journal of the American Statistical Association*, **65**, 1338–1349.

Substantive Implications of Longitudinal Design Features: The National Crime Survey as a Case Study

David Cantor

1. INTRODUCTION

Longitudinal surveys offer a number of substantive and methodological features that, on their face, are extremely attractive. Substantively, a panel survey allows the study of social and physical dynamics that cannot be inferred from a cross-sectional design. Methodologically, it allows for the accurate measurement of how exogenous and endogenous phenomena interrelate over the short and long term by minimizing the problems associated with asking retrospective questions.

What is often overlooked when assessing the need for a longitudinal survey, however, is how the complications of conducting such a survey can seriously threaten or negate many of its analytical advantages. The decision to conduct a longitudinal survey affects every component of the survey design. These components include the basic sample structure, administration of the survey, data base structure, estimation, and analysis. The decision to conduct a longitudinal survey, therefore, dictates the basic framework of the entire design.

While this statement may seem obvious to many, it is commonly overlooked when considering design options at the planning stage of a study. When operating on a fixed budget, there are trade-offs between a panel design and directing the money into collecting the information in some other way (e.g., conducting more cross-sectional interviews). When deciding between a panel and cross-sectional design, then, careful thought must be given as to how operational complications might affect the quality of the longitudinal data.

The purpose of this paper is to illustrate how three key operational features of a longitudinal survey affect the substantive utility of the information derived from it. These features include: (1) interview spacing, (2) respondent selection, and (3) mode of interview. While only one of these is uniquely applicable to a longitudinal design (spacing), all three have specific repercussions that are not present with a cross-sectional survey. The discussion below will show how these three features can affect the quality and utility of data from a panel survey.

The discussion will try to remain applicable to almost any type of longitudinal survey. This is an extremely difficult task, however. There are many different types of longitudinal surveys and each has its own unique operational and substantive features. To facilitate this discussion, therefore, much of the material presented below will use the National Crime Survey (NCS) as a "case study." This will give the discussion a point of reference from which it can draw a few ready examples.

The NCS is a victimization survey sponsored by the Bureau of Justice Statistics (BJS) and administered by the U.S. Bureau of the Census. Its purposes are to measure the amount of crime in the United States and provide a data base to understand the causes and consequences of crime from a victim's perspective. The survey is structured as a rotating panel design where housing units are contacted by interviewers every six months. Every person in the unit who is aged 12 years and over is interviewed as part of the survey. A housing unit stays in sample for a total of 7 household visits (3 years).

The NCS is a useful example to draw from because it is undergoing active consideration to be redesigned as a longitudinal survey that follows persons. As a result of a six-year redesign program directed by Albert D. Biderman at the Bureau of Social Science Research and sponsored by BJS, a number of analytic advantages were identified that lead to the conclusion that following persons would greatly benefit the utility of the data. There are several key problems, however, that must be resolved prior to converting the design. It is around these problems that many of the examples discussed below will be drawn.

2. INTERVIEW SPACING

Probably no single feature of a longitudinal survey affects its design and response structure more than the length of time between interviews. Determining this feature not only influences the amount of change that can be observed on relevant variables but also influences several other important aspects of the survey, including its error structure and cost. While the precise spacing depends on the substantive content of the study, there are certain generic, oftentimes competing, considerations when determining this aspect of the survey design. As discussed below, very little information is available to guide survey designers in balancing these competing demands when determining an optimal spacing between interviews.

Two primary considerations can be identified across all types of surveys when setting the length of time between follow-up interviews. The first is the amount of time it takes to expect meaningful change and/or occurrence in the variables that are of substantive interest. Different topics can lead to substantially different interview spacing periods. For example, epidemiological studies typically require a number of years for a follow-up period in order to allow for physical and mental development. Ten years, in this instance, is not considered unusually long and could represent a minimum amount of time required to detect significant changes in biological circumstances. Alternatively, detecting significant changes in employment or financial patterns might require a relatively short time interval, since changes in status with respect to these variables can occur quite quickly. An individual may exit and enter the labor force several times in a period of less than a year.

When trying to detect the consequences of particular events, it is important to obtain at least one measure of the outcome variable shortly after the event occurs. In measuring change in psychological phenomena, like depression, for example, the most dramatic changes occur shortly after a significant event. In those cases where the timing of the event cannot be precisely predicted (e.g., death of a family member, victimization) it is important to space the interviews over a short enough time period so that the survey can measure the outcome variables soon after the event for as many people as possible.

The second primary consideration when determining the length of time between interviews is the budget of the survey itself. If the investigator wishes to study a phenomenon over a fixed period of time, then interviewing more often will require more money per observation unit. In several government surveys, such as the Survey of Income and Program Participation (SIPP) and the National Medical Care Expenditure Survey (NMCES), annual estimates of particular phenomena are required. It is therefore important that the reference periods cover all months of the year, preferably with the same respondents. If the interview spacing is three months, this would require at least four interviews per person to cover the entire year.

As a practical matter, sponsors of a study will provide enough money to conduct the survey for fixed periods of time, and once this is determined, the interview spacing is determined as a natural consequence of what the study can afford. In this sense, therefore, determining interview spacing is a natural extension of determining the "optimal" sample size. In the latter case, investigators typically examine the anticipated award amounts and then work backwards to determine sample sizes sufficient to conduct the study.

While both cost and substance play primary roles in determining the spacing between interviews, a number of other factors may be factored into the equation. The extent to which these other factors can be actively considered when formulating the design depends on the substance of the study and the amount of information available to the investigator. In almost all cases, however, the repercussions of particular follow-up intervals affect the quality of the data and usually have substantial impacts on the results.

2.1. Telescoping and Omissions

One reason for conducting a panel survey is to control for recall problems when asking retrospective questions. A longitudinal design affords an opportunity to control for telescoping errors[1] (Neter & Waksberg, 1964; Sudman & Bradburn, 1972) by defining the beginning of the reporting period as the last time the respondent was interviewed. This provides a specific boundary that can be used by both the interviewer and respondent when deciding whether certain events have occurred in the time period. When previous interviews are used as a bound, then interview spacing, which coincides with the recall period for the respondent, can have dramatic effects on the nature of the omission and telescoping problems that will be encountered. Specifically, the longer the spacing, the more telescoping and omission errors that occur.

Whether the interview spacing is "too long" depends on a number of factors, most importantly the type of event that is of interest.[2] The more significant or salient the event is to the respondent, the more likely it is remembered and dated properly. Somewhat related to salience is the absolute number of times an event occurs. Frequent events are relatively less likely to be remembered and dated properly than those that occur less often.

According to at least one recall model (Sudman & Bradburn, 1972), the extent to which the bounding procedure affects external telescoping will vary by the interview spacing. The longer the interval, the fewer events the respondent should be able to remember at the beginning of the period, and therefore the less likely an event is to be misdated as occurring in the period. This has implications for panel surveys that use a very short recall period that are a fraction as long as the interview spacing. For example, the Current Population Survey (CPS) asks about employment that occurred during the week prior to the interview, even though the previous interview occurred at least one month in the past (U.S. Bureau of the Census, 1978). While the respondent might be omitting fewer spells of employment because of the short recall period, he/she might also be telescoping in proportionately more events than would be the case if the recall period was more tightly bounded.

If the analyst has a substantial budget at his/her disposal, then a methodological study could be designed, probably in the form of a reverse/forward record check (Cannell & Fowler, 1965; Cash & Moss, 1972; Cohen & Cohen, 1985; Mathiowetz & Duncan, 1984), to determine the optimal period. Typically, however, educated guesses are made with respect to what a reasonable period might be.

The NCS is designed as a rotating panel survey with six months between interviews. Before establishing this spacing, a number of studies attempted to examine how omissions and telescoping varied by the recall interval (Penick & Owens, 1976). These studies, primarily in the form of reverse record checks that sampled police records, seemed to indicate that for very serious events, a relatively high rate of omission and telescoping occurred for periods of longer than 6 to 12 months.[3] The present interview spacing of 6 months was chosen as a compromise between these methodological concerns and budgetary constraints.

Even after a very carefully planned agenda of research, it is still unclear what the "optimal" interview spacing for the NCS is. What is clear from this research is that interview spacing plays a critical role in the types of substantive results the survey produces. Analysis of a field experiment that varied the recall period within the ongoing NCS seems to show that significantly higher victimization rates are obtained when comparing 3-month recall periods to a 6-month recall period and when comparing 6-month to a 12-month recall period (Bushery, 1981). Perhaps more important from an analytical perspective is that there seems to be a relationship between recall patterns and selected respondent characteristics (Cantor, 1985; Kobilarcik et al., 1983). This is illustrated in Table 1, which displays estimates of annual victimization rates by 3- and 6-month recall periods. As can be seen, the rates are substantially higher when using the shorter period. Table 2 displays ratios of these rates between specific demographic subgroups. Note the substantial increase in violent crime reporting for the blacks when using the shorter period. This is especially significant when studying victimization because of the theoretical importance of race in current theories of victimization (Cantor, 1985). For more discussion of this topic, see the Kalton et al. and the Bailar papers in this volume.

2.2. Panel Effects

Interview spacing is also related to the types of panel effects that can be expected. As used here, panel effects refer to the extent that previous interviews influence the results of subsequent interviews. These effects can take two basic forms. First, there are panel effects in which interviews actually influence the behavior of the respondent. A behavioral effect might occur if the respondent directly benefits as a result of being a survey participant. This has been hypothesized to be the case in health examination surveys, for example, if the first exam reveals a malady that would not have normally been discovered if the survey had not been taken. As a result of the survey, the respondent might see a doctor and get treatment for the problem before the next round of interviewing. While this is extremely fortunate for the patient, it complicates the ability of the survey to make unbiased cross-sectional estimates of levels of occurrence. In this case, more frequent interviewing increases the likelihood of this kind of behavioral effect.

Behavioral effects are also a problem in many types of evaluation

TABLE 1. ANNUAL VICTIMIZATION RATES BY RECALL INTERVAL
(PER 100 PERSONS 12+ YEARS OLD)

| | Recall Interval | | |
Type of Crime	6 Months	3 Months	Diff.
Total personal crimes	12.85	15.49	− 2.64**
Crimes of violence	3.46	4.29	− 0.83**
Crimes of theft	9.39	11.20	− 1.81**
Total household crimes	23.00	26.38	− 3.83**
Burglary	8.53	9.68	− 1.15*
Larceny	12.70	15.09	− 2.39**
Auto theft	1.78	2.07	− 0.29

Source: Bushery, 1981.

*Significant at the 10% level.
**Significant at the 5% level.

studies where interviews are performed both before and after a particular program is instituted or event occurs. For example, panel surveys of voting behavior, which interview prospective voters both before
and after an election, have been hypothesized to influence the number
of respondents that actually go out to vote (Traugott & Katosh, 1979).
This might occur if the first interview highlights the election as being
particularly important for the survey respondent.

For phenomenon of this kind, it is important when trying to gauge a
respondent's attitude about the election (e.g., candidate preferences,
probability of voting) to interview as close to the election as possible.
The accurate measurement of these variables is critically related to
when they are taken with respect to the election, since many people
really do not make up their minds until the last few days before an election. Alternatively, the interview should be spaced adequately before
the election in order to minimize any panel effect that might occur.

For the crime victimization surveys, respondents may increase crime
avoidance behavior or become more frightened of becoming a victim if
there is substantial contact with the survey over a very short period of
time (Office of Management and Budget, 1986).

A second and more complicated type of panel effect is related to
respondent conditioning. This occurs when earlier interviews affect the
respondent reaction to subsequent survey contacts, independent of any
real behavioral change. The most common type of respondent conditioning effect is the tendency of respondents to gradually decrease the
number of positive responses to questions over time, resulting in a
decline in incidence and prevalence measures of the phenomena being
studied (Bailar, 1972; Cohen & Burt, 1985). This decline in rates has

TABLE 2. RATIO OF PERSONAL VICTIMIZATION RATES ACROSS
DEMOGRAPHIC SUBGROUPS BY RECALL INTERVAL

Relative Demographic Subgroup	Recall Interval	Total Personal Crimes	Crimes of Violence	Crimes of Theft
Black/total	3 mos.	1.14	1.71	0.94
	6 mos.	1.00	1.19	0.93
Male/female	3 mos.	1.34	1.83	1.19
	6 mos.	1.35	1.96	1.18
12–24/25–49 yrs.	3 mos.	1.87	2.27	1.72
	6 mos.	1.65	2.14	1.49
12–24/50+ yrs.	3 mos.	4.58	7.68	3.84
	6 mos.	4.25	7.16	3.59
25–49/50+ yrs.	3 mos.	2.45	3.46	2.23
	6 mos.	2.58	3.35	2.41

Source: Kobilarcik et al., 1983.

been observed with the NCS. Figure 1 displays victimization rates for the NCS for the second through seventh contacts of the housing unit. As can be seen, there is a noticeable decline between most contacts, especially the second and third.

The extent to which interview spacing might influence these patterns depends on what accounts for this phenomena. The most common explanation for this phenomenon is "respondent burden"—that is, respondents get tired of participating in the survey and quickly learn, from previous experience, that positive answers tend to lead to a longer interviewing session. If this is the case, then, shorter interview spacing or frequent interviewer contact could accelerate the decline in occurrence rates. Another consequence of this effect is that shorter interview spacing may lead to higher unit nonresponse.

While intuitively clear, there is very little evidence to either support or deny this tendency. Frankel and Sharp (1984) found that when asked to define respondent burden, respondents did not seem particularly concerned with the number of times the interviewer returned for follow-ups. More direct evidence on this stems from several studies that varied the follow-up interval for a short period of time. A study that examined linking the National Survey of Family Growth (NSFG) to a follow-up survey did not find any dramatic decline in response rates as the follow-up interval varied (Mathiowetz et al., 1986). Sudman and Lannom (1984) did not find a significant decline in response rates when comparing monthly to bimonthly intervals. Similarly, work with the NCS (Bushery, 1981) did not find any significant decline in response rates when switching from a 6 to a 3 month follow-up period. These

FIGURE 1. ANNUAL VICTIMIZATION RATES BY TIME IN SAMPLE

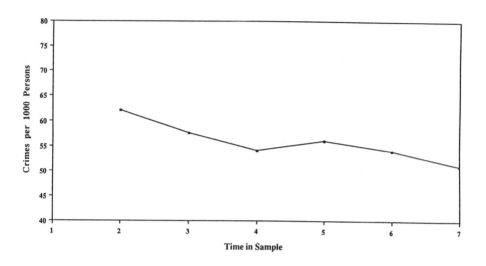

Source: Roman & Sliwa, 1982.

studies contrast with work on selected health surveys that have found a
significant increase in attrition when comparing monthly and bimonthly
follow-ups (Yaffee & Shapiro, 1981).

Another explanation for the drop in rates exhibited in Figure 1 is
that the respondent is actually learning what the survey is about over
time (Biderman et al., 1986 pp. 217–230; Cannell et al., 1981). Under
this response model, respondents are actually very eager to report
something to the interviewer whenever possible. Upon the first administration, when the topic of the survey is not entirely clear, the
respondent reports anything remotely qualified. After several contacts
and more detailed questions, the respondent obtains a better understanding of the survey and thus restricts answers in ways consistent
with this understanding. If this were the case, then more frequent
interviewing should have, at worst, a benign effect on response quality.
It actually might improve responses if shorter spacing leads to a better
recall of what was reported during the previous interview.

Given the competing nature of these two explanations, it is important
to determine what accounts for these response patterns when designing
and analyzing a panel survey. The respondent burden model views frequent interviews as competing with a shortened recall period, while the

respondent learning model does not place as much emphasis on the negative aspects of frequent contact. This author's speculation is that the applicability of each explanation probably varies by the sensitivity of the subject matter and length of the interview. With frequent contacts, sensitive and lengthy interviews are more likely to annoy the respondent than interviews that are shorter and less sensitive. SIPP, for example, has a lengthy questionnaire and deals with the sensitive topic of income, so perhaps the short follow-up interval partially accounts for its nonresponse rate. Topics like victimization, which many people are very eager to talk about, might be a different matter.

Understanding respondent conditioning goes beyond determining the proper interview spacing and gets to the very heart of using and interpreting the survey results. This is most clearly seen when trying to formulate adjustments for responses assumed to have a significant error component. The type of adjustment one uses will depend on which response model is assumed. Are later panel interviews less biased than earlier ones? At which interview are responses best? These types of questions are critical to the utility and adjustment of panel data.

When formulating adjustments for the NCS, the problem is attenuated because of the confounding of respondent conditioning and the telescoping that is assumed to occur in an unbounded interview. On the NCS it is assumed that unbounded interviews, which typically elicit victimization rates 30%–40% higher than bounded interviews, contain more error than bounded ones because of telescoping problems. Consequently, the entire first round of NCS interviews, which are unbounded, are not used for analysis purposes. As part of the NCS redesign, adjustments were formulated to allow incorporation of these data into the NCS analytical data base. Because of the respondent conditioning phenomenon referred to above, however, this proved to be a difficult task, since it is unclear to what degree the drop in reports between the first (unbounded) and second (bounded) interviews is due to telescoping or conditioning. Since each type of effect can be observed only when occurring simultaneously, it is difficult, short of conducting extensive field experiments, to rationalize inclusion of adjusted first-interview data unless one can show that the adjustment actually makes the responses less biased. This, of course, cannot be done without assuming that unbounded interviews are clearly inferior to the bounded ones (an assumption that the Census Bureau is not currently willing to accept.)[4]

2.3. Tracking Mobile Respondents

The length of the follow-up period affects the number of people who move between interviews and must be traced and found again. This, in turn, influences how many people are lost to follow-up (LTF) and contributes to the overall nonresponse rate. While overall mobility over a fixed time period is invariant to inter-interview length, the number LTF should decrease with a shorter inter-interview period. It is easier to follow movers if attempts to recontact the respondent are made very close to when they had moved. Shorter periods, therefore, reduce the number of "cold trails" vis-a-vis longer periods. Since mobility is highly related to most major demographic and social characteristics, it is extremely important to try to minimize the number LTF as much as possible. Mobility, for example, is one of the highest known correlates of crime patterns. Neighborhoods that exhibit high residential mobility usually have very high crime rates (Shaw & McKay, 1972). Similarly, for victimization, those who are the hardest to trace seem to be subject to the highest risk (Biderman & Cantor, 1984; Nelson, 1984; Reiss, 1978). See, for example, Table 3, which presents victimization rates at the first unbounded interview on the NCS by whether the respondent had moved out by the second time the housing unit was contacted. One can see that those most likely to have moved out by the next interview are disproportionately located in the cells with the highest victimization rates. In addition, within each age group, those who move out have substantially higher rates than those who do not.

As is evident from the Burgess paper in this volume, if funds are available, methods of tracking mobile populations are extremely efficient. Once the tracing part of the budget is fixed, the analyst must consider whether the potential losses of highly mobile respondents are particularly troublesome.

For many types of longitudinal surveys, geographical restrictions are placed on where respondents can be followed. This is common in evaluation studies, for example, where there is rarely money or time to set up procedures to follow people all over the country. This is also the case for surveys conducted in person and surveys that require elaborate field procedures to collect data. For example, health examination surveys have extremely high interviewing costs and cannot substitute a telephone or mail interview for most data collection. Typically, one or two very expensive medically trained examiners (e.g., doctors or nurses) collect the data using expensive testing equipment. This is why many epidemiological surveys are centered around a population clearly defined geographically, such as a small town—e.g., Framingham, Massachusetts (Gordon et al., 1959) or Tecumseh, Michigan (Napier, 1962).

Losses due to mobility are especially important to studies that either cannot afford to conduct any tracking, or those that return to the same area over time rather than follow individuals. Studies that evaluate neighborhood programs, such as crime control activities, are commonly of the latter type (Rosenbaum et al., 1986). In a study of this kind, conflicts typically exist between setting the interview spacing to minimize losses due to mobility and measuring change in the outcome variables of interest.

TABLE 3. PERSONAL CONTACT VICTIMIZATION RATES BY MOBILITY STATUS AT NEXT INTERVIEW AND AGE (PER 1,000 PERSONS 12+ YEARS OLD)[a]

Age	Mobility Status at Next Interview		Percentage Moved
	Moved	Did Not Move	
12–15	57.0 (3,138)	46.9 (14,440)	18.0%
16–19	78.2 (5,640)	50.7 (11,456)	33.0
20–24	67.7 (7,443)	41.1 (11,853)	38.6
25–34	54.2 (8,414)	30.9 (24,794)	25.3
35–49	27.7 (6,208)	16.0 (29,991)	17.1
50–64	17.5 (4,842)	10.6 (28,183)	14.7
65+	10.2 (3,516)	8.5 (19,483)	15.2

[a]Population bases are in parentheses. Rates were computed using NCS unbounded incoming rotation groups. Determination of mobility status was estimated across panel waves. The mobility estimates have not been corrected for matching errors.

2.4. Migration and Bounding Status

Some designs allow for significant migration of respondents between panel waves. The SIPP, for example, follows respondents originally selected into the sample wherever they move, under certain restrictions. If these "key" sample persons move into a new household, every person in that household is interviewed. Once a key person moves out of that household, however, nonkey individuals are not followed. Designs that return to areas or structures, rather than following people, will gain and lose a significant number of people throughout the life of the survey. Cox and Cohen (1985) have labeled these designs as longitudinal dwelling unit designs. The NCS, for example, returns to the same housing structure every six months, regardless of who lives in the structure. When this is the case, interview spacing is directly related to the amount of migration that occurs between waves. This can complicate the analysis stage because respondents will have been subject to a different number of previous interviews.

This is an extremely difficult problem for the NCS because of the large impact previous interviews have on subsequent ones. Currently, any person who moves into a sampled housing unit after the initial contact is interviewed as if a bounding interview had already taken place. As a result, a significant proportion of interviews at any particular time in sample are with the first-time, unbounded, respondents. Families or individuals who move into a housing unit after the first housing unit contact, for example, are unbounded. These responses are used in aggregate rate estimation and included in public use data sets, even though their responses are inflated by 30–40%. The validity of the estimates from the NCS, therefore, are affected in at least three ways by the number of people moving into sample:

(1) Over a one-year period about 18% of the NCS interviews are unbounded due to in-movers (Biderman & Cantor, 1984). If these responses are inflated by 40%, then an extremely crude calculation indicates that crime estimates are, on average, inflated by 7% $[1.4(.18) + 1(.82) = 1.07]$.

(2) Estimates of year-to-year change could be unduly influenced by changes in factors that effect mobility rates (e.g., the economy). This is especially troubling given the relationship between mobility and crime.

(3) Analysis of Census and public use data, which currently do not have indicators of bounding status, are suspect because of the high correlation between mobility and victimization. This includes both methodological and substantive analyses (Bushery, 1981; Cantor, 1985).

If shorter intervals were used on the NCS, these problems would be minimized somewhat. For example, changing the follow-up interval to three months would reduce the number of unbounded interviews due to mobility by about one-third (Cantor, 1985). Alternatively, if the interval were increased to 12 months, the number of unbounded interviews contributing to NCS analyses would substantially increase, probably by about 50%. Adjustment of these unbounded interviews is desirable, but is subject to the interpretation problems referred to in the section on panel effects. Ignoring these interviews is obviously impossible, since it would seriously bias the representative nature of the sample.

It is interesting to note that the impact of interview spacing on the NCS would not be affected by changing the design to one that followed movers in the same way SIPP does. Nonkey individuals would still provide unbounded data at the interview that immediately followed the move.

3. RESPONDENT SELECTION

Respondent selection, as used here, refers to the rules and procedures that a survey institutes to select an appropriate reporter for the events of interest. The rules and procedures implemented on a panel survey can affect its analytical utility because of the potential for changes in respondent between waves. These changes can affect the quality of the data the respondents provide and estimation of the overlap of retrospective reports between waves. When choosing a particular respondent selection design, therefore, it is important to consider the implications of respondents switching between various statuses (e.g., proxy to self-respondent).

Three types of respondent status categories are typically relevant when conducting a household survey:

(1) *Self-respondent:* Answers questions referring to his/her own behavior, experiences, and attitudes.

(2) *Proxy respondent:* Answers questions for the person selected in the study.

(3) *Household respondent:* Answers questions referring to characteristics, behavior, experiences, and attitudes of a household as a unit (e.g., family demographics, family economic and legal needs, purchases of household property, characteristics of family relationships).

In this section we discuss the implications of using these three types of respondents in a panel survey.

3.1. Response Quality

There has been a great deal of research assessing the quality of different types of respondents for reporting various events. Much of this research has been undertaken in the reporting of health status and expenditures (Cannell et al., 1963; Cannell & Fowler, 1965; Kovar & Wright, 1973). This work has primarily compared self-respondent reports to proxy reports and has generally supported the idea that self-respondents are better reporters than proxy reporters. One notable exception to this research is that done by Mathiowetz and Groves (1985), who find exactly the opposite in an experiment that simulates the conditions of the Health Interview Survey.

Depending on the respondent rules, a panel survey could result in confounding changes between interviews in respondent status with changes in substantive phenomena. This would frequently happen, for example, if proxy rules are used on the basis of some type of age criterion distinguishing children and adults. The NCS, which collects information for every person aged twelve or over in sample housing units, used proxy respondents for persons who are 12 and 13 years old during most of its collection history. As the 13-year-olds turned 14, however, their respondent status changed and so, probably, did the quality of data they provided.[5] Reiss (1982), for example, presents analyses showing that changes between proxy and self-respondent status between ages 13 and 14 are marked by an unusually large increase in victimization reporting. He argues that this change is largely due to changes in respondent status.

A second problem that arises with the use of different respondent rules is in designs where the person acting as the proxy or household respondent also participates in the survey as a self-respondent. In this case, the exposure to the proxy/household questions may in some way affect how he/she responds as a self-respondent. This reaction could take many forms. It might act as an extra burden and decrease the person's motivation to participate in the survey. Alternatively, acting as a proxy/household respondent might "warm up" or remind the respondent about relevant phenomena and improve survey performance.

This last type of effect seems to be present in the household respondent (HR) design of the NCS. In the NCS, the HR, selected to be a knowledgeable adult aged 18 or over, is asked to report on characteristics of the household and on thefts against common household property. Residential burglary, for example, is considered a crime against the entire household and is usually expressed as a rate per household rather than per person. An HR is used to collect this household information because it minimizes duplicate victimization in aggregate estimates of event counts. If more than one member of the household reports a "household" event, these duplicate reports must be reconciled by the interviewer prior to sending the interviews for processing; the use of a HR design minimizes the interviewer's work.

One problem with this design for conducting a panel survey is that HRs have been found to report substantially more personal victimizations than self-reporters (Biderman et al., 1985). This is explained by the fact that the HR is asked an extra set of victimization screen questions that seem to help in the recall of crime events. Since it is quite common for HRs to switch between panel contacts, these changes can seriously confound the measurement of individual changes in victimization experience.

The magnitude of this problem can be seen from Table 4, which displays aggregate personal crime rates for the second and third NCS interview by respondent status at each interview. As can be seen, a switch between HR and self-respondent (SR) status is associated with a large drop in rates for both "contact" and "no-contact" crimes. For the no-contact crimes, this drop is much larger than any drop associated with time in panel. For the contact crimes, the respondent status effect is as large as the time in panel effect. The seriousness of this effect is compounded after noting that respondent status is correlated with characteristics associated with victimization.

One obvious way of solving the problem of switching respondent status between panel waves is to try to develop protocols that keep the proxy/household respondent consistent. When using proxy respondents

TABLE 4. SIX-MONTH PERSONAL VICTIMIZATION RATES REPORTED AT THE SECOND AND THIRD INTERVIEWS BY STATUS CHANGE SEQUENCE OF HOUSEHOLD RESPONDENT AND TYPE OF CRIME[a]

Interview Status Sequence	Contact Crimes		No-Contact Crimes	
	Interview No. 2	Interview No. 3	Interview No. 2	Interview No. 3
HR(2) & SR(3)[b]	13.6	9.3	30.9	18.6
	(23,334)		(23,334)	
SR(2) & HR(3)[b]	10.6	11.5	18.9	30.7
	(23,334)		(23,334)	
T(2) & T(3)[c]	13.0	11.4	31.2	28.4
	(147,146)		(147,146)	

Source: Biderman et al., 1985.

[a]"No-contact" crimes exclude all thefts involving auto parts. This table refers only to respondents interviewed during the first three times the housing unit was contacted. The population also excludes one-person households.

[b]HR(i) = household respondent at interview i; SR(i) = self-respondent at interview i.

[c]Includes respondents who did not change status between interviews.

for the mentally and/or physically impaired this should not be difficult to implement. Proxies are typically close relatives or friends who are frequently available for interview. For other types of respondent rules, however, this might be impractical. When proxy status is defined by age, it is impossible to prevent changes as the respondent crosses the age boundary. This solution is also impossible to implement in surveys where selection to be the HR is done using random procedures. For surveys like the NCS or CPS, where "the most knowledgeable" respondent is used as a proxy/household respondent, trying to maintain consistency could add substantially to costs. If nothing else, it will increase the number of contacts the interviewer must make to interview the correct respondent.

3.2. Eliminating Reporting Overlap across Panel Waves

Whenever the reporter for a particular respondent or household can change between panel waves, there will be a problem of eliminating overlap between the two reporters' responses. This is a problem when retrospective questions are used and the previous interview is used to mark the beginning of the recall period. If respondents change between interviews, then the bounding control for both the interviewer and respondent is much less concrete, making it more difficult for telescoped incidents to be detected.

The situation becomes more complicated when a respondent moves from one household to another. Household phenomena that occurred prior to the move could be reported by the mover in the new household. Alternatively, if the mover is chosen as a household respondent in the new household, then reporting on household phenomena will not be particularly accurate. In these particular situations, therefore, it is important to ask the respondents where they were living at the time the events of interest occurred. When choosing the appropriate household respondent, it is also important to decide whether people moving into the household are eligible. If so, rules must be devised to decide whether to include events occurring before the move took place.

The NCS has a relatively unusual complication associated with its household respondent design. As briefly mentioned, one of the goals of the NCS is to estimate the number of crime events that take place in the United States. This results in offering statistics comparable to the major crime indicator released by the Federal Bureau of Investigation, which represents criminal events reported to the police. In order to do this, however, the NCS transforms each individual report of victimization to counts of crime events. Since each event can have multiple victims who are eligible to report it on the survey, the weighting scheme must somehow provide for this multiple reporting.

For crimes in which the victim actually sees the offender, this is accomplished by using a multiplicity estimator that divides the final sample weights by the number of other people the respondent reports were victimized at the same event. Assuming that there is little error in these reports and that every person has approximately the same probability of reporting the event, this estimator is unbiased.

This procedure, however, cannot be used for crimes where the victim and offender never see each other. These "no-contact" crimes include all types of personal thefts, such as larceny, burglary, and motor vehicle theft, and actually make up the bulk of the crimes reported on the survey. The strategy used to estimate event counts for these crimes is to have the interviewer eliminate all duplicate reports within the same household while in the field. The assumption here is that most of these crimes will be restricted to the household being interviewed. If one can eliminate these reports ahead of time, then there are no multiplicity problems to worry about when estimating events.

There are a number of problems with this procedure (Biderman et al., 1986), but several that are especially troublesome with respect to a longitudinal design. First, a longitudinal design makes it almost impossible for the interviewer to perform "unduplication" procedures for respondents who move into new households. In the vast majority of these cases, different interviewers will be involved in the pre-move and post-move households and it simply would not be possible to reconcile duplicate events across reporters. Second, if an entire household breaks up between interviews and is scattered into different locations, it is extremely difficult to ensure that any of these sample persons would be chosen as a HR. This would decrease the likelihood that any of the household crimes that occurred against the old household will ever get reported, since HRs are the only ones that are specifically asked about these types of crimes.

One way to solve these problems is to resolve the overlap between victimization reports without having the interviewer do it in the field. For the NCS, this could be achieved by eliminating the HR design. All respondents would be given the same set of questions (Biderman et al., 1986). Along with this uniform procedure, a computer algorithm would be developed that matched incidents, once the reports are processed. For the reporting of victimizations, an algorithm would be developed that depended on several key characteristics, such as the type of crime, month of occurrence, amount of property loss (in the case of property crime) and whether the incident was reported to the police or not. Since none of these items are ever reported with 100% accuracy, some margin of error would have to be allowed in judging a match between different reporters. For other types of survey events that are reported more accurately, such as hospitalizations, fewer criteria might be required.

Matching potential duplicate events on the computer would solve the problem of resolving the overlap between different proxy respondents over time. For the purposes of estimating event counts, however, it is less than ideal because it does not allow for the fact that not all potential reporters for certain victimizing events necessarily reside within the same household. Many larcenies, for example, may involve the theft of a number of items that belong to multiple households in a particular crime incident. This event could theoretically be reported by members of different households if they all happen to fall into the NCS sample. Resolving duplicates within households, therefore, does not allow for the possibility of multiple reporters between different households.

Although this multiplicity problem exists with the present NCS design, transforming the NCS to a longitudinal survey that follows movers would accentuate the problem. For movers, no-contact crime victims would be distributed across multiple households, even though the event was originally restricted within the same household. So, for example, a burglary, which is by definition restricted to one household, could be represented in multiple households if a household splits or breaks up between panel waves.

A second, related, solution to this problem would be to formulate a multiplicity estimator that can be used as an additional adjustment to the final weights. Such a weight would be based on the probability that anyone affected by the incident could report the event, rather than simply downweighting by the number of people that actually do report it (Czaja et al., 1986), as the present "unduplication" procedure now effectively does. There are theoretical problems with this solution, however, because it involves developing rather complex response models for the probability of reporting events.

3.3. Respondent Conditioning Effects

If the proxy/household respondent changes over time, there also could be differential response patterns due to different conditioning effects. For example, if a new proxy is used in the middle of a panel, then there might be telescoping effects associated with unbounded interviewing. Similarly, experience with the survey across respondents will vary at any time in sample as particular individuals build up differential experience as a proxy/household respondent. These differentials could be quite substantial if the proxy/household respondent is also acting as a self-respondent.

There is very little evidence to suggest that respondent conditioning effects associated with respondent selection are particularly large. This, however, is mostly due to the lack of research in this area and the fact that so little is known about the mechanisms that seem to cause respondent conditioning in the first place.

4. MODE OF DATA COLLECTION

As with respondent status, the mode of administration of a panel survey has implications for operational features that could ultimately affect its substantive utility. These include: (1) initial and follow-up cooperation, (2) the ability to effectively track survey respondents and maintain high response rates, and (3) response quality and reliability. In this section these three areas are reviewed with respect to the use of self-administered, telephone, and in-person modes of data collection.

4.1. Initial and Follow-up Cooperation

The initial mode of administering a panel survey is important because it is related to both the response rate at the first interview and establishment of rapport with the respondent. Mode has been shown to be related to response rates on many cross-sectional surveys (Dillman, 1978; Oksenberg et al., 1986) and establishes a set of baseline measures for subsequent interviews. In general, as the mode of initial contact deviates further from being in person, the lower the expected response rates. Low response rates at the beginning of the survey automatically restrict the number of respondents that will be eligible for participating in all interviews. Partly for this reason, many panel surveys use a personal interview as the initial mode of contact.

The initial mode of contact is also generally believed to be important in maintaining response rates for subsequent interviews waves (Sudman & Lannom, 1980, 1984). This stems from the belief that the further one gets away from an initial in-person contact, the harder it is to establish the importance and credibility of the study. For tracking purposes, "cold contact" could make the respondent more hesitant to respond to questions about future intentions on mobility and to give the names of friends/relatives who can be contacted in case of a move.

There is very little evidence that either corroborates or disproves the extent to which "cold contact" surveys affect follow-up cooperation. It is clear, however, that for the average survey, using telephone procedures could produce comparable, although slightly lower, response rates than those conducted in person. Similarly, mail questionnaires can also produce respectable response rates (for example, 60%–80%) if the proper follow-up procedures are used and the questionnaire is not particularly complex (Dillman, 1978). If a cheaper mode than in-person interviewing is used, such as centralized telephoning, the savings can be used to develop procedures to compensate for much of this "credibility gap." Careful preparation of the respondent through the use of advance letters and providing "800" telephone numbers have been found to increase the response rates significantly (Dillman, 1978). Sudman and Lannom (1984) show, for example, that even though in-person contacts produce higher response rates than do "leave with" diaries or telephone contacts, the latter two methods can be comparable to the in-person visits with the proper preparation, incentives, and callback procedures. The concern with response rates when using a telephone or self-administered mode of data collection, therefore, is probably not very different from a similar concern when using these modes for any type of cross-sectional design.

4.2. Tracking and Response Rates

Trying to maintain a high response rate on a panel survey is partly determined by how effectively sampled respondents can be tracked between interviews. The number of respondents that are LTF places an upper bound on the maximum response rate. The mode of interview is important when tracking respondents because it typically determines the types of tracking techniques that are used. Personal interview surveys will use the mail and telephone whenever possible, but will typically place a heavy reliance on the interviewer to do much of the tracking in the field. Similarly, self-administered mail-back and telephone surveys will use the mail to track respondents, but will predominantly use the telephone to confirm the location of particular respondents. These types of surveys rarely send people into the field to obtain this information. Large-scale telephone surveys are increasingly using centralized telephone facilities that have separate interviewing and tracking personnel.

The relative effectiveness of tracking operations dominated by either personal or telephone methods is not clear-cut. Both modes can be extremely effective in tracking respondents (Call et al., 1982). The primary difference between the two probably lies in the types of populations that are the hardest to contact by each mode and their respective costs. In-person field methods have the advantage of not depending on the respondent and/or the respondent's close friends/relatives to have a listed telephone number. As discussed in the paper by Burgess in this volume, tracking by telephone depends, in part, on having friends/relatives of the respondent keep the survey informed on the location of the respondent. If these people are hard to contact by telephone, then this limits the success of finding the respondent.

Assuming a centralized design, tracking by telephone has the advantage of being considerably easier to manage the transfer of cases between different locations. This can be expensive when conducting telephone interviews that have a decentralized set of field offices. It could also place severe restrictions on whom the survey can follow if the administration of the survey requires that it be done in person (e.g., on an examination survey). If someone moves out of a geographic area such as a Primary Sampling Unit, the respondent is essentially lost to follow-up.

4.3. Response Quality

There are three basic concerns when considering how collection mode affects the validity or reliability of longitudinal data: (1) differences in panel effects, (2) differences in the asking of retrospective questions, and (3) changes in mode during the survey.

First, in single-mode designs, panel effects may differ by mode. Duncan et al. (1984) point to evidence that telephone surveys have been found to produce less detailed answers to open-ended questions and to be perceived by the respondent to be longer (Benus, 1975; Groves & Kahn, 1979). This could have implications for the respondent burden aspects of a longitudinal survey. If a telephone survey is considered more burdensome, then it might be important to compensate in other areas, such as the length of the interview or interview spacing.

Second, response quality in answers to retrospective questions might differ by mode. For several different types of phenomena, it is generally believed that personal interviews and self-administered forms are superior to telephone interviews. When conducting an in-person interview, the interviewer has more flexibility when trying to promote respondent recall. It is common for personal interviews to use visual memory aids to facilitate recall. The interviewer might also be able to get a better sense of pace when observing how the respondent is reacting to the recall tasks. In a self-administered form, the respondent has much more time and resources to try to recall and accurately report events. Diaries, for example, result in more accurate reporting of events that are difficult to remember (Sudman & Lannom, 1980, 1984).

One point that is often overlooked when conducting telephone surveys is that use of this mode tends to isolate the respondent from others who might be present at the time the survey is conducted. On the NCS, for example, it was found that during personal interviews, other people were actually present 70% of the time (Martin et al., 1986). This increases the amount of coaching, whether helpful or detrimental, that respondents receive from others. The presence of others can inhibit the reporting of sensitive items. On the other hand, when asking retrospective questions, members of the same household might help the respondent remember and determine events that might be eligible for the survey. This often happens in the NCS, where it is quite common for the entire family to be interviewed at the same time.

Using a self-administered questionnaire to report many different events over a particular reference period poses unique problems of constructing a form with room to report multiple events, and to be able to associate a specific set of detailed questions to each event in an economical way. The NCS redesign, for example, initiated work on a self-administered victimization form, but could not construct a form that clearly communicated how to skip to more detailed questions without including a separate set of forms for each possible report. Including separate sets of questions for more than one or two events made the form too long to expect adequate response rates.

A third impact that mode has on response quality is when changes between interviews might confound changes in substantive phenomena. In on-going longitudinal surveys, changes in mode are probably inevitable if the original interview is done in person. Once the credibility of the survey is established, it is a cost-saving measure to use the less expensive mode. This is the case, for example, for the National Medical Care Expenditure Survey (NMES), the follow-ups to the National Health and Nutrition Examination Surveys I (NHANES I), the National Longitudinal Survey (NLS), the Panel Survey on Income Dynamics (PSID), and the NCS.

Analysis of the NCS indicates that for victimization reporting, this type of switch does make a difference when examining change across interviews. In the late 1970s, the NCS instituted a maximum telephone procedure where every even interview was conducted by

FIGURE 2. ANNUAL VICTIMIZATION RATES BY TIME
IN SAMPLE AND INTERVIEW PROCEDURE

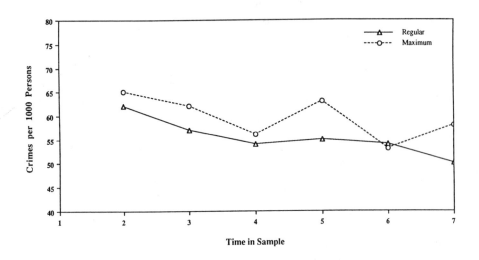

Source: Roman & Sliwa, 1982.

telephone whenever possible. Roman and Sliwa (1982) evaluated the
impact of this procedure on aggregate victimization rates. While they
did not find any effect of this procedure on the aggregate rates, the
response patterns across waves were found to deviate from the normal
time-in-sample decline referred to above. Namely, at later times in
sample, rates went up when switching from the telephone to personal
modes (Figure 2). Similarly, Biderman et al. (1985) examined, on an
individual basis, persons who switched between personal and telephone
modes. This analysis found that switching from the personal to
telephone modes had a statistically significant effect on reporting rates.

There is no clear explanation for why changes between the two
modes would produce such patterns. One possibility is that personal
interviews will produce more reported events because the conditions
under which the interview takes place promote better recall and higher
reporting. A second is that these patterns are a result of the telephone
interview not serving as good a bounding interview as the personal one.
When switching between telephone and personal interviews, therefore,
there might be more telescoping of incidents occurring during the personal interview.

TABLE 5. SUMMARY OF NCS DESIGN FEATURES, LONGITUDINAL COMPLICATIONS, AND IMPLICATIONS FOR NCS ANALYSES

Design Feature	Complication	Effects on NCS Analysis
Interview spacing	Recall effects	Shorter periods produce more reports of victimization
	Panel effects	Shorter periods increase respondent crime avoidance behavior
	Tracking respondents	Longer periods increase number of respondents lost to follow-up
	Migration	Longer periods increase the number of unbounded interviews
Respondent selection	Response quality	Switch household respondents between waves; change the number of personal victimization reports.
		Switch from proxy to self-reporters when turning 14 years old
	Eliminating overlap between moves	Proxy/household reporters' knowledge and abilities change between panel waves
		Eliminating the overlap between household crimes
	Respondent conditioning	Differential bounding when proxy/household respondent changes between waves
		Increases in respondent burden for selected respondents
Mode of interview	Tracking	Initial contact response rates will vary by mode
		Cooperation with follow-up efforts vary by mode
	Response quality	Respondent burden relationship to mode
		Self-administered forms are difficult to design for multiple-victimization reporters
		Switching between telephone and personal

Given the inevitability of switching modes during the life of a panel, it is important to plan for these changes ahead of time. Designing the survey so that changes in mode can be evaluated will facilitate the understanding of how analyses may be affected by such an operation.

5. SUMMARY AND CONCLUSIONS

This discussion pointed to three specific design features of a panel survey that can have substantial effects on the substantive products of the survey. Table 5 summarizes this discussion by listing how each of the three features may affect the analytic utility of NCS longitudinal data.

The NCS serves as a good example for this discussion because even though it is a panel survey, most of its detailed analysis is from a cross-sectional perspective. Thus, none of the analyses attend to such important problems as migration of persons into and out of sample and the consequent effects of previous interviewing experience. Thus, while the longitudinal nature of the survey creates a number of nonuniformities in response patterns, the average user largely has to ignore them in the analysis.

Once the survey is redesigned, and if a panel design that follows movers is adopted, many of these problems will be dealt with by capitalizing on the experience of other panel surveys such as the SIPP. However, design problems unique to the NCS will still have to be solved in order for it to continue to effectively generate a valid data structure and produce annual estimates of crime events. These problems include:

(1) Determining the optimal interview spacing. This will influence the recall properties, panel effects, and migration of persons into the sample.

(2) Determining a valid way to collect and estimate event counts for no-contact crimes.

(3) Determining how cold-contact interviews affect response rates.

(4) Determining how mode of interview might affect estimates of individual level and change.

To accomplish this, more research will be required to examine the causes of the various response patterns that have been discussed throughout this paper. The need for this type of research, however, is not unique to the NCS. Very little is currently known about most of the questions raised above and a better understanding of these response phenomena is essential to the design and utility of all types of panel surveys. Unless this occurs, survey designers will continue making design decisions on such crucial features as interview spacing, respondent selection, and mode of interview on the basis of relatively little understanding of response patterns in longitudinal surveys.

REFERENCES

Bailar, B. A. (1972), "The Effects of Rotation Group Bias on Estimation from Panel Surveys," *Journal of the American Statistical Association*, **70**, 23–30.

Benus, J. (1975), "Response Rates and Data Quality," in *Five Thousand American Families—Patterns of Economic Progress, Vol. 2: Analyses of the First Six Years of the Panel Study of Income Dynamics* (G. T. Duncan & J. N. Morgan, eds.), Survey Research Center, Institute for Social Research, The University of Michigan, Ann Arbor.

Biderman, A. D., & Cantor, D. (1984), "A Longitudinal Analysis of Bounding, Respondent Conditioning, and Mobility as Sources of Panel Bias in the National Crime Survey," *Proceedings of the Section on Survey Research Methods*, American Statistical Association, 708–713.

Biderman, A. D., Cantor, D., & Reiss, A. J. (1985), *Household and Secondary Respondents: A Quasi-Experimental Analysis of Personal Victimization Reporting by Household Respondents in the National Crime Survey*, Bureau of Social Science Research, Washington, DC.

Biderman, A. D., Cantor, D., Lynch, J. P., & Martin, E. (1986), *Final Report of*

Research and Development for the Redesign of the National Crime Survey, Bureau of Social Science Research, Washington, DC.

Biderman, A. D., & Lynch, J. P. (1981), "Recency Bias in Data in Self-Reported Victimization," *Proceedings of the Social Statistics Section*, American Statistical Association, 31–40.

Bushery, J. M. (1981), "Recall Bias for Different Reference Periods in the National Crime Survey," *Proceedings of the Section on Survey Research Methods*, American Statistical Association, 238–243.

Call, V. R. A., Otto, L. B., & Spencer, K. I. (1982), *Tracking Respondents: A Multi-Methods Approach*, Lexington Books, Lexington, MA.

Cannell, C. F., Fisher, G., & Bakker, T. (1963), *Reporting of Hospitalization in the Health Interview Survey* (PHS Publication No. 1000, Series 2, No. 6), U.S. Government Printing Office, Washington, DC.

Cannell, C., & Fowler, F. (1965), *Comparison of Hospitalization Reporting in Three Survey Procedures* (Vital and Health Statistics Series, No. 2, DHHS Publication No. (PHS) 81–3281), National Center for Health Statistics, Hyattsville, MD.

Cannell, C., Miller, P. V., & Oksenberg, L. (1981), "Research in Interviewing Techniques," in *Sociological Methodology 1981* (S. Leinhard, ed.), Jossey Bass, San Francisco, pp. 389–437.

Cantor, D. (1985), "Operational and Substantive Differences in Changing the NCS Reference Period," *Proceedings of the Social Statistics Section*, American Statistical Association, 128–137.

Cash, W. S., & Moss, A. J. (1972), *Optimum Recall Period for Reporting Persons Injured in Motor Vehicle Accidents* (Vital and Health Statistics Series 2, No. 50), U.S. Public Health Service, National Center for Health Statistics.

Cohen, S. B., & Burt, V. L. (1985), "Data Collection Frequency Effect in the National Medical Care Expenditure Survey," *Journal of Economic and Social Measurement*, 13, 125–151.

Cohen, S. B., & Cohen, B. B. (1985), "Data Collection Frequency Effect in the National Medical Care Utilization and Expenditure Survey," *Proceedings of the Section on Survey Research Methods*, American Statistical Association, 511–516.

Cox, B. G., & Cohen, S. B. (1985), *Methodological Issues for Health Care Surveys*, Marcel Dekker, New York.

Czaja, R. F., Snowden, C. B., & Casady, R. J. (1986), "Reporting Bias in Sampling Errors in a Survey of a Rare Population Using Multiplicity Counting Rules," *Journal of the American Statistical Association*, 81, 411–419.

Dillman, D. A. (1978), *Mail and Telephone Surveys: The Total Design Method*, John Wiley and Sons, New York.

Duncan, G. J., Juster, T. F., & Morgan, J. N. (1984), "The Role of Panel Studies in a World of Scarce Research Resources," in *The Collection and Analysis of Economic and Consumer Behavior Data in Memory of Robert Ferber* (M. A. Spaeth & S. Sudman, eds.), Bureau of Economic Research, University of Illinois, Champaign, pp. 301–328.

Frankel, J., & Sharp, L. M. (1984), "Dimensions and Correlates of Respondent Burden: Results of an Experimental Study," *Proceedings of the Fourth Conference on Health Survey Research Methods* (DHHS Publication No. (PHS) 84–3346), National Center for Health Services Research, U.S. Public Health Service, U.S. Department of Health and Human Services, Washington, DC, pp. 205–212.

Gordon, T., Moore, F. E., Shurtleff, D., & Dawber, T. R. (1959), "Some Methodologic Problems in the Long-Term Study of Cardiovascular Disease: Observations on the Framingham Study," *Journal of Chronic Diseases*, 10, 186–206.

Groves, R. M., & Kahn, R. L. (1979), *Surveys by Telephone: A National Comparison with Personal Interviews*, Academic Press, New York.

Jabine, T. B., Straf, M. L., Tanur, J. M., & Tourangeau, R. (eds.) (1984), *Cognitive Aspects of Survey Methodology*, National Academy of Sciences Press, Washington, DC.

Kobilarcik, E. L., Alexander, C. H., Singh, R. P., & Shapiro, G. M. (1983), "Alternative Reference Periods for the National Crime Survey," *Proceedings of the Section on Survey Research Methods*, American Statistical Association, 197–202.

Kovar, M. G., & Wright, R. A. (1973), "An Experiment with Alternate Respondent

Rules in the National Health Interview Survey," *Proceedings of the Social Statistics Section*, American Statistical Association, 311–316.

Martin, E., Groves, R. M., Matlin, J., & Miller, C. (1986), *Report on the Development of Alternative Screening Procedures for the National Crime Survey*, Bureau of Social Science Research, Washington, DC.

Mathiowetz, N., & Duncan, G. J. (1984), "Temporal Patterns of Response Error in Retrospective Reports of Unemployment and Occupation," *Proceedings of the Section on Survey Research Methods*, American Statistical Association, 652–657.

Mathiowetz, N., & Groves, R. M. (1985), "The Effects of Respondent Rules on Health Survey Reports," *American Journal of Public Health*, **75**, 639–644.

Mathiowetz, N., Northrup, D., Sperry, S., & Waksberg, J. (1986), *Analysis of Field Trials to Test and Evaluate Alternative Strategies for Linking the NSFG to the NHIS* (Report prepared for the National Center for Health Statistics, Hyattsville, MD).

Napier, J. A. (1962), "Field Methods and Response Rates in the Tecumseh Community Health Study," *American Journal of Public Health*, **52**, 208–216.

Nelson, J. F. (1984), "Modeling Individual and Aggregate Victimization Rates," *Social Science Research*, **13**, 352–372.

Neter, J., & Waksberg, J. (1964), "A Study of Response Errors in Expenditures Data from Household Interviews," *Journal of the American Statistical Association*, **59**, 18–55.

Office of Management and Budget (1986), *Federal Longitudinal Surveys* (Statistical Policy Working Paper No. 13, National Technical Information Service PB86-139730).

Oksenberg, L., Coleman, L., & Cannell, C. F. (1986), "Interviewer Voices and Refusal Rates in Telephone Surveys," *Public Opinion Quarterly*, **50**, 97–111.

Penick, B. K. E., and Owens, M. (1976), *Surveying Crime*, National Academy of Sciences Press, Washington, DC.

Reiss, A. J. (1978), *The Residential Mobility of Victims and Repeat Victims in the National Crime Survey* (Data Report No. 5, Reports on Analytical Studies in Victimization by Crime, prepared under LEAA Grant No. 75 55–99–6012 and 6013), Institute for Social and Policy Research, Yale University, New Haven, CT.

Reiss, A. J. (1982), *Victimization Productivity in Proxy Interviews*, Institute for Social and Policy Studies, Yale University, New Haven, CT.

Roman, A. M., & Sliwa, G. A. (1982), *Final Report on the Study Examining the Increased Use of Telephone Interviewing in the National Crime Survey (NCS)* (Memorandum dated August 9, 1982), U.S. Bureau of Census, Washington, DC.

Rosenbaum, D. P., Lewis, D. A., & Grant, J. A. (1985), *The Impact of Community Crime Prevention Programs in Chicago: Can Neighborhood Organizations Make a Difference?*, Center for Urban Affairs and Policy Research, Northwestern University, Evanston, IL.

Shaw, C., & McKay, W. (1972), *Juvenile Delinquency in Urban Areas* (rev. ed.), University of Chicago Press.

Sudman, S., & Bradburn, N. (1972), "Effects of Time and Memory Factors on Response in Surveys," *Journal of the American Statistical Association*, **68**, 805–815.

Sudman, S., & Lannom, L.B. (1984), "Methods of Collecting Health Behavior and Expenditure Data," in *The Collection and Analysis of Economic and Consumer Behavior Data in Memory of Robert Ferber* (M. S. Spaeth & S. Sudman, eds.), Bureau of Economic Research, University of Illinois, Champaign, pp. 329–356.

Sudman, S., & Lannom, L. B. (1980), *Health Care Surveys Using Diaries* (DHHS Publication No. (PHS) 80–3279), National Center for Health Services Research.

Traugott, M. W., & Katosh, J. P. (1979), "Response Validity in Surveys of Voting Behavior," *Public Opinion Quarterly*, **43**, 359–377.

U.S. Bureau of the Census (1978), *The Current Population Survey: Design and Methodology* (U.S. Bureau of the Census Technical Paper No. 40), U.S. Government Printing Office, Washington, DC.

Yaffee, R., & Shapiro, S. (1981), "Medical Economics Survey Methods Study: Cost Effectiveness of Alternative Sample Survey Strategies," in *NCHSR Research Proceedings Series: Health Survey Research Methods Third Biennial Conference* (Department of Health and Human Services Publication No. (PHS) 81–326), U.S. Government Printing Office, Washington, DC.

NOTES

[1]Telescoping refers to misdating the occurrence of an event. Omissions refer to failing to report an event that actually occurred.

[2]Recently there has been a great deal of collaborative activity between survey researchers and the cognitive science community that is trying to apply psychological memory models to survey design principles (Jabine et al., 1984). Discussion of these more detailed theories of recall, however, is beyond the scope of this paper.

[3]Biderman & Lynch (1981) discuss a number of problems with these studies that limit their utility for evaluating the accuracy of respondents reports.

[4]The NCS redesign project ultimately recommended adjusting the first-interview data on the basis of increasing the reliability of the annual estimates. The adjustment used was formulated so as to simply make unbounded interviews "look like" bounded ones. No reference to which is "more correct" was made.

[5]The NCS no longer uses proxies for 12- and 13-year-old respondents.

Major Issues and Implications of Tracing Survey Respondents

Richard D. Burgess

1. INTRODUCTION

In planning a panel or longitudinal survey of persons, or of groups of persons (households or families), the question of tracing, or tracking, respondents who have moved must arise. At the one extreme, tracing may not be an important issue simply because the number of "movers" is very small. At the other extreme, tracing may pose such difficulties that it is necessary to use a survey methodology that, although otherwise not as suitable, will substantially reduce or eliminate the need to trace.

Even if some survey respondents have moved, tracing need not be part of the survey methodology. Sampled units requiring tracing can be dropped from the survey, or substituted or adjusted for as a form of nonresponse. Such options, however, will necessitate making assumptions that may not be supportable. Not infrequently the characteristics of "movers" will be different from those of the remainder of the population. For some respondents the reason for moving may be related to the subject matter of the survey.

Including a thorough tracing apparatus in a survey may be costly and will extend the length of data collection and processing. At the same time, it will not necessarily ensure that a high proportion of traces attempted will result in a contact with the sampled person or household. Cases not traced, despite all efforts, may represent an important uncertainty in estimates from the survey.

In this paper some of the major issues and implications in planning and carrying out a tracing operation are discussed. General methods and matters related to the extent to which a survey organization is able to trace people are discussed in Section 2. In Section 3 various tracing operations and the use of information sources are discussed. Some considerations for planning and costs are also discussed.

The verification of tracing and matching is discussed in Section 4. Some methods used, or which may be used, to deal with cases for which tracing or matching have been unsuccessful are discussed in this section.

In Section 5 major tracing operations for a coverage and response quality evaluation study associated with the 1986 Census of Population of Canada are discussed. The various legal and other constraints placed on these operations and the particular methodology used are presented. Some results of similar tracing operations from the 1981 Census of Population are also discussed in this section. Finally, a few brief summary remarks are given in Section 6.

2. TRACING ENVIRONMENT

For a given survey, the success of tracing respondents is largely controlled by three factors: (1) the general method of tracing; (2) the length of time between the two survey dates in question; and (3) the survey organization's ability to efficiently obtain access to reliable information necessary to trace the survey respondents. There will be, of course, a relationship between these three factors. That relationship, however, will be unique to each type of survey organization and perhaps to each individual organization.

2.1. Methods of Tracing

Survey respondents can be traced, in one of (or a combination of) three ways—retrospective, forward, or reverse tracing.

2.1.1. *Retrospective Tracing.* What can be termed "retrospective tracing" is a procedure by which survey respondents are traced only as the need arises (Hogan, 1983). In the case of a longitudinal survey, this means that an attempt would be made to contact each respondent at the address of the previous interview. If the respondent has not moved (and is identifiable), then the survey interview is conducted as required. If the respondent has moved (or is not properly identifiable), then the respondent must be traced. This may be done by asking the current residents of the whereabouts (or identity) of the survey respondent, by asking neighbors, and/or by various other procedures.

This type of tracing appears to make little distinction between tracing for a panel survey and tracing for a follow-up survey or for a one-time survey of persons selected from some form of list or frame.

Retrospective tracing must be used at the first stage of a panel survey if a sample of persons (as opposed to dwellings) is selected from a list for which addresses are not current, complete and correct. For later stages of the panel survey there may be advantages to collecting, at the time of the interview, information that can be used for later tracing. There will, of course, be some cost and response burden associated with this "forward tracing" of all respondents.

2.1.2. *Forward Tracing.* There are various methods that can be termed "forward tracing" (Hogan, 1983). All of these have some procedure built in whereby a current or more recent address is obtained for each survey respondent prior to conducting the survey (or some iteration of the survey). For a panel or longitudinal survey, forward tracing may be particularly appropriate.

At the initial interview the respondent could be asked to send a forwarding address if he/she moves, or to provide the name and address of a person who would know his/her whereabouts. He/she could be asked for an identification number (e.g., a Social Insurance Number), assuming the survey organization has access to administrative records that contain the identification numbers and up-to-date addresses.

If the intersurvey period is of sufficient length, then tracing at regular intervals within this period could be done. If the intersurvey period is five years, for example, it may be more cost-effective to trace a person five times over five years than once after five years. Tracing may be retrospective in nature, and/or information obtained from the initial interview may be used. It may entail the use of administrative files, city directories, interviews, etc. There may or may not be contact with the respondent. The object, however, would be to maintain a periodically up-to-date address for each survey respondent.

In general, forward tracing is more timely, in terms of the survey completion·date, than retrospective tracing. Comparative costs will depend upon the nature of the survey, the survey organization, and other factors.

2.1.3. *Reverse Tracing.* In very specialized situations it may be possible to use what will be termed "reverse tracing." It is useable if the first stage of the panel or longitudinal survey is a census. It cannot be used beyond the first iteration or second stage of a sample survey.

In such an application a sample of persons with known addresses is selected from a current list of persons. It may be a sample of dwellings with persons within the households forming the sample. For purposes of longitudinal comparison, the respondent is asked where he/she resided at the time (previous) of the reference census. Matching will be to that stated address, or to some set of addresses if there is some uncertainty or ambiguity.

This technique is potentially useful only to survey organizations that have access to micro-level data from a census, or to a census of a subset of the population. Some amount of follow-up will be required if there are discrepancies between the census and survey records. The use of this methodology assumes that such discrepancies (for example, different names or household composition) will be infrequent.

2.2. Length of Intersurvey Period

If there is only a short time between stages or iterations of a survey, then forward tracing may be cost-effective only for surveys of very particular population groups or for certain groups (e.g., young adult students) within the survey sample. As the intersurvey period increases, tracing becomes a more important requirement, and more difficult and time consuming if done retrospectively. Expenditure of resources on forward tracing thus becomes increasingly cost-effective.

Reverse tracing will suffer from errors of memory as the intersurvey period increases. Typically, several previous addresses would have to be requested and some follow-up undertaken.

2.3. Tracing Information

The objective of tracing is for the most part directed to determining the current address of the survey respondent. In doing this a survey organization will be restricted by the survey schedule and budget, and the organization's own infrastructure, its potential sources of information, how it may use information obtained, its self-imposed constraints, and the cooperation of its sources.

The survey schedule and budget are an obvious constraint on tracing. The infrastructure of the survey organization will have a direct impact on the cost and timeliness of tracing, and/or the methods used. The geographic dispersion of people who have moved may make it impossible for some survey organizations to carry out tracing. Not only may people who move great distances be difficult to trace, it may be beyond the resources (and perhaps even the purposes) of a survey to trace nationally, or even beyond a limited region.

The survey organization must establish a set of sources of tracing information that are available to it—forwarding addresses provided by respondents, relatives, contacts established through the respondent, neighbors, employers, directories, administrative files, etc. It must establish criteria for the use and limits of use of these sources under various circumstances. It must work within its legal powers— stemming from the organization itself, or from the laws or acts under which it operates. Some sources may be available only to certain survey organizations. Some will not be available to any outside organization (e.g., Census and Income Tax records). The survey organization must also work within the perhaps unwritten guidelines of what it believes to be proper and to be in its own long-term best interest. This is critical to the continued and increased cooperation of respondents and information sources.

Access to information may imply or require confidential treatment of the data provided. There is an implied requirement if the information is provided because of the nature or reputation of the survey organization or of the survey. If there is a legal or other obligation (on the part of the source or of the respondent to the source) to provide information, then any corresponding obligation on the part of the survey organization to use data obtained on a confidential basis will require that such data not be divulged in tracing. Data collected at an earlier stage of the panel survey itself may also be confidential.

Confidential data that provide an exact location for a respondent are usually not a constraint on tracing for the survey organization. Typically, restrictions must be imposed when interviewing a neighbor, relative, another person with the same surname, etc. In requesting information on the respondent's address it may first be necessary to identify the respondent more specifically than giving his/her name. Stating the respondent's age, sex, and occupation may seem innocuous, but if the source of such data is a census or income tax record (and no publicly available source has been accessed), then these data cannot be used in the interview.

Further, confidentiality usually applies to the individual. Members of the family, for instance, may not be aware of certain facts about the respondent. In conducting a panel or follow-up survey of unwed or single mothers, as an example, the interviewer should realize that the child may have been placed for adoption and the mother subsequently married. It would be quite inappropriate to contact the husband in the course of tracing and divulge the existence of the child.

The survey organization must, of course, guard against obtaining or using information it should not have. In the course of an inquiry a person may provide information that he/she should not. Even when formal arrangements for access can be made it may be tempting to simply deal informally with a colleague, friend, or sympathetic contact. Such methods may prove embarrassing. It is not uncommon, for example, for a respondent to ask how he/she was traced.

Sources will provide different degrees of cooperation to different survey organizations. The extent of cooperation may change over time as laws, public concerns regarding privacy, and the political and economic climates change. To obtain cooperation and to maintain its reputation (for all of its survey functions and needs) the survey organization must deal with the concerns for privacy and the sensitivities of respondents and contacts. In the context of tracing, even legal methods with the most benign purposes may seem insidious to some respondents or to persons being contacted.

To reduce any negative reaction to tracing, it should be possible to inform the respondent of a panel survey that tracing will be undertaken, if necessary. Nonetheless, tracing a respondent is not so important that any and all legal means be used. Some of the people being traced may have separated or divorced, may have died, may have been adopted, or may have run away or gone to jail in the intersurvey period. Persons contacted for purposes of tracing such respondents might be upset by the contact. If pursuing an interview is likely to offend, or is unlikely to lead to reliable information about the respondent, then the interview (and perhaps tracing of the respondent) should be discontinued. It is impossible to set rules for all eventualities. Clearly tact and discretion must be used, but these are subjective.

3. TRACING OF SURVEY RESPONDENTS

The method or methods of tracing to be used, whom to trace, the sources of tracing information used (among those established by the survey organization) and the order in which they are used, and when, should be decided (within the constraints discussed in Section 2.3) on the basis of cost and timeliness. Much will depend upon the nature of the given survey.

3.1. The Nature of the Survey

The nature of the survey—the population of interest, the sample design, the subject matter—may be the key factor in deciding whether to undertake tracing and, among other factors, will be critical to the success of tracing.

Typically a more mobile population will present more difficulties in tracing. In a general, however, population mobility by itself will not make tracing difficult or infeasible. Specific segments of the population of movers may present difficulties. Tracing for a longitudinal survey of persons living in urban slums, for example, is more likely to prove infeasible than for a similar survey of recent medical school graduates.

The potential impact of different sample designs on tracing success will depend upon the population of the survey. For the urban slum population example the sample design may have little impact on the feasibility and cost of tracing. For the medical school graduate population example the sample design would be an important consideration. For this population, primary sample units (medical schools) would be directly linked to tracing information sources (registrar, alumni associations, social clubs, etc.) and the sample units (graduates) would tend to maintain some communications with the tracing information source and with other persons in the sample (Clarridge et al., 1978).

If the longitudinal survey includes subject matter of a very personal, confidential, or sensitive nature, the probability of tracing may be reduced. A respondent who moves may be less willing to provide the survey organization with a forwarding address or to identify contact persons. Information sources may be less cooperative and useful ones less numerous and accessible.

3.2. Whom to Trace

Whether the respondent should be traced directly or someone else traced prior to contacting the respondent will also depend upon the nature of the survey. It will also depend upon the characteristic of the particular respondent (or sample unit) and the nature of the information sources being used. The probability of obtaining useful information for tracing from any given type of source will vary from one respondent to another. Some survey respondents will be difficult to trace directly, regardless of what information sources are available.

In order to facilitate tracing, a network of persons associated with the respondent should be identified in the course of sample selection, of interviewing various contacts and respondents, and in obtaining information from various sources. The value of establishing a preliminary network at the outset of the survey is two-fold. First, it permits identification of those respondents for whom such associations are limited or nonexistent and who thus might be considered for forward tracing or for subsampling. Second, some of the persons identified as part of the network will not have moved in the intercensal period and thus can be contacted without having to be traced.

As noted in Section 2.1.2, respondents can be asked during the initial interview to provide the names and addresses of persons who would know their whereabouts. At the same time the names of all adult members of the respondent's household should be requested (if this is not already an aspect of the survey). For persons living outside of a family unit the names and addresses of parents and/or of adult brothers and sisters will be useful. For older persons the names of sons, daughters, nieces, nephews, etc., should be requested.

Many information sources identify persons who have some close association with the respondent—the source file used to select the sample, administrative and civil records, directories, voter lists, etc. The names and any available characteristics of such persons should be recorded for possible use in tracing. Some of these associations will be discovered during tracing. The association may only be due to the address. The use of a telephone directory is perhaps the most obvious example of how such an association might be identified.

At the outset or at any stage of tracing it is necessary to decide whom to trace. A child, for example, is unlikely to be listed in a telephone directory. There may be little value in tracing a child directly (unless using school or birth records, for example). In this case, people contacted during tracing may be more cooperative and able to provide useful information if tracing is largely restricted to locating parents and guardians.

In many cases it will be useful to trace the respondent and at least one other person associated with them simultaneously or to commence tracing the other person(s) if a few quick attempts to trace the respondent are unsuccessful. This will be particularly helpful for respondents who are single adults. A daughter may marry, change her surname, and move. If a parent can be traced, then he/she probably will be able to provide the address of the selected person and any name change.

3.3. Sources of Information and Their Use

Identifying useful sources of information for tracing is in part a matter of ingenuity and perseverance in dealing with individual cases and circumstances. However, some basic sources that can be used for tracing with a variety of surveys are listed in Table 1. (Access to some of these sources may not be universal.) Some of the sources listed can provide two types of information—official records and the knowledge of individuals within the organization about local residents. The post office and police are examples of such sources. The post office also provides a third form of information in that some people can be traced through the forwarding of mail (Crider et al., 1971; Eckland, 1968).

The sources of information used for tracing and the order used will depend upon the nature and constraints of the survey and its size, the survey collection methodology, the probability of obtaining useful information for any given respondent from the various sources, and the cost.

Tracing techniques can be classified into three types: (1) interviewing techniques; (2) manual-based administrative and other record searches; and (3) computer-based administrative record searches and matching.

The first of these can be, in the extreme, an exercise of the interviewer's ingenuity or caprices. The second is meant to include any systematic manual procedure of checks in directories or administrative lists or records. The distinction of the third is that it is computer based.

Most tracing work would use the first and second techniques together. The third technique would typically be used in conjunction with the first two. Different techniques could be used for different subgroups of the population together with forward or retrospective tracing.

If the survey is conducted by face-to-face interviewing, then for some respondents the need to trace will be identified as a result of an attempt to interview (some amount of forward tracing may already have been conducted). In such cases it may be most efficient at this stage to use interviewing techniques; that is, at this time ask the current occupants and neighbors for information on the whereabouts of the respondent.

For mail or telephone surveys, for large surveys, and for cases where it is unlikely that current occupants or neighbors will know the whereabouts of the respondent, techniques other than a purely interviewing technique should be used at the outset.

The first stage of tracing should be one that is likely to be successful for a large proportion of cases to be traced and for which the unit cost is low. Different information sources could be used for different subgroups of the survey sample. Administrative, civil, or institutional records might be appropriate initial sources, particularly if an identification number is available and usable for matching.

Telephone and city directories are some of the most useful sources of

TABLE 1. SOME BASIC TRACING INFORMATION SOURCES

Reference:	*Government:*
Telephone directories	Post office
Telephone directory assistance	City/municipal records
City/suburban directories	—Land titles
—Numerical, reverse, street	—Assessment/tax records
	Motor vehicle/driver license records
General Public/Companies:	Civil records
	Department/agency records
Neighbors	Voter lists
Apartment managers	
Banks and trust companies	*Others:*
Credit bureaus	
Real estate companies	Utility companies
Social clubs, cultural centers	School, college, university records
Landlords	Subscription/mailing lists
Employers	

tracing information (Booth & Johnson, 1985; Clarridge et al., 1978; Crider et al., 1971; Eckland, 1968). These are particularly appropriate sources for retrospective stages of tracing and for tracing operations that are organized into stages or operations according to the source being used. Directories can be searched quickly and almost all households within the given geographic area are represented. Their use is likely to provide (at low cost) information that will lead to a respondent; thus, they can be used for all cases to be traced. Telephone directories can be used to locate the respondent household, relatives of the respondent, or persons identified prior to or during tracing as possibly knowing the whereabouts of the respondent. City directories can be used to identify neighbors and sometimes other household members and employers. For purposes of tracing, both current telephone and city directories and those current for the previous stage of the survey or of frame creation can provide useful information (e.g., if the respondent was not listed in the telephone directory at his/her original survey address, but another household member was, then use of a current directory should include a search for this other household member).

Some sources will provide useful information only for a subset of cases or only at high cost. To use these sources it is necessary to have some prior information indicating that the use of such sources has some meaningful chance of being productive. Information would be requested from a university only if there were evidence that a person in the sample attended that school. A telephone directory for a distant city would be searched only for those respondents for which there was evidence that they had moved to that city.

The value of some sources may vary from one geographic area to another. A rural post office, for example, may be a more useful local source than an urban post office. Some postal areas may provide better services in the forwarding of mail.

3.4. Planning/Costs

Unless circumstances for the survey are such that tracing is a trivial matter, a decision on whether to trace survey respondents, and how, should be an integral part of planning at the earliest stages of the survey. The viability of the survey and efficient use of resources may depend upon the tracing methodology.

A requirement to trace can be a critical factor in deciding issues as basic as the survey population, the sample design and the overall survey methodology. For the survey of persons living in urban slum areas, for example, it might be necessary to exclude dwellings and buildings occupied mainly by transient persons and/or to include only families with one or more children. It may be necessary to include only persons registered for welfare or some form of community support. Such restrictions may severely modify the scope and objectives of the survey.

The survey of medical school graduates is one where, from a tracing, cost, and timeliness perspective, a cluster sample design (as a first-stage sample) may be preferable to a design not involving cluster sampling.

The problems of tracing using an existing but out-of-date frame or list may force the survey organization to create its own list or to conduct a large preliminary survey so as to identify (a sample of) the population of actual interest.

To decide which combinations of forward and retrospective tracing and techniques to use, it is necessary to have reliable information on the mobility of the population of interest. For some surveys, useful information on the mobility of the population of interest could be obtained at the first stage of the survey. Some knowledge of the correlation between mobility and the characteristics of interest to the survey could be critical as well. Such factors as the cost and timeliness of tracing various groups within the population should be considerations, not only in making a decision on tracing, but also in the sample design and allocation.

Each of the three tracing techniques cited in Section 3.3 has different resource and possibly cost implications. The first two (interviewing methods and manual checks) are demanding of human resources. These techniques are not difficult to plan and would be used unless there is a long-term need for a tracing methodology for a given population, and/or a very large volume of tracing to be done. In these circumstances, the third technique (computer-based) could be used if feasible. This technique will be very demanding of computer-related resources.

Computer-based tracing techniques may be complex. Processing costs may be high. Planning for such techniques should consider the cost and timeliness, and expertise and skill requirements of manual versus computer methods. Whether a system is to be used on more than one occasion is an important consideration as well.

The availability (prior to tracing) of an identification number for each respondent would permit a computer search of an appropriate administrative file. Nonetheless, some problems and exceptions should be anticipated. More important, how the information obtained as a result of the search is used may not always be straightforward. Typically it will be necessary to know whether the address obtained corresponds to a place of residence and whether the respondent has moved. Are these to be determined manually or by computer?

Comparison of addresses to determine whether a respondent has moved requires matching algorithms or · record-linkage software. Matching respondent records to an administrative file using name, characteristics, and possibly address—when an identification number cannot be used—also requires such software. The cost of development of such software would be prohibitive except for large and/or often repeated surveys. Even if this software is available in generalized form, the need for supporting expertise and software may currently place computer techniques for tracing beyond the means of many surveys.

4. VERIFICATION AND CASES NOT TRACED

There is no guarantee that tracing will always be successful or that the correct person will be the one traced. If cases are improperly traced or not traced, then there will be some effect on the quality of survey results. To minimize these forms of error it is necessary to incorporate a decision and verification methodology for cases to be considered matched and a weight adjustment or imputation methodology to deal with cases not traced (Platek, 1985).

4.1. Verification

The effect of different matching algorithms or rules has been studied for various surveys and the results in some cases have been dramatic—in terms of the matching rates and the results of the studies (Cooley & Cox, 1981; Miller & Groves, 1985; U.S. Department of Commerce, 1964). The uncertainty or potential for error introduced to a longitudinal survey may be particularly great if the matching is, in effect, being carried out by the interviewer in the course of the survey. In such cases the rules to determine what constitutes a match must be clear and straightforward. The results of such tracing or matching should be subjected to subsequent verification.

In order to decide that the respondent has been identified or traced, it is necessary to have some set of data for the respondent that is the same for each of the two or more stages or iterations of the survey to be linked. These data must permit, with very high probability, the correct linking of the survey records. At the same time matching data must not include variables that, if different, will constitute part of the survey results. (For example, if the survey is measuring change in occupation, then similarity or dissimilarity of occupation should not be a factor in deciding whether the correct person has been traced.)

Difficulties will arise if the matching data are similar but not identical, and/or if there is some nonresponse or erroneous response. Matching may be less certain for common names, for small households, and for persons who have moved.

If the respondent or the household has not moved in the intersurvey period, then (assuming identification of the housing unit is unambiguous) verification of the match or of the respondent's status (and existence) should be straightforward. Some nonresponse or discrepancies are tolerable, but they will introduce uncertainty.

If a respondent has moved, verification may be more difficult and the criteria for acceptance of a match (or of the trace) typically will be more stringent. The name and basic demographic characteristics of the respondent alone generally will not be sufficient to accept the trace. There should be some information that links the traced person to the original address or the known respondent to the traced address. An identification number, names of other household members (if they move with the respondent), a forwarding address provided by the respondent or other contact, and/or a confirmation through the interview should in most cases provide this link. One must also guard against spending time and resources in tracing the wrong person, especially to the extent that this contributes to the number of cases not traced.

The use of directories and administrative files (without using an identification number) can present some problems as information gaps may occur—i.e., what links the respondent with the address given on the file or in the directory? There is no guarantee that a record or entry identified with the same name and characteristics will represent the respondent. If the file or directory is large or if the respondent is not listed or recorded, then it is likely that someone else with name and characteristics similar to the respondent will be listed.

4.2. Not-Traced Cases

It is reasonable in surveying most groups of the population to expect to contact or trace 80% to 90% of all respondents even over very extended intersurvey periods (Booth & Johnson, 1985; Clarridge et al., 1978; Crider et al., 1971). Complete success should be considered an unreasonable target, although something approaching it for many surveys should not be. Nonetheless, even one respondent not traced must be dealt with, just as any form of nonresponse—as a "not stated" category, adjusted or imputed for, or ignored (by publishing percentages only).

Not-traced persons for a longitudinal survey represent another form of nonresponse, however. Some respondent characteristics (age and sex, for example) will be known from previous interviews. Unless there is an identification problem it will also be known that the respondent has moved (or has died).

An adjustment of the weights of traced cases to compensate for the not traced seems the most appropriate course of action for a longitudinal survey. Imputation would be reasonable if there are only a few not-traced cases or if few data are being collected or published for the survey. The steps taken in tracing or attempting to trace a respondent, whether the respondent was listed in a source (traced or not), as well as standard characteristics, should be used in defining weighting groups. The weighting groups should be formed to minimize any distortion of geographic distributions and to minimize any bias introduced by the tracing methods used.

A potential for bias exists if traced and not-traced persons within the weighting group have greatly different probabilities of being traced by the methods employed, or if there is some correlation between the survey estimates and the probability of a respondent being traced. Consider, for example, a survey on change of educational attainment. If a register of practicing doctors is used as a tracing information source, then (assuming this list is complete and up to date) traced persons found on that list ideally should not be included in the same weighting or imputation group as persons not traced and not found on that list. If such a respondent was found on a register of university graduates but still not traced, then the weighting group that includes that respondent could be defined accordingly to include only persons found on that list (if numbers permit)—traced and not-traced. In general nonmovers and not-traced persons (who are movers or deceased) should not be included in the same weighting or imputation group (see Kalton & Kasprzyk, 1986).

5. A CASE STUDY

Tracing of respondents is used extensively in coverage and response evaluation studies for the Census of Population in Canada. While the purposes of these studies are not typical of panel surveys, the studies represent valid examples of tracing scenarios. The largest of these studies is the Reverse Record Check (RRC). This study, with a sample of approximately 36,500 persons, has been conducted as part of the Census of Canada since 1966. The objective of the RRC is to provide various estimates of population and household undercoverage. The study is also used to produce estimates of response quality based upon longitudinal comparisons of data for a sample of persons as well as to do some longitudinal analysis.

The largest proportion of the sample is selected from the previous Census. These addresses, therefore, must be updated over five years (with a Census in Canada every five years). Between 60% and 65% of the selected persons are expected to undergo some degree of tracing. Almost 95% of these are expected to be traced.

The survey organization in this case is Statistics Canada, a federal government agency mandated to collect data and information for statistical purposes. It has the authority to obtain information from a broad range of public and private sources, including administrative files of federal and provincial government departments. It has eight permanent regional offices across Canada with experienced interviewers who can participate in tracing.

The same Act that provides Statistics Canada with the authority to collect or obtain information prescribes penalties for respondents who refuse to provide the requested information. This vested authority provides the legal basis for all of the activities necessary to locate a particular respondent in order to satisfy the requirements of a statistical activity. Nonetheless, the ability to trace a respondent, given this legitimate purpose, still rests largely upon cooperation—the cooperation of other government departments and of members of the general public. The Act also guarantees that micro-data and information collected under authority of the Act not be divulged by Statistics Canada staff, with penalties prescribed for so doing. The cooperation of other government departments in many cases would and could not be forthcoming without the authority and guarantee of confidentiality provided by the Act. However, persuasion rather than legal authority has been the most effective force in securing formal access to records in a timely and efficient manner.

The cooperation received by Statistics Canada from the general public is, no doubt, based upon many historical and cultural factors. Of perhaps greatest importance is the well publicized confidentiality guarantees of the Act. In requesting information from the general public for RRC, there have been few complete refusals. (It is not known how much misinformation is deliberately given. The results of tracing indicate that this possibility could influence the outcome of no more than 5% of cases for which tracing is required). This level of cooperation may have depended upon the nature of the survey involved. Surveys that, from the respondent's perspective, have a less clear purpose, and/

or involve a controversial subject matter may have less success.

5.1. Overview of the Reverse Record Check

There are two key elements of the RRC methodology (Statistics Canada, 1984): (1) the construction of an independent list of persons among whom are included all persons who should be enumerated in the Census and (2) the tracing/follow-up operations undertaken to identify those persons on the list who should have been enumerated, and to determine their usual place of residence as of Census Day. Conceptually, these two elements will permit matching to the Census to determine the accuracy of the enumeration.

The independent list is constructed on a sample basis from four frames:

(1) A sample of persons enumerated in the previous Census (for the 1986 Reverse Record Check, the 1981 Census is the source of this sample).

(2) A sample of intercensal births, selected from vital statistics records (available within Statistics Canada).

(3) A sample of intercensal immigrants, selected from records of Employment and Immigration Canada.

(4) A sample of persons missed in the previous Census—which is the sample of persons determined to have been missed by the previous (1981) Reverse Record Check.

Given the nature of the constructed list, certain selected persons will not be enumerated in the 1986 Census for valid reasons—e.g., intercensal deaths and emigrants. Further, the address information obtained at the time of sample selection will be out of date and thus will not represent an individual's usual place of residence for the 1986 Census. Tracing operations are therefore conducted to determine the status of the selected persons (i.e., to determine which selected persons should not be enumerated) and the Census Day usual place of residence of the selected persons who should have been enumerated.

5.2. Constraints on Tracing for Census-Related Studies

The policy for tracing in studies associated with the 1986 Census of Population requires that tracing "be employed in a discrete, ethical and authorized manner" (Statistics Canada, 1986). This policy is directed to three major concerns: (1) confidentiality of data, (2) the privacy and sensitivities of individuals and families, and (3) the reputation of the Census and of Statistics Canada.

Microdata obtained under authority of the Act may not be divulged in tracing. Data obtained during tracing must also be kept confidential. The interviewing process for tracing is, under these circumstances, very restricted. Address information may be used in tracing as a point of attempted contact. For any contacts with the general public, only the selected person's name and publicly available data may be used.

If the selected person is under the legal age of majority, then a parent/guardian of the selected person must be interviewed. There may be exceptions, for example, when the selected person is sixteen years old and can provide reliable information. The parent/guardian, however, must be aware of the nature of the contact, and its purpose, prior to the interview. If the selected person is an intercensal birth, then a parent will be the person traced. When dealing with the public, any direct or indirect references to the child, or to a child, which suggest a parental relationship for an identifiable person may only be made to the person known to be the parent. If the vital statistics record or a (government) administrative file indicates the child has been adopted, then the "birth parents" are not to be contacted (Statistics Canada, 1986).

In using administrative files, only data needed for tracing are requested and accessed. Telephone numbers obtained from administrative files can be used only if they can be determined to be "listed." In current practice no attempt is made to obtain information·from the administrative files of police or credit agencies, or from financial institutions. (In small communities and rural areas police offices may be contacted to determine the whereabouts of selected persons. These contacts are made to take advantage of the familiarity of officers with their communities and not for reference to police records). All contacts with the general public (including selected persons) are made by Statistics Canada employees.

5.3. Tracing for the Reverse Record Check

Prior to 1986, tracing for the RRC had been, for the most part, retrospective. For 1986, prior to Census Day, forward tracing without contact (so as not to condition respondents and thus bias the results of the study) was used (Burgess & Charron, 1985). A change was made for several reasons, including reducing the proportion of cases not traced and reducing the period of time required for tracing by eight months.

For the 1981 RRC, 3.4% of the persons in the sample were not traced. While this rate may be acceptable for many longitudinal survey purposes, it resulted in uncertainty in an estimate of undercoverage that was approximately 2%.

For a more typical longitudinal survey, timeliness might have been the more important of the two issues. For 1981, tracing operations commenced in July 1981 (Census Day was June 3) and ended as scheduled late in August of 1982. Results were available some eight months after publication of the Census population counts and eight months after what would have been expected for a survey of comparable size that did not require tracing of respondents. Reducing the duration of tracing to a large extent could be achieved by assigning more resources to this task. The use of forward tracing is, however, a much more efficient method when the proportion of cases to be traced is high (see Section 5.6).

5.3.1. *Initial Updating of Addresses.* For most selected persons, family and household members—as of the date of the previous Census, or at the time of birth or immigration of the selected person—are known. Tracing is of the selected person, and/or of up to two other related household members (to be called contact persons). The first step matched selected persons and all other household members to an administrative data file containing names and addresses, which had originally come from Income Tax records. The reference date of this file was April 1981, just prior to the 1981 Census. The volume of records required that matching be done by computer.

The second step consisted of address matching to confirm possible matches or to distinguish among possible matches. If an address did not match but would otherwise have been considered a definite match based upon name and characteristics, it remained a match.

The Income Tax based administrative data file was chosen because it was expected to contain at least one person record for more than 90% of Canadian households. It contained the necessary data for matching. It is updated annually, and each record has an identification number that remains the same for each person from year to year. The 1981 year was chosen as it provided addresses close to Census Day 1981. Many matches, therefore, could be confirmed based upon address. It was also less expensive to match within geographic areas (provinces) rather than matching nationally, as would have been necessary had a file for a more recent year been used.

The identification numbers for all persons matched to the administrative data file were used to extract the corresponding addresses from the 1985 version of this file—the most current version of the file available prior to Census day.

This updating was not considered adequate for births and immigrants. It was not completely adequate for selected persons who were age 15–19 at the time of the 1981 Census or 65 and over in 1986. These groups would have a low probability of being traced using only the Income Tax based administrative data file. Manual matches to three other administrative files were carried out for the births (to a Family Allowance file), immigrants (to a Social Insurance Number file), and those aged 65 and over (to an Old Age Security file). Overall, at least 90% coverage was expected for all but the immigrant cases.

It was expected that many persons aged 15–19 at the time of the 1981 Census would have subsequently moved away from their 1981 address. To update the address for these selected persons a manual match to the 1985 file was carried out. More than 50% of these cases were assigned updated addresses as a result.

5.3.2. *Searching and Subsequent Tracing.* For a typical longitudinal survey the next stage would be an attempt to contact and interview the selected persons. For the RRC this was not necessary. The next stage was carried out after completion of the Census collection. At regional processing sites, where completed Census questionnaires were sent for coding and data capture, there was a searching/matching operation. The updated addresses were used for this purpose. It was expected that at least 70% of all selected persons would be found enumerated at this stage. This compares with 36% of all selected persons enumerated at this stage of the 1981 RRC, for which there was no forward tracing. (Actual results for 1986 were not known at the time of writing.)

For selected persons not identified at this stage on a 1986 Census questionnaire, a further stage of tracing is carried out at the eight Statistics Canada regional offices. In many cases a contact person or other household member will be enumerated during the searching operation at the processing site. If this were the case, then that person is telephoned as the initial step of regional office tracing. If no such person had been found enumerated during the earlier searching, then the initial step is an attempt to obtain a telephone number for the selected person or a contact person. Telephone directories in the area of the last known address are searched and telephone directory assistance services used.

Information obtained during this initial step is followed up until the selected person is located or until continuing to follow up on the information seems inefficient. (For example, names of potential relatives or employers might be obtained and they would be contacted.) If the selected person is not contacted, then the next step will be to identify and telephone neighbors, using city directories. In turn this may lead to landlords, employers, relatives, etc.

At some point the RRC manager in the regional office must decide whether to continue tracing through local means or to commence searching provincial administrative files. The administrative files are searched unless there is reliable information that some other source is likely to provide information on the status or location of the selected person. Administrative files relating to health care, drivers' licences, or motor vehicles are searched. If necessary, cases are transferred between regional offices.

As a last resort all persons with the same surname in the local telephone directory are telephoned. This might require some publicly available data—no matter how old—for the selected person or contact person. This is not done for very common names. (Persons listed in the telephone directory with the same surname and initials are telephoned earlier in tracing.) It is not uncommon for 50 or more telephone calls to be made in an attempt to interview one selected person.

Telephone tracing and interviewing are the norm. Face-to-face contacts are carried out as necessary. Cases not resolved are referred to the head office for assistance—additional administrative file searching—or for a decision on whether to continue tracing. For the 1986 RRC this work continued into January 1987.

5.4. Verification of RRC Tracing

Attempts to locate most selected persons use, as a starting point, an address, or at least a specific area, where the selected person was known to have resided.

For the manual matching for the 15- to 19-year-olds there is clearly a potential gap in tracing information. There may be a gap in the tracing for births, immigrants, persons 65 years and over, and difficult to trace cases. This is particularly true for the intercensal births and immigrants. (While a typical panel or longitudinal survey would not be concerned with intersurvey births etc., the same records could be the starting point for sample selection for such surveys). Two types of problems historically occur for these latter cases. There is a high rate (perhaps 10%) of mismatches for births that are detected upon contact with the household or by matching to the Census questionnaire. This results in some recycling through the tracing system and perhaps an increase in the number of cases not traced because of insufficient time.

The major difficulty for the immigrant selected persons is with names. Name spelling and order vary from one document or source to another. Sufficient information to verify that the correct household has been located is not necessarily available. The same types of problems occur for the Census portion of the sample, although for a smaller proportion of cases.

While various checks are performed, where this is feasible, there is no definitive method of ensuring that all traces (or matches) are correct.

5.5. RRC Tracing Results

Results for the 1986 RRC are not available at the time of writing, but results for the 1981 study provide information on what can be expected for this study and the methods used. The basic results of tracing and interviewing for the study are given in Table 2. Table 3 provides information on the method or source of information that resulted in a case requiring tracing being traced.

TABLE 2. 1981 RRC FINAL CLASSIFICATION OF SELECTED PERSONS

Final Classification	Census		Birth		Immigrant		Missed		Total	
	Cases	%	Cases	%	Cases	%	Cases	%	Cases	%
Traced	29,761	97.1	3,211	92.3	1,392	96.1	807	96.1	35,171	96.6
Enumerated	27,541	89.8	3,096	89.0	1,113	76.8	696	82.9	32,446	89.1
Deceased	1,056	3.5	33	0.9	5	0.3	26	3.1	1,120	3.1
Emigrated/abroad	299	1.0	34	1.0	111	7.7	24	2.8	468	1.3
Missed	865	2.8	48	1.4	163	11.3	61	7.3	1,137	3.1
Not Traced	895	2.9	267	7.7	57	3.9	33	3.9	1,252	3.4
Total	30,656	100.0	3,478	100.0	1,449	100.0	840	100.0	36,423	100.0

TABLE 3. 1981 RRC CASES TRACED BY SOURCE[a]

Method/Source	Census		Birth		Immigrant		Missed		Total	
	Cases	%	Cases	%	Cases	%	Cases	%	Cases	%
Telephone/city directory	11,006	35.9	—	—	—	—	—	—	11,006	30.2
Federal government admin. data[b]	1,900	6.2	3,079	88.5	384	26.5	209	24.9	5,572	15.3
Mail out[b]	—	—	8	0.2	398	27.5	332	39.5	738	2.0
Interview	3,781	12.3	124	3.6	600	41.4	266	31.7	4,771	13.1
Not traced	895	2.9	267	7.7	57	3.9	33	2.9	1,252	3.4
Tracing not required	13,074	42.6	—	—	—	—	—	—	13,074	35.9
Total	30,656	100.0	3,478	100.0	1,449	100.0	840	100.0	36,423	100.0

aTraced means information obtained led to final classification other than "not traced."
bMail out was not used for 1986 as it was believed that direct contact could be made by telephone for cases traced.

5.6. RRC Tracing Costs

Consideration of the costs associated with tracing for the 1986 RRC, given in Table 4, will be useful in comparing tracing costs for a manual approach (a combination of manual record and directory searches and interviewing techniques) with one that depends on a manual and computer approach. Tracing costs would vary, however, from one sample allocation to another, from one survey to another, and from one survey organization to another. The actual costs are therefore not important in themselves. The relative size of costs and their nature—fixed or variable—are worth examination.

The cost of tracing at the regional office stage for the 1986 RRC (the cost per case is for only 30% of the selected persons) includes the cost of completing a questionnaire. Among these cases will be the more difficult to trace cases, and those that will not in the end be traced. The costs for items 4, 5, 6, and 7 are mostly variable costs. The other components contain a large element of fixed cost. The cost of systems preparation, or development, does not include software for record linkage. This was already available as a generalized package. The cost of human resources to prepare supporting software, to test weights for matching or linkage, to program edits, etc., are included in this item.

The cost of tracing without a significant element of computer-based tracing would not be different given the sample size and allocation. A reduction in costs for items 1 and 2 of Table 4 would result in a corresponding increase for items 6 and 7. If the sample size had been increased there would have been some savings for the computer approach. Since the systems were developed for 1986, to the extent that they are reusable, costs in 1991 should be lower for the computer approach by a minimum of 25%.

6. CONCLUSION

Developing and organizing a tracing operation can be a challenging and interesting undertaking. Tracing activities and quality, however, can be difficult to monitor and control. The operations have to be planned, developed, and organized in such a way that the needs of the survey are met within cost and timeliness constraints. More important, confidentiality and other requirements of the survey organization must be respected. In deciding to undertake a survey, in setting specific objectives, in planning and designing the sample and questionnaires etc., the needs and limitations of tracing should be considered.

How tracing is conducted and what sources of information will be used will depend upon the nature of the survey and of the survey organization. While there are no prescribable methods and techniques to cover all situations, there are basic operations (e.g., telephone inter-

TABLE 4. APPROXIMATE TRACING COSTS FOR
THE 1986 REVERSE RECORD CHECK

Tracing Component	Percentage of Selected Persons Involved	Approximate Cost Per Selected Person Involved	Approximate Total Cost
1. Systems preparations	100%	$ 4	$146,000
2. Computer processing	100	6	219,000
3. Procedures preparation	100	2	73,000
4. Head office manual	20	5	37,500
5. Processing site searching	100	2	73,000
6. Regional office	30	23	252,000
7. Head office searching	30	12	131,400
Total	100	26	931,900

viewing) and information sources (e.g., the respondents themselves, telephone and city directories) that will be common to almost all survey tracing.

Historically, for most surveys at least 80% of respondents can be contacted if tracing is used. Higher rates of success in excess of 90% should be expected, however, even for extended intersurvey periods. The potential for bias as a result of not tracing all survey respondents will be greater to the extent there is correlation between the subject matter of the survey and population mobility.

The potential impact of not-traced cases on survey results should be an integral part of analysis and description of the quality of the survey results.

REFERENCES

Booth, A., & Johnson, D. R. (1985), "Tracing Respondents in a Telephone Interview Panel Selected by Random Digit Dialing," *Sociological Methods & Research*, **14**(1), 53–64.

Burgess, R. D., & Charron, A. (1985), *Proposal for Tracing in the 1986 Reverse Record Check* (Unpublished document), Statistics Canada.

Clarridge, B. R., Sheehy, L. L., & Hauser, T. S. (1978), "Tracing Members of a Panel: A 17-year Follow-up," in *Sociological Methodology*, Jossey-Bass, San Francisco, pp. 185–203.

Cooley, P. C., & Cox, B. G. (1981), "An Automated Procedure for Matching Record Check and Household-Reported Health Care Data," *Proceedings of the Section on Survey Research Methods*, American Statistical Association, 418–423.

Crider, D. M., Willits, F. K., & Bealer, R. C. (1971), "Tracing Respondents in Longitudinal Surveys," *Public Opinion Quarterly*, **35**, 613–620.

Eckland, B. K. (1968), "Retrieving Mobile Cases in Longitudinal Surveys," *Public Opinion Quarterly*, **32**, 51–64.

Hogan, H. (1983), "The Forward Trace Study: Its Purpose and Design," *Proceedings of the Section on Survey Research Methods*, American Statistical Association, 168–172.

Kalton, G., & Kasprzyk, D. (1986), "The Treatment of Missing Survey Data," *Survey Methodology*, **12**, 1–16.

Miller, P. V., & Groves, R. M. (1985), "Matching Survey Responses to Official Records: An Exploration of Validity in Victimization Reporting," *Public Opinion Quarterly*, **49**, 366–380.

Platek, R. (1985), "Nonresponse Adjustment Procedures in Sample Surveys: Discussion," *Proceedings of the First Annual Research Conference*, U.S. Department of Commerce, Bureau of the Census, 462–468.

Statistics Canada (1984), *1981 Census of Population—Public Use Reverse Record Check Tape User's Guide*.

Statistics Canada (1986), *1986 Census of Population—Policy Decision Record* (Working paper).

U.S. Department of Commerce, Bureau of the Census (1964), *Evaluation and Research Program of the U.S. Censuses of Population and Housing, 1960: Record Check Studies of Population Coverage* (Series ER 60 No. 2).

Collection and Design Issues: Discussion

Stanley Presser

A key element in the data collection plan for longitudinal surveys is the treatment of nonrespondents. In many panel surveys, nonrespondents to earlier waves are not recontacted in later waves. As no attempt is made to reinterview nonrespondents in this design, a noninterview at one point in time automatically becomes a noninterview at all future waves.

The cumulative nature of this attrition has a clear implication for response rates. Consider a survey with a within-wave response rate of 94% that does not return to nonrespondents: By the tenth wave, there will be observations on barely half the original sample.

Given the increase in potential nonresponse bias, why exclude non-respondents from later waves? The practice is based on the assumption that providing an interview is very highly related across waves. On its face this seems plausible. Individuals who are not found at one point in time are more likely not to be located at another time. Similarly, those who refuse an interview at one wave are probably more apt to refuse at other waves.

There is evidence, however, that the size of these relationships is more modest than generally assumed. Consider the experience of the Wisconsin High School Class of 1957 study, which consisted of three waves conducted in 1957, 1964, and 1975. In 1975 the study was able to locate 99% of those who had provided an interview in 1964, compared to 86% of those who had been nonrespondents in 1964 (Clarridge et al., 1978). Thus although tracing success was correlated over time, most of those for whom there were no wave 2 data were found in wave 3.

How many of these 1964 nonrespondents who were located in 1975 provided an interview? Unfortunately, Clarridge et al. (1978) are silent on this issue. But relevant data are available from a similar study. Wave 3 of the National Longitudinal Survey of the High School Class of 1972 returned to nonrespondents from earlier waves. Of those who had provided an interview at wave 2, 95% were interviewed in wave 3. By comparison, 52% of the wave 2 nonrespondents were interviewed at wave 3 (Bailey, 1976). Again, there was a correlation over time, but the yield from the prior wave's noninterviews was substantial.

Moreover, after a nonrespondent became a respondent the chances of retaining the individual in succeeding waves were excellent. Of those interviewed at wave 3 who had been nonrespondents at wave 2, 88% provided an interview at wave 4 (Wisenbaker, 1981). Thus the payoff to returning to prior wave nonrespondents persisted over time.

No doubt there are limits to this strategy in a panel with many waves. In wave 4 of the National Longitudinal Survey of the High School Class of 1972, individuals who had never provided an interview (who had been nonrespondents to each of the three prior interviews) were recontacted. The response rate? 13% (Wisenbaker, 1981). Obviously one can reach a point of diminishing returns.

In addition, one needs to ask about the quality of the data provided by individuals who were previously nonrespondents. In the National Longitudinal Survey of the High School Class of 1972, the completion rate on key items was somewhat lower for this group than for those who had responded to all prior waves (Wisenbaker, 1981). This may have been due in part to the fact that the survey was conducted largely in self-administered fashion by mail. Interviewer-administered surveys may be less prone to the problem.

Overall, though, the evidence reviewed here suggests that nonrespondents from at least some prior waves should be included at later waves.[1] In order to locate individuals who were not interviewed at the last wave, the kinds of tracing discussed by Burgess are essential.

Many applications require what Burgess calls "forward tracing." The farther apart interviews are spaced, the more important this becomes. Burgess suggests asking respondents in the initial interview for a forwarding address in the event they move. Other survey researchers recommend gathering more extensive information including names, addresses, and telephone numbers of persons likely to know the respondent's whereabouts. This strategy should be considered even in one-time cross-section surveys, to facilitate later, unplanned reinterviews.

Burgess also notes that a useful supplement to collecting this kind of information is communication with sample members between interviews. The Panel Study of Income Dynamics, to take one example, pays respondents $5.00 if they provide their current address on a postcard that is mailed out prior to each wave of interviewing (Survey Research Center, 1972). The National Longitudinal Survey of the High School Class of 1972, to take a second example, included a newsletter reporting on results of the study with its between-wave address card (Bailey, 1976).[2]

Of course, even when these techniques are employed, addresses for some sample members will still be incorrect. Tracing then resembles detective work, involving the creative use of various sources of information. The sources used in the Canadian Reverse Record Check (presented in Burgess' Table 1) constitute a comprehensive list. As Burgess suggests, however, privacy laws will restrict the use of some of these sources to government surveys.

One source with few access problems that covers a large fraction of most populations is the license registry of state motor vehicle bureaus. It was a useful tracing tool on a recent Institute for Social Research study (Jennings & Markus, 1984). Exactly how useful, however, is unclear, because records of the number of respondents located through different channels were not kept. Regrettably, such information is rarely published, even in technical reports.

A much-needed model of reporting practice on this score is provided by the Wisconsin High School Class of 1957 study. Clarridge et al. (1978) present a table showing the yield of their eight different tracing sources (alumni records, the post office, etc.) for both the total sample and 11 subgroups (based on respondent characteristics such as sex and education). If more studies documented and reported exactly how tracing was conducted and with what results, we would be in a better position to evaluate what works and what does not. (This is true for other aspects of data collection as well.)

In carrying out tracing operations it is important to bear in mind Burgess' cautions about confidentiality and public reaction. Clarridge et al. convey this in vivid terms:

> We sometimes discovered that people did not want to be found. Some owed money for child support and others moved frequently, leaving a trail of unpaid bills. Sometimes sources would refuse to give us information because they wanted to protect the respondent from harassment or because the respondent had a serious problem such as alcohol or mental illness. (Clarridge et al., 1978, p. 196)

There will be times when the needs of survey research must give way to other, nonresearch needs.[3]

The point made above that little is usually published about the details of field work is related to how little we know about the effects of various design features. There is a split in the survey world between the operations people, who know a great deal about the conditions under which data are produced, and the researchers, who analyze data but have limited, if any, first-hand experience with field work. The two groups' differing recruitment patterns and interests, as well as the way operations jobs are structured, mean that those who know the most about data collection are unlikely to carry out research about what they do.

The result is that survey analysts not only have scant hands-on experience with these matters; they also have few research reports to turn to for enlightenment. For example, some practitioners say that matching respondents to the same interviewers across waves increases efficiency, but this is based on impressionistic evidence. Like the justification for many data collection practices, it seems plausible, yet it would be lovely to know for sure. And that requires systematic research.

David Cantor provides a useful way to think about some of this much-needed research. He organizes his discussion in terms of the effects on data quality of three survey decisions: spacing between waves, respondent selection, and mode of administration. Data quality, in turn, is seen in terms of three factors: response rates, recall problems, and panel effects or conditioning.[4]

There are difficult trade-offs here. Reducing the interval between waves may make it easier to trace respondents but harder to get cooperation from those who are located. Likewise, shortening the between-wave interval may reduce recall errors but increase conditioning effects.

The trade-offs involve other factors as well. For example, with fixed resources, decreasing the between-wave interval and thereby increasing the number of waves implies a reduction in sample size. Conversely, greater precision of the results (a larger sample) could be purchased by increasing the interval between waves. Thus a given decision might decrease sampling error but increase potential nonresponse bias, while at the same time reducing conditioning and increasing telescoping.

How are researchers to make informed judgments about these trade-offs? Even if there were extensive research on the issues, such judgments would not be easy. But the paucity of relevant studies means we are frequently operating in the dark.

REFERENCES

Bailey, J. P., Jr. (1976), *National Longitudinal Study of the High School Class of 1972: Second Follow-up Survey, Final Methodological Report*, Research Triangle Institute, Research Triangle Park, NC.

Clarridge, B. R., Sheehy, L. L., & Hauser, T. S. (1978), "Tracing Members of a Panel: A 17-year Follow-up," in *Sociological Methodology 1978* (K. F. Schuessler, ed.), Jossey-Bass, San Francisco, pp. 185–203.

Jennings, M. K., & Markus, G. B. (1984), "Partisan Orientations over the Long Haul: Results from the Three-Wave Political Socialization Panel Study," *American Political Science Review*, **78**, 1000–1018.

Kalton, G. (1986), "Handling Wave Nonresponse in Panel Surveys," *Journal of Official Statistics*, **2**, 303–314.

Kalton, G., McMillen, D., & Kasprzyk, D. (1986), "Nonsampling Error Issues in the Survey of Income and Program Participation (SIPP)," *Proceedings of the Second Annual Research Conference*, U.S. Bureau of the Census, 147–164.

Survey Research Center (1972), *A Panel Study of Income Dynamics: Study Design, Procedures, and Available Data: Vol. 1*, Inter-University Consortium for Political and Social Research, Institute for Social Research, The University of Michigan, Ann Arbor.

Wisenbaker, J. M. (1981), *Factors Related to Third Follow-up Survey Responses*, National Longitudinal Study Sponsored Reports Series, Research Triangle Park, NC.

NOTES

[1]Although this will create missing data problems for certain longitudinal analyses, it should be noted that the missing data problem is more severe (albeit less visible) when nonrespondents are not included at later waves. Moreover, the data from the waves on either side of the missing wave may be used as the basis for imputation. (For a discussion of the ways to handle the analysis of data with wave nonresponse, see Kalton, 1986.)

[2]When resources do not allow the adoption of these strategies for an entire sample, they could be used for a subsample targeted on the basis of mobility probability.

[3]Much recent social change (e.g., rising crime rates and increasing rates of labor force participation among women) have made survey research more difficult. The growing tendency of women to keep their last name after marrying, by contrast, has made tracing easier.

[4]Cantor's discussion is presented in the context of the National Crime Survey. For a similar discussion with examples from the Survey of Income and Program Participation, see Kalton et al. (1986).

Coverage and Classification Maintenance Issues in Economic Surveys*

Michael J. Colledge

---

1. INTRODUCTION

This review paper was motivated by the requirement to examine coverage and classification procedures during the course of a major redesign initiative at Statistics Canada, the Business Survey Redesign Project. The paper focuses on procedures and developments at Statistics Canada that are relevant to the design and maintenance of panel survey frames.

In fact, there are very few panel surveys at Statistics Canada. Even the Labour Force Survey is not a panel survey in the strict sense of the word, although there is considerable sample overlap between successive months. Statistics Canada does, however, have many repeated surveys that, for both statistical and operational reasons, have a substantial overlap from occasion to occasion and, de facto, a longitudinal component. Although these surveys could, in principle, have been designed as panel surveys, for practical reasons they were not. Thus the premise upon which this paper depends for its relevance is that the issues of coverage and of classification are closely related for all types of repeated surveys, panel or otherwise.

The scope of this review has been essentially confined to existing or planned subannual economic surveys at Statistics Canada. This restricted class of surveys provides ample scope to display issues and problems arising from the creation and maintenance of frames for a dynamic universe. There are occasional references to practices in other statistical agencies, notably the U.S. Bureau of the Census, for the purposes of contrast and illustration of alternative methods.

Writing a paper of this type presents semantic difficulties, because many terms have several slightly different meanings. Even the term "panel survey" itself can be ambiguous. Three terms that must be defined explicitly at the outset are those appearing in the title: "coverage," "classification," and "economic." "Coverage" refers to the set of units constituting the target population and "classification" refers to the descriptive values applied to each unit for the purposes of sampling. However, the distinction is not clear-cut, as the industrial classification of a unit can render it in-scope or out-of-scope for a particular survey—

80

i.e., the classification can affect coverage. For this reason, classification issues are considered along with coverage. Together, coverage and classification define the survey "frame."

"Economic" is used in contradistinction to "social." At Statistics Canada, economic surveys are often referred to as "business surveys." All units of commercial economic production—including manufacturers, builders, carriers, etc.—are referred to as "businesses." Indeed, the term "business" is often extended to the economic activities of professionals, institutions, and governments. This usage has been avoided as far as possible in this paper; however, it is unavoidably embedded in some Statistics Canada nomenclature such as the "Business Register" and "Business Survey Redesign Project."

The paper is structured along the following lines: Section 2 provides perspective to the review by outlining the environment at Statistics Canada in which subannual surveys are designed and conducted. Sections 3 and 4, respectively, deal with the centralized and specialized procedures by which frame data are established and maintained. In Section 5 the results of the review are summarized in terms of generic problems and alternative solutions that have been or will be explored during the course of the Business Survey Redesign Project. Section 6 summarizes the underlying themes and future direction.

A reader who wants to get directly to the heart of the paper can skim Sections 2 through 4 and focus on Sections 5 and 6.

2. ECONOMIC SURVEY ENVIRONMENT AT STATISTICS CANADA

2.1. Program of Economic Surveys

The program of economic surveys at Statistics Canada includes surveys of financial, industrial, commodity, employment, capital expenditure, taxation, etc., statistics, collected on monthly, quarterly, annual and occasional bases.

The program is distributed organizationally over a number of "subject matter" divisions. In all there are roughly 300 surveys (depending upon the definition of survey), of which about 125 are subannual and the balance, annual or occasional. There is no economic census.

In general terms the function of annual surveys is to collect "structural" information, i.e., the values of as broad a range of data items at as fine a geographic and industrial breakdown as resources and response burden permit. The function of subannual surveys is to collect information concerning change over time for a smaller number of key data items at a coarser level of geographic and industrial detail.

A program of such wide scope needs centrally organized efforts to standardize and integrate both systems and data. Three major factors contributing toward this goal are the System of National Accounts (Statistics Canada, 1975), to which many surveys supply data; the Standard Industrial Classification (Statistics Canada, 1980); and the central provision of frame and income tax information for survey use. In principle, the same frame data base should be used by all surveys, as illustrated in Figure 1.

2.2. Characteristic Features of Economic Surveys

Economic surveys have certain distinctive features that must be taken into account in any design or review. Three factors of particular importance in the context of frame construction and maintenance are as follows.

The target populations are often difficult to define precisely, involving consideration of several different but interrelated sets of units. To be more specific, in the world outside a statistical agency there are units corresponding to administrative processes, e.g., payroll deduction; there are legal units, e.g., incorporations under federal and provincial charters; and there are operating units corresponding to the way economic entities organize their operations and keep their accounts. Within the statistical agency there are sets of target statistical units, and there are reporting units that bridge the gap between the statistical targets and the real world capacity to report. The process, termed "profiling" at Statistics Canada, of delineating and relating all the various units in these organizational structures is complex. It is fundamental to ensuring good coverage.

The target populations for economic surveys are usually very heterogeneous in terms of size. For example, the largest 12% of corporations account for over 80% of corporate gross revenue. This serves to emphasize the importance of profiling as well as the need to classify and stratify by size.

Classifying the target populations by industry is nearly always essential for efficient survey sampling and data collection, and for interpreting the resulting statistics. Yet industrial codes are rarely available as data items from the respondents themselves. They have to be assigned, based on respondents' descriptions of their economic activities.

2.3. Policies and Guidelines

A number of current policies and guidelines have particular influence upon survey design at Statistics Canada. These include:

- A requirement to alleviate response burden, especially on small businesses.
- A requirement to maximize use of administrative data in order to reduce costs and response burden. These data can be employed both for coverage and classification purposes and to supplement

or replace collection of data by direct survey.

- A requirement to respond to new demands for data but within a climate of budgetary constraint. This has promoted a commitment to develop and use standardized systems and procedures in all future survey designs or enhancements. In particular, the current policy focuses all major economic survey redevelopment effort through the Business Survey Redesign Project. This initiative is often referred to locally as the Infrastructure Project because the primary thrust is to upgrade the "infrastructure" that provides frame and income tax data for use by economic surveys (Cain et al., 1984).

3. CENTRALIZED FUNCTIONS FOR COVERAGE AND CLASSIFICATION

3.1. Introductory Remarks

To facilitate a review of the coverage and classification maintenance procedures for subannual economic surveys it is necessary to indicate, at the outset, the centralized services available to each survey. In principle, as previously noted (see Figure 1), frames for all subannual surveys should be provided from a single source. In practice the present situation at Statistics Canada is rather more complicated, as illustrated in Figure 2. Coverage and classification maintenance activities are divided between individual, survey-specific systems and three centralized functions, namely the Business Register, the Corporate Tax Data Base, and the Tax Record Access Program. The following subsections outline the objectives and performance of these functions insofar as they relate to subannual surveys. Incidentally, this description was not included within Section 2, as the functions cannot be regarded as an unchanging part of the environment. Indeed, they are presently being extensively redesigned as part of the Business Survey Redesign Project.

3.2. The Business Register

The Business Register system in its present form was implemented in 1972. It was accompanied by a policy statement (Statistics Canada, 1973) defining the five principal objectives of the operation: control of economic survey coverage; control of classification; implementation of a common identifier; maintenance of an inventory of surveys for which each unit is in-scope/in-sample; and tracking changes in units over time.

Thus the original intention was that the Business Register should provide a complete, up-to-date list containing frame data for all economic surveys, although in practice it has not achieved this role (Moser et al., 1980).

FIGURE 1. CENTRAL PROVISION OF FRAME DATA: IDEAL ARRANGEMENT

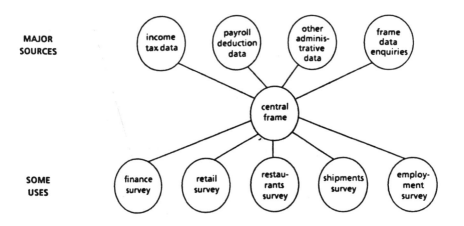

FIGURE 2. CENTRAL PROVISION OF FRAME DATA:
ACTUAL ARRANGEMENT (SIMPLIFIED)

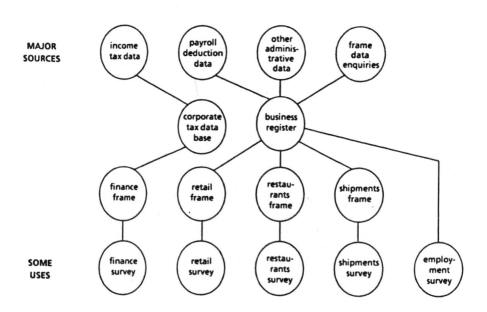

The Business Register Master File currently contains a set of about 1.3 million legal entities, statistical and payroll deduction account records (Statistics Canada, 1985b). It is based primarily on payroll information from Revenue Canada.

The essential features of payroll deduction data in this context are as follows:

(a) The data refer exclusively and exhaustively to "employers" required by Revenue Canada to remit payroll deductions.

(b) Data available monthly in machine-readable form provide updates to employer account name, address, and remittance status; hence they can indicate additional or changed economic production but are not by themselves adequate for delineation or classification of statistical or reporting units.

(c) Data available annually in machine-readable form include total earnings and number of employees paid for each payroll deduction account during the year.

(d) Copies of registration questionnaires sent by Revenue Canada to new account holders are forwarded to the Business Register. However, a new account cannot be taken as a definitive indication of the birth of a new economic unit since existing units can open additional accounts. Completed questionnaires for perhaps 70% of all new accounts are received within four months of the account being opened. The data are generally sufficient for the identification and approximate classification of small unit births, but must be supplemented when classification information is inadequate or when the birth of a large unit is indicated.

In rough terms, payroll deduction data generate 180,000 potential births and 1,000,000 updates per annum and this information is available via the Business Register Master File to all surveys.

The Business Register is also the focus for a range of nature- and structure-of-business enquiries, including: (a) mail cards requesting size, industrial classification or contact details; (b) telephone interviews conducted by Regional Office staff to collect or verify the coverage and classification details for small economic entities with simple unit structures; and (c) personal interviews by Regional or Head Office staff to establish the structural coverage and classification details for large, multi-unit entities.

All frame data collected by the Business Register, with the exception of detailed structural information for large units, are stored in machine-readable form on the Business Register Master File and are available to survey operations. To facilitate use of the information, frame update files are generated monthly for each survey or group of surveys.

3.3. The Corporate Tax Data Base

The primary objective of the Corporate Tax Data Base is to provide income tax data for the Annual Surveys of Financial and Taxation Statistics (Statistics Canada, 1984), and to supplement data collected annually under the Corporations and Labour Unions Reporting Act. However, de facto, the data base provides a frame of corporations.

Data are acquired from Revenue Canada in three forms: (a) a machine-readable file containing five key financial variables, for all corporate tax filers; (b) a duplicate copy of the complete tax return for all large corporate tax filers; and (c) a photocopy of the complete tax return for a small sample of the other corporate tax filers.

All corporate returns appearing for the first time are assigned a standard industrial code by Statistics Canada. This code is verified in subsequent years only for those returns for which copies are received, i.e., (b) and (c) above.

The Corporate Tax Data Base is thus a potential source of coverage and classification information for all subannual surveys. However the data base is not systematically linked to the Business Register and in consequence it is used only by subannual surveys of financial statistics.

3.4. The Tax Record Access Program

The principal function of the Tax Record Access Program (Darcovich, 1982) is to obtain income tax data for small economic entities to supplement the data collected by annual surveys. The program is complementary to the Corporate Data Base in the sense that it focuses exclusively on small units and it includes acquisition of data from personal tax returns indicating self-employed income as well as from corporate tax returns.

Personal tax data are obtained from Revenue Canada in two forms: (a) a machine-readable file containing minimal financial data for all individuals reporting self-employed income and (b) microfilm of the complete tax return for a selected sample.

Jointly with the Corporate Tax Data Base the personal tax files thus provide complete coverage of all economic entities liable to file income tax returns. However, as in the case of the corporate data, the files are not systematically linked to the Business Register. They are not presently used by subannual surveys.

4. REVIEW OF SUBANNUAL SURVEY COVERAGE AND CLASSIFICATION METHODS

4.1. Introductory Remarks

In the previous section the general facilities available for coverage and classification were described. This section contains a review of the frame maintenance procedures actually used by specific subannual economic surveys at Statistics Canada. Since there are many subannual surveys and they are very diverse in objectives, size and methods, a group of fourteen has been purposively selected for more extensive investigation. The group includes four employment surveys, two financial surveys, and eight surveys of production statistics, as follows:

(a) *Employment*: Employment, Payrolls and Hours; Federal Government Employment; Provincial Government Employment; Local Government Employment.

(b) *Financial*: Industrial Quarterly, Financial Institutions.

(c) *Production*: Passenger Bus and Urban Transit; Restaurants, Caterers and Taverns; Retail Trade; Department Stores; New Motor Vehicle Dealers; Wholesale Trade; Current Shipments, Inventories and Orders; Business Conditions.

Each of these surveys involves substantial direct data collection, i.e., none of them is based purely on tabulation of information from administrative sources. Although this sample is not representative of all subannual surveys in the program, it serves to illustrate most of the important frame maintenance procedures and issues.

The target populations of the 14 surveys are all different, each having, in addition to a general goal, its own particular set of inclusions and exclusions. No survey covers all units of economic production. The closest to total coverage is provided by the Survey of Employment, Payrolls and Hours, for which the target population is (nearly) all employers. Each of the production surveys covers a subset of industries, e.g., manufacturing, retail trade, etc. The sampled population sizes range from 750,000 for the Survey of Employment, Payrolls and Hours to 50 for New Motor Vehicle Dealers.

Among the 14 surveys there are three basic sample designs: a census (4 cases); a cut-off census of large units only (2 cases); and a sample survey (8 cases). For each member of the last group, sampling is stratified simple random, the largest units being sampled with certainty. Sample size determination and allocation schemes vary from one survey to another.

The survey designs lead to a need for classification of the units on the survey frame by industry (11 cases), by geography to at least provincial/territorial level (11 cases), and by size.

Data collection frequencies coincide with publication frequencies except for the government employment surveys where collection is monthly and publication quarterly. Data collection is by mail questionnaire (12 cases) or telephone interview (2 cases). Response rates (after follow-up and before replacement of refusals) vary from 30%−80%. They tend to be much worse for small units, particularly in the Industrial Quarterly Survey, where many financial data items are collected (Statistics Canada, 1986c).

4.2. Sources of Coverage and Classification Data

Sources Used for Initial Frame Creation. For each survey reviewed, the initial construction of the frame was based on an administrative source or a previous annual or periodic census. Frames for the Manufacturing Shipments, Inventories and Orders Survey and the Business Conditions Survey were (and still are) derived from the Annual Census of Manufacturers for which the frame has been put together over fifty years from a variety of sources. One survey, the Retail Trade Survey uses an area frame (described in more detail later). This frame was derived from a now superseded version of the Labour Force Survey area frame.

In all instances the initial frame data have been substantially overlaid by subsequent updating information. Thus the original sources of frame data are largely irrelevant except to illustrate that each survey had a different starting point. This diversity of historical beginnings is one of the factors that complicate the goal of standardizing concepts, systems and procedures across all subannual surveys.

Data Sources for Coverage Updates. The sources of coverage updating information available to a subannual survey may be classified into six groups: (1) administrative and commercial lists; (2) data collected by survey questionnaire/interview, i.e., during the course of routine survey operations; (3) data collected as a result of special procedures for contact of surveyed units, e.g., for the purposes of explaining the survey to newly sampled units, for follow-up of nonresponse, etc.; (4) data collected by other surveys; (5) data collected by direct contact initiated by the Business Register; and (6) data collected for units in the area frame.

The administrative sources available are essentially the same as those for initial frame creation, each being regularly updated by the corresponding administrative process and hence providing the basis for identifying births and deaths. The relationship between the source and a survey target population, together with the timeliness and frequency of source updating, are the factors that determine the utility of each source.

By far the most significant source, administrative or otherwise, of potential births and deaths for subannual surveys is the payroll deduction system. Of the 14 surveys reviewed, 9 treat this source, directly or indirectly, as the major source of births and as an indicator of potential deaths. The major problem experienced in using payroll deduction data stems from difficulty of relating the survey target statistical units to payroll deduction accounts. The relationship can be complex for large units, thereby limiting the value of data from this source.

There are many other administrative and commercial sources with potentially useful data for coverage maintenance. Some are relevant to several surveys, others are survey-specific. Examples are telephone subscribers, hydro account-holders, trade journals, business and union registration lists, government releases concerning youth employment programs or departmental reorganizations, financial magazines, etc. The number of sources that can be effectively employed is limited by the need to unduplicate the information, and in this regard there are two basic constraints. First, units are defined to suit each administrative or commercial process and hence may well differ from one source to another. Second, even where units coincide, there is no commonly used business identifier, thus unduplication has to be based primarily on a name and address comparison. The net result is that, in practice, no survey makes extensive use of more than one basic source.

The routine survey data collection process itself, though generally incapable of detecting births, is a valuable source of data on deaths for a survey. Of course, death identification is confined to units currently in sample, and its use is complicated by difficulties in distinguishing dead units from active nonrespondents.

The requirement to establish good respondent relationships with a unit newly selected for survey, or to follow-up nonrespondents, or to check unusual data values, leads each survey to establish procedures for respondent contact in addition to those associated with routine data collection. Such contacts are conducted as part of the survey operation or may be delegated to the Business Register. In either case, the data gathered are an important source of information about deaths, organizational structure changes, and, occasionally, births.

For any given survey, data collected by other surveys (annual, subannual or occasional) are a potentially rich source of unit deaths, and to a lesser extent births. The channel of communication between surveys is the limiting factor in the use of such information. In principle, all frame data obtained by one survey operation of potential relevance to another should be fed back to the Business Register for distribution. In practice this goal has yet to be fully achieved, with the consequence that none of the surveys reviewed is able to take full advantage of such data.

Another major source of coverage information is the program of nature-of-business/structure-of-business enquiries conducted by the Business Register, usually in response to an administrative indication of a frame change or to a request by a survey operation for investigation (Statistics Canada, 1985b). In conjunction with survey operations, this is the most important source for detecting the coverage changes that may follow an internal structural reorganization of a large economic entity.

Area frames are the final source of coverage information considered in this review. In fact there is only one such frame in current use for economic surveys at Statistics Canada. It is part of the monthly Retail Trade Survey. The procedure for creation and use of an area frame in conjunction with a survey list frame is along the following lines. The country is divided into geographic areas. A sample of areas is selected and all the economic units within the sampled areas are enumerated. The results of this enumeration are unduplicated against the survey list frame and those units not found on the list are surveyed. On a regular, usually rotational basis, the areas are re-enumerated to detect births, deaths and classification changes. Thus an area frame can provide coverage for units excluded from the list for any reason, for example, time lag in updating the list, being out-of-scope for the administrative process on which the list is based, etc. An area frame is a safety net. However it is also relatively expensive to maintain compared with a list frame, and it implies cluster sampling, which is less efficient.

Data Sources for Classification Updating. The same data sources as are available to surveys for coverage updating also provide classification updates, but with two major reservations. First, the payroll deduction process is not a source of industrial reclassification information. It can provide geographic and size updates, but industrial information is acquired only through the initial registration form and is thus related only to new account holders. Second, the data collected by subannual surveys themselves generally provide no basis for industrial reclassification except where a questionnaire that is clearly inappropriate to the unit to which it was sent is returned with a comment to that effect from the respondent. Thus the options for detecting industrial classification change are substantially fewer than for detection of changes in size or geography, or for births and deaths.

Data Sources for Contact Information Updating. The set of data sources as previously outlined can also provide updates of name, address, "for attention of," and other reporting arrangements. The major source, however, is the survey process itself, sometimes via the routine questionnaire, more usually via special contact procedures for newly sampled or nonresponding units.

4.3. Systems for Processing Coverage and Classification Data

Systems for processing frame data can be classified into two categories, centralized and survey-specific. All the subannual surveys reviewed have their own autonomous frame data maintenance systems. The degree to which these surveys use the centralized systems described in Section 3 varies from virtually not at all (for Federal and Provincial Government Employment Surveys) to near total dependence (for the Passenger Bus Survey).

In principle it does not matter whether a survey makes more or less use of a centralized system, so long as all the available frame data are processed effectively and, where relevant, are communicated to other surveys. In practice, the bigger the role played by the centralized system the more efficient and effective data communication between surveys is likely to be. This was one of the basic reasons for initiating the Business Survey Redesign Project.

4.4. Procedures for Coverage and Classification Maintenance

Procedures for Births. There are basically two alternative approaches adopted by the surveys reviewed:

(1) Births are added to the frame and subjected to sampling as soon as they are detected and classified. Thus the frame is updated with births between successive survey occasions.

(2) Births are not added to the frame until a subsequent annual update, with the exception that very large and significant births may be added immediately.

Procedures for Deaths. Procedures for handling data concerning the death of a unit may be classified according to the nature and source of the information, as follows:

(a) Death of a unit sampled with certainty, detected by the survey process itself;

(b) Death of a unit sampled with probability less than one, detected by the survey process itself;

(c) Death of a unit not in sample, detected by a source independent of the survey process;

(d) Death of a unit in sample and not responding, detected by a source independent of the survey process;

(e) Death of a unit in sample and responding, detected by a source independent of the survey process.

Based on operational considerations, a unit death detected by the survey itself (i.e., categories (a) and (b)), or consistent with survey observation (i.e., category (d)), will result in the immediate extraction of the unit from the data collection process. For some surveys the unit is removed immediately from the survey frame as well. In other cases, a decision regarding deletion of the unit is deferred until an annual update and, in the interim, zero values are ascribed. For only one survey (the Survey of Employment) is an adjustment made to deal with the potential bias in the selection of future samples caused by updating the frame with information based on a weighted sample unit, category (b).

Death information that is inconsistent with survey observation, category (e), does occur. In the Survey of Employment, for example, there are currently 2,000 in-sample units representing 20,000 population units that are responding to the survey questionnaire and yet, according to the administrative (payroll deductions) source, are dead. This sort of inconsistency is suggestive of duplication on the survey frame or of discrepancies in identification between units on the survey frame and another information source. In most of the surveys under review, this situation occurs infrequently or is not recognized. Procedures for handling it are ad hoc. Sometimes the information is considered too insignificant to be worth the expense of investigating and is ignored.

Procedures for Changes in Unit Structures. Of the surveys reviewed only one (the Industrial Quarterly Survey) has a formal procedure for detecting changes in organizational structures and any unit additions to or deletions from the survey frame that are thus required. Collection of this information is coordinated with the annual update of the survey frame. For the remaining surveys, procedures for detecting and responding to changes in structures are ad hoc. In general, as most structure changes refer to large units that are in-sample with certainty, there are no particular problems posed by updating the sample and frame immediately.

Procedures for Handling Duplication of Units. Notwithstanding efforts to remove duplicate units from survey frames prior to sampling, evidence of duplication does occur in all surveys with substantial target populations. There are no formal procedures for dealing with this situation. In most cases it is handled by eliminating the duplicate units from both the sample and the frame, notwithstanding the bias in future sampling that may result.

Procedures for Changes in Classification. Data concerning changes in classification can be categorized according to their impact in terms of survey stratification (pre- or post-sampling) as either: (i) not involving a stratum change or (ii) involving a stratum change. A unit with a classification change of type (i) can be updated without risk of bias. If the unit is in-sample, estimates (if any) for the domains associated with the two classifications will be subjected to corresponding changes in value. If the unit is not in sample there will be no effect at all on the estimates.

Data concerning changes of type (ii) can be considered as constituting a potential death in one stratum and a birth in another. They can thus be further categorized into four groups similar to those previously defined in (a), (b), (c), and (d) for data concerning deaths. In fact, a change to a stratum that is out-of-scope for the survey is, in effect, a death.

Review of the 14 surveys indicated no consistent pattern of treatment. For the manufacturing surveys, classification changes are held until the annual update. For the Survey of Employment they are held until the annual update or until the unit rotates out of sample if scheduled to do so. In exceptional cases, where the change is considered to be a correction, it may be implemented immediately.

Procedures for Changes in Contact Information. In every survey, to avoid respondent irritation and to reduce costs, changes in contact information are effected as soon as possible, given the facilities provided by the updating system.

5. PROBLEMS AND SOLUTIONS

Based on the detailed investigation of 14 subannual surveys summarized in the previous section, and on a more cursory review of many others, a wide range of generic coverage and classification problems has been identified. These problems are presented below under five broad headings. Each is briefly described together with its immediate consequences and the possible, alternative solutions being considered at Statistics Canada in the context of the ongoing Business Survey Redesign Project (BSRP).

5.1. Coverage—Large Units

Problem 1: For large economic entities the target populations and reporting arrangements are difficult to establish. The consequence is a risk of undercoverage and of duplication of coverage.

Problem 2: Different types of data are available at different levels from different sets of accounts within an organization. There is a risk of refusal or poor quality information if data are requested at an inappropriate level. Intercompany transactions can cause unwanted duplication of data. For example unconsolidated reports from two companies within an enterprise may result in double counting of revenues.

These two problems are illustrated in Figure 3, which indicates the structural organization of a hypothetical medium-sized business enterprise. It comprises three chartered companies owned or controlled by a fourth, the head of the enterprise. It is structured into three operating divisions, each of which has two or more plants, warehouses, outlets, etc. (depending upon the nature of the business). These operating units may at one time have been related directly to the legal units— for example, division X may originally have been belonged to company A before the latter was acquired by the enterprise. However, there is now no direct linkage; operating and legal units are quite separate. In addition, the enterprise has defined a third and different set of administrative units for payroll deduction purposes. It is within this context that a broad range of data items—employment, sales, production

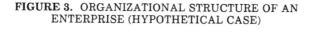

FIGURE 3. ORGANIZATIONAL STRUCTURE OF AN
ENTERPRISE (HYPOTHETICAL CASE)

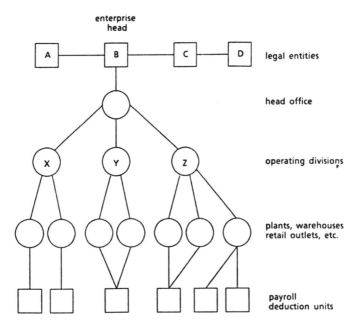

statistics, capital investment intentions, financing, etc.—must be collected.

The solution to both these problems adopted by the BSRP is to define a comprehensive information model (Statistics Canada, 1985a) that recognizes the real world complexity of economic entities. Five distinct types of units are included within the model:

(a) Legal—for example, incorporations under federal or provincial charter.

(b) Administrative—for example, payroll deduction accounts.

(c) Operating—for example, divisions, profit centers, plants, etc., corresponding to the way in which the entity organizes itself and keeps its operating accounts; the legal, administrative and operating units jointly define the view the entity has of itself.

(d) Statistical—the target units for statistical measurement purposes, i.e., the statistical agency's view of the entity.

(e) Reporting—providing the linkage between the statistical target units and the real world economic operating units.

The statistical units are themselves further classified in terms of a four-level hierarchy, each corresponding to capacity to report economic data, as follows:

(a) Statistical enterprise—the highest level of unit, associated with an autonomous operating entity, capable of reporting all forms of economic data about itself.

(b) Statistical company—a subdivision of the statistical enterprise, corresponding to an entity capable of providing an unconsolidated financial report, e.g., a division of an enterprise.

(c) Statistical establishment—a subdivision of the statistical company, roughly corresponding to a profit centre, able to report production statistics and the components of value added.

(d) Statistical location—a subdivision of the statistical establishment, corresponding to an individual plant, warehouse, retail outlet, etc., capable of reporting revenue and, possibly, employment.

Accompanying the definition of an information model is the requirement for a comprehensive "structure profiling" program to build and maintain all these lists of units and their structural relationships, i.e., "structures," as the basis for provision of survey frames.

Problem 3: Changes of structure occur quite frequently and there. are a bewildering number of alternative ways in which the corresponding frames can be updated. In particular, changes of legal structure are commonplace and do not necessarily reflect any significant change in the operating or corresponding statistical structures. If the frame is updated directly in terms of legal or administrative units, the consequence is a spuriously high incidence of apparent deaths and births, and an attendant risk of incomplete or duplicated coverage.

The BSRP solution is to extend its information model to include a set of "standard events" that take place in the real world and are relevant to structure maintenance (Armstrong et al., 1986). Any indication of a change is treated as a "signal" in response to which the corresponding structure is "reprofiled" to see which, if any, of the standard events has occurred. Updating is always in terms of these events.

In the absence of signals, structures are profiled on a periodic basis, at a frequency depending upon their significance and their propensity to change.

Problem 4: Structure profiling is a complex operation. It is time-consuming and can put a heavy demand upon skilled human resources.

The BSRP solution is to restrict structure profiling to a relatively few large units. For the remainder of the target population (i.e., the small units), statistical and administrative entities are equated on a one-to-one basis. In other words, all four levels of statistical units are presumed to coincide within one another and with a unit in an administrative list. The boundary between "large" and "small" units in this context is a compromise between the high cost associated with structure profiling and the statistical impurity resulting from its absence. Within the set of large units, profiling procedures are matched to the size of unit—e.g., free-format, personal interviews for the largest and fixed-format, telephone interviews for the smallest. Maximum use is made of self-profiling questionnaires along the lines of the U.S. Bureau of the Census Company Organization Survey (Konschnik et al., 1985).

5.2. Coverage—Small Units

Problem 5: There is a time lag associated with detection of births, assignment of industrial codes, and making new units available for sampling. The net consequence is undercoverage.

During the course of the BSRP a number of ways of addressing this issue have been considered:

    (a) Classify only a sample of births and use a two-phase sample design;

    (b) Classify and survey only larger units, using other sources to estimate for smaller units;

    (c) Speed up processing by focusing upon systems' bottlenecks, using automated coding procedures, etc.;

    (d) Use an area frame to pick up a sample of births (discussed later—see Problem 9).

A two-phase sample design can be effective not only in cutting down the time lag but also in reducing costs for given sampling variance. It is used, for example, in the U.S. Bureau of the Census Business Surveys (Konschnik et al., 1985). However, the benefits of the technique depend upon being able to use a first-phase sample that is substantially smaller than the target population. Its size is determined by the requirement that there be sufficient units for reliable survey estimates, also, possibly, for sample rotation. Thus, the more numerous and diverse the surveys using the same first-phase sample the larger it has to be, and the less statistically and operatively attractive the technique becomes. The approach has been studied during the BSRP (Foy, 1986) but has not been adopted, primarily on account of the Survey of Employment requirements (for which the U.S. Bureau of the Census has no equivalent).

Restriction of direct subannual survey data collection to larger units, coupled with estimation for smaller units based on annual survey data, is being seriously entertained in the redesign of the Industrial Quarterly Survey.

While automated industrial coding procedures cannot yet match human coding in terms of precision (Estevao & Tremblay, 1986), they have the virtue not only of being much faster but also cheaper and more consistent. An automated procedure is presently used by the Business Register as the first stage in coding registration questionnaires for new payroll deduction accounts. Enhanced systems with wider application are being developed.

Problem 6: There is a time lag in detecting the deaths of units on the frame, or sometimes even a complete failure to do so. Typically this will occur when the frame is based on an administrative source. Units become inactive or dead in terms of economic production, yet are retained on the administrative list because the administrative process is not complete—e.g., payments are outstanding. The consequence is operational inefficiency and reduced effective sample size when dead units are sampled and their inactive status determined. Overcoverage will occur if the death of a unit is not detected and data are imputed as if the dead unit were a nonrespondent.

Three approaches are being considered in the BSRP to deal with this problem:

(a) Information concerning possible deaths is extracted from the administrative source underlying the frame in as exhaustive and as timely a fashion as possible. This implies using administrative data items to deduce that, in terms of economic production, a unit is inactive or dead before the unit is deleted from the administrative list.

(b) Each unit newly entering a survey sample is subject to an "initial contact" prior to data collection to ensure the unit is active. (The purpose of this contact is also to establish whether the unit is in-scope for the survey, check the industrial classification, explain the objectives of the survey and to establish reporting arrangements).

(c) When a dead unit is detected in sample, whether by the initial contact process or as a result of nonresponse follow-up, the information is fed back to the frame maintenance process for possible use by other surveys.

Problem 7: A unit is misclassified on the frame as being in-scope for the survey when in fact it does not belong to the target population. The effect is much the same as for a dead unit, as just discussed—i.e., inefficiency and/or overcoverage.

Various ways of coping with this problem are as follows:

(a) The classification of all units on the frame is reviewed on a periodic basis—for example, along the same lines as the Bureau of Labor Statistics SIC review procedure (Hostetter, 1983)—preferably with a frequency dependent upon propensity of each type of unit to a change of classification.

(b) Each newly sampled unit is subjected to the "initial contact" described in Problem 6(b).

(c) On detection of the misclassification through the survey process the unit is removed immediately from the sample and the classification on the frame is corrected. This may require application of an adjustment to sample and population counts to avoid bias (see Problem 23).

(d) The procedure adopted by the U.S. Bureau of the Census for business surveys when an out-of-scope unit is detected in the sample is to retain the unit in sample until the next reselection and to impute data based on responses from units having the same nominal classification (Garrett et al., 1986). The justification for this approach is that it will provide compensation for the undercoverage of units misclassified as being out-of-scope (see Problem 8).

(e) Conceptually, after detecting a misclassified unit in the sample, the ideal procedure is to transfer the unit with its sampling weight to the appropriate stratum, which, in the case of an out-of-scope unit, would be the corresponding subannual survey for which the unit was in-scope. In practice this is impossible, at Statistics Canada anyway, first, because the set of subannual surveys do not jointly provide full coverage of all economic units and, second, because of operational complexity. It is, however, a goal that will become closer to practical reality with future standardization of the subannual survey program.

Problem 8: A unit is misclassified on the frame as being out-of-scope for a survey when in fact it does belong to the target population. This is the complement of Problem 7. It results in undercoverage.

The ideal but impractical solution was outlined in Problem 7(e). The approach described in Problem 7(d) of compensating for such undercoverage by inclusion of out-of-scope units will not be adopted by the BSRP as there is no firm basis for presuming that under- and over-coverage effects balance for any particular survey. Efforts will be focused instead on classification maintenance and measurement. Undercoverage due to misclassification will be estimated as part of the frame quality assurance program. Such estimates may be footnoted in survey publications.

Problem 9: Multiple sources, usually administrative, from which a complete frame could be constructed and maintained do not always refer to the same type of units or, where the units are identical, the unit identification numbering schemes are different. The result is duplication, the extent of which can be substantial if large numbers of units are involved and unduplication procedures are based on automated matching of names and addresses, etc. There is also scope for undercoverage as a result of false matches.

The BSRP approach is to use only one basic source of frame data for small units and to estimate for units not covered by that source. The alternative is to use multiple frame techniques (for example, see Bankier, 1986), which require only sampled units to be unduplicated. This approach has not been adopted by the BSRP because of its complexity. It would imply either multiple frames for each subannual survey separately or the maintenance of a multiple frame "master sample" meeting the needs of all subannual surveys and implying a two-phase design. The area frame presently used in conjunction with the Retail Trade Survey list frame is not being continued (Statistics Canada, 1986b). This is in contrast to the approach adopted for the Retail Trade Survey at the U.S. Bureau of the Census (Konschnik et al., 1985).

Problem 10: For simplicity of frame construction and maintenance, administrative and statistical units may be equated for sampling purposes (Problem 4). However, there are instances where several administrative units may actually relate to the same statistical entity. This implies duplication.

The solution being adopted by the BSRP is to establish, with the "initial contact" previously described in Problem 6, the number of administrative units that refer to the same statistical entity, and to divide the sampling weight by the number of duplicates. This approach avoids the necessity of linking all duplicate units on the frame, a process that is virtually impossible without direct contact.

Problem 11: As noted above, administrative and statistical units may be equated for sampling purposes. However, there are cases where a single administrative unit would actually cover several separate statistical entities if the structure were to be properly established. The consequence is a lack of precision resulting from the aggregation of data from separate units into a single classification cell.

A BSRP approach to this problem is to establish at an "initial contact" the reporting arrangement that will allow for separate collection and tabulation of data, even though there is only one sampling unit.

5.3. Classification—Small Units

Problem 12: Uncertainties and errors in the industrial classification of administrative units used for statistical purposes result in misallocation of units by industry. Though there are some compensating effects, i.e., misclassifications may tend to balance out, there is also evidence of systematic bias. Special cases of this problem where misclassification causes a unit to be erroneously in-scope or out-of-scope for a survey were previously considered in Problems 7 and 8.

The BSRP approach is along the following lines:
  (a) Establish the level of industrial precision that can reasonably be achieved using the administrative data source.
  (b) Establish the level of industrial precision necessary for an efficient subannual survey design.
  (c) Compare the precision required (b) with that readily available (a) and, in the event of a discrepancy, err on the side of reduced sample efficiency—i.e., cut back the classification detail and use domain estimates.
  (d) Standardize and improve industrial coding procedures; introduce

automated and computer assisted coding where effective.

An alternative to domain estimation, item (c) above, is to have a two-phase sample design with a more detailed classification being given to units selected in the first phase, as previously mentioned in connection with Problem 5. It has not been adopted for the BSRP, first, because it would require a major redesign effort, and second because the first phase sample would have to be very large to accommodate the separate needs of the full range of subannual surveys, hence giving limited scope for savings in comparison with classifying the whole frame to the more detailed level of precision.

Problem 13: The administrative process providing data to enable initial industrial classification of new units provides no mechanism for classification maintenance. The result is a gradual decay in the quality of industrial codes over time and consequent increase in misallocation of units by industry.

The BSRP approach to this problem is fourfold:

(a) Establish the average rates of change of industrial activity and review the classification of all units on the frame periodically, at a frequency matched to the propensity for change, using an inquiry procedure that takes into account the current classification and facilitates self-classification.

(b) Maximize the use of classification data feedback from other reliable sources, especially annual surveys, by introduction of a centralized frame data base.

(c) Use any other classification data as the trigger for a self-classification enquiry (item (a) above).

(d) Cooperate in classification maintenance with the administrative body responsible for the source data.

Problem 14: It is frequently impossible to assert at what point in time a unit changed its industrial classification. Furthermore, an industrial classification may not be precise anyway. It may be based on the larger of two nearly equal revenue-producing activities that may oscillate in significance during the course of a year or two. (See Jabine, 1984, for a definitive account of industrial coding problems.) Classification by size can be equally difficult to pin down to a particular date. Geographic classification is usually much easier, being less subject to short term variation, although there can be gradual migration, for example, from one location to another. The net result is that, if classification changes are applied to the frame as soon as they are detected, there will be artifacts in the estimates resulting from frame maintenance activities, having little or no relationship to what has actually occurred during the preceding month or quarter.

The approach being considered in the BSRP is to define more precisely than previously the rules for assigning and updating unit classification, and in particular:

(a) Require that industrial and size codes be based on a 12-month period.

(b) Consider holding all classification changes detected during the course of processing for a periodic update, at which time all accumulated revisions of both frame and survey data will be made, and users of the data informed what has happened.

(c) Ensure by use of "resistance rules" that classification "flip-flops" do not occur from year to year.

Problem 15: Correction of classification errors, or introduction of genuine changes in classification but with a long time lag, may lead to artifacts that distort the estimates.

This is linked to the previous problem. The approach being considered in the BSRP is to build facilities for backwards revision of all subannual series. Such revision would be coordinated with the "annual update" of frame and survey data just described.

Problem 16: Size classification boundaries based on dollar values are eroded over time by inflation, with the result that sampling designs become inefficient.

Within the BSRP, systems are being designed so that it is relatively easy to respecify boundaries, recompute sample sizes and allocations, and reselect samples. Consideration will be given to the idea of building inflation indices into size boundaries.

5.4. Response Burden, Collection, and Rotation

Problem 17: Subannual surveys imply a monthly or quarterly response burden that sometimes causes vociferous complaints, especially from small businesses who do not benefit greatly from the resulting statistical series. The BSRP is addressing this problem by considering procedures along the following lines:

(a) Make an "initial contact" with all units newly entering a survey sample to explain the purpose of the survey, validate classification, etc. as previously noted (Problem 6(e)).

(b) Arrange that units rotate out-of-sample wherever this is feasible—i.e., whenever units are not sampled with certainty, and subject to availability of suitable replacement units. The goal is to have a period in-sample of one year and to ensure that once a unit has rotated out of sample for a particular survey it will not, in general, be surveyed again for at least one further year. The Survey of Employment provides an example (Schiopu-Kratina & Srinath, 1986).

(c) Monitor the response burden collectively imposed by all subannual surveys, with the ultimate goal of coordinating sampling and rotation across these surveys, for example using synchronized sampling as practiced by the Australian Bureau of

Statistics (Hinde & Young, 1984).

(d) Maximize the use of administrative data in place of survey data wherever this is technically and operationally feasible.

(e) Restrict sampling to larger units and estimate for smaller ones based on annual survey, administrative, or other auxiliary data (see Problem 5).

Problem 18: The respondent does not keep the survey information requested on a subannual basis or finds it difficult to access. The possible consequences are nonresponse, complaints, and poor quality data. High incidence of nonresponse or partial response will tend to cause bias.

The BSRP approach is threefold:

(a) Review all subannual survey objectives in conjunction with businesses' bookkeeping practices and thus to match data requests to respondent capacity to report and to the reference periods for which accounts are maintained.

(b) Introduce follow-up of all nonrespondents; reducing sample size if necessary to make this operationally feasible.

(c) Use administrative or other sources to impute for persistent refusals.

Problem 19: It is difficult to distinguish the death of a unit from nonresponse, with consequent bias if data are imputed for a dead unit under the assumption that it is active.

The BSRP approach is to aim for complete follow-up, as just noted.

Problem 20: Units new to the sample are more likely to be nonrespondents or to provide poor quality or incomplete data.

The BSRP approach is to ensure that every new unit is subject to the initial contact previously described and that there is no rotation over and above what is necessary for response burden reasons.

5.5 Sampling and Estimation

Problem 21: Units in economic survey populations tend to be very heterogeneous. Population totals are dominated by the contributions of a small proportion of the largest units. If these factors are not taken into account, designs are inefficient and give rise to many outliers.

The BSRP approach is to stratify by size, to sample the largest units with certainty and the remainder using a power or other size dependent allocation (Statistics Canada, 1986a).

Problem 22: Recently born units and older units tend to have different characteristics—for example, births are more likely to become inactive or dead. This implies births must be sampled or estimates will be biased.

The BSRP approach is to stratify and sample births separately. However having been once subjected to sampling as a birth, a unit is not regarded on a birth for sampling on future occasions.

Problem 23: In order that sample selection be as efficient and effective as possible it is desirable to maximize the use of feedback from survey sample units to update the frame with respect to deaths and changes in classification. However, as a subannual survey sampling process invariably ensures a systematic carry-over of units from one occasion to the next, there is statistical dependency between updating based on survey feedback and future sample selection. The result is bias unless suitable precautions are taken.

Three possible approaches regarding treatment of deaths have been considered by the BSRP:

(a) Ignore the bias on the grounds that it is insignificant. This is not practical for subannual economic surveys at Statistics Canada for which the frame currently contains some 20% dead units (Estevao & Tremblay, 1985).

(b) Retain the dead units on the frame until their deaths are detected by a process that is statistically independent of survey sampling, e.g., an administrative process. This approach is the one most likely to be adopted. It avoids bias but implies that frame maintenance is not as effective as it might be.

(c) Delete the dead units from sample and frame but keep a running estimated total of the number of dead units in the population and adjust the sampling weights accordingly, as developed by Srinath in the context of the Survey of Employment (Schiopu-Kratina & Srinath, 1986).

In principle, an adjustment procedure similar to (c) above for dead units could be applied to allow unbiased frame classification updates from sample feedback. Various methods have been proposed but the benefits have to be weighed against the complexities they entail. Furthermore, unlike detection of dead units in the sample that will occur with high probability (albeit with time lag), detection of misclassification by a subannual survey is far from certain. Thus, an approach similar to (b) is favored for application.

Problem 24: Subannual estimates and corresponding annual estimates are discrepant, giving an apparent of lack of data integrity.

There are many legitimate reasons why subannual survey values, when added or averaged (as appropriate) over a year, should not equal the annual values. Not the least of these is that the subannual and annual frames will be different. However if annual estimates are explicitly designed to have smaller mean square error, then it is worth considering their use to benchmark subannual estimates. This is the approach being adopted by the BSRP.

Problem 25: The period for which the respondent reports does not coincide with the subannual reference period. The result is bias due to seasonal and growth factors.

The BSRP approach is to allow respondents to report data according to their accounting practices in order to minimize response burden (Problem 19), and then to adjust data to the reference period. The adjustment procedures have yet ⬎) be determined. Adjustment may be at micro or aggregate level.

Problem 26: Small units in the target population are not surveyed either because they are not on the frame (Problem 9) or following a conscious decision to exempt them from direct data collection (Problem 5).

The undercoverage corresponding to a decision, voluntary or otherwise, not to collect data by survey from small units can be defined out of existence by re-identifying the target population in terms of larger units only. This is hardly a very satisfactory approach (Hogan et al., 1986). The alternative being actively pursued in the BSRP is to attempt to estimate for units not covered by a subannual survey using annual survey or administrative data.

6. GENERAL THEMES AND FUTURE DIRECTION

From the preceding review of frame maintenance issues in subannual economic surveys, certain themes and design principles emerge that may be applicable to repeated or longitudinal economic surveys in general. They are summarized in the following paragraphs, illustrated by examples drawn from the ongoing Business Survey Redesign Project.

The first theme is standardization. In an environment of resource constraint but increasing demand for a wider range of more coherent data, standardization of concepts, methods, and systems is vital. This inevitably involves some simplification and force-fitting of individual surveys, which cannot then be as fine-tuned to their parochial objectives. But the benefits are reduced development and maintenance costs and enhanced capacity for data integration.

The starting point for standardization is definition of concepts and objectives. The BSRP has developed a comprehensive information model and set of classification standards. Subannual survey goals are being reviewed within the context of the whole program of economic surveys. The main thrust is to reduce data item collection on a subannual basis, to focus on estimates of change, leaving annual surveys as the source of comprehensive structural data. This simplifies frame maintenance.

A bank of standard statistical procedures is a prerequisite for data integration, coordination of response burden, and efficient system development. The BSRP is developing general strategies for both annual and subannual economic surveys covering not only coverage and classification procedures but all other aspects of survey design.

A flexible centralized system for frame creation and maintenance is vital for communication of data from administrative sources and between individual surveys. The BSRP is constructing a new Central

Frame Data Base that will be the focal point for receiving, processing, storing, and disseminating all frame data; for generating mailing lists; and for mail-out and follow-up. It will also be the vehicle for coordinating the acquisition and use of administrative data. In addition to the data base, generalized systems are being developed for sampling, data capture, automated coding, edit and imputation, automation, etc.

The second theme is special treatment for large units. Given a very heterogeneous target population with a small number of exceedingly complex large units and a large number of individually insignificant small units, it is appropriate to give special treatment to the former. In the context of frame maintenance this implies recognition of the various levels within an organization at which particular data are available, and the need to relate organizational structures to statistical targets. The BSRP is introducing a comprehensive program for delineation and maintenance of a four-level hierarchy of statistical units and reporting arrangements together with linkages to the corresponding real-world legal, administrative, and operating structures for the larger units.

The third theme, complementary to the second, is to simplify the treatment given to the vast mass of small units. There are often neither the resources nor the need to handle small units in the same way as large ones. The approach adopted by the BSRP is to equate statistical and administrative units for sampling purposes, accepting that a certain degree of statistical impurity will result. One administrative source (payroll deduction accounts) will, in essence, provide the frame of small units for all subannual economic surveys. Estimates for any units not covered by this frame will be derived from statistical models. For sampled units only, there will be some recognition of differences between administrative and statistical units in the handling of partnerships and of multiple economic operations. To ensure currency of the estimates, new units will be introduced into the frame and subjected to sampling as soon as they are detected. By contrast, to limit the incidence of artifacts due to frame maintenance activities, changes in unit classifications may be held back until a periodic mass update.

The fourth and final theme is the need to alleviate response burden, especially on smaller units. Respondents are a statistical agency's most valuable data source. High nonresponse rates can produce results of poor equality. For large units the BSRP approach is comprehensive organizational structure profiling, as already mentioned, during the course of which the objectives and benefits of the whole survey program can be described to respondents. For small units the policy is to make an initial contact with all new entrants to a subannual sample to explain the survey objectives as well as to validate classification data and establish reporting arrangements. Whenever possible, small units are promised they will rotate out-of-sample at the end of a fixed period. Requests for data are matched to bookkeeping practices and respondents are encouraged to report in accordance with their accounting reference periods.

REFERENCES

Armstrong, G., Monty, A., & Wood, S. (1986), *Definitions of Standard Events* (Business Survey Redesign Project Working Paper), Statistics Canada, Ottawa.

Bankier, M. (1986), "Estimators Based on Several Stratified Samples with Applications to Multiple Frame Surveys," *Journal of the American Statistical Association*, 81, 1074–1079.

Cain, J., et al. (1984), *Infrastructure Development, Objectives, Policy and Strategy* (Business Survey Redesign Project Working Paper), Statistics Canada, Ottawa.

Darcovich, N. (1982), *Current Procedures and Proposals for the Acquisition and Usage of Income Tax Data* (Business Register Division Working Paper), Statistics Canada, Ottawa.

Estevao, V., & Tremblay, J. (1985), *A Report on the Quality of Data in the BRMF* (Business Register Working Paper), Statistics Canada, Ottawa. ˙

Estevao, V., & Tremblay, J. (1986), *An Evaluation of the Assignment of Standard Industrial Codes from PD-20 Data* (Business Survey Redesign Project Working Paper), Statistics Canada, Ottawa.

Foy, P. (1986), *Évaluation préliminaire de l'emploi d'un échantillon maître des comptes PD dans la partie non-intégrée du CFDB* (Business Survey Redesign Working Paper), Statistics Canada, Ottawa.

Garrett, J., Hogan, H., & Pautlier, C. (1986), "Coverage Concepts and Issues in Data Collection and Presentation," *Proceedings of the Second Annual Research Conference*, U.S. Bureau of the Census, 329–334.

Hinde, R., & Young, D. (1984), *Synchronized Sampling and Overlap Control Manual* (Working Paper), Australian Bureau of Statistics, Canberra.

Hostetter, S. (1983), "The Verification Method as a Solution to the Industry Coding Problem," *Proceedings of the Section on Survey Research Methods*, American Statistical Association, 499–504.

Jabine, T. B. (1984), *The Comparability and Accuracy of Industry Codes in Different Data Systems*, National Academy Press, Washington, DC.

Konschnik, C., Monsour, N., & Detlefsen, R. (1985), "Constructing and Maintaining Frames and Samples for Business Surveys," *Proceedings of the Section on Survey Research Methods*, American Statistical Association, 113–122.

Moser, C., Ruggles, R., Popkin, J., Martin, M., & Waksberg, J. (1980), *Methodology Review of Statistics Canada* (Consultants Report), Statistics Canada, Ottawa.

Schiopu-Kratina, I., & Srinath, K. P. (1986), *The Methodology of the Survey of Employment, Payrolls and Hours* (Business Survey Methods Division Working Paper), Statistics Canada, Ottawa.

Statistics Canada (1973), *Policy for the Implementation and Use of the Business Register and for the Conduct of Business Profiling* (Business Register Policy Paper), Statistics Canada, Ottawa.

Statistics Canada (1975), "National Income and Expenditure Accounts," *Catalogue 13-549E*, Vol. 3, Statistics Canada, Ottawa.

Statistics Canada (1980), "Standard Industrial Classification 1980," *Catalogue 12-501E*, Statistics Canada, Ottawa.

Statistics Canada (1984), "Corporation Financial Statistics," *Catalogue 61-207*, Statistics Canada, Ottawa.

Statistics Canada (1985a), *Business Statistics Survey Frame Model* (Business Survey Redesign Project Working Paper), Statistics Canada, Ottawa.

Statistics Canada (1985b), *Business Register Reference Manual* (Business Register Working Paper), Statistics Canada, Ottawa.

Statistics Canada (1986a), *Overview of the Methodological Framework: Wholesale/Retail Trade Surveys Redesign* (Business Survey Redesign Project Working Paper), Statistics Canada, Ottawa, rev. April 1986.

Statistics Canada (1986b), *Frame Options for the Monthly Wholesale Retail Trade Survey Redesign* (Business Survey Redesign Project Working Paper), Statistics

Canada, Ottawa.

Statistics Canada (1986c), "Industrial Corporations Financial Statistics," *Catalogue 61-003 Quarterly*, Statistics Canada, Ottawa.

NOTE

*In preparing this paper the author has drawn upon the experience of many others at Statistics Canada, in particular, colleagues associated with the Business Survey Redesign Project. The author would like to thank G. Brackstone, M. Hidiroglou, N. Chinnappa, C. Patrick, M. P. Singh, and the referees for their helpful comments. Although the paper has been subject to review within Statistics Canada, the opinions expressed in it are those of the author and not necessarily those of the agency.

A Probability Sampling Perspective on Panel Data Analysis

Ralph Folsom, Lisa LaVange, and Rick L. Williams

---

1. INTRODUCTION

This paper focuses on methods of estimation and inference based on the probability sampling distribution of panel survey data. Three areas of data analysis are discussed and methods from each that are appropriate for use with panel survey data are presented. The presentation is not intended to represent an integrated treatment of panel survey data analysis; rather, the methods lie in areas where the authors have particular interest and recent experience. The first section focuses on the changing nature of the target population during the life of a longitudinal survey. In this dynamic population environment, the definition of longitudinal households and families becomes an issue. Section 2 reviews the alternative longitudinal household definitions specified for the 1977 National Medical Care Expenditure Survey (NMCES), the 1980 National Medical Care Utilization and Expenditure Survey (NMCUES), and the ongoing Survey of Income and Program Participation (SIPP). Longitudinal household weights are developed using methods designed for sampling frames with multiple links between the sampling and analysis units. Section 2 concludes with a discussion of time-adjusted estimation of panel survey means, rates, and distribution quantiles (e.g., rates per person day).

The third section of the paper deals with the estimation of superpopulation model parameters associated with the general linear multivariate model (GLMM). National Crime Survey applications of the repeated-measurements model with missing data are presented. Efficient sample design-weighted estimators are also derived for the regression coefficients and variance component parameters in a polynomial growth curve model. Covariance matrix estimation for the growth curve model is accomplished employing a design-weighted EM algorithm adapted from Laird and Ware's (1982) results. The paper's final section addresses survival and hazard function estimation from a probability sampling perspective.

108

2. DESCRIPTIVE ANALYSIS

This section concentrates on descriptive parameter definition and estimation for panel survey populations. Issues relating to longitudinal household definition and weighting are addressed. Time-at-risk notions are used to develop descriptive parameters for dynamic panel survey populations.

2.1. Longitudinal Population Definitions

When defining the target population for a longitudinal survey, one must deal with the consequences of birth, death, and mobility during the life of the panel. Goldstein (1979) enumerates the following four alternative population definitions that explicitly incorporate the target population dynamics:

(1) The universe including all eligible individuals who were living in the specified geographic area at the beginning of the survey and who are still eligible and living in the area at each subsequent measuring occasion. The target area may be defined as the entire nation or a geographic subregion. Eligibility may exclude full-time military and institutionalized individuals.

(2) The group above plus all eligible individuals who are born into or move into the target area after the first measuring occasion.

(3) All eligible individuals living in the area of interest at the first occasion, wherever they happen to live at subsequent occasions.

(4) The initial population or cohort who reside in the specified area and are eligible at the first measurement occasion (3) plus all additions due to births, geographic in-migration, and eligibility status changes during the survey reference period.

Goldstein's first population minimizes practical problems associated with following out-migrants and fosters the collection of complete survey responses. If, however, important subsets of the relevant universe are lost by out-migration from the eligible target group, then the survey population may not provide sufficient coverage of the relevant inference universe. For example, restricting the survey population to exclude persons who move into institutions may unduly bias the survey statistics if one's inference goals include the institutionalized subpopulation. Cost considerations often dictate a universe definition that is less inclusive than desirable. For panel surveys whose objectives include the estimation of monthly service utilization rates, program eligibility rates, and benefit recipiency rates, it is desirable to include new entrants who join the inference population due to birth, changing eligibility status, or geographic in-migration.

Goldstein's fourth definition is the most inclusive, representing an ideal that few longitudinal surveys attempt to achieve. The two previous national medical care expenditure surveys, the 1977 NMCES and the 1980 NMCUES, excluded individuals from the target population during their stays in institutions. The SIPP survey population definition also excludes institutionalized segments. The 1987 National Medical Expenditure Survey (NMES) followed persons from the household universe into institutions. A sample of persons institutionalized on January 1, 1987 was also selected and followed throughout the year, including those who moved back into the household population. While such multiple frame samples substantially complicate data collection and estimation with attendant magnification of survey costs, the increase in relevant population coverage can be particularly beneficial for panel surveys.

One group of population in-migrants that is typically missed by surveys that follow an initial sample of household members is those persons whose eligibility status changes during the survey reference period (military to civilian and institution member to household member) and who establish single-person households. These in-migrants are not linked to the initial cohort of households (population 3). The NMCES, NMCUES, and SIPP surveys all exclude this subpopulation of in-migrants. Research needs to be conducted into the cost effectiveness of strategies for eliminating this source of coverage bias. At a minimum, one would need to follow a probability subsample of the initial survey dwelling units after the original household occupants had moved away. In the following subsection, we examine issues relating to longitudinal household definition.

2.2. Longitudinal Household Definition

While birth, death, and migration clearly impact on person-level analyses of panel data, household- and family-level analyses are particularly disrupted by deaths of members, by separations and divorces, and by marriages. During the life of a panel survey, family and household type units are created, disappear altogether, and undergo critical changes in type. Such changes in household type cause difficulties for simple descriptive analyses focused on monthly rates of program participation or mean levels of benefit recipiency. Longitudinal family unit concepts are discussed in the context of the 1977 NMCES by Horvitz and Folsom (1980). Rules were considered for linking families and single-person households to produce longitudinal units over the 12-month survey reference period. A family unit analysis strategy was proposed based on estimation theory developed for sampling frames with multiple links between the initial sampling units and the analysis units.

The family unit linkage and estimation strategy implemented for NMCES is described by Bentley and Folsom (1981). The linkage rule employed caused a family unit to be dissolved with one or more new units potentially being created when a household head or spouse change occurred. If the household head (householder) died or the householder's spouse died, the original family was dissolved. Remaining household members continuing to reside together were reconstituted into a new family or single-person household. Divorces and separations similarly caused a change of head or spouse, resulting in the dissolution of the original unit and the creation of new units from the constituent parts. With the NMCES linkage rules, marriages clearly changed householder and spouse relationships resulting in the death of two previous households and the birth of a new merged unit. Additions of non-householder or nonspouse family members due to births or returns of adult children who had previously lived independently did not lead to changes in family unit definitions.

The NMCES rules for longitudinal family definition can be viewed in terms of breaking links when household type changes occurred. With households classified as married-couple families, single-male householder families, single-female householder families, single-person male households, and single-person female households, the NMCES rules result in a linkage break when household type changed. There is an exception to this characterization of the NMCES linkage rules as being keyed to a household type change. Single-householder families and single-person households were not reconstituted when the addition of a child or the loss of all children changed the unit's family versus single person status.

The provisionally adopted SIPP longitudinal household definition described by Citro et al. (1986) also keys on changes in the Bureau's family typology. The SIPP and NMCES rules are in this respect quite similar. The differences occur when single-householder families become single-person households and vice versa. SIPP reconstitutes these units while NMCES left them intact. For the 1980 NMCUES survey, family linkage rules were formalized by Dicker and Casady (1982). To insure that the linkage rules were unique, Casady introduced the notion of principal predecessor families and principal successor families. Interval families were defined as units containing the same set of members for a continuous interval of time during the reference year. Two interval families were linked if they shared a principal successor and principal predecessor relationship. The NMCUES rule for linkage was that a principal successor family should receive the *majority of the family members*. A principal predecessor family must correspondingly donate the majority of members to a merged unit. A family birth or death occurred when there was respectively no principal predecessor or principal successor family. These birth and death situations included cases with equal-sized splits and mergers; that is, cases where ties occurred. For two interval families to be linked required that they had a reciprocal relationship, one constituting the principal predecessor and the other the principal successor.

Dicker and Casady had the objective of forging family linkages of longer durations than the type-change-based rules of NMCES and SIPP. For a strictly longitudinal mode of analysis where change variables and state transitions are of interest, one does not ideally want to dissolve the analysis unit every time an important structural change occurs. On the other hand, analysts familiar with household unit tabulations from cross-sectional surveys want to condition their analyses by family type. These first priority descriptive analyses of household units are not served well by a rule that tolerates type changes in the same unit. It would appear that one could serve both purposes by creating subfamilies of common type within the longer duration units. If one carried this compromise to the point of creating a separate set of subunit weights, then the result would amount to an unappealing "do it both ways and take your pick" solution. In the next section, we describe a multiplicity weighting strategy that places family unit estimation methodology in a familiar context.

2.3. Longitudinal Household Weighting

Frame multiplicity estimation methods can be adapted to provide design unbiased analysis weights for linear longitudinal household statistics. The multiple links to a longitudinal household can be counted in terms of the initial cohort of households or in terms of persons who were population eligibles at the beginning of the survey reference period. By applying the specified longitudinal household or family definition at the population level over the lifetime of a particular panel survey, one can identify the associated universe of longitudinal family units. We will depict a member of this longitudinal household universe by the symbol f_j with j ranging up to a total of J. Such a longitudinal unit exists for a period of time and includes a set of individuals who belong to the unit for some fraction of its lifetime. Single-person households can have lifetimes with missing intervals due to ineligible periods of institutionalization.

Initial sampling units in a panel survey of households are typically housing units, mailing addresses, or telephone numbers. Initial survey reporting units are defined by the data collection protocol. For the NMCES and NMCUES surveys where area household sampling was employed, the wave 1 reporting units (RUs) were defined as groups of one or more individuals residing in the same dwelling unit who were related by blood, marriage, adoption, or foster parent relationship. Units that reported a college student member living away from home were assigned two reporting unit IDs. These college students were traced and interviewed as separate reporting units. To define wave 1 families, college student RUs were combined with their base family RU. This population of wave 1 family units was conceived of as the universe of all such civilian families and single-person households existing on the beginning date of the survey reference period (January 1, 1977 for NMCES). Suppose that the symbols f_{oi} denote the time-zero family units existing in the eligible survey population with $i = 1, \ldots, I$. The survey-eligible persons belonging to family f_{oi} at time zero can be similarly depicted by the symbols p_{oik}.

If the probability of selecting the dwelling unit that contains family f_{oi} is denoted by π_{oi} and the associated unbiased design weight by $w_{oi} = \pi_{oi}^{-1}$, then the unbiased multiplicity weight for a longitudinal family f_j can be specified as

$$wf_{js} = \sum_{i \in s} w_{oi} H_{ij}/H_{+j} \tag{1}$$

where H_{ij} is a one-zero indicator variable that assumes the value one if the f_{oi} family contributes one or more members to longitudinal family unit j. The summation in equation (1) extends over all base families f_{oi} included in the area sample s of dwelling units. The H_{+j} quantity in equation (1) sums the H_{ij} over all f_{oi} families in the population and represents the "base family" multiplicity for the jth longitudinal unit. Note that these H_{ij} indicator variables are defined relative to the entire panel survey time period. The associated weights wf_{js} are similarly defined to be constant over the entire survey reference period.

Note that if H_{+j}, the population count of distinct f_{oi} families that contribute members to longitudinal unit f_j, can be determined accurately from the survey participants and the associated H_{ij} links are accurately recorded, then the condition for linear unbiasedness is satisfied; that is, the weights in equation (1) have the expected value of unity over repeated area household samples. This result follows simply from the definition of the wf_{js} as follows:

$$E_s \{wf_{js}\} = \sum_s P(s)wf_{js} = \sum_{i=1}^{I} H_{ij}/H_{+j} = 1 \tag{2}$$

with the summation extending over all possible samples s and $P(s)$ representing the probability of selecting sample s.

For the unbiasedness condition derived in equation (2) to be achieved in practice, the biasing effects of survey nonresponse must be negligible. Both initial wave 1 family nonresponse and subsequent attrition adversely affect longitudinal family weighting. Initially responding family units f_{oi} that drop out in a later wave may subsequently spawn new longitudinal units that are not observed. Assuming that the w_{oi}^* weights have been suitably adjusted for nonresponse and attrition so that the associated bias is negligible, then

$$wf_{js}^* = \sum_{i \in s(j)} w_{oi}^* / H_{+j} \tag{3}$$

is a negligibly biased weight for longitudinal family unit f_j with $s(j)$ denoting the set of f_o families in sample s that contribute members to family f_j .

The weighting procedure applied for the NMCES and NMCUES longitudinal families was a person-level version of the multiplicity estimation strategy described above. If we denote the person p_{oik} weight by $W_{oik} = w_{oi}$ for individuals residing at time t_o in family f_{io}, then it is possible to define longitudinal family weights in terms of nonresponse adjusted person weights W_{oa}^* where a is a single person-level index used in place of the nested ik label. At the person level, one requires a multiplicity indicator, say g_{aj}, that assumes the value 1 when an initial population member p_{oa} spends some time during the panel's life as a member of longitudinal family f_j . The total number of time t_o eligible persons that ever belong to f_j during the panel's life is g_{+j}, the person-level multiplicity.

The decision to employ the associated person-level multiplicity estimator was based on the presumption that multiplicity counting at the person level would be more accurate than at the initial household or base family level. While unique person-level IDs were maintained throughout these surveys, it was not easy to link IDs assigned to new reporting units and associated longitudinal families back to the base families f_{oi} that spawned them. Given sufficient attention to designing prospective RU linkage procedures, this reason for preferring the person-level multiplicity weight can be minimized.

A second reason for preferring the person-level multiplicity counting rule was the feeling that more accurate person-level population counts were available for nonresponse and attrition adjustments. On the other hand, there is reason to believe that the age × sex × race/ethnicity poststrata applied at the person level do not adequately account for differential nonresponse patterns across household types. For example, single-person households are more likely to be missed due to not being at home on any of the interviewer's contact attempts. Ideally, one should include household type as a poststratification variable in the person-level nonresponse adjustments.

Acknowledging this potential shortcoming, the person-level nonresponse adjusted weight for longitudinal family f_j becomes

$$wp_{js}^* \equiv \sum_{a \in s(j)} W_{oa}^*/g_{+j}$$

(4)

with $s(j)$ denoting the set of wave 1 sample persons with $g_{aj} = 1$. In the following section, time-at-risk type weight adjustments are introduced for longitudinal descriptive statistics. These adjustments are particularly relevant for longitudinal family units whose survey period time durations are more variable than those of persons.

2.4. Longitudinal Descriptive Estimation

A difficulty when analyzing panel data is that population members move into and out of the target population during the study. Thus, the size of the population changes over time. This complicates the definition of such descriptive quantities as the annual mean or median. Various definitions are discussed in this section.

Assume that we are studying a fixed time period T (say one year) consisting of M distinct time points (days). A population of N units is followed such that on any given day a particular unit may be in or out of some eligible target group. Further assume that we desire to estimate the mean number of events per population unit that occur while the units are in the eligible group. For example, we may wish to estimate the mean number of physician visits per year for the civilian noninstitutionalized population of the United States. In this situation, members of the U.S. population move into and out of the civilian noninstitutionalized population on a daily basis and hence the size of the population changes every day.

In order to study this situation, first define Y_i to be the total number of events for unit i while eligible during T and

$$\alpha_{ij} = 1 \text{ if person } i \text{ is eligible at time point } j$$

$$= 0 \text{ otherwise.}$$

The total days that unit i is eligible is $\alpha_{i+} = \Sigma_j \alpha_{ij}$ and the proportion of T that unit i is eligible is $\alpha_i = \alpha_{i+}/M$. Next, the size of the eligible population at time j is $\alpha_{+j} = \Sigma_i \alpha_{ij}$. Finally, note that the average size of the eligible population over the target period T is

$$\alpha_+ = \sum_i \alpha_i = \sum_j \alpha_{+j}/M.$$

We will assume that events occur at some underlying instantaneous rate λ over T and that

$$E_m(Y_i) = \lambda \alpha_i$$

(5)

where E_m denotes expectation under the model. This model for Y_i imposes a simple dependence between the number of events associated with unit i and the proportion of the reference period T that unit i is in the eligible target population.

One definition of the eligible target population mean is the overall mean

$$\bar{Y} = \sum_i Y_i/N .$$

(6)

Notice that the expectation of $\bar{Y}$ under the model (5) is

$$E_m (\bar{Y}) = \sum_i E_m(Y_i)/N = \sum_i (\lambda\alpha_i)/N = \lambda\alpha_+/N = \lambda\bar{\alpha}$$

(7)

where

$$\bar{\alpha} = \sum_i \sum_j \alpha_{ij}/MN .$$

This is a function of the underlying rate with which events occur and the proportion of the total number of unit-days ($\bar{\alpha}$) which are actually realized in the population. This definition is appropriate if interest is directed toward the joint effects of population mobility and the occurrence rate λ.

On the other hand, if interest is solely directed toward the underlying event rate, λ, then the following time-adjusted definition of the population mean is appropriate

$$\bar{Y}^* = \sum_i Y_i/\alpha_+ .$$

(8)

The model expectation of $\bar{Y}^*$ is

$$E_m(\bar{Y}^*) = \sum_i E_m(Y_i)/\alpha_+ = \sum_i \lambda\alpha_i/\alpha_+ = \lambda.$$

The time-adjusted mean can be rewritten as a weighted average of the form

$$\bar{Y}^* = \sum_i Y_i^*\alpha_i/\alpha_+ = \sum_i Y_i^*\alpha_i^*$$

(9)

where $Y_i^* = Y_i/\alpha_i$ and $\alpha_i^* = \alpha_i/\alpha_+$. The value Y_i^* will be referred to as the annualized value for unit i.

A consistent estimate of the mean per person-time from a sample of size n is

$$\bar{y}^* = \sum_k w_k Y_k / \sum_k w_k \alpha_k$$

where w_k is the sampling weight for sample unit k. This is in the form of a simple ratio estimate and can be easily linearized for variance estimation.

In order to study definitions of the population median, we first define the ordered values

$$Y_{(1)} \le \ldots \le Y_{(N)}$$

and

$$Y^*_{(1)} \le \ldots \le Y^*_{(N)}$$

where the $Y_{(.)}$ represent the ordered population values and the $Y^*_{(.)}$ represent the ordered annualized population values. Also, let $\alpha^*_{(i)}$ be the α^* value associated with $Y^*_{(i)}$. Notice that $Y_{(i)}$ and $Y^*_{(i)}$ are not necessarily associated with the same population unit.

The overall population median will be defined as

$$Y_{.5} = Y_{([N/2])} \tag{10}$$

where [.] denotes the integer part and the .5 subscript symbolizes the $p = 0.5$ quantile of the Y variate population distribution. This is the median of the unadjusted population values. The time-adjusted population median is defined to be

$$Y^*_{.5} = Y^*_{(m)} \tag{11}$$

where m satisfies

$$\sum_{i=1}^{m-1} \alpha^*_i < \frac{1}{2} \le \sum_{i=1}^{m} \alpha^*_i \, .$$

This is the weighted median of the annualized population values.

To study the differences between these two definitions, we consider their associated cumulative distribution functions (CDFs). The overall population CDF is given by

$$F(x) = N^{-1} \sum_{i=1}^{N} I[Y_i \le x] \tag{12}$$

where $I[z \le x]$ is equal to 1 if $z \le x$, and zero otherwise. The time-adjusted population CDF is

$$F^*(x) = \sum_{i=1}^{N} \alpha_i^* \, I \, [Y_i^* \le x].$$

$$(13)$$

Notice that the overall median $Y_{.5}$ is the median of F and that the time-adjusted median $Y_{.5}^*$ is the median of F^*.

Next, consider the means of the two population CDFs, F and F^*. The mean of F is

$$\text{Mean}(F) = \sum_{(i)=1}^{N} \{F[Y_{(i)}] - F[Y_{(i-1)}]\} Y_{(i)}$$

$$= \sum_{(i)=1}^{N} \frac{1}{N} Y_{(i)} = \bar{Y}$$

$$(14)$$

where the quantity in curly brackets is the "jump" in F at $x = Y_{(i)}$ and $F[Y_{(o)}]$ is defined to be zero. Thus, we see that $\bar{Y}$ and $Y_{.5}$ are the mean and median of the population CDF F. Likewise, the mean of F^* is

$$\text{Mean}(F^*) = \sum_{(i)=1}^{N} \{F^*[Y_{(i)}^*] - F^*[Y_{(i-1)}^*]\} Y_{(i)}^*$$

$$= \sum_{(i)=1}^{N} \alpha_{(i)}^* Y_{(i)}^* = \bar{Y}^*$$

$$(15)$$

Again, $F^*[Y_{(o)}^*]$ is defined to be zero. In this case, we see that $\bar{Y}^*$ and $Y_{.5}^*$ are the mean and median of the time-adjusted population CDF F^*.

3. MODEL-BASED MULTIVARIATE ANALYSIS

In this section, classical techniques of longitudinal data analysis are reviewed for application to panel survey data. The general linear multivariate model (GLMM) for repeated measurements is presented with an extension for missing data often occurring in a rotating panel design. A reduced version of this model is illustrated using data from the National Crime Survey (NCS). Efficient estimation of the parameters in the classical growth curve model (GCM) is also discussed.

3.1. Repeated Measurements

3.1.1. *Model Definitions.* The GLMM is a tool for analyzing data arising from an experiment in which p measurements are taken on n subjects. The p measurements are univariate random variables that might represent p distinct characteristics or one characteristic measured at p time points. If Y denotes the $(n \times p)$ data matrix where y_{ij} represents the jth measurement on the ith subject, the usual model is given by

$$E_m(Y) = XB$$

$$\text{Var}(i\text{th row of } Y) = \Sigma \qquad (16)$$

where E_m denotes expectation under the model, X is a $(n \times q)$ model matrix, and B and Σ are the $(q \times p)$ and $(p \times p)$ matrices of unknown model parameters, respectively. The assumptions implicit in the GLMM formulation are vectors of complete observations on each person and identical design matrices for each of the p measurements. A unique vector of model parameters b_j is then estimated for the jth time point, $j = 1, \ldots, p$.

Various extensions of the GLMM to allow for missing data as well as different design matrices for different time points have been proposed and are reviewed in LaVange (1983). The most general version of the GLMM with incomplete data allows for each subject to be measured at arbitrary time intervals rather than at a proper subset of the original p time points. Estimation of the model parameters in this setting requires assumptions about the covariances among measurements on a subject, namely that they can be expressed as functions of an estimable set of variance component parameters. The full generality of this model is typically not needed for rotating panel designs, since panel members are interviewed at prespecified intervals.

The general incomplete model (GIM) originally proposed by Kleinbaum (1973) has been used in conjunction with longitudinal clinical studies in which different groups of subjects were rotated into the sample for a series of measurements, each spanning a subset of the reference period (e.g., Rao & Rao, 1968; Woolson et al., 1978), and is appropriate for any type of rotating panel design. This model formulation consists of first grouping subjects according to like response times. Let $j = 1, \ldots, u$ index these groups and define the group-specific model as

$$E_m[Y_j] = D_j B K_j$$

$$\text{Var}(i\text{th row of } Y_j) = K_j^T \Sigma K_j \qquad (17)$$

where

Y_j is the $(n_j \times m_j)$ data matrix,
B is the $(q \times p)$ matrix of model parameters,
D_j is the $(n_j \times q)$ design matrix,
K_j is the $(p \times m_j)$ incidence matrix defining the m_j time points
 $\quad (m_j \leq p)$ at which the subjects in group j were interviewed.

To combine over the u groups, let y_j denote the vector of columns of Y_j, i.e., Y_j "rolled out" by columns. We have the following model holding for $y^T = [y_1^T, \ldots, y_u^T]$:

$$E_m(y) = Z\,\gamma \qquad\qquad (18)$$

$$\mathrm{Var}_m(y) = \Omega$$

where $\gamma^T = [b_1^T, \ldots, b_p^T]$, or B rolled out by columns,

$$Z = \begin{bmatrix} K_1^T \otimes D_1 \\ \cdot \\ \cdot \\ \cdot \\ K_u^T \otimes D_u \end{bmatrix},$$

$$\Omega = \begin{bmatrix} K_1^T\,\Sigma\,K_1 \otimes I_{n_1} & & 0 \\ & & \\ 0 & & K_u^T\,\Sigma\,K_u \otimes I_{n_u} \end{bmatrix},$$

$N = \sum_{j=1}^{u} n_j m_j$, and $\otimes$ denotes the left-hand Kronecker product.

To see how this model can be applied to a rotating panel design, consider the NCS design. Each sample of about 70,000 dwelling units is divided into six rotation groups, with each rotation group being introduced into the sample every 6 months. Dwelling units are interviewed once every 6 months for 3–1/2 years, with 1/6 of a rotation group or one panel interviewed each month. Persons living in the dwelling units are asked to provide information concerning victimizations that occurred during the 6-month reference period prior to each interview. Thus the subsample of dwelling units contributing to victimization statistics for a particular reference month varies from month to month. Table 1 gives the rotation chart for January 1978 through June 1979. Keeping in mind the 6-month reference period per interview, victimizations occurring in January 1978 were reported by households in exactly 42 panels, corresponding to panels 2–6 of group 3 and panels 1–6 of groups 4–6 in samples J03/J04, and to panels 1–6 of groups 1–3 and panel 1 of group 4 in samples J05/J06. Households in 42 panels contribute to February 1978 data, but only 6/7 or 86% of these overlap with those contributing to January 1978 data.

To illustrate the GIM for this particular panel design, consider the problem of fitting linear models to vectors of monthly victimization rates for the 12 months of 1978. In the absence of missing data due to non-response, the data groups j can be defined from the rotation chart in

Table 1. For example, panel 2 of group 3 in samples J03/J04 would contribute data for only one response time, January, and would constitute one data group, while panel 2, groups 5–6 in J03/J04 and panel 2, groups 1–3 in J05/J06 would provide data for all 12 response times, reported in three separate interviews, thereby constituting another data group. A total of 22 data groups would need to be defined, each contributing from 1 to 12 months of data.

3.1.2. *Estimation.* For simple random sampling, the best linear unbiased estimator of B in the GLMM, assuming normality for each row of Y, is given by the usual ordinary least squares estimator. However, in the presence of missing data, efficient estimation of γ (and hence of B) requires knowledge of the covariance matrix Σ. If Σ is known, the generalized least squares estimator is optimal, namely

$$\hat{\gamma} = (Z^T \Omega^{-1} Z)^{-1} Z^T \Omega^{-1} y \qquad (19)$$

where y, Z, γ, and Ω are defined in equation (18). In practice, Σ is usually not known and best linear unbiased estimates do not exist for B. However, Kleinbaum (1973) has shown that if a consistent estimator of Ω is substituted in equation (19), γ is a best asymptotically normal estimator.

Large-scale panel surveys such as the NCS or SIPP are typically the result of a complex sample design. For example, each NCS sample of about 70,000 households was selected as a stratified multistage cluster sample. Counties or groups of counties were selected at the first stage, followed by enumeration districts and segments of households. Unequal probabilities of selection were used at each stage. The effects of designs such as that used for the NCS, namely, unequal probabilities of selection and clustering, must be taken into account when applying classical longitudinal data analysis techniques. The normality assumption as well as the assumption of independent observation vectors (rows of Y) are violated in both models presented (GLMM and GIM). As a result, the ordinary least squares and generalized least squares estimators of B and γ do not have the desirable properties mentioned above. Methods for estimating linear model parameters for complex survey data are readily available through software packages such as RTI's survey regression package, SURREGR (Holt, 1977). SURREGR produces design-consistent estimators of B and the corresponding variance-covariance matrix using the Taylor Series linearization method. Wald test statistics are computed for hypotheses of interest about the parameters in B. SURREGR accepts as input a vector of y values and a design matrix; thus it is suitable for the vector version of the GIM given in equation (18). To fit the GLMM with SURREGR, the columns of Y in equation (16) would need to be "rolled out" in a manner similar to that used for the GIM.

While the estimates of B produced via SURREGR are design-

TABLE 1. NCS ROTATION CHART, JANUARY 1978–JUNE 1979

Year and Month	Sample								
	J03/J04				J05/J06				
1978									
January	13	14	15	16	11	12	13		
February	23	24	25	26	21	22	23		
March	33	34	35	36	31	32	33		
April	43	44	45	46	41	42	43		
May	53	54	55	56	51	52	53		
June	63	64	65	66	61	62	63		
July		14	15	16	11	12	13	14	
August		24	25	26	21	22	23	24	
September		34	35	36	31	32	33	34	
October		44	45	46	41	42	43	44	
November		54	55	56	51	52	53	54	
December		64	65	66	61	62	63	64	
1979									
January			15	16	11	12	13	14	15
February			25	26	21	22	23	24	25
March			35	37	31	32	33	34	35
April			45	46	41	42	43	44	45
May			55	56	51	52	53	54	55
June			65	66	61	62	63	64	65

Key: kl = *k*th panel of *l*th rotation group.

consistent for the finite population parameters, they are not efficient in the presence of missing data since they ignore the covariance structure. If Σ were known, then the original vector of observations y could be transformed by the inverse square root matrix $\Omega^{-1/2}$ so that the following model holds:

$$E_m(\Omega^{-1/2}y) = \Omega^{-1/2}Z\gamma \qquad (20)$$
$$\text{Var}_m(\Omega^{-1/2}y) = I$$

The transformed vector of observations $\Omega^{-1/2}y$ could then be analyzed using SURREGR to produce design consistent estimates of B that are also efficient. If Σ is not known, an alternative approach would be to produce maximum-likelihood estimates of the variance and covariance elements in Σ and use these to transform y as above. A two-step iterative procedure for estimating variance components by fitting a linear model to the vector of squares and cross-products of residuals that is discussed in Anderson (1973) is applicable in this setting.

3.1.3. *Reduced Models.* An advantage of repeated-measurements analysis of panel survey data is the range of hypotheses that can be tested once the model has been fitted. The model parameters B include effects of various regressors for each of the p response times. By generating the appropriate contrast matrices C and U, hypotheses of the form H_0: $CBU = 0$ can be tested to determine trends in B over time. The design-consistent estimates of B and Var(B) produced by SURREGR can be used to compute a Wald test statistic for the various hypotheses of interest. Note that in the GIM setting (equation 18), the contrast $H\gamma$ is equivalent to CBU if $H = U \otimes C^T$.

A disadvantage of the model framework for repeated-measurements analysis is that different effects are estimated for each time point. If the number of time points and/or the number of independent regressors is large, this could result in a large number of model degrees of freedom. An alternative approach is to fit a model to the vector of time-specific observations that includes the time dimension in the set of regressors. This enables the user to selectively interact time with the other regressors and to model the trend in time using, for example, polynomial effects, as in growth curve analyses.

This approach was taken in an analysis of NCS victimization rates that included survey measurement effects, respondent characteristics, time period indicators, and their selected interactions. The goal of the study was to produce predicted victimization rates that were adjusted for the effects of known survey measurement errors such as recall bias and forward telescoping of events.

If y_k denotes the vector of monthly victimization counts for person k, where $k = 1, \ldots, n$ and $y^T = [y_1^T, \ldots, y_n^T]$, the following model was assumed:

$$E[y] = Xb \qquad (21)$$

where X is the $(n \times q)$ design matrix and b is the $(q \times 1)$ vector of unknown model parameters. This model formulation differs from classical repeated-measurements models in that the same set of effects b is assumed for each response time. However, by including time period indicators in X, trends over time can be tested for any effect with which time is interacted. A further advantage of this modeling approach is that reduced polynomial time trends can be included in the model that sufficiently characterize the effects over time, thereby reducing the number of degrees of freedom required for the time period interactions.

For the NCS analysis, it was decided that quarterly time period indicators were sufficient for capturing trends in victimizations during a reference period of length one and one-half years. Reference quarter indicators were included in the model so that predicted quarterly rates could be produced. In addition, these quarterly indicators were interacted with reporting subgroup variables such as race and sex in order to produce subgroup-specific predicted rates for comparison with previously published data.

To illustrate this approach to modeling effects over time, consider two NCS measurement effects that were thought to vary with time, the effects due to recency and time in panel. Recency was defined as the time lag in months between the victimization occurrence and the interview. Including recency effects in the model partially accounted for potential biases due to memory loss and internal telescoping of events. Time in panel was defined as the number of interviews for each person. Although each sampled dwelling unit was interviewed exactly seven times, the NCS sample was not balanced with respect to person time in panel due to movers and missed interviews. Including effects for time in panel in the model partially accounted for potential biases due to forward telescoping of events at the first, unbounded interview and conditioning or interview fatigue over subsequent interviews. In order to minimize the number of degrees of freedom required to form interactions with reference quarter, orthogonal polynomial effects were defined and tested for significance. It was determined that a third-degree polynomial provided an adequate fit for main effects due to time, and a linear trend in time was sufficient for forming interactions with other variables. Both recency and time in panel were found to interact significantly with time. For example, the recall bias in reporting crimes as measured by the first- and sixth-month recency effect was much less in the first quarter of 1978 than in the third or fourth quarters of 1978. Similarly, the increase in reported victimizations associated with unbounded interviews was less in the first and second quarters than in the fourth quarter.

The brief discussion given here of the NCS analysis serves to illustrate a repeated-measurements analysis applied to a rotating panel design that uses a reduced form of the classical model. The model was sufficient for characterizing the dynamic nature of important survey measurements effects while allowing for enough detail to both test for trends in effects over time and produce time-period-specific predicted victimization rates. A detailed discussion of this modeling approach applied to the NCS data can be found in LaVange and Folsom (1985).

Another variation of the classical repeated-measurements model that is particularly suitable for panel data such as the SIPP data involves transforming the $(n \times p)$ data matrix Y into a $[n \times (p - 1)]$ matrix of successive differences. The outcome of interest is then the change from month to month of an individual's status with respect to the original outcome variable, such as dollars received from various sources. For rotating panel designs, transformations can be defined for each missing-data group. The model parameters then refer to effects with respect to the change in status from month to month rather than to the effects on a particular month's expected value. The covariance structure reflects the correlations among the monthly status changes of a panel member.

Estimation techniques discussed in Section 3.1.2 are appropriate for this reduced model formulation.

3.2. Growth Curve Models

Building on Kleinbaum's method of dealing with missing-data patterns in the repeated-measurements design, we consider a polynomial growth curve model with stochastic coefficients and missing time points. We envision a panel survey with an outcome variable Y_{it} reported for up to p time points, say p months. Due to panel rotation and nonresponse, survey respondent i provides $m_i \le p$ monthly reports.

3.2.1. *The Stochastic Coefficient Model.* We initially state the stochastic coefficient model in the following repeated-measurement form

$$Y_i = K_i \, \mu_i + \epsilon_i \qquad (22)$$

where μ_i $(p \times 1)$ is the full vector of p monthly outcome means for individual i and K_i $(m_i \times p)$ is Kleinbaum's incidence matrix of ones and zeros. If outcome Y_{it} corresponds to the jth reference month, then the tth row of K_i has a one in column j and zeros elsewhere. The vector ϵ_i $(m_i \times 1)$ can be viewed as a vector of variable measurement errors. In this formulation, the μ_i vectors of individual means are assumed to be stochastic. We let Σ $(p \times p)$ depict the between-individual covariance matrix among the elements of μ_i. The within-individual variable measurement errors ϵ_i are assumed to have zero mean vector, to be uncorrelated across months, and to have common variance over repeated measurements; that is, $E(\epsilon_i \epsilon_i^T) = \sigma^2 I_{m_i}$.

With these within- and between-individual variance assumptions, the total covariance matrix for Y_i is

$$\mathrm{Cov}(Y_i) = K_i \, \Sigma \, K_i^T + \sigma^2 I_{m_i} \qquad (23)$$

3.2.2. *The Polynomial Reparameterization.* The full $(p - 1)$ degree polynomial growth curve model is obtained by reparameterizing equation (22). If we define Q as the $(p \times p)$ matrix with a leading column of common elements $(1/p^{1/2})$ and the remaining $(p - 1)$ columns consisting of orthonormal polynomial contrast vectors, then $Q^T = Q^{-1}$ and the full set of polynomial regression coefficients for individual i is defined by

$$\theta_i = Q^T \mu_i \,.$$

Letting $Q_i = K_i \, Q$ define the subset of $m_i \le p$ rows of Q corresponding to the ith individual's nonmissing months, then

$$Y_i = Q_i \theta_i + \epsilon_i \qquad (24)$$

The appeal of the polynomial growth curve model in equation (24) is its potential to provide a more parsimonious description of the individual growth curves. To achieve this parsimony, one must test for the non-significance of higher-order polynomial terms.

Before moving to a reduced-degree within-person polynomial, we consider the between-person model for the full $(p \times 1)$ vector of coefficients θ_i. We begin by letting a_i $(1 \times r)$ denote a vector of individual characteristics with a leading element $a_{i1} = 1$ for all persons i. The γ_t quantity will denote an associated $(r \times 1)$ vector of between-individual regression coefficients for predicting the mean of the tth polynomial contrast θ_{it}. With these definitions, the between-individual model becomes

$$\theta_i = (I_p \otimes a_i)(\gamma_1^T, \ldots, \gamma_p^T)^T + \eta_i$$

$$\equiv A_i\,\gamma + \eta_i \tag{25}$$

where $\otimes$ denotes the Kronecker product and η_i is the $(p \times 1)$ vector of between-individual sampling errors in the θ_i coefficients. Recalling that $\theta_i = Q^T\mu_i$, one notes that the η_i errors have covariance matrix

$$E(\eta_i\,\eta_i^T) = \Delta\,(p \times p) = Q^T\,\Sigma\,Q, \tag{26}$$

where Σ $(p \times p)$ is the between-individual covariance matrix among the μ_i monthly means. While it would be tempting to reduce the variance component model dimensionality by assuming that Δ is diagonal, this is not reasonable in light of the Δ representation in equation (26).

Combining equations (24) and (25), the full model for Y_i becomes

$$Y_i = Q_iA_i\gamma + Q_i\eta_i + \epsilon_i \tag{27}$$

Note that the expanded design matrix, D_i $(m_i \times rp) = Q_iA_i$, has rows $D_{it} = Q_{it} \otimes a_i$. Since Q_{it} and a_i have leading elements that are uniformly $(1/p^{1/2})$ and 1 respectively, D_{it} includes the time polynomial main effects in Q_{it}, the individual characteristic main effects in a_i and all the associated first-order interactions.

With the combined error term in equation (27) denoted by $\nu_i = Q_i\eta_i + \epsilon_i$, the full model can be recast as

$$Y_i = D_i\gamma + \nu_i \tag{28}$$

with

$$E(\nu_i\,\nu_i^T) = Q_i\,\Delta\,Q_i^T + \sigma^2I_{m_i}$$

3.2.3. Fitting Reduced Models. In the context of the full model in equation (28), the hypothesis that the higher-order polynomial coefficients θ_{it} with $t > q$ all have zero expectations is equivalent to

$$H_0: E(\theta_{it}) = a_i\gamma_t = 0 \text{ for all } t > q$$

or equivalently, with ϕ_r denoting an $(r \times 1)$ null vector,

$$H_0: \gamma_t = \phi_r \text{ for all } t > q. \tag{29}$$

A robust large sample test for the appropriateness of a degree $(q - 1)$ polynomial can be performed simply using the sample design-weighted estimator for γ; namely,

$$\gamma_w = (\sum_{i=1}^{n} w_i \, \boldsymbol{D}_i^T \, \boldsymbol{D}_i)^{-1} (\sum_{i=1}^{n} w_i \, \boldsymbol{D}_i^T \, \boldsymbol{Y}_i) \tag{30}$$

with w_i denoting the design weight for individual i. The SURREGR program (Holt, 1977) described in Section 3.1.2 can be used to calculate γ_w and provide a design-consistent estimate of its sampling covariance matrix. Other programs also available include SUPER CARP (Hidiroglou et al., 1979), NASSREG (Mohadjer et al., 1986), and REPERR (Survey Research Center Computer Support Group, 1981). The F-transformed version of the T^2 or Wald statistic employed in SURREGR and NASSREG provides a robust simultaneous test for the hypotheses in equation (29).

Having established that $E(\boldsymbol{\theta}_{it})$ is nonsignificant for $t > q$, the reduced model can be fitted using efficient weighted least squares methods. We first define $\boldsymbol{X}_i \, (m_i \times q)$ as the first q columns of the ith individual's matrix of polynomial contrasts; that is, $\boldsymbol{X}_i$ is comprised of the first q columns of $\boldsymbol{Q}_i = \boldsymbol{K}_i \boldsymbol{Q}$. Given this definition of $\boldsymbol{X}_i$, the reduced version of the design matrix $\boldsymbol{D}_i$ is $\boldsymbol{G}_i \equiv (\boldsymbol{X}_i \otimes \boldsymbol{a}_i)$. The associated reduced vector of between-individual regression coefficients will be denoted by $\boldsymbol{\xi}^T = (\gamma_1^T, \ldots, \gamma_q^T)$. This allows us to state the reduced model

$$\boldsymbol{Y}_i = \boldsymbol{G}_i \, \boldsymbol{\xi} + \boldsymbol{\nu}_i \tag{31}$$

where $\boldsymbol{G}_i$ has dimensions $(m_i \times rq)$. If $\boldsymbol{\Omega}_i$ is used to denote the covariance matrix for $\boldsymbol{\nu}_i$ displayed in equation (28), then with $\boldsymbol{\Omega}_i$ known, the design-weighted GLS estimator for $\boldsymbol{\xi}$ is

$$\hat{\boldsymbol{\xi}}_{GLS} = (\sum_{i=1}^{n} w_i \, \boldsymbol{G}_i^T \, \boldsymbol{\Omega}_i^{-1} \boldsymbol{G}_i)^{-1} (\sum_{i=1}^{n} w_i \, \boldsymbol{G}_i^T \, \boldsymbol{\Omega}_i^{-1} \, \boldsymbol{Y}_i) \tag{32}$$

To employ the efficient GLS estimator in equation (32), one requires a design-consistent estimator for $\boldsymbol{\Omega}_i$. In the next section, design-weighted pseudo maximum-likelihood estimators for $\boldsymbol{\Delta}$ and σ^2 are developed under the added assumption that the $\boldsymbol{\eta}_i$ and $\boldsymbol{\epsilon}_i$ error vectors are independent multivariate normal variates.

3.3. Variance Component Estimation

To obtain estimates for $\boldsymbol{\Delta}$ and σ^2, we will assume that the individual effects $\boldsymbol{\eta}_i$ and the within-individual measurement errors $\boldsymbol{\epsilon}_i$ of model (27) are multivariate normal and independent of each other. In its reduced form, namely

$$Y_i = G_i\xi + Q_i\eta_i + \epsilon_i ,$$

the resulting growth curve model is equivalent to Laird and Ware's (1982) "conditional-independence model." This model states that the m_i monthly responses on individual i are independent, conditional on η_i and ξ. Laird and Ware show how the EM algorithm can be exploited to yield maximum-likelihood (ML) and restricted maximum-likelihood (REML) estimates for ξ, Δ, and σ^2. To apply the EM algorithm to the growth curve setting, note that if it were possible to observe η_i and ϵ_i as well as Y_i, then ML estimation of Δ and σ^2 would be straightforward. Therefore $(Y_i, \eta_i, \epsilon_i)$ are considered the "complete" data and Y_i the "incomplete" data. The EM algorithm then consists of iterating between ML estimation of Δ and σ^2 based on observing the "complete" data, and estimation of the unobservable η_i and ϵ_i. The second step is accomplished by equating η_i and ϵ_i to their expected values conditional on Y_i, $\hat{\xi}$, $\hat{\Delta}$, and $\hat{\sigma}^2$.

The EM algorithm is iterative, and requires starting values for ξ, Δ, and σ^2. Since a good set of starting values can substantially improve the rate of convergence, it is advisable to choose one's starting values carefully. Laird et al. (1987) suggest a method for calculating starting values that should perform well for growth curve applications.

Given a set of starting values, the sample design weighted EM algorithm provides the following pseudo ML estimate for ξ at iteration $(g + 1)$:

$$\xi_{g+1} = [\sum_{i=1}^{n} w_i \, G_i^T \, W_{ig} \, G_i]^{-1} [\sum_{i=1}^{n} w_i \, G_i^T \, W_{ig} \, Y_i]$$

with

$$W_{ig} = (Q_i \, \Delta_g \, Q_i^T + \sigma_g^2 \, I_{m_i})^{-1}.$$

The corresponding individual effects $\eta_i \, (g + 1)$ are estimated by

$$\eta_i \, (g + 1) = \Delta_g \, Q_i^T \, W_{ig}(Y_i - G_i \, \hat{\xi}_{g+1}) . \tag{33}$$

We similarly define the conditional variance functions:

$$tv_{ig} \equiv \mathrm{tr} \, \mathrm{Var} \, [\epsilon_i \mid Y_i; \, \xi_g; \, \Delta_g; \, \sigma_g^2] = \sigma_g^2[m_i - \sigma_g^2 \, \mathrm{tr}(W_{ig})] ,$$

and
$$\tag{34}$$

$$V\eta_{ig} \equiv \mathrm{Var} \, [\eta_i \mid Y_i; \, \xi_g; \, \Delta_g; \, \sigma_g^2] = [\Delta_g - \Delta_g Q_i^T W_{ig} Q_i \Delta_g] .$$

where $\mathrm{tr}(W_{ig})$ denotes the trace of the matrix $W_{ig} \equiv \Omega_{ig}^{-1}$.

In terms of the predicted individual effects in equation (33), the predicted measurement errors are

$$\epsilon_i(g + 1) = \left[Y_i - G_i\, \xi_{g+1} - Q_i\eta_i(g + 1) \right] . \tag{35}$$

Based on the definitions in equations (33), (34), and (35), the sample design-weighted estimate for σ_{g+1}^2 is

$$\sigma_{g+1}^2 = \sum_{i=1}^{n} w_i\left[\epsilon_i(g + 1)^T\, \epsilon_i(g + 1) + tv_{ig}\right]/\hat{m}_+ \tag{36}$$

with $\hat{m}_+ \equiv \sum_{i=1}^{n} w_i m_i$.

The $(g + 1)$ step estimator for Δ is

$$\Delta_{g+1} = \sum_{i=1}^{n} w_i\left[\eta_i(g + 1)\eta_i(g + 1)^T + V\eta_{ig}\right]/\hat{w}_+ \tag{37}$$

with $\hat{w}_+ \equiv \sum_{i=1}^{n} w_i$.

Examining the definitions in equation (34), one observes that $tv_{ig} \geq 0$ and $V\eta_{ig}$ is nonnegative definite given proper starting values $\tilde{\sigma}_0^2$ and $\tilde{\Delta}_0$. With this realization, it is clear that the solutions σ_{g+1}^2 and Δ_{g+1} are properly constrained to the parameter space. Therefore, the convergent solutions σ_*^2 and Δ_* will be properly restricted to the parameter space where $\sigma^2 > 0$ and Δ is nonnegative definite. If one were to settle for a fixed g-step solution, these estimates are also guaranteed to satisfy the parameter space restrictions.

3.4. Design-Consistent Estimation and Robust Inference

Godambe and Thompson's (1986) theory of optimal estimating functions provides a formal rationale for the sample design-weighted estimation of superpopulation model parameters. For the growth curve models dealt with in the previous section, the application of sample design weighting was a straightforward extension of the EM algorithm.

When the superpopulation model of equation (27) holds, the convergent solution for the superpopulation parameter ξ, say $\hat{\xi}_*$, has the following covariance matrix over repeated sampling from both the superpopulation and the finite population

$$E_m E_s(\hat{\xi}_* - \xi)\,(\hat{\xi}_* - \xi)^T = E_m\, \text{Var}_{s|m}(\hat{\xi}_* - \hat{\xi}_N)(\hat{\xi}_* - \hat{\xi}_N)^T$$
$$+ \text{Var}_m(\hat{\xi}_N - \xi)(\hat{\xi}_N - \xi)^T , \tag{38}$$

where the first term is the average sampling covariance of $\hat{\xi}_*$ and ξ_N is the universe-level version of equation (32). Note that the second term in equation (38), representing the model variance of ξ_N, is of order no larger than N^{-1} and is negligible compared with the first term, which is of order no larger than n^{-1}. A consistent estimate of the leading term in equation (38) is formed from the linearized vectors

$$g_{i*} \equiv G_i^T \, W_{i*} (Y_{i*} - G_i \, \hat{\xi}_*) \tag{39}$$

The covariance matrix for $\hat{\xi}_*$ is then approximated by calculating the sampling variance for

$$g_{s*} = \sum_{i \in s} w_i g_{i*} \, .$$

If this covariance matrix estimator is depicted by $\mathrm{cov}_s(g_{s*})$, then

$$\mathrm{cov}_s(\hat{\xi}_*) = J_{s*}^{-1} \mathrm{cov}_s(g_{s*}) J_{s*}^{-1} \tag{40}$$

where J_{s*}^{-1} is the inverse of the sample design weighted left-hand-sides matrix defining $\hat{\xi}_*$.

Design-consistent estimators for the sampling variances and covariances of σ_*^2 and Δ_* can also be specified simply employing Binder's (1983) results. Consider, for example, the upper triangular vector $\Delta_*(K \times 1)$ with $K = [p(p-1)/2 + 1]$. Let $\eta\eta_{i*}$ denote the corresponding upper triangular portion of $\eta_{i*} \eta_{i*}^T$ and similarly define $V\eta_{i*}$. With this notation, the linearized vector corresponding to g_{i*} in equation (39) is

$$d_{i*} = [\eta\eta_{i*} + V\eta_{i*} - \Delta_*] \, . \tag{41}$$

The analog for J_{s*}^{-1} is, in the case of Δ_*,

$$H_{s*}^{-1} \equiv I_K(1/\hat{w}_+)$$

If one chooses to stop the EM iterations after a fixed number of steps, say $g = 3$, the linearized variates in equations (39) and (41) are computed based on the $g = 3$ step solutions; namely, W_{ig}, ξ_g, $\eta\eta_{ig}$, $V\eta_{ig}$, and Δ_g . Considering that the EM algorithm has a reputation for slow convergence, one might be well advised to accept a solution based on $g = 2$ or 3 steps. For large samples, such a solution is likely to achieve most of the efficiency of the fully convergent pseudo ML estimator.

4. SURVIVAL ANALYSIS

4.1. Introduction

Survival analysis methods are often applicable to panel survey data. This is immediately apparent for such longitudinal studies as the National Health and Nutrition Examination Survey Epidemiologic Follow-up Study, which relates baseline measurements to subsequent morbidity and mortality. Survival analysis is commonly used when studying morbidity and mortality. It is less commonly used to study such activities as program participation. As an example, RTI used survival methods to study the effects of participation in the AFDC program caused by changes in the eligibility rules under the Omnibus Budget Reconciliation Act of 1981. Survival analysis was used to determine if recipients removed from the AFDC program because of rule changes modified their employment behavior over time to regain their AFDC benefits.

Survival analysis requires the estimation of the distribution function of the time from study entry (e.g., baseline survey date) until the occurrence of a specified event (e.g., death or termination of program participation). Letting T be the time until failure (or some other event), the survival function we wish to estimate is

$$S(t) = P(T \geq t) \quad 0 < t < \infty.$$

The associated hazard function, which is important in some survival analysis methods, is the "instantaneous rate of failure" given by

$$\lambda(t) \equiv P[\text{failure at time } t \mid T \geq t]$$

$$= -d \log S(t)/dt.$$

The estimation of the survival function or the hazard function is complicated by the fact that in most studies individuals are followed for varying lengths of time and that the duration of the study is usually short relative to the maximum lifetime remaining for a subject entering the study. A number of parametric models are available for the analysis of survival data. However, the usual methods of choice are nonparametric or quasi-parametric methods which make minimal assumptions about the underlying survival distributions.

This paper discusses three methods for conducting a survival analysis using sample survey data. The first method is a contingency table approach similar to that presented by Koch et al. (1972) for implementing the actuarial method of life table estimation. A second method is the Kaplan–Meier (1958) product limit estimator of the survival function. Both of these methods allow direct subgroup comparisons by estimating separate survival functions for each group and then comparing the estimated functions. Reduced models can be fitted to the estimated survival functions via weighted least squares for additional study and summarization. However, neither of these two methods facilitates covariate adjustments prior to making subgroup comparisons. The proportional-hazards model is presented as a third method, which allows covariate-adjusted analyses. The design-consistent version of this method is a pseudo ML procedure and relies upon Binder's (1983) work for variance estimation and inference. The proportional-hazards model for simple random samples was studied in detail by Chambless and Boyle (1985).

4.2. Contingency Table Approach

The general approach is taken from Koch et al. (1972). They present contingency table methods for estimating survival rates via the actuarial approach of Berkson and Gage (1952). Koch et al. assume that the rows of the contingency table are independent and that the observations in a row follow a multinomial distribution. We depart from these assumptions and assume that the data were obtained from a possibly stratified, clustered and unequally weighted sample from a finite population of individuals. This induces additional correlations among the entries in the contingency table. In the description below, the notation and development of Koch et al. has been presented whenever possible.

The basic situation considers the case where a single decrement is involved and where withdrawals must be considered. It is assumed that the population is grouped into r aggregates of individuals for which separate survival distributions are to be studied. Within each group, an individual unit can contribute to one or more of t exposure-periods depending upon the number of periods until death or withdrawal. Thus, there are rt group × exposure-period combinations. Next, define π_{ij1} to be the proportion of the population in group i who are alive at the beginning of exposure-period j and survive to the end of period j. Similarly, π_{ij2} is the proportion of the population in group i who are alive at the beginning of period j and die during that period. Finally, π_{ij3} is the proportion of the population in group i who are alive at the beginning of period j, survive for part of period j, and are withdrawn. Since these three events are mutually exclusive and exhaustive, we have $\pi_{ij1} + \pi_{ij2} + \pi_{ij3} = 1$, for all i and j.

Let n_{ijk} be the sample number of individuals that contribute to cell ijk and let $n_{ij} = n_{ij1} + n_{ij2} + n_{ij3}$ be the number of individuals alive at the beginning of period j for group i. Also define w_{ijkl} to be the sampling weight or inverse selection probability for the lth sample member that contributes to cell ijk.

A consistent sample estimate of π_{ijk} is

$$
p_{ijk} = \left[\sum_{l=1}^{n_{ijk}} w_{ijkl} \right] \bigg/ \left[\sum_{k=1}^{3} \sum_{l=1}^{n_{ijk}} w_{ijkl} \right] .
\tag{42}
$$

For the sampling situation equivalent to the case that Koch et al. considered (i.e., stratified simple random sampling), the sampling weights are constant, at least within each group, and equation (42) simplifies to $p_{ijk} = n_{ijk}/n_{ij}$.

It is convenient to define the column vectors

$$\pi'_{ij} = \left(\pi_{ij1}, \pi_{ij2}, \pi_{ij3}\right)$$
$$p'_{ij} = \left(p_{ij1}, p_{ij2}, p_{ij3}\right)$$
$$\pi' = \left(\pi'_{11}, \pi'_{12}, \ldots, \pi'_{rt}\right).$$

and

$$p' = \left(p'_{11}, p'_{12}, \ldots, p'_{rt}\right).$$

A key step in this analysis is the estimation of the covariance matrix of p. In Koch et al., the multinomial covariance matrix of p_{ij} was estimated by

$$V(p_{ij}) = [\mathrm{Diag}(p_{ij}) - p_{ij}p'_{ij}]/n_{ij}$$

and the covariance matrix of p by the $(3rt \times 3rt)$ block-diagonal matrix with the $V(p_{ij})$'s on the diagonal and zeros elsewhere. These estimators are appropriate for the case they considered. For a complex finite population sample of cases, a more complicated covariance matrix estimator is required. For most designs, appropriate covariance matrix estimators are available using either a linearization or sample reuse approach—e.g., SURREGR (Holt, 1977) and NASSREG (Mohadjer et al., 1987).

The estimated vector of survival rates can now be obtained as a special case of the log-linear function

$$s = \exp\{K \, [\log(Ap)]\} \tag{43}$$

where A and K are known matrices that will be specified later and $\exp$ and $\log$ are vector operators that transform the elements of the argument vector into a vector of exponentials or natural logarithms. Grizzle et al. (1969) have shown that

$$V_s = D_s \, K \, D_a^{-1} \, A \, V \, A' \, D_a^{-1} \, K' \, D_s \tag{44}$$

is a consistent estimate of the variance matrix of s where D_a and D_s are diagonal matrices with the elements of the vectors $a = Ap$ and s on their main diagonals, respectively. When V is a design-consistent estimator for the covariance matrix of p, the V_s matrix in (44) is similarly design-consistent for $\mathrm{Cov}(s)$.

To construct s, we will assume that units withdrawn in a given exposure-period were exposed to risk for an average of half of the period. Define the $(2rt \times 3rt)$ block-diagonal matrix A by $A = I_{rt} \otimes A^*$ where I_{rt} is the rt-dimensional identity matrix and

$$A^* = \begin{bmatrix} 1 & 0 & 0.5 \\ 1 & 1 & 0.5 \end{bmatrix} .$$

Also let K be the $(rt \times 2rt)$ block-diagonal matrix $K = I_r \otimes K^*$ where K^* is the $(t \times 2t)$ matrix

$$K^* = \begin{bmatrix} 1 & -1 & 0 & 0 & \cdots & 0 & 0 \\ 1 & -1 & 1 & -1 & \cdots & 0 & 0 \\ \cdot & \cdot & \cdot & \cdot & \cdot & \cdot & \cdot \\ \cdot & \cdot & \cdot & \cdot & \cdot & \cdot & \cdot \\ \cdot & \cdot & \cdot & \cdot & \cdot & \cdot & \cdot \\ 1 & -1 & 1 & -1 & \cdots & 1 & -1 \end{bmatrix} .$$

Using these definitions of A and K in conjunction with equation (43) yields a vector of estimated survival rates with the following structure

$$s' = (s'_1, s'_2, \ldots, s'_r)$$

where $s'_i = (s_{i1}, s_{i2}, \ldots, s_{it})$. Here s_{ij} is the estimated j-period survival rate for group i given by

$$s_{ij} = \prod_{k=1}^{j} (p_{ik1} + p_{ik3}/2)/(p_{ik1} + p_{ik2} + p_{ik3}/2) .$$

The population vector of survival rates is given by

$$S = \exp\{K[\log (A\pi)]\} .$$

Hypotheses concerning S of the form

$$H_o: CS = 0 \text{ vs. } H_1: CS \neq 0$$

can be evaluated using the quadratic form

$$Q = (Cs)' (C'V_s C)^{-1} (Cs) .$$

For large samples, Q approximately follows a chi-square distribution with rank (C) degrees of freedom.

4.3. Kaplan–Meier Product Limit Estimator

Another commonly used method of estimating the survival function is the product limit estimator based upon the work of Kaplan and Meier (1958). For this situation, assume that the sample consists of n units followed over time. These units will fail or withdraw during the study or will survive the follow-up period. Let $t_1 < t_2 < \ldots < t_m$ be the distinct time points at which failures occur. The risk set R_i $(i = 1, \ldots,$

m), corresponding to failure time t_i, consists of all the sample members who have not yet failed or withdrawn just prior to t_i. That is, R_i contains the sample members who are at risk of failure at time t_i.

The product limit estimator of the survival function is a discrete-valued function taking values

$$s(t_j) = \prod_{i=1}^{j} [1 - d(t_i)/n(t_i)] = \prod_{i=1}^{j} [1 - q(t_i)] = \prod_{i=1}^{j} p(t_i)$$

where $d(t_i)$ estimates the number of population units that fail at time t_i and $n(t_i)$ estimates the population size of the risk set R_i. Explicitly

$$d(t_i) = \sum_k w_k \gamma_k(t_i)$$

and

$$n(t_i) = \sum_k w_k \delta_k(t_i)$$

where w_k is the sampling weight for unit k, $\gamma_k(t_i) = 1$ if unit k fails and 0 otherwise, and $\delta_k(t_i) = 1$ if unit k is a member of R_i, and 0 otherwise.

The design-consistent variance estimator for $s(t_i)$ (conditional on $t_1 < t_2 < \ldots < t_m$) can be obtained from the linearized values (for unit k)

$$z_k[s(t_i)] = \sum_{i=1}^{j} p_i^*(t_j) \, z_k \, [p(t_i)]$$

where

$$p_i^*(t_j) = \prod_{\substack{i'=1 \\ i' \neq i}}^{j} p(t_{i'})$$

and

$$z_k[p(t_i)] = \frac{w_k[\gamma_k(t_i) - q(t_i)\delta_k(t_i)]}{n(t_i)}.$$

The associated covariance matrix of $s = [s(t_i), \ldots, s(t_m)]'$, the column vector of survival proportions, can be estimated using the linearalized values above. The estimation of s and its covariance can easily be extended to separate subgroups of the population. Armed with such estimates, reduced models can be fitted to the survival proportions using weighted least squares, and relevant hypotheses concerning the shape of s and subgroup differences can be assessed.

4.4. Discrete Proportional Hazards Model

As a final example, we consider the discrete proportional hazards model, which allows covariate adjustment to be made for population comparisons. The brief description below follows that given in Kalbfleisch and Prentice (1980) and is a generalization of the work of Chambless and Boyle (1985) for a simple random sample.

Consider a set of $m + 1$ disjoint time intervals $A_i = [a_{i-1}, a_i)$, $i = 1$, $\ldots$, $(m + 1)$ with $a_0 \equiv 0$ and $a_{m+1} \equiv \infty$. For panel survey applications these intervals A_i are likely to be days or months. We assume that only the interval of failure or withdrawal is recorded for each sample unit. The discrete proportional hazards model is characterized by the hazard function

$$\lambda_i(z) = 1 - (1 - \lambda_i)^{\exp(z'\beta)} \qquad i = 1, \ldots, (m + 1)$$

where λ_i is the baseline $(z = 0)$ hazard for interval A_i, z is a p-dimensional vector of covariates and β is the vector of covariate coefficients.

The weighted log-likelihood for this model from a sample of size n is

$$\hat{L}(\tau, \beta) = \sum_{k=1}^{n} w_k \sum_{i=1}^{m} \{\gamma_{ki} \log[1 - \exp(-\exp(\tau_i + z_k'\beta))]$$

$$- \alpha_{ki} \exp(\tau_i + z_k'\beta)\}$$

where w_k is the sampling weight for unit k, $\tau_i = \log[-\log(1 - \lambda_i)]$, $\gamma_{ki} = 1$ if unit k fails in interval A_i and 0 otherwise, and $\alpha_{ki} = 1$ if unit k survives interval A_i and 0 otherwise.

The associated weighted score statistic is the vector of partial derivatives

$$\hat{U}(\theta) = \left\{ \frac{\partial \hat{L}}{\partial \theta_j} \right\}$$

with $\theta = [\tau', \beta']$ and θ_j is the jth element of θ. The elements of $\hat{U}(\theta)$ are

$$\frac{\partial L}{\partial \tau_i} = \sum_{k=1}^{n} w_k[\gamma_{ki} b_{ki} - \alpha_{ki} h_{ki}]$$

and

$$\frac{\partial L}{\partial \beta_j} = \sum_{k=1}^{n} w_k \sum_{i=1}^{m} [\gamma_{ki} z_{kj} b_{ki} - \alpha_{ki} z_{kj} h_{ki}]$$

where β_j and z_{kj} are the jth elements of β and z_k respectively,

$$h_{ki} = \exp(\tau_i + z'_k\beta)$$

and

$$b_{ki} = h_{ki} \exp(-h_{ki})[1 - \exp(-h_{ki})]^{-1}.$$

The estimate of θ is found by solving the equations $\hat{U}(\hat{\theta}) = 0$ for $\hat{\theta}$. Binder (1983) has shown that the variance of θ can be estimated by

$$\hat{V}(\hat{\theta}) = (\hat{J}^{-1}) \text{ Var}[\hat{U}(\hat{\theta})](\hat{J}^{-1}),$$

where $\hat{J} = \partial\hat{U}(\hat{\theta})/\partial\hat{\theta}$. Noting that $\hat{U}(\hat{\theta})$ has the linear form

$$\hat{U}(\hat{\theta}) = \sum_{k=1}^{n} w_k U_k(\hat{\theta}),$$

a design-consistent covariance matrix estimator for $\hat{U}(\hat{\theta})$ is obtained by viewing $\hat{U}(\hat{\theta})$ as a vector of survey outcome variables reported without error by sample member k. Probability sampling theory provides the associated design-consistent estimator for $\text{Var}[\hat{U}(\hat{\theta})]$.

5. DISCUSSION

In the preceding sections, we have shown how sample design weighting and associated design-consistent covariance matrix estimation can be accomplished for a variety of superpopulation models commonly used to analyze panel survey data. The paper began with a review of our contributions to longitudinal family definition and estimation. Following a discussion of time adjustments for descriptive statistics, three sections were devoted to repeated-measurement and polynomial growth curve models. The final section presented three methods for estimating survival functions.

In addition to the utility of these specific results, we hope that our paper will stimulate wider interest in methods of estimation and inference that focus on the probability sampling distribution of survey statistics. In this regard, we believe that sample design consistency is a valuable robustness property for estimates of superpopulation model parameters and their covariance matrix estimators. The analytical tools are available to achieve this desirable property and we would admonish panel data analysts to take advantage of them.

REFERENCES

Anderson, T. W. (1973), "Asymptotically Efficient Estimation of Covariance Matrices with Linear Structure," *The Annals of Statistics*, 1, 135–141.

Bentley, B. S., & Folsom, R. E. (1981), *Family Unit Analysis Weighting Methodology for the National Medical Care Expenditure Survey* (RTI Final Report No. 1320–11F, Contract No. HRA 230–76–0268), National Center for Health Services Research, Rockville, MD.

Berkson, J., & Gage, R. P. (1952), "Survival Curve for Cancer Patients Following

Treatment," *Journal of the American Statistical Association*, **47**, 501–515.

Binder, D. A. (1983), "On the Variances of Asymptotically Normal Estimators from Complex Surveys," *International Statistical Review*, **51**, 279–292.

Citro, C., Hernandez, D., & Herriot, R. (1986), "Longitudinal Household Concepts in SIPP: Preliminary Results," *Proceedings of the Bureau of the Census' Second Annual Research Conference*, 598–611.

Chambless, L. E., & Boyle, K. E. (1985), "Maximum Likelihood Methods for Complex Sample Data: Logistic Regression and Discrete Proportional Hazards Models," *Communications in Statistics, Part A: Theory and Methods*, **14**, 1377–1392.

Dicker, M., & Casady, R. J. (1982), "A Reciprocal Rule Model for Defining Longitudinal Families for the Analysis of Panel Survey Data," *Proceedings of the Social Statistics Section*, American Statistical Association, 532–537.

Godambe, V. P., & Thompson, M. E. (1986), "Parameters of Superpopulation and Survey Population: Their Relationships and Estimation," *International Statistical Review*, **54**, 127–138.

Goldstein, H. (1979), *The Design and Analysis of Longitudinal Studies*, Academic Press, London.

Grizzle, J. E., Starmer, C. F., & Koch, G. G. (1969), "Analysis of Categorical Data by Linear Models," *Biometrics*, **25**, 489–504.

Hidiroglou, M. A., Fuller, W. A., & Hickman, R. D. (1979), *Super Carp Manual*, Statistical Laboratory, Iowa State University.

Holt, M. M. (1977), "SURREGR: Standard Errors of Regression Coefficients from Sample Survey Data" (rev. April 1982 by B. V. Shah), Research Triangle Institute, Research Triangle Park, NC.

Horvitz, D. G., & Folsom, R. E., Jr. (1980), "Methodological Issues in Medical Care Expenditure Surveys," *Proceedings of the Section on Survey Research Methods*, American Statistical Association, 21–29.

Kalbfleisch, J. D., & Prentice, R. L. (1980), *The Statistical Analysis of Failure Time Data*, Wiley, New York.

Kaplan, E. L., & Meier, P. (1958), "Nonparametric Estimation from Incomplete Observation," *Journal of the American Statistical Association*, **53**, 457–481.

Kleinbaum, D. G. (1973), "Testing Linear Hypotheses in Generalized Multivariate Linear Models," *Communications in Statistics*, **1**, 433–457.

Koch, G. G., Johnson, W. D., & Tolley, H. D., (1972), "A Linear Models Approach to Analysis of Survival and Extent of Disease in Multidimensional Contingency Tables," *Journal of the American Statistical Association*, **67**, 783–796.

Laird, N. M., Lange, N., & Stram, D. (1987), "Maximum Likelihood Computations with Repeated Measures: Application of the EM Algorithm," *Journal of the American Statistical Association*, **82**, 97–105.

Laird, N. M., & Ware, J. H. (1982), "Random-Effects Models for Longitudinal Data," *Biometrics*, **38**, 963–974.

LaVange, L. M. (1983), *The Analysis of Incomplete Longitudinal Data with Modeled Covariance Structures* (Institute of Statistics Mimeo Series No. 1449), University of North Carolina, Chapel Hill.

LaVange, L. M., & Folsom, R. E. (1985), "Regression Estimates of National Crime Survey Operations Effects: Adjustment for Nonsampling Bias," *Proceedings of the Social Statistics Section*, American Statistical Association, 7–10.

Mohadjer, L., Morganstein, D., Chup, A., & Rhoads, M. (1986), "Estimation and Analysis of Survey Data Using SAS Procedures WESVAR, NASSREG, and NASSLOG," *Proceedings of the Section on Survey Research Methods*, American Statistical Association, 258–263.

Rao, M. N., & Rao, C. R. (1968), "Linked Cross-Sectional Study for Determining Norms and Growth Rates," *Sankhya*, **28**, 237–258.

Survey Research Center Computer Support Group (1981), *OSIRIS IV User's Manual* (7th ed.), Institute for Social Research, The University of Michigan, Ann Arbor.

Woolson, R. F., Leeper, J. D., & Clarke, W. R. (1978), "Analysis of Incomplete Data from Longitudinal and Mixed Longitudinal Studies," *Journal of the Royal Statistical Society, A*, **141**, 242–252.

Weighting Issues for Longitudinal Household and Family Estimates*

Lawrence R. Ernst

1. INTRODUCTION

In the last few years several surveys have had as one of their goals to tell us what happens to households and families over time. These include the National Medical Care Expenditure Survey (NMCES), the National Medical Care Utilization and Expenditure Survey (NMCUES), and the Survey of Income and Program Participation (SIPP). The process of obtaining longitudinal estimates for households and families presents some important questions that either are not encountered or are easier to answer for cross-sectional estimation or longitudinal person estimation. The following are three key questions:

1. Since the composition of households and families can and does change over time, which changes should allow the unit to be considered still continuing and which mark the dissolution of the unit?

2. What operational rules should be used to determine which households, families, and individuals are to be followed over time, and what retrospective questions should be asked of individuals who join the sample after the beginning of the panel?

3. What weighting procedures should be employed to obtain weights that yield unbiased estimates, and how should the weights be adjusted to reduced the variances and biases of the estimates?

In this paper the focus is on the third question, but as will become clear in Section 4, in order to obtain unbiased estimates, the right combination of weighting procedure, longitudinal household or family definition and operational procedures is required.

In Section 2, we state the assumptions that are made in this paper and fix notation and terminology. In Section 3, we explain why a common type of weighting procedure used in sampling to obtain unbiased estimates—weighting by the reciprocal of the probability of selection—does not, in practice, generally work for longitudinal household and family estimation, and we present a class of weighting procedures that can accomplish this task. In Section 4, we explain the difficulties that can arise in applying these weighting procedures because the information necessary to determine the weight, the continuity of a household or family, or some of the subject-matter data needed in the estimates may not be collected under the assumed operational procedures. Also presented are conditions that, if satisfied by a longitudinal household or family definition, are sufficient for there to exist a weighting procedure that avoids these difficulties. Finally, in Section 5, some thoughts on the adjustment of the weights used to produce unbiased estimates are discussed. This discussion focuses on procedures for adjusting for non-interview and for controlling estimates for key demographic variables to independent estimates, since these procedures may have to be handled differently for longitudinal household and family estimation than for either cross-sectional estimation or longitudinal person estimation.

Many of the ideas in this paper were originally developed in Whitmore et al. (1982), Ernst (1983, 1985, 1986), and Ernst et al. (1984). In particular, this author wishes to acknowledge the work of his co-authors Hubble and Judkins in the last-mentioned paper, which has been incorporated into the present paper.

2. PRELIMINARIES

In order to keep the discussion from becoming overly complex, a fixed set of design and operational procedures is assumed throughout this paper and will now be described. The set of procedures chosen is motivated by the procedures used in NMCES, NMCUES, and SIPP. The presentation in this paper is, for the most part, with respect to households only, not families. Since a family is a subset of a household it should be relatively easy to make the necessary modifications for family estimation. Also, in Sections 2–4 it is assumed that no data are missing or in error, and that there is perfect frame coverage. Modifications and adjustments of the estimation procedures because of the unrealistic nature of these assumptions are considered in Section 5.

We take a month to be the basic unit of time. For each month t, H_t denotes a universe of cross-sectional households and P_t the set of persons residing in a household in H_t. For example, H_t might be the set of all households in the United States that contain civilian persons. The initial sample at month B is a probability sample of members of H_B. A person in a chosen household is known as an *original sample person*. There are several rounds of interviewing, with each interview covering the month or months since the previous interview; E denotes the final interview month for the sample panel. For each month t, the set of individuals to be interviewed are all original sample people in P_t plus all other people in P_t residing with an original sample person. These latter people are referred to as the *associated sample people*. Note that associated sample people are interviewed only for months in which they reside with an original sample person. When associated sample people are initially interviewed they are asked only enough retrospective information to ascertain whether they were in P_B .

A longitudinal household universe, U, is a collection of disjoint subsets of $\bigcup_{t=B}^{E} H_t$ such that each member L of U is of the form $L = \{h_b ,$ $h_{b+1} , \ldots , h_e\}$, where $h_t \in H_t$ and where b and e are the beginning and ending months, respectively, of the longitudinal household (the meaning that b and e have throughout this paper). The definition of a specific household universe consists of two parts. (Often "household" is referred to in this paper without an indication of whether it is longitudinal or cross-sectional, as has just been done. Unless cross-sectional is obvious from the context, longitudinal household should be assumed.) The first part is the household definition itself, which we consider a set of rules that for any $h_t \in H_t$ specifies which $h_{t+1} \in H_{t+1}$, if any, is eligible to be in the same member of U, or, more descriptively, which household h_{t+1} at month $t + 1$ is eligible to be the continuation of h_t at month t. (If there is no such h_{t+1} then $t = e$, that is t is the last month for the household.) The following are six examples of such rules. Each of the first five constitutes a household definition by itself, or they can be used in combination (in some cases redundantly) in a household definition; the sixth rule does not constitute a household definition alone.

1. *Same householder:* h_t and h_{t+1} have the same householder. (The householder of a household, in the Census Bureau terminology, is the first adult person listed on the household roster. According to the survey instructions this person should be an owner or renter of the housing unit.)
2. *Same spouses:* If h_t is a married-couple household, then h_{t+1} is also a married couple with the same husband and wife; otherwise h_t and h_{t+1} have the same householder.
3. *No change:* h_t and h_{t+1} have the same household members.
4. *Reciprocal majority:* The majority of members of h_t are in h_{t+1} and the majority of members of h_{t+1} are in h_t .
5. *Reciprocal plurality:* h_{t+1} contains more household members in h_t at month t than any other household in H_{t+1}, and h_t contains more household members in h_{t+1} at month $t + 1$ than any other household in H_t .
6. *Household type:* Either h_t and h_{t+1} are both married-couple households, both other family households, or both nonfamily households.

To illustrate these definitions consider the following example. A man, M, resides with his grown child, C_1 , in the interval $[B, t_1)$, while a woman, W, resides with her three grown children, C_2 , C_3 , C_4 , in the same time interval. At month t_1 , M and W marry, with M, W, and the four children residing together in the interval $[t_1 , t_2)$. At month t_2

all four children leave, with each child residing alone in the interval $[t_2 , E]$ while M and W continue to live together in this interval. In Table 1, the longitudinal households that arise from these six people are indicated for each of the five household definitions consisting of one of the rules alone, other than *household type*. Since for each of these definitions there are four households consisting of one of the four children with period of existence $[t_2 , E]$, these four households are omitted from the table. In the table each row constitutes a single household, with each column indicating the household members during the specified interval. An asterisk indicates that the household was not in existence during the interval.

To illustrate the effect of the *household type* rule, if it is combined with any of the other rules but the *no change* or *reciprocal majority* rules, then for this example the result would be the same as for the *same spouses* rule alone. *Household type* together with *no change* or *reciprocal majority* is the same as *no change* alone.

NMCES and NMCUES essentially used the *same spouses* and *reciprocal majority* rules, respectively, as household definitions. No household definition has been officially adopted yet for SIPP, although a definition combining the *same spouses* and *household type* rules is currently the leading candidate. It is not the purpose of this paper to compare any of these rules from a social science point of view. They will, however, serve to illustrate the discussion in Sections 3 and 4.

The second part of the definition of a specific household universe places restrictions on the form each member of U must take. For example, there might be the requirement that each household be in existence for at least two months. Another example is the requirement that $b = B$ and $e = E$, that is that the universe is restricted to households in existence for the entire life of the panel. Finally, there may be the restriction that U consists of a cohort of households in existence at month B, the *initial longitudinal households*, plus a set of households formed after month B, the *subsequently formed longitudinal households*, "generated by" the set of initial households. This last restriction only is assumed in this paper, for, as will be explained in Section 3, this assumption is necessary to obtain unbiased estimates.

To illustrate the vague concept "generated by," if the household definition is *same householder* then the set of subsequently formed households generated by the initial households might be all households whose householder at month b is in P_B ; if the household definition is *no change*, the concept might be the set of households with at least one member in P_B at month b, or, alternatively, the set of households for which all members at month b are in P_B .

TABLE 1. ILLUSTRATION OF HOUSEHOLD DEFINITIONS

Definitions	Intervals		
	$[B, t_1)$	$[t_1, t_2)$	$[t_2, E]$
Same householder	$\{M,C_1\}$	$\{M,W,C_1,C_2,C_3,C_4\}$	$\{M,W\}$
	$\{W,C_2,C_3,C_4\}$	*	*
Same spouses	$\{M,C_1\}$	*	*
	$\{W,C_2,C_3,C_4\}$	*	*
	*	$\{M,W,C_1,C_2,C_3,C_4\}$	$\{M,W\}$
No change	$\{M,C_1\}$	*	*
	$\{W,C_2,C_3,C_4\}$	*	*
	*	$\{M,W,C_1,C_2,C_3,C_4\}$	*
	*	*	$\{M,W\}$
Reciprocal majority	$\{M,C_1\}$	*	*
	$\{W,C_2,C_3,C_4\}$	$\{M,W,C_1,C_2,C_3,C_4\}$	*
	*	*	$\{M,W\}$
Reciprocal plurality	$\{M,C_1\}$	*	*
	$\{W,C_2,C_3,C_4\}$	$\{M,W,C_1,C_2,C_3,C_4\}$	$\{M,W\}$

*Household not in existence during the time interval.

3. OBTAINING UNBIASED ESTIMATES

To motivate the approach to obtaining unbiased estimates for longitudinal households to be presented in this paper, we will first explain why weighting by the reciprocal of the probability of selection is not in general feasible for this purpose, and hence the need to consider alternatives.

Let $X = \sum_{i=1}^{N} x_i$ be a parameter of interest, where x_i is the value of the characteristic for the ith unit in a population of size N. Typically in survey work to estimate X, a sample would be drawn in such a manner that the ith unit has a known positive probability of being chosen, p_i, and X would then be estimated by

$$\hat{X} = \sum_{i=1}^{N} w_i x_i ,$$

(1)

where

$$w_i = 1/p_i \text{ if the } i\text{th unit is in sample,} \tag{2}$$

$$= 0 \text{ otherwise.}$$

Unfortunately, for longitudinal household estimation such an approach is generally not practical, as is illustrated by the following example. Consider any subsequently formed household under the *no change* definition. Such a household would be in sample if and only if at least one household member is an original sample person and consequently, to use equations (1) and (2) as an estimator it would be necessary to determine the probability of this event. It is operationally impossible to determine this probability, since it is necessary to determine the first-round household for each member of the current household, including associated sample persons, and then compute the probability that at least one of these first-round households was selected.

Fortunately though, it is not necessary that w_i satisfy equation (2) in order that equation (1) be unbiased. If fact, if w_i is any random variable associated with the ith unit in the population satisfying

$$E(w_i) = 1, \tag{3}$$

then $E(\hat{X}) = X$. Thus, defining unbiased household weighting procedures reduces to defining random variables w_i satisfying equation (3).

All of the weighting procedures that have been proposed for obtaining unbiased household estimates that satisfy equations (1) and (3) have been of the following form. Associate with the jth individual in P_B, a weight, w_j', as follows. Let p_j denote the probability that the jth individual's household is in sample at month B, and then let

$$w_j' = 1/p_j \text{ if the individual's household is in sample at month } B, \tag{4}$$

$$= 0 \text{ otherwise.}$$

Then for the ith household in U, the longitudinal household universe, associate a set of constants α_{ij}, where j ranges over all people in P_B, with α_{ij} independent of w_j' and

$$\sum_j \alpha_{ij} = 1. \tag{5}$$

Finally, let the weight w_i of the ith household be

$$w_i = \sum_j \alpha_{ij} w_j'. \tag{6}$$

Clearly, any set of w_i's that satisfies equations (4)–(6) also satisfies equation (3). Furthermore, and these are the crucial differences, while equation (2) in general requires knowledge of the joint probability of selection for some members of H_B, equations (4)–(6) only require knowledge of individual probabilities of selection, and while equation (2) requires that the probability of selection be known for some members of H_B not in the initial sample, equations (4)–(6) do not.

The most common examples of estimators satisfying equations (4)–(6) are those for which we have a subset S_i of the ith members of the household such that

$$\alpha_{ij} = 1/m_i \text{ if the } j\text{th individual is in } S_i \cap P_B , \qquad (7)$$

$$= 0 \text{ otherwise,}$$

where m_i is the number of individuals in $S_i \cap P_B$. Equation (6) is thus the arithmetical average of the weights of the people in $S_i \cap P_B$. Note that equation (7) requires that $S_i \cap P_B$ be nonempty for each household in U.

Below are five examples of weighting procedures that have appeared previously in the literature (Ernst, 1983; Ernst et al., 1984) and that satisfy equations (4)–(6). The first three examples are of the form of equation (7) and also yield weights that do not vary with the interval for which the estimates are made. The fourth and the fifth examples are provided to illustrate weighting procedures that lack the latter property and the former property, respectively. The procedures are:

1. *Beginning date of household:* S_i is the set of all household members at month b.

2. *Householder weight:* S_i is the singleton set consisting of the householder at b.

3. *Average of spouses' weights:* S_i consists of the householder and spouse at b if the householder is a married-couple household at b; otherwise the householder at b is the only member of S_i.

4. *Beginning date of interval:* S_i is the set of all household members at the beginning of the time interval of interest.

5. *Average of monthly weights:* For the ith longitudinal household let d_i be the number of months that the household contains at least one member of P_B, and for each month t let m_{it} be the number of such individuals. For the jth person in P_B let T_{ij} denote the set of months that this individual is in the ith household. Then let

$$\alpha_{ij} = \frac{1}{d_i} \sum_{t \in T_{ij}} \frac{1}{m_{it}} .$$

Note that this yields the same weights as would be obtained by averaging the weights of the individuals in P_B who are in the household each month and then averaging the result over all months that the household contains a member of P_B.

To illustrate these procedures, consider the example of Section 2. Assume that M and C_1 are original sample people with probability of selection 1/2400, and that the woman and her children are not original sample people. Then for the household in the third row under the *same spouses* definition in Table 1, the weight would be 800 for the *beginning date of household* procedure, 2400 for the *householder weight* procedure, and 1200 for the *average of spouses weights* procedure. For the *beginning date of interval* procedure with beginning month t, the weight is 800 if $t \in [t_1, t_2)$ and 1200 if $t \in [t_2, E]$. If additionally it is given that the durations of $[t_1, t_2)$ and $[t_2, E]$ are 12 and 8 months, respectively, then the weight is 960 for the *average of the monthly weights* procedure.

Corresponding to each of these weighting procedures is a largest universe for which the procedure is defined. For any procedure of the form of equation (7) it is all households for which $S_i \cap P_B$ is nonempty. For example, for the *beginning date of household* procedure it is all households that at b contain at least one member of P_B, while for the *householder weight* procedure it is all households for which the householder at b is in P_B. For the *average of monthly weights* procedure it is all households that contain a member of P_B for at least one month. Each weighting procedure can of course also be used for any smaller universe. These examples illustrate why it is necessary, if strictly unbiased estimates are desired, to restrict the universe to a cohort of initial households plus a set of subsequently formed households generated by the initial households, as was indicated in Section 2.

In the next section we will show that some of the weighting procedures that have just been defined in combination with some of the household definitions in Section 2 can overcome some practical problems in obtaining unbiased estimates.

4. OVERCOMING PROBLEMS ASSOCIATED WITH OBTAINING UNBIASED WEIGHTS

Although estimators obtained using weighting procedures satisfying equations (4)–(6) avoid the difficulties arising from the use of equation (2), some of the information needed to compute equation (1) is generally not available for many combinations of weighting procedures and household definitions because the information is not collected under the assumed operating procedures. (This is true even though it is assumed in this section, as in the previous two sections, that there is complete response and that there is perfect frame coverage.) This unavailable information can result in the following four problems for a household:

1. *Unknown weight:* The weight associated with the household is not known because it depends on uncollected information about the household before it entered or after it left sample.
2. *Missing subject-matter data:* The household has a positive

weight but subject-matter data are incomplete because the household existed before entering or after leaving sample.

3. *Period of existence* (problem 1): The household has a positive weight and yet its period of existence is unknown because it cannot be determined whether the household existed before entering or after leaving sample.

4. *Period of existence* (problem 2): The household has a positive weight and yet its period of existence is unknown because for some month t it cannot be determined if a household $h_{t+1} \in H_{t+1}$ is the continuation of a household $h_t \in H_t$ even though both h_t and h_{t+1} are in sample.

Note that the *unknown weight* problem is equivalent to not knowing some of the w_i's in (1); the *missing subject-matter data* problem is equivalent to not knowing some of the x_i's when $w_i > 0$; and the two *period of existence* problems are equivalent to not knowing the set of units to be used in equation (1).

Table 2 presents the problems that exist for each combination of the five household definitions in Table 1 and the five weighting procedures described in Section 3. In this table "all but last" means all the problems except *period of existence* problem 2.

To illustrate, let us use the example in Section 2 to verify that all four problems arise for the combination of the *reciprocal plurality* definition and the *beginning date of household* procedure, as asserted in Table 2. In this illustration it is also assumed, as in Section 3, that M and C_1 are the only original sample people, with probability of selection $1/2400$.

Focus on the household that for the interval $[t_1, t_2)$ consists of M, W, C_1, C_2, C_3, C_4. According to Table 1, this household is the continuation of the household in existence since time B consisting of W, C_2, C_3, C_4, and hence has weight 0. However, since the woman and her three children are not original sample people, we would not know, under the assumed operating procedures, their living arrangements prior to t_1, and hence do not have the information necessary to determine continuity from $(t_1 - 1)$ to t_1. For example, a second possibility is that these four people each resided in a different household prior to t_1, in which case the household consisting of all six people at t_1 is the continuation of the household with M and C_1 prior to t_1, and hence has weight 2400. A third possibility is that the woman resided with one child at month $(t_1 - 1)$, in which case t_1 is the beginning month for the household consisting of all six people at t_1, and this household then has weight 800. Thus, since there are at least three possibilities for the weight of this household, the *unknown weight* problem occurs. Furthermore, because of the possibilities it cannot be determined if the six person household at month t_1 is a continuation of the M, C_1 household at $(t_1 - 1)$, even though both households are in sample, and hence *period*

TABLE 2. PROBLEMS

	Procedures				
Definitions	Beginning Date of Household	Householder Weight	Average of Spouses Weights	Beginning Date of Interval	Average of Monthly Weights
Same householder	All but last	None	All but last	All but last	All but last
Same spouses	All but last	None	None	All but last	All but last
No change	None	None	None	None	None
Reciprocal majority	All but last	All but last	All but last	All but last	All but last
Reciprocal plurality	All	All	All	All	All

of existence problem 2 occurs.

As for the other two problems, if the longitudinal household that consists of all six people for $[t_1, t_2)$ has a positive weight, then these problems can also occur because the living arrangements of the woman's three children at month t_2 are unknown. For example, if instead of all three children living separately, as is assumed in Table 1, the woman's children are all living together at t_2, that would constitute the continuing household at t_2, but since it would not be in sample, the *missing subject-matter data* problem would occur. Another possibility is that exactly two of these children are living together at month t_2, in which case, instead of the household that we are focusing on continuing to E, $(t_2 - 1)$ is its last month of existence. Consequently, *period of existence* problem 1 occurs.

The same example can also be used to illustrate that the indicated problems can occur for the other entries in Table 2 for which there are some problems. We now proceed to explain how all four problems can be avoided simultaneously for the right combinations of household definition and weighting procedure, as listed in Table 2.

The key result of this section is that if the following two conditions are satisfied by the household definition, then there exists a weighting procedure that avoids all the problems.

Condition 1. For each month t and each $h_t \in H_t$ there exists a nonempty subset h_t^* of h_t that depends only on h_t, such that $h_{t+1}^* = h_t^*$ is necessary for h_{t+1} to be the continuation of h_t.

Condition 2. The determination of whether $h_{t+1} \in H_{t+1}$ is the continuation of $h_t \in H_t$ depends only on h_t and h_{t+1}.

If conditions 1 and 2 are satisfied and if a weighting procedure of the form of equation (7) is used with $S_i = h_b^*$, then, as we will show shortly, all four problems are avoided.

First, however, let us give an intuitive explanation of these conditions and demonstrate that they are satisfied for the entries in Table 2 for which no problems are listed. Condition 1 states that when a household is formed there is some known subset of the household members, h_b^*, that will remain in the household throughout its period of existence. Then, using equation (7) with $S_i = h_b^*$ averages the weight of all people in this set who are in the cross-sectional universe at month B. For the *no change* definition, for example, all people in the household at month b must remain in the household throughout its period of existence. Consequently, condition 1 is satisfied with $h_t^* = h_t$. Therefore equation (7) can be used with $S_i = h_b$, which is the *beginning date of household* procedure. Furthermore, if condition 1 is satisfied for a set of people, it is also satisfied for any subset. In particular, for the *no change* definition, condition 1 is satisfied with h_t^* the set consisting of the householder (and spouse if a married-couple household) at month t, which leads to the *average of the spouses weights* procedure. It is also satisfied with h_t^*, the singleton set consisting of the householder at month t, which yields the *householder weight* procedure.

If there is more than one choice of h_t^* satisfying condition 1, the best choice may be the largest possible h_t^*, since this would result in the largest number of households with positive weights, which would in general result in an estimator given by equation (1) with the greatest precision. Consequently, for the *no change* definition, the *beginning date of household* procedure would be preferred over the *average of the spouses weights* procedure and the *householder weight* procedure. Also note that for the *no change* definition, the *beginning date of interval* and *average of monthly weights* procedures yield the same weights as the *beginning date of household* procedure.

As for the other definitions in Table 2, the largest h_t^* satisfying condition 1 for the *same spouses* definition is the set corresponding to the *average of spouses weights* procedure, which is consequently the preferred procedure for this definition. Condition 1 is also satisfied for this definition by the singleton set corresponding to the *householder weight* procedure. For the *same householder* definition, the only h_t^* satisfying the condition consists of the householder at t, that is only the *householder weight* procedure can be used. For the *reciprocal majority* and *reciprocal plurality* definitions, no h_t^* satisfies condition 1, since it is the quantity of the people present rather than any specific set of people that determines continuity for these definitions.

Condition 2 amounts to saying that if h_t and h_{t+1} are both in sample then it can be determined whether h_{t+1} is the continuation of h_t. Thus, if this condition is satisfied the *period of existence* problem 2 is avoided.

Each of the definitions in Table 2 except *reciprocal plurality* satisfies condition 2, as can be ascertained by a careful reading of these definitions. As for *reciprocal plurality*, the example presented earlier showed that not enough information is available to determine if the six-person household at month t_1 is a continuation of the household consisting of M and C_1 at $(t_1 - 1)$ even though both households are in sample, illustrating that condition 2 is not satisfied for *reciprocal plurality*.

For more complicated definitions in which more than one of the six rules of Section 2 must be satisfied, the following is true about these conditions. Condition 1 is satisfied if and only if either the *no change*, *same householder*, or *same spouses* rule is in the definition. Condition 2 is satisfied if the *reciprocal plurality* rule is not in the definition. It is also satisfied if *reciprocal plurality* is present in combination with either the *no change* or *reciprocal majority* rules since *reciprocal plurality* is then redundant.

We now establish that if conditions 1 and 2 are satisfied then all four problems are avoided with the stated weighting procedure. In fact, only condition 1 is required for all but *period of existence* problem 2. Clearly this is true for the *unknown weight* problem. The *missing subject-matter data* problem is avoided since if a household $L = \{h_b, \ldots, h_e\}$ has a positive weight then there exists an original sample person that is a member of h_b^* and hence a member of h_t^* for each month t of L's existence, and therefore L is in sample throughout its period of existence. Similarly, *period of existence* problem 1 is avoided since the household cannot be in existence before entering sample or after leaving sample, because this original sample person would not have been a member.

Condition 1 alone and this weighting procedure do not guarantee that *period of existence* problem 2 does not occur. For example, a definition combining *same householder* and *reciprocal plurality* does not avoid this problem, even though condition 1 is satisfied with the h_t^* that leads to the *householder weight* procedure. However, as noted previously, condition 2 guarantees that *period of existence* problem 2 is avoided.

Finally, we remark that in general, conditions 1 and 2 do not assure that complete information is available for each household that ever appears in sample. In fact, only the *no change* rule assures that. What these conditions together with the specified weighting procedure do guarantee is that all households that are not in sample for the entire period of existence are zero-weighted and, consequently, that the missing information is not needed.

5. WEIGHTING ADJUSTMENTS

To obtain final weights for use in producing estimates for household surveys, weights that result in unbiased estimates are typically subjected to one or more adjustment procedures to reduce the variance of the estimates and reduce the bias resulting from undercoverage and non-response. For example, at the Census Bureau these adjustments generally include a noninterview adjustment, an adjustment to reduce variability between primary sampling units, and an adjustment to independent estimates of key demographic characteristics of the analytic unit.

Adjustments of the weights for a sample of longitudinal households would generally incorporate the same basic concepts as adjustments for a cross-sectional sample, but they are subject to additional complications arising from the time element. In this section we outline one approach to weighting adjustments for longitudinal households. Other approaches are certainly possible.

Our approach is motivated in part by the nature of our assumed universe, that is a cohort of initial households plus a set of subsequently formed households generated by the initial households. The procedure envisioned consists of the following three steps.

Step 1: Adjust the weights of the set of initial sample households as one adjusts a cross-sectional sample at month B, through a series of adjustments culminating in an adjustment to independently derived estimates of the number of cross-sectional households with specific characteristics in existence at B.

Step 2: Next, adjust the weights of the subsequently formed sample households to carry over to these households the adjustments to the initial sample households. This includes adjusting for noninterviews among subsequently formed households that result from noninterviewed initial sample households.

Step 3: Finally, adjust the weights of subsequently formed sample households to account for the other category of noninterviews among these households, namely, noninterviews that result from original sample people in interviewed initial sample households who later become noninterviews.

The distinction between the adjustments in steps 2 and 3 is that the adjustment in step 2 is for subsequently formed households that contain people from initial sample households who became noninterviews while they were in their initial household. In step 3, the adjustment is for subsequently formed households that contain people in initial sample households who became noninterviews after their initial household ended. For the rare case when a subsequently formed household contains both types of people, the adjustment is a combination of steps 2 and 3.

Before detailing these steps, we digress to present observations on two general aspects of the adjustments that affect more than one of these steps. The first of these is the question of the number of points in time for which the sample estimates should be controlled to independent demographic estimates. The proposed procedure envisions doing this only at month B. One fundamental reason for not proposing adjustment at more than one point in time is that the household universe that we are considering excludes subsequently formed households not generated by initial households. Consequently, the number of households in the universe at any time after month B would not agree with independent cross-sectional estimates that include such households. Furthermore, even if appropriate controls are obtained at more than one point in time, there are difficulties in attempting to obtain agreement with these controls. One approach to obtaining this agreement is to group the households in each cell according to their pair of beginning and ending months and to then apply a different weighting factor to each such group. The values for these factors can be determined by considering them as variables in a mathematical programing problem. This approach is described in detail by Judkins et al. (1984). However, in certain situations no solution would be possible unless some weighting factors are allowed to be very large or negative, and sometimes not even then. For example, agreement cannot be reached when controlling to independent estimates at two months t_1 and t_2 when some cells have the identical set of units in both months, but the control totals are different.

The second item of discussion is whether the final weights for households should vary with the interval for which estimates are produced, even when the weights used to produce unbiased estimates do not. This is also an issue in longitudinal person estimation, and in fact was first discussed by Kobilarcik (1985). It arises in the noninterview adjustment problem because a considerable number of sample households are interviewed for some but not all of their period of existence. If one final weight is used for each household, then such households would have to be considered noninterviews in the noninterview adjustment process (unless data were imputed for the missing time periods) and the data collected from these households would not be used directly in the estimation. In contrast, the use of final weights that vary with the time interval for which estimates are to be made allows the use of households that are interviewed for some, but not all, rounds to be used in estimates for time intervals throughout which they are interviewed. This would be accomplished by allowing the set of households considered noninterviews in the noninterview adjustment process to vary with the time interval, which would cause the noninterview adjustment factors and final weights to vary also.

A drawback to the use of more than one final weight for a household is the increase in operational complexity. Even if more than one final weight is produced, the number of different intervals for which distinct final weights are obtained may have to be limited to keep the processing problem from becoming unmanageable. To simplify this problem, assume that the noninterview pattern for each household and person is nested, that is, noninterview for one month implies noninterview for all subsequent months. (Then for any actual case for which the noninterview pattern is not nested, either missing interviews would be imputed or interview data subsequent to the first noninterview month would not be used in the estimation.) It would then be appropriate to obtain weights for a limited number of intervals, say, of the form $[B, t_i]$, $i = 1$, $\ldots, k$. Then if estimates are desired for an interval $[t, t']$, the weights associated with the smallest of the intervals $[B, t_i]$ containing $[t, t']$ would be used in the estimation.

We now detail the suggested three steps of the adjustment process. Assume that a single final weight for the interval $[B, E]$ will be obtained for each sample household. However, by simply considering t as the final month of the panel, this process would apply equally well for any interval of the form $[B, t]$, and hence can also be used if the final weights vary with the interval for which estimates are to be produced.

5.1. Step 1: Adjustment of Weights of Initial Sample Households

Conceptually, the adjustment procedures for initial sample households are similar to the adjustments for a cross-sectional survey at month B, and we consequently highlight here only aspects for which there may be important differences. Noninterview adjustment is one such area. At the Census Bureau, for example, a single noninterview adjustment is generally used in the household surveys for cross-sectional estimation. In this adjustment the analytic units are partitioned into adjustment cells defined by demographic characteristics of the unit. The weight of each interviewed unit is multiplied by a noninterview adjustment factor, namely, the sum of the weights of the interviewed plus the noninterviewed units in the cell divided by the sum of the weights of the interviewed units; and the noninterviewed units are then zero-weighted, thereby redistributing the weights of the noninterviewed units in each adjustment cell to the interviewed units.

For the initial sample households, however, we propose that the noninterview adjustment be performed in two steps: In the first step, the weights of units not interviewed at B would be redistributed to all other units in the cell; in the second step, the weights after the first adjustment of units interviewed at B, but not interviewed for their entire period of existence, would be redistributed to the initial households interviewed for their entire period of existence. The reason for proposing two such adjustments here is that this permits using a selection of variables in forming adjustment cells from the extensive data collected in previous interviews from those households interviewed at B, instead of being restricted to the limited information that typically would be available for households not interviewed at all. (The concept of using two noninterview adjustments in this context applies equally well to longitudinal person estimation and, in fact, was first developed for SIPP longitudinal person estimation [Kobilarcik & Singh, 1986].)

The first noninterview adjustment presents no unusual difficulties, and in fact can be done precisely as a noninterview adjustment at B just as in a cross-sectional survey. However, there are at least two complications that arise in the second adjustment. First, if a household had month t as its first noninterview month, then it might not be known whether the household actually continued to exist at month t. This information is important since such a household can obviously be a noninterviewed household only if it continued to exist at month t. Imputation may be necessary to make this determination. Second, it may be desirable to redistribute the weights of households with first noninterview month t only to interviewed households still in existence at month t by computing noninterview factors F_{tC} that vary not only with the adjustment cell C, but also with t. To compute F_{tC}, $B < t \le E$ first let I_{tC} denote the weighted count in cell C of interviewed households with period of existence $[B, t]$ (using the weights after the first noninterview adjustment) and let N_{tC} denote the weighted count of noninterviewed households in cell C with first noninterview month t. (Note that $N_{BC} = 0$ because of the first noninterview adjustment.) Then let

$$F_{tC} = 1 + \sum_{i=B}^{t} \left(N_{iC} \Big/ \sum_{j=i}^{E} I_{jC} \right).$$

Application of this factor redistributes the weights of all noninterviewed households in cell C with first noninterview month t to all interviewed households in existence at month t. Furthermore, the sum of the weights of all interviewed households in cell C after this adjustment can now be obtained. In the derivation that follows, an interchange of order of summation is involved. This is valid since the summation over all pairs (i, t) for which $B < i < t < E$ can be written as $\sum_{t=B}^{E} \sum_{i=B}^{t}$ or $\sum_{i=B}^{E} \sum_{t=i}^{E}$. The sum is

$$\sum_{t=B}^{E} F_{tC} I_{tC} = \sum_{t=B}^{E} \left[1 + \sum_{i=B}^{t} \left(N_{iC} \middle/ \sum_{j=i}^{E} I_{jC} \right) \right] I_{tC}$$

$$= \sum_{t=B}^{E} I_{tC} + \sum_{i=B}^{E} \left[N_{iC} \left(\sum_{t=i}^{E} I_{tC} \right) \middle/ \sum_{j=i}^{E} I_{jC} \right]$$

$$= \sum_{t=B}^{E} I_{tC} + \sum_{i=B}^{E} N_{iC} \, ,$$

which as desired is the sum of the weights before this adjustment of all households in cell C, both interviewed and noninterviewed.

Another important question that arises in the weighting adjustment for initial longitudinal households is what to use as a source for the independent controls. Among United States national surveys, the Current Population Survey (CPS) estimates have been the choice in NMCUES, and also for SIPP longitudinal person and cross-sectional estimation. It should be noted that until recently CPS household and family estimates appropriate to use as controls were produced only for March estimates in conjunction with the Annual Demographic Supplement, and hence this was the month for which NMCUES estimates were controlled. Since the advent of SIPP, these estimates have been produced monthly to provide controls for SIPP cross-sectional estimates, and they can also be used as controls for longitudinal household estimates.

Some necessary imperfections in the CPS household control totals should be noted. Although CPS estimates of the total numbers of persons in given age-race-sex categories are themselves controlled to independent demographic estimates that have no sampling variability, there are no such controls for household estimates. Consequently, such key CPS household estimates as the number of households with the householder in a specific age-race-sex category, or the number of households of a given size or type, or even the total number of households are subject to sampling variability and unknown biases. Despite this drawback, it is felt that adjusting to CPS estimates would be worthwhile in reducing sampling variability and many biases because of the large size of the CPS sample and the relative reliability of CPS estimates.

As for the specific variables to use in the control process, this would of course depend on the needs of the particular survey. The variables used in NMCUES are presented in Whitmore et al. (1982).

5.2. Step 2: Carry-over of Weight Adjustments for Initial Sample Households to Subsequently Formed Sample Households

Undercoverage affecting the set of initial households also affects the subsequently formed households generated by the initial households. To compensate for this in a simple fashion, the weights for the set of sample subsequently formed households can be adjusted by modifying equation (4), so that the final weight of each original sample person's initial household is used in place of the reciprocal of the probability of selection. The motivation for this adjustment is that for a subsequently formed sample household that contains original sample people from a single initial sample household, as is nearly always the case, the ratio of the weight after this adjustment to the weight used in the unbiased estimates would appropriately be the same as the ratio of the final weight to the weight used in the unbiased estimates for the corresponding initial sample household.

5.3. Step 3: Additional Noninterview Adjustments for Subsequently Formed Sample Households

Even if all original sample people who were members of interviewed initial sample households continue to be interviewed throughout the life of the panel, there would still generally be for each $t \in [B, E]$ a set of noninterviewed households formed at month t resulting from the noninterviewed initially formed households. This set of noninterviews is compensated for by the adjustment in step 2. However, in practice, there is also a set of noninterviewed households formed at month t, denoted N'_t, whose noninterview status results from later noninterviews among original sample households. Additional noninterview adjustments are required to compensate for this latter set. These adjustments present some significant complications that are not found in longitudinal person estimation. To illustrate, consider the case of a sampled initial household with final weight w that moves at month t and is not followed. Prior to the move the household contained five people but no information is available concerning the composition after the move. Then, at one extreme, each of these five people might be living alone at month t, in which case the initial household generates five new households at month t. At the other extreme these five people might remain together, in which case there are no new households at month t generated by the initial household. Furthermore, the weight of any new households would in general not be known. For example, with the *average of spouses weights* procedure, if one of these people is living alone at month t, the weight of this newly formed household after step 2 would be w. However, if that person instead forms a two-person household by marrying an associated sample person, the corresponding weight would be $w/2$. Finally, if the person becomes part of a household in which the householder and spouse, if present, are associated sample persons, then the household would be zero-weighted. Thus, in addition to the problem of missing subject-matter data, noninterviewed households in N_t' entail the additional problems of determining the number of noninterviewed analytic units and their weights. These problems would likely be handled by some form of imputation procedure.

Once this imputation is performed, it is proposed that the household weights for the sample set of subsequently formed households be adjusted to compensate for noninterviews in N_t', for any $t > B$, through a sequence of noninterview adjustments computed by using recursion on t as follows. First partition N_t' into two subsets, N_t'' and N_t''', the former consisting of noninterviewed households that would have been interviewed had all original sample people interviewed at month $(t - 1)$ continued to be interviewed throughout the life of the panel, and the latter consisting of the remaining households in N_t'. Adjustments for noninterviewed households in N_t'' can be made by redistributing their weights to interviewed households. However, the members of N_t''' would be unknown and compensation for them must be made by prior noninterview adjustments. Since any member of N_t''' must have resulted from noninterviewed households that were in N_i'' for some $i < t$, the noninterviewed households in N_t''' can be adjusted for by carrying over the adjustments for N_i'' for each i into households formed after month i.

The following is the suggested noninterview adjustment procedure to compensate for noninterviews in N_t''. For each month $t > B$, a noninterview adjustment factor f_{tL} would be applied to each member L of the set of interviewed households formed at t, denoted by I_t, as follows. For each month $i \in (B, t)$ any $L_i \in I_i$ would have previously received a noninterview adjustment factor f_{iL_i} to compensate for noninterview in N_i'. From these factors, each original sample person interviewed at month $(t - 1)$ would be assigned a month $(t - 1)$ adjusted weight

$$w \prod_{i=B}^{t-1} g_{iL_i} , \tag{8}$$

where w is the final weight of the person's initial household, L_i is the person's household for month i, and

$$g_{iL_i} = f_{iL_i} \text{ if } L_i \text{ was formed at month } i,$$
$$= 1 \text{ otherwise.}$$

Thus, a noninterview adjustment factor would be applied to each original sample person for each month after B that the person was a member of a newly formed interviewed household. Then to compute

f_{tL} , first compute an adjusted household weight w_{tL*} (but not the final weight) for each household $L^* \in I_t \bigcup N_t''$, using equations (6) and (4), with expression (8) substituted for the reciprocal of the probability of selection of the person's initial household. This is where the recursion occurs. The factor f_{tL} is then the sum of the weights w_{tL*} of all households $L^* \in I_t \bigcup N_t''$ in the same adjustment cell as L divided by the sum of the same weights of all households only in I_t in this adjustment cell. The final household weight for L would then be the product $f_{tL} w_{tL}$. Note that this final weight is also the weight that would be obtained from equations (6) and (4) with each person's month t adjusted weight replacing the reciprocal of the probability of selection of the person's initial household. Thus, the final weight for L includes the factor f_{tL} to compensate for noninterviewed households in N_t'' , factors f_{iL_j} to account for noninterviews in N_t''' that result from noninterviews in N_i'' , and the weight after step 2 that accounts for noninterviewed households formed at month t resulting from noninterviewed initial households.

In practice, there would be at least one major difficulty, in addition to the general complexity, in computing the f_{tL} factors using the method just outlined. The set I_t for a fixed t may often be too small to form adjustment cells containing a sufficient number of cases. Consequently some compromise may be necessary to the principle that members of N_t'' should have their weights distributed only to members of I_t .

REFERENCES

Cox, B. G., & Cohen, S. B. (1985), *Methodological Issues for Health Care Surveys*, Marcel Dekker, New York.

Ernst, L. R. (1983), *Preliminary Thoughts on Unbiased Longitudinal Household Estimation* (U.S. Bureau of the Census report).

Ernst, L. R. (1985), *The Impact of Conditions in a Longitudinal Household Definition on the Ability to Produce Unbiased Longitudinal Estimates in SIPP* (Report Census/SRD/RR-85/16), U.S. Bureau of the Census, Statistical Research Division.

Ernst, L. R. (1986), *SIPP Longitudinal Household Estimation for the Proposed Longitudinal Household Definition* (Report Census/SRD/RR-86/5), U.S. Bureau of the Census, Statistical Research Division.

Ernst, L. R., Hubble, D. L., & Judkins, D. R. (1984), "Longitudinal Family and Household Estimation in SIPP," *Proceedings of the Section on Survey Research Methods*, American Statistical Association, 682–687.

Judkins, D. R., Hubble, D. L., Dorsch, J. A., McMillen D. B. & Ernst, L. R. (1984), "Weighting of Persons for SIPP Longitudinal Tabulations," *Proceedings of the Section on Survey Research Methods*, American Statistical Association, 676–681.

Kobilarcik, E. L. (1985), *Overview of SIPP Longitudinal Weighting* (U.S. Bureau of the Census report).

Kobilarcik, E. L., & Singh, R. P. (1986), "SIPP: Longitudinal Estimation for Persons' Characteristics," *Proceedings of the Section on Survey Research Methods*, American Statistical Association, 214–219.

Whitmore, R. W., Cox, B. G., & Folsom, R. E. (1982), *Family Unit Weighting Methodology for the National Household Survey Component of the National Medical Care Utilization and Expenditure Survey* (Research Triangle Institute report).

NOTE

*The views reflected are not necessarily those of the Census Bureau nor do they necessarily represent Census Bureau statistical policy or practice.

Statistical Design and Estimation: Discussion

Martin R. Frankel

These are three very interesting papers, each dealing with issues that are important in attempting to develop the sample selection and estimation strategy to be applied in panel surveys. Before addressing some of the specific issues covered by the papers, some general comments may be in order.

First, when we consider some of the sample design issues (i.e., frame preparation, sample selection, and estimation) that must be addressed for any panel survey, I often think we are in a situation similar to that of a person who exists in two dimensions and is suddenly exposed to the third. Whatever the issues are that must be faced, addressed, and dealt with in a cross-sectional survey, these issues seem to grow by at least one order of magnitude and take us into the next dimension of complexity when we deal with panels.

What strikes me as encouraging about these papers is that while they are obviously grounded in actual survey experience, they are all able to operate at a level of generality that goes beyond any single survey. This means that in the area of panel surveys we have made some progress in developing general rules, principles, and even theory that builds on specific solutions.

In single-point-in-time surveys, the standard approach in design-based sampling has been to define a population by defining individual elements and then defining an aggregation rule. Included in either the definition of elements or in the aggregation rule is some mention of time. Typically, this time-based definition involves a single instant in time, for example, all households within the United States as of November 20, 1986, or all business establishments of a certain type as of January 1, 1987.

When we begin to consider panel surveys we must deal with the notion of a population of elements moving through time. Our population definition will usually involve at least two time points—typically endpoints. Our population definition may include any element in the population that exists as an element at any time within the two time endpoints or it may be more restrictive and require existence for some duration within the endpoints. In either case, our design must recognize that over time individual elements may be born (enter the popula-

160

tion for the first time), they may die (leave the population forever), and they may enter and leave the population temporarily. Even more complex, however, is the fact that our definition may have to deal with the changes in the fundamental make-up or composition of the element itself over time.

The paper by Michael Colledge deals with some of the issues that must be faced in trying to create and maintain a sampling frame of business establishments. While it is true that Statistics Canada does not presently conduct many surveys that may be classified as panel surveys, the issues that must be faced when dealing with the maintenance of a sampling frame for cross-sectional surveys of business establishments are the same issues that would have to be faced if we were dealing with panel surveys of business establishments.

In dealing with frames, the approach taken by Colledge is to be commended. He does not take the position that we often hear in statistics— "if the model does not fit the real world, then assume that the real world will see the error of its ways and change." Instead, he deals with many of the actual complexities that must be recognized as business organizations move through time—changing their business activities, locations, sizes, and possibly the entire nature of their business.

Both the paper by Folsom et al. and the paper by Ernst are written by individuals who have had first-hand exposure to major panel survey efforts. The Folsom et al. paper is authored by individuals who have worked with the very real and complex problems associated with the National Medical Care Expenditure Survey (NMCES) and the National Medical Care Utilization and Expenditure Survey (NMCUES), while the Ernst paper is based on real-world experience with the Survey of Income and Program Participation (SIPP).

I was impressed by Ernst's start in the task of formally specifying some of the rules that might be applied in defining population members over time. By specifying these rules in a mathematical fashion, it is then possible to apply a somewhat more formal and rigorous examination of available options for producing unbiased estimates, nonresponse-adjusted estimates, and other poststratification estimates.

The author points out that in the attainment of unbiased estimation, the weight w_i does not necessarily have to satisfy the inverse probability condition. Rather, it need only satisfy a condition based on expected value. Since the probability of selection is not required in order to produce an unbiased estimate, this points out that the traditional definition of probability sampling (known, nonzero probabilities for all elements in the population) may be somewhat too restrictive. Perhaps we need to modify our working definition of a probability sample.

The paper by Folsom et al. covers a good deal of ground, some of which is quite new. One of the major impacts of increased availability of panel data is the fact that new and more meaningful parameters may be defined over a dynamic population and estimated from a panel survey.

Several years ago statisticians at the Research Triangle Institute were among the first to give explicit recognition to the fact that means, ratios, and proportions calculated from panel data involved denominators that, even in a parameter sense, might take on fractional values. This is often the result of the fact that the population that is considered in-scope or at-risk may change on a day-by-day, hour-by-hour, or minute-by-minute basis. Rather than having a single snapshot at one instant, we have a parameter that is defined over time.

I was pleased to see what I believe is a worthy next step in the process of explicit recognition that the parameters appropriate for panel analysis are somewhat different, and indeed in many ways more enlightening, than the parameters that are usually applied to cross-sectional data.

I found that the GLMM, growth curve, and survival analysis approaches add to the analytical battery of methods that are available for the analysis of panel data. These parameters, as defined in the paper, are not simply enhancements of cross-sectional methods. Rather, the time dimension is given a formal place in the parameter. I believe that as we begin to analyze panel survey data in ways that use parameters that give more explicit recognition to the dimension of time, we will be faced with a need to modify the way we structure and store our data. I note with interest that several papers in this volume deal with the topic of data base management for panel surveys.

Data Base Strategies for Panel Surveys

Pat Doyle

Data base design is an intuitive and artistic process and there is no algorithm to guide users in the selection of the best strategy (Kroenke, 1983). An approach that is optimal for a specific institution to support a specific set of surveys and applications can be impractical or highly inefficient at another institution for a different set of users and uses. For example, there are options that penalize retrieval in order to optimize the file updating process and the reverse. There are still other options that restrict the application to specialized hardware that may be expensive to acquire.

This paper attempts to provide potential users of complex panel data with a discussion of methods to approach the problem of data organization and access. No single approach is identified as being optimal for all users and all uses. Each institution must choose the system and file organization that best meets the needs of the users while functioning within its own fiscal and physical constraints.

1. OVERVIEW

The data base strategies described in this paper address the problem of creating a master research data base from data collected in complex panel surveys. These surveys are defined as a function of the types of studies supported by the data that can be classified as follows:

Cohort studies are analyses of individuals drawn from a sample of persons with a particular set of characteristics, such as the sample of men aged 45 to 59 that formed the basis for the National Longitudinal Survey of the labor market experience of men (Center for Human Resource Research, 1981). The surveys on which these studies are based are longitudinal, allowing the observation of changes to the initial cohort from which the sample was drawn.

Attribute-based studies are also individual-based longitudinal studies. However, the sample of individuals to be followed is derived from a sample of addresses (or households) or groups of related persons within common dwellings (families) at a particular point in time. The analyses focus on individuals but describe attributes of families or households to which they belong.

Longitudinal aggregate studies are analyses of aggregate groups of individuals. The samples on which these studies are based are the same as those that form the basis of the attribute-based studies but the units of analysis are aggregate units defined longitudinally.[1]

Complex panel surveys support the second and third types of studies. Data base strategies for surveys that support only cohort studies are much easier to develop than strategies for the more complex surveys. Furthermore, options for organizing and accessing these are adequately addressed elsewhere.[2]

Development of data base strategies for the more complex surveys is clearly difficult and numerous approaches can be taken. Useful data strategies vary according to the design of the panel survey and according to the specific uses of the data. Section 2 provides background on the survey design issues and the types of applications that complicate the data base design.

The complexity of the surveys that motivate the discussions in this paper are difficult to describe in a paper of reasonable length. Hence, a hypothetical example was developed and used to illustrate the design features and uses discussed in Section 2 as well as subsequent discussions of data base design strategies. The hypothetical example is described in Section 3.

Section 4 presents alternative data base strategies. The discussion is subdivided into two parts, the first of which addresses issues related to data organization. Various data base designs are presented and classified according to the survey design features and uses of panel data presented in Section 2 and are illustrated using the hypothetical example. The second part of Section 4 provides an overview of the various options for accessing complex panel data. The discussion does not focus on one preferred option recognizing that different users have different constraints affecting their selection or development of an access system. The report is summarized in Section 5.

2. ISSUES AFFECTING THE COMPLEXITY OF DATA BASE STRATEGIES

Surveys that support attribute-based and longitudinal aggregate studies have several common design elements. The samples are based on groups of people defined at a particular point in time rather than on individuals with a particular characteristic. These surveys attempt to follow all members of the original sample for an extended period of time but often lose some portion thereof due to attrition, death, or movement beyond specified boundaries. These surveys also allow new individuals to enter the sample subsequent to the first interview for reasons such as birth, marriage, adoption or cohabitation. Specific elements of the designs of these surveys that impact on data base strategies are discussed below, followed by a description of the variety of uses that motivate the data base design strategies presented in Section 4.

2.1. Aspects of Survey Design That Complicate Data Base Strategies

This paper focuses on selected survey design elements that have an important effect on data base strategies. The first element is the time unit for which characteristics are measured as contrasted with the interviewing cycle. It is desirable from a number of perspectives to use an interviewing cycle that coincides with the basic time unit imbedded in the survey design. However, that is often not practical, particularly if the basic time unit is the month or some shorter period. In a perfect world, the interviewing cycle should not matter, but in practice it does, for several reasons. One reason important to the current discussion is that the use of an interviewing cycle longer than the basic time unit often results in roadblocks to the formation of useful data structures. One such roadblock is the tendency of the data producer to measure some characteristics as of the interview date rather than collect repeated measures for each basic time unit.

The second element is the separation of the survey instrument into questions to be repeated at every interview for each respondent and questions to be administered less frequently or perhaps only once. Information collected in this manner includes baseline statistics collected at first interview as well as topical data introduced in interviews subsequent to the first. This design feature is chosen at least in part to reduce respondent burden while increasing the number of characteristics measured. It is also used to avoid repetition when characteristics are not expected to change very often, if at all. The principal problem with this design element is the absence of the temporal dimension around which the majority of the data items in the survey are organized.

Another survey design element that complicates the data base design problem is the use of an inappropriate response unit for selected questions. For example, some surveys will collect information once per address (in a survey whose original sample was address based) rather than once per resident. Clearly, there are variables for which the most efficient response unit is the address—state of residence being a prime example. These are variables that specifically pertain to the address and hence everyone residing at that address can be unambiguously treated as having that characteristic. However, there are other variables collected once per address for which this condition cannot be met. This questionnaire design feature results in ambiguity as to how the address attribute applies to the individual and hence complicates the data base designer's attempts to create useful data structures.

The final element is the method chosen to identify and track individuals over time. It is absolutely essential to uniquely identify individuals in a panel study. This would seem to be a fairly simple objective that could be met through the assignment of a unique sequential number to each person as they enter the survey. However, two recent surveys have failed in this regard.[3] It appears the reason for these failures was the use of an identification scheme that was too complicated and intended to do more than simply link individuals over time. The implications of the lack of a foolproof method of linking observations are obvious. Either the data base designer must create such a variable, which could be expensive, or the analyst must simply accept that errors exist in the data base.

2.2. Uses Can Influence Data Base Design

Panel studies can support a variety of research topics, many of which have conflicting data needs. As a result, optimization of the data base to support one set of analyses may confound the user's ability to extract pertinent information for a study of a different type. The variety of uses that can have conflicting demands on the data structure are illustrated below. This is followed by a discussion of the status of the current debate over the problem of defining households, families, or other aggregate units longitudinally. This is a conceptual issue of considerable import to the data base designer because the preferences of the users in this regard affect the choices for organization and access of the data.

2.2.1. *Conflicting Uses of Panel Data.* Complex panel surveys can support both cross-sectional and longitudinal applications. The former includes analysis of net flows requiring a time series of cross-sectional data sets as well as the analysis of distributional characteristics of the population at a single point in time. Longitudinal studies are exemplified by the need to examine the behavior of each observation unit over time. Analysis of the number of units who enter and leave public welfare programs and the reasons for these transitions are examples of the type of longitudinal studies for which complex panel surveys are uniquely suited.

Studies relying on complex panel data will have different requirements for data elements. Some will use only information collected in a uniform manner in every round of interviewing, while others will require integration of data collected less frequently. The integration of data collected periodically is a substantively difficult access issue due to the absence of repeated measures needed to account for changes over time.

Uses of panel data will vary with the desired units of analysis and panel surveys vary in the number of different types of aggregate units that can be supported. Data strategies for surveys focused primarily on households, families and individuals as the units of observation and of analysis are more straightforward than strategies for surveys that support a variety of nonnested aggregations of individuals. Data strategies also vary depending on whether attributes of aggregate units are measured directly. Surveys that directly collect most of the attributes of specific aggregate units such as families or households can be dealt with more easily than surveys that collect vast amounts of data for each individual in anticipation that a variety of aggregate units could be formed and that most of the attributes of those units will be newly generated for each application.

Solenberger et al. (this volume) provide a clear and comprehensive review of the data base strategies that can be applied to the complex panel surveys focused primarily on families and individuals and for which the attributes of families are collected at the family level. This review extends that discussion to surveys that support dynamic formulation of a variety of aggregate units and that collect most of the information at the individual level in order to provide the flexibility to newly generate attributes of aggregate units for each research project. Facilitation of the latter type of analysis adds a new dimension to the already complex problem of accessing and managing panel data.

Uses of panel data can also be classified according to the time and resource constraints imposed on each research project. Ad hoc studies that must be completed quickly (within twenty-four hours) or that have limited budgets will require access to previously developed analysis files or study-specific variable constructs that can be used directly with little or no data base development work. Other projects with longer time horizons and larger budgets can support the needed study-specific file development.

2.2.2. *What Is the Unit of Analysis?* There is a controversial issue surrounding the uses of complex panel data that impacts on the choice of a data base design. This issue centers around the appropriate unit of analysis for longitudinal studies concerned with the behavior of social or economic units such as families or households. There are essentially three schools of thought:

- A fixed definition of longitudinality can be established for households or families as was done for the National Medical Care Expenditure Survey (Folsom, 1980) and the National Medical Care Utilization and Expenditure Survey (Dicker & Casady, 1982).
- The concept of longitudinal households is futile (Duncan & Hill, 1985). Instead, longitudinal studies should be based on the person, with household and family characteristics being treated as attributes.
- The appropriate definition of longitudinality varies depending on the analysis issue and hence flexibility to alter the definition

should be maintained in both the survey and data base designs. Followers of the third school of thought do not rule out attribute-based studies, but they do maintain that there are some issues that cannot be adequately addressed with the person as the unit of analysis.

It is important for the data base designer to understand the distinction between these approaches to longitudinal analysis. Surveys with fixed longitudinal unit definitions contain clear time unit structures that can and should be used in the design of the data base. Surveys that permit flexibility in defining units longitudinally but that will be used exclusively to support attribute-based studies also have clear time unit structures and these are based on the individual. On the other hand, surveys for which the unit of analysis for longitudinal studies will vary, do not have clear time unit structures. Furthermore, imposition of a specific structure on the data base will complicate applications relying on a different structure.

There are two issues to consider in dealing with users belonging to the third school of thought. First is the maintenance of flexibility in the actual definition of longitudinality to be employed. This is discussed in Section 4. The second is the creation of units that exist for less than the full reference period (part period units). The existence of part period units poses some difficulties that have not been entirely resolved. For example, Duncan and Hill (1985) perceive that these part period units are not longitudinal and are hence always excluded from the universe of study.[4] They present this as one reason why defining households longitudinally is futile. However, not all researchers will exclude these, as illustrated by the research conducted by Citro et al. (1986) and Carr et al. (1984).

Duncan and Hill assume concepts of longitudinal households are futile because they "obscure the nature of household composition changes and obfuscate attempts to describe the experiences of the population over time." Presumably, these roadblocks exist because it is assumed that part period units are either excluded from the analysis or examined in the absence of any knowledge of what went on before or what will go on after they exist. This need not and in fact should not be the case. The data base designer should ensure that once part period units are formed, they can be linked to their predecessors and successors. This link provides the analyst with the requisite data on the reasons for unit formation and dissolution.

The uses of panel data illustrated in this section have a direct effect on the optimum strategy for data base design. Therefore, one key to successful development of a research data base is the knowledge of the intended uses of the data. The uses, however, may be so varied that one research data base design will not be sufficient for all applications. Section 6 describes the effect of these intended uses on the selection of a data base design strategy.

3. A HYPOTHETICAL EXAMPLE

In order to clarify subsequent discussions of data base strategies, a hypothetical sample unit has been constructed and used as the basis of examples. This hypothetical sample unit, displayed in Figure 1, is from a complex panel survey similar in size and scope to the Survey of Income and Program Participation (SIPP).[5] In the interest of brevity, very little data are actually presented for the hypothetical sample unit. However, the reader should note that SIPP and other complex panel surveys provide repeated measures of an extremely large number of characteristics. This richness of data complicates the data base design effort immensely and should not be overlooked in considering the methods presented in this paper.

The survey from which the hypothetical sample unit is drawn is a nationally representative multipurpose longitudinal survey of approximately 50,000 people in the United States. The initial sample consists of all adults in a cross-section sample of addresses in existence at the time of the first interview (Wave 1). For these adults Wave 1 measures monthly income, demographics and labor force participation for the four-month period preceding the date of interview. In addition, the presence of children is determined and selected items pertaining to the address are collected. In Figure 1 the first adult is identified as the husband, age 30, who worked the full period and who received $10 in benefits each month. This adult reported tenure and school lunch participation for the address. The second adult is the wife, age 30, who received no income but did report the presence of a child age 10. The third adult is an other relative reported to be working the whole period.

Subsequent waves of this survey are administered every four months at every address containing an adult interviewed in Wave 1. Some observations are lost if they move out of the United States, die or refuse to participate while other observations are added to the sample if they move in with one or more of the Wave 1 adults. The other relative in the example moved out of the country subsequent to Wave 1 and hence was dropped from the sample.[6]

The second and subsequent waves of the survey collect repeated measures of the information included in Wave 1. This is referred to as core data since the questions do not fluctuate across time. These waves also include new questions pertaining to a specific topic (and hence are referred to as topical data). In Figure 1, the vehicular assets collected in Wave 2 is an example of topical data collected therein. From the Wave 2 data in that figure, it can be seen that the intact household splits between Waves 1 and 2 and that the other relative has left the sample. There was also a change in benefit receipt, tenure and school lunch participation.

FIGURE 1. HYPOTHETICAL SAMPLE UNIT

Response Unit	Response
Wave 1:	
Address 11	Composition = Husband, wife, child, other relative Tenure = Owned School lunch participation = Yes, one child participating
Husband	Age = 30 each month Benefits = $10 each month Working = Yes each month
Wife	Age = 30 each month Benefits = $0 each month Working = No each month Child's age = 10 each month
Other relative	Age = 15 each month Benefits = $0 each month Working = Yes each month
Wave 2:	
Address 11	Composition = Wife and child all 4 months, and other relative and husband in months 1, 2 Tenure = Owned School lunch participation = No Topical data = 1960 Chevy
Wife	Age = 30 each month Benefits = $0 in months 1 and 2 = $40 in months 3 and 4 Working = No in each month Child's age = 10 each month
Other relative	This person was not interviewed since she left the sample. She was imputed to have worked and to have no benefits in all months.
Address 21	Composition = Husband in month 3 and 4 Tenure = Rented School lunch participation = No Topical data = 1980 Ford
Husband	Age = 30 each month Benefits = $10 in months 1 and 2 = $0 in months 3 and 4 Working = Yes each month

The hypothetical example illustrates four survey design features that complicate the data base design problem:

- School lunch participation is collected once per address without sufficient information to associate it with the child or children actually benefitting from the program (an example on an inappropriate response unit)
- Tenure and school lunch are only determined every four months

so insufficient data are available to pin point the timing of fluctuations that occur between interviews

- Vehicular data are only collected in the second interview and hence ownership of vehicles in prior periods is unknown
- Less than the full amount of information is collected for the other relative because of sample attrition.

Furthermore, the hypothetical survey can support the range of uses illustrated in the preceding chapter. For example:

- Two different analytic units, persons and households, can be used in analysis
- Most of the information is collected for each individual in anticipation that household attributes such as total income will be generated for each application.

There are numerous data base strategies that can be used for this example. The selection of the optimum approach varies depending on the intended uses and on the available software and resources. Alternative approaches and guidelines for the selection of an approach are presented in the next section.

4. DATA BASE STRATEGIES

Information that can be logically characterized as a two-dimensional array does not pose a problem for retrieval and storage. Furthermore, all users of social science data can comprehend and access information stored in this fashion. Complex panel data, however, do not fit logically into this simple strategy because observations can come and go and relationships among individuals change over time as do their economic circumstances. As a result, users need a strategy for storage and retrieval of the information that explicitly accounts for the dynamic nature of the underlying population. This strategy should take into account the design features and the uses of the survey discussed in Section 2 as well as the constraints imposed on the user by the computer installation or by funding restrictions.

This discussion assumes the user is familiar with a few terms used frequently to describe data base design alternatives, access, and storage methods. The terms listed below are defined briefly in the text that follows. Readers requiring additional explanation are referred to basic textbooks on data base design and access, such as Kroenke (1983).

Terminology:

Data Organization:
  Hierarchical
  Network
  Relational
  Special case: rectangular

Access Method:
  Sequential
  Direct

Access Software:
  Procedural languages
  Statistical packages
  DBMS (hierarchical, network, relational)

4.1. DATA ORGANIZATION

Cross-sectional and longitudinal applications are both possible and likely uses of complex panel data. However, optimizing the data base for one may penalize its use for the other. Furthermore, the tools required to analyze the data cross-sectionally are distinctly different from the tools required to analyze the data longitudinally. Institutions requiring both uses may need to have more than one master research file for a single survey.

4.1.1. *Cross-Sectional Applications.* For purposes of this discussion assume the objective of the study is an analysis of the net change in the number of persons in households with household attribute X over time, as illustrated in Table 1. There are two fundamentally different ways to approach the design of a data base for a complex panel survey to support this application. The structure can be oriented around the questionnaire or around the application itself. (Assume the questionnaire is administered to each adult rather than once per household.) The two options apply regardless of the choice of a data model such as hierarchical or relational. It even applies if the design is constrained by the need to have fixed-length rectangular files. The following discussion is subdivided according to the complexity of the design, noting the two alternative approaches in each section questionnaire image and application specific.

Multi-level files: Organization of the data around the questionnaire suggests the first design in Figure 2, while organization of the data around the application suggests the second design in that figure. These could be built using a hierarchical scheme implied in Figure 2 where the sort sequence is important and the data are essentially in one file. Alternatively, they could be a series of files within a relational or network system.[7]

Note that in the questionnaire image format, there is one record per individual with the basic time unit (month) within interviewing cycle (wave) as an attribute and that there is one record per address interviewed. Also note that because persons can relocate within a wave, there can be more than one address record per individual thus destroying a natural hierarchical structure.

The questionnaire image design has four features that complicate the

TABLE 1. NUMBER OF PERSONS IN
HOUSEHOLDS WITH ATTRIBUTE X BY MONTH

Attribute X	Month			
	1	2	3	4
X1				
X2				
X3				

production of the desired table:

(1) Each physical person record must be counted more than once.

(2) Persons can enter and leave the sample each month but the questionnaire is designed to collect all four months of data for every person so each person record must be examined four separate times to determine if he or she is in the sample.

(3) There is more than one choice for the appropriate record from which to extract address-specific attributes such as tenure.

(4) The lack of a true hierarchical relationship complicates the creation of attribute X if it does not already exist on the household record.

Of course these problems are easily resolved if sophisticated software packages or highly skilled programmers are available. However, not all users of surveys like the hypothetical example have that luxury. Furthermore, in many cases the problems would be solved by implicitly creating the alternate file structure described above principally because it is expressly suited for this application.

The general point emphasized here is that if the interviewing cycle is longer than the basic time unit, use of the interviewing cycle as a method of subdividing the data will confound the users' ability to directly analyze net flows relying on the basic time unit.

Single-level files: The choices for data organization can be expanded to consider those users who do not have access to software that can easily handle a data structure with more than two dimensions. For such users the data can be organized in a rectangular format. This means to create a single-level file with one record per unit of analysis (persons in the hypothetical example) replicating all information for aggregate units like households and padding the record to account for missing information from the lower levels (such as time-varying personal characteristics that are not collected for the other relative in the sample unit after Wave 1). It is well recognized that this approach is highly inefficient and computationally expensive. However, even if some redundancy of data is involved, this simplistic structure can be ef-

FIGURE 2. DATA BASE DESIGN ALTERNATIVES
FOR CROSS-SECTION APPLICATIONS

Option 1: Questionnaire Image

Wave
   Address (Tenure, School Lunch)
     Person (Address, Benefits, Working, and Sample by Month)
For example—Wave 2:

Address Records

Address	Tenure	School Lunch
11	Own	No
21	Rent	No

Person Records

| | Address | | | | Benefits | | | | Working | | | | Sample | | | |
Month	1	2	3	4	1	2	3	4	1	2	3	4	1	2	3	4
W	11	11	11	11	0	0	40	40	N	N	N	N	Y	Y	Y	Y
C	11	11	11	11	0	0	0	0	0	0	0	0	Y	Y	Y	Y
O	11	11	0	0	0	0	0	0	Y	Y	Y	Y	Y	Y	0	0
H	11	11	21	21	10	10	0	0	Y	Y	Y	Y	Y	Y	Y	Y

Option 2: Application Oriented

Month
   Household (Tenure, School Lunch)
     Person (Benefits, Working, Sample)
For example—Month 3 of Wave 2 (equivalent to Month 7 of the longitudinal survey):

Household Record

Address	Tenure	School Lunch
11	Own	No

Person Record

ID	Benefits	Working	Sample
W	40	N	Y
C	0	O	Y

Household Record

Address	Tenure	School Lunch
21	Rent	No

Person Record

ID	Benefits	Working	Sample
H	0	Y	Y

fective. It is preferred in situations where storage and access costs are low relative to labor costs or when the only available software cannot efficiently handle more complex files.

A rectangular version of the hypothetical survey could be structured in two ways as illustrated in Figure 3. Here PID represents the system of uniquely identifying each observation and HID represents each household-month. The principal difference between these two structures is that the first file contains one record per interview cycle (wave) for each person in the sample and the second contains one record for each basic time unit (month) for each person.

The first file is patterned after the questionnaire design and therefore suffers from the same problems of double counting records in the production of the table as the original questionnaire image file. Furthermore, generation of attribute X if it does not already exist on the file can be a challenging task because individuals can belong to multiple addresses within a collection cycle (or wave) thus prohibiting the use of a simple grouping of records based on a single address identifier. However, for household attributes stored explicitly on the file it is easier to extract the appropriate value of attribute X from a rectangular file than from the original questionnaire image file.

The second rectangular file is more uniquely suited to the application at hand and is quite simple to process in any system. However, it is slightly larger than the first rectangular file due to the replication of interview-specific information and the person identifiers.

The family: Thus far the existence of families or other aggregate units within households has been ignored. For some panel studies this issue does not arise because the interview unit is actually a group of related people, i.e., a family, and other aggregate units cannot be formed due to lack of information. However, in many of the surveys conducted in the United States by the Census Bureau, the tradition is to use the physical dwelling to define the basic interview unit and to define family groupings within that dwelling. Residents of a common address are defined to be a "household" whether or not they are related. Groups of related persons within households are often referred to as families. This leads to conceptual problems in part because not all members of a family may be residing together at the time of the interview and also in part because the method of defining families is based on each person's relationship to one person (currently called the reference person, formerly called the head). This also confounds the data base design effort when the desire is to use a questionnaire image format for a panel survey. Consider the hypothetical example: The problem here is that the composition of families within addresses can change drastically at any point between the interviews. This happened to the wife's family, which changed from an intact husband–wife unit to a single-parent unit between Waves 1 and 2. Depending on the analysis, the wife's family

FIGURE 3. ALTERNATIVE STRATEGIES FOR CROSS-SECTIONAL FILES IN RECTANGULAR FORMAT

Option 1: Questionnaire Image—Wave 2:

Month =		Household												BEN				Person							
		1			2			3			4			1	2	3	4	WRK 1	WRK 2	WRK 3	WRK 4	SAM 1	SAM 2	SAM 3	SAM 4
WAV	PID	HID	TEN	SCH	HID	TEN	SCH	HID	TEN	SCH	HID	TEN	SCH	BEN	BEN	BEN	BEN	WRK	WRK	WRK	WRK	SAM	SAM	SAM	SAM
2	W	11	Own	N	11	Own	N	11	Own	N	11	Own	N	0	0	40	40	N	N	N	N	Y	Y	Y	Y
2	C	11	Own	N	11	Own	N	11	Own	N	11	Own	N	0	0	0	0	0	0	0	0	Y	Y	Y	Y
2	O	11	Own	N	11	Own	N	0	0	0	0	0	0	0	0	0	0	Y	Y	Y	Y	Y	Y	0	0
2	H	11	Own	N	11	Own	N	21	Rent	N	21	Rent	N	10	10	0	0	Y	Y	Y	Y	Y	Y	Y	Y

Option 2: Application-Oriented—Month 7 (equivalent to Month 3 of Wave 2 in the cross-sectional survey):

MON	PID	HID	TEN	SCH	BEN	WRK	SAM
7	W	11	Own	N	40	N	Y
7	C	11	Own	N	0	0	Y
7	H	21	Rent	N	0	N	Y

FIGURE 4. OPTIONS FOR INCORPORATING THE FAMILY LEVEL

Option 1:	Option 2:
Wave	Month
Address	Household
Family 1	Family
Family 2	Person
Family 3	
Family 4	
Person	

may or may not have been the "same" before and after the split, thus destroying a natural hierarchical relationship between addresses, families, and persons. Therefore, the data base designer is forced to either impose a longitudinal definition of families on the data base or insert multiple family records per address with a file structure similar to option 1 of Figure 4. The former is not recommended and the latter is quite an illogical treatment of time. On the other hand, the family level does fit very nicely within the two structures organized around the basic time unit (option 2, Figure 4) because of the natural hierarchical relationship that exists between households, families and persons at a specific point in time.

Other complicating factors: The list of attributes of panel surveys that can complicate the analysis of net flows certainly has not been exhausted at this point. The desire to analyze program units that are not directly related to families within addresses confounds the problem immensely.[8] In fact, even the file structure organized by the basic time unit is not amenable to this situation simply because the relationships between families and selected types of program units or between units for different programs do not follow a logical hierarchical path.[9]

A second factor is the survey design feature whereby selected information is collected only irregularly, like the vehicular data in the hypothetical example. The basic problem with information collected sporadically is that it does not correspond to the temporal dimension around which all of the other data are organized. In essence, the survey provides one or perhaps two snapshots and leaves it to the user to determine how these data can be integrated with the data collected in every interview.

The final factor to note is the special type of missing information common among complex panel surveys. The sample is not static and, therefore, the observations to be examined cross-sectionally vary across time. The data base designer should ensure that in each cross-section application, only individuals observed at the relevant point in time are included. Furthermore, those individuals must be properly weighted to account for the missing observations. The latter implies that sample members will have multiple weights, one for each basic time unit.

4.1.2. *Longitudinal Applications.* The tools to be used to analyze gross flows and to analyze the dynamics of the population are quite different than those used for analysis of cross-sectional data. The optimum data structures are different as well. From a data management perspective, the principal problem to solve is the treatment of time.[10] In the literature on design of data structures for very large statistical data bases there is little discussion of the temporal dimension (Doyle et al., 1987). Furthermore, that which does exist tends to view the problem as defining a data structure appropriate for a current snapshot of the population with associated historical information. This is similar to the view traditionally taken for the Michigan Panel Study on Income Dynamics (PSID).[11] The discussion that follows is not restricted to this approach. Instead, the suggested options support that view in addition to other means of analyzing the population longitudinally.

Individual-based studies: The software available for survival analysis such as the RATE model (Tuma, 1979), and LIMDEP (Greene, 1986) require that the data be logically organized into spells but that the physical representation be single-level rectangular files.[12] If these tools are to be used to analyze the data, then the data should be arrayed into a master file in such a manner as to facilitate extraction of rectangular files with one record for each spell of interest that can then be used as input to the relevant software.

In choosing the data base design, it is also useful to consider whether the discrete events directly measured by the survey, such as working, will be the primary objects of study. If so, then the designer need only be concerned with efficient access to the directly measured events. If, on the other hand, users will often require the description of concurrent events such as receipt of benefits while working, then it will be necessary to choose a design that facilitates construction of nested spells as recodes of directly measured spells. Data base design considerations should also include whether the users will want to examine time-varying covariates or quantitative flows, such as changes in benefits during a spell of program participation.

Logically the information to support longitudinal analysis of individuals could be arrayed in one of three ways, as illustrated in Figure 5. Option 1 essentially describes all characteristics of individuals at each point in time whereas the other two options organize the information into spells or events. The spell-oriented structure could be designed either to store information as characteristics by time unit (option 2) or to record the characteristics of spells such as beginning and end dates (option 3).

Each of the approaches has advantages and disadvantages as can-

FIGURE 5. STRATEGIES TO SUPPORT
INDIVIDUAL-BASED LONGITUDINAL ANALYSIS

Alternatives		Illustration for Wave 1, Months 1 and 2
Option 1:	Person Time Data	Husband Month = 1 Benefits = 10 Working = Y Sample = Y Month = 2 Benefits = 10 Working = 10 Sample = Y
Option 2:	Person Event Time	Husband Benefits Month 1 = 10 Month 2 = 10 Working Month 1 = Y Month 2 = Y Sample Month 1 = Y Month 2 = Y
Option 3:	Person Event Characteristics	Husband Benefits Start and end dates, level Working Start and end dates Sample Start and end dates

didates for organization of the master file. The first option facilitates cross-sectional analysis if one of the objectives is to support both types of studies with one integrated file. This option is also well suited to the treatment of individuals present in the sample for less than the full reference period (assuming users conducting longitudinal studies will want to access the less than complete information available for these individuals). In this case, records can be suppressed where no information exists, hence conserving storage costs. Furthermore, the absence of the record has explicit meaning, i.e., the observation is not in the sample at that point in time and therefore no information exists.

Both the first and second options are more convenient if the objective of the analysis is to study an event defined by interactions of measured events, as they make the necessary recoding more straightforward.

Similarly, they can allow flexibility in what defines an event. For example, the user can easily choose between a discrete event, such as receipt of benefits, or quantitative flows, such as a change in benefits during the periods of participation.

On the other hand, the third approach is clearly the most effective use of storage space when the reference period of the survey contains a large number of discrete time points, as fewer variables are needed to describe each event. This method can also be used to reduce storage costs through reduction of the information on each spell to the universe of persons who experienced each event (which is often considerably smaller than the full sample).

Households as attributes: To support attribute-based studies, the inclusion of household characteristics at each time point does not fundamentally alter the basic choices for conceptual organization of the data. (Note that the topic at hand is the household as an analytic unit, rather than a physical address. These two concepts are very different in a longitudinal context.)[13] The primary issue is how to store the household-level information. One approach is to treat household characteristics as attributes of persons, i.e., replicating household data for each individual (see option 1 of Figure 6). This maintains a simple logical view of the data, but it is a highly inefficient method of storage and it can complicate retrieval.

An alternative approach would be to retain household characteristics in a separate file or separate level within the master research data base (option 2 of Figure 6). Then they can be accessed as needed for the analysis. This method reduces storage costs and facilitates retrieval in comparison to the first alternative. Furthermore, with this method it is easier to create new household characteristics as needed for specific applications. As a result, this approach also gives the designer the option of limiting the amount of information retained for households to just those characteristics that cannot be generated by aggregating over individuals. For example, variables like total household income by source need not be created and stored in the data base. Instead, income of individual members would be summed whenever the application requires the household total. This minimizes duplication of information and therefore storage requirements. Obviously, the designer should consider the trade-offs between storage and retrieval costs when using this option. For example, if every application requires the same variables to describe total household income then it is not necessarily cost efficient to make every user create them. On the other hand, if each user will describe total household income differently, then it may not be cost effective to retain all or even any of the desired variations in the master research data base. The designer should also be aware that user-created aggregate measures may have consistency problems if the fields to be aggregated are not carefully documented.

FIGURE 6. STRATEGIES TO SUPPORT LONGITUDINAL
ANALYSIS OF HOUSEHOLD CHARACTERISTICS

Alternatives	Illustration for Month 1, Wave 1		
Option 1: Person Time Data (household)	Husband Month = 1 Benefits = 10 Working = Y Sample = Y Address = 11 HH Income = 10 Tenure = Own School lunch = Y		
Option 2: Person Time Time Household Data	Husband Month = 1 Benefits = 10 Working = Y Sample = Y Address = 11	Month = 1 Address = 11 Tenure = Own School Lunch = Y	
Option 3: Time Household Person	Month = 1 Address = 11 Tenure = Own School Lunch = Y Husband Benefits = 10 Working = Y Sample = Y		

Another approach to the treatment of household characteristics is to
maintain the hierarchical structure proposed for cross-sectional analysis
(option 3 of Figure 6). This maintains the logical hierarchical relationship between persons and households within each time unit and
facilitates the creation of new household characteristics as needed.
However, depending on the access software, it can complicate construction of spells for individuals. Therefore, it is not recommended if a
large majority of the applications are individual-based longitudinal
studies.

Households as units of analysis: As noted previously, it is not recommended that a universal definition of the longitudinal analysis unit be
imposed on the master research data base. Instead, users should have
the option of employing a suitable definition should that be called for in
the analysis. Before discussing data base design alternatives, it is useful to explain the requirements for maintenance of flexibility in defining
households longitudinally.

Construction of longitudinal households requires three basic steps.

First, the user needs to define "sameness" over time (or continuity); for example, the user may define sameness in terms of the head and spouse (while present). Second, the user needs to construct a measure of composition of the household at each point in time; for example, this could simply be a list of the head and spouse and selected characteristics. Third, the user needs to compare these variables across-time. As long as the determinants of longitudinality remain the same, the unit will continue to exist. When the unit no longer meets the requirements of the definition of sameness then the unit will disappear and perhaps one or more new units will be formed.

When the constructed composition measures differ, the user will have created longitudinal units that do not exist for the full reference period. The existence of part-period units can pose two types of conceptual difficulties for the researcher. The first is the problem of how units existing for varying lengths of time can be compared. This is not a data base design problem. The second is the need to ascertain the reasons for unit formation and dissolution, which cannot necessarily be determined based on the part-period unit alone. As discussed in Section 2, the data base designer should ensure that once part-period units are formed, they can be linked to their predecessors and successors thus providing the analyst with the reasons for formation or dissolution of the unit.

Returning to data base design strategies, procedures to be used in the first step of defining units longitudinally suggest maintenance of households as a separate file within the master research data base organized around the basic time unit (option 2 or 3 of Figure 6). The problem to consider is the extent to which detailed information on the household members should be retained in the household characteristics file. The solution to this problem is a function of the access software and other intended uses of the file.

Logically, the household characteristics file would be organized as in option 3. This facilitates flexibility in creating the variables needed for each longitudinal unit definition. However, if most uses of the data revolve around longitudinal analysis of individuals, then the main data base should not be physically organized in this fashion. Instead, the master file would be more efficiently organized using option 2, where most of the data are organized around the individual but a separate household characteristics file is maintained. Some information would be duplicated across these two files. The extent of the duplication depends on whether the access software can easily allow the user to construct household characteristics needed to develop the appropriate longitudinal unit definition.

On the other hand, if the longitudinal applications are principally household-based and the users will be performing a multitude of cross-sectional analyses, then the data base designer can minimize retrieval costs by employing the logical time–household–person organization depicted as option 3 as the main approach to physical data base design.

Some issues to consider: As discussed previously, analysis of units other than persons and households is complicated by the variety of ways in which individuals can be grouped—tax units, program units, families, just to name a few. If it is clear to the data base designer which units are of most interest, then they can be explicitly dealt with in the same manner as households. However, at least for the hypothetical survey, the only thing that is clear is that all units that can be formed will be of interest to at least one user. Hence, it would be rather difficult to accommodate all applications or to plan to reorganize the data each time a different unit of analysis is desired. In lieu of that, the optimum approach is to ensure that among the person characteristics retained, the master research file should include explicit representation of unit composition for all units measured in the survey for all points in time.

Another issue to consider is that the discussion of data base design alternatives has assumed that the data producer has provided the data base designer with panel data in a form that supports longitudinal analysis. In particular, it was assumed that the appropriate sample weights have been provided and that all editing and imputations have been performed longitudinally. Data base designers should exercise caution in applying these data structures to panel data that have been processed cross-sectionally. Users should understand that cross-sectional processing can introduce artificial transitions over time and that the use of a longitudinal data base scheme to store what is essentially a series of cross-section data sets will not solve these inherent problems.

4.2. Data Access Options

There are a number of options for processing complex panel data, all of which can be applied to at least some of the methods of organizing data discussed in the previous section. The selection of which approach or combination of approaches to take for panel data can be difficult if insufficient budget or insufficient options for storage restrict the choices that can be made. In order to assist users in the selection of access systems, some of the trade-offs between various options are discussed below.

Based on a review of the literature on very large statistical data bases, Doyle et al. (1987) note that conventional data base management systems—i.e., those developed principally for commercial applications—are not optimum for statistical analysis of large complex panel surveys. They are instead optimized for extracting a few data elements from a small number of observations and for repeated updates to the data base. In fact, one of the primary motivations behind high-level normalization of data often discussed in conjunction with relational data models is the optimization of the process of file updating rather than efficient access. As Kent (1983) pointed out, high-level normalization can penalize retrieval. In contrast to commercial applications, applications using complex panel data will often require the extraction of a large number of variables for a large number of observations and will less frequently require updates to the data base once the data have been cleaned. Hence, it is not necessarily cost efficient to select a data model for which retrieval is penalized in order to optimize the updating process.

There is also some evidence that the traditional sequential approach to accessing large social science data sets may in fact be the most efficient approach. For example, based on a comparison of direct access techniques used by conventional data base management systems, access systems retrieving data organized in transposed format, and sequential access systems, Teitel (1985) concludes that:

- Sequential access is better than conventional DBMS access techniques for statistical queries
- Traditional sequential access techniques are optimal for retrieval of most variables on most observations (like creating an analysis file from complex panel data)
- Access methods designed for data stored in transposed format are better than standard sequential access methods for accessing few variables for most observations.[14]

Aside from the performance issues, there are practical considerations to make in selecting an access system. In particular, the data base designer needs to consider the total cost of a specific system (the preceding refers only to the cost of execution), the ease of use, and the physical and monetary limitations imposed by the institution. These are discussed below for each of the three main approaches to data access.

4.2.1. *Data Base Management Systems (DBMS).* Even though the available evidence suggests that conventional data base management systems may not employ the optimal access methods for analysis of complex panel data, they do offer other advantages. Many have "user friendly" methods of extraction of information, that is, they are supposedly designed to allow novice computer users instant results. Relational DBMSs seem particularly appropriate to the problem of accessing complex panel data. The main advantage of this technology is the ability to subdivide the data according to either the commonality of the universe for collection or the substance of the information and then dynamically reformulate the information according to the needs of the research. Relational technology is currently being used as the vehicle to manage and access SIPP data at the University of Wisconsin. David (1985) describes the advantage of relational technology as "[making] it possible to store data according to clear semantic principles and to facilitate aggregation of data . . . on several principles of analysis . . . ."

There are drawbacks, though. First, acquiring a DBMS can be expensive if the institution does not already subscribe to one or is not eligible for acquisition at noncommercial rates. Second, since DBMSs rely on direct access methods they require that the information be stored on conventional disk or mass storage devices. If such devices are either expensive or not abundant at the computer installation, the use of a DBMS may not be feasible.[15] Third, "user friendly" is not always synonymous with easy to use. While relational data base systems provide great flexibility in analysis, they are often confusing to users in that a thorough knowledge of relational theory and of the data structure is needed before work can begin. Fourth, none of the DBMSs in existence today are truly relational (Codd, 1985). Of course, the debate begun by Codd focuses on the application of the term "relational" to a DBMS rather than to the actual merits of the system (Curtis, 1986). In other words, the fact that a "relational" system does not meet all of Codd's original qualifications does not mean the system is not useful as a data management tool.

Another potential drawback is that many DBMSs, particularly those developed more than five years ago, tend to place impractical limits on the data base, such as restricting the number of variables or record size to an unrealistically low level (at least in the context of social science research). Thus, a fairly straightforward design with say three files (or tables in relational terminology) quickly becomes a system of thirty or more files by subdividing each logical unit into smaller physical files that fit within the limitations of the system. Furthermore, if the research task requires more than the maximum number of variables, the user is forced to extract multiple files and merge them using software external to the data base management system.

It is certainly true that some subdivision of the data in the very large surveys is efficient and in fact recommended. However, it is important that this subdivision be guided by the substance of the data rather than the physical limitations of the system. Furthermore, not all classes of applications routinely require large extracts. If relatively few such applications are anticipated at a particular institution, then this restriction should not be a deterrent to the selection of a particular DBMS.

The final drawback to note is that DBMSs typically contain very few statistical functions. This means that many of these systems cannot be used to perform basic descriptive analysis of complex panel data. Instead, they serve principally as a data management tool producing analysis files for input to other software systems. In the recent past many DBMSs have added interfaces to statistical packages to avoid this intermediate file creation step.

The complications described above arise with DBMSs principally because they were not designed for social science research. Relational technology could be more appropriately applied to the problems of such research if that was the objective. In fact, RAPID (Hammond, 1983) is a potentially useful tool for this purpose and is currently being investigated as a potential system to access the SIPP longitudinal research file. Another potentially useful system is the one currently under development at the Institute for Social Research (ISR) at The University of Michigan (Solenberger et al., this volume). This is an expansion of OSIRIS IV to employ direct access technology for the PSID. While this is not a relational system, it does employ many of the important features of relational technology. More important, however, the system is also being specifically designed to support the construction of spells and descriptions of spell characteristics.

Statistical packages: An alternative to DBMSs is the use of so called statistical packages. For purposes of discussion this title refers to sequential access systems that allow the user to compute basic statistics or manipulate data using a set of predetermined commands. Because these systems rely on sequential access they eliminate the need for large amounts of disk storage. Some of the more powerful statistical packages cannot be classified as "user friendly," but a fairly wide range of applications can be conducted by relatively junior programmers or research assistants. They can be costly to acquire or to access, but they tend to be considerably less expensive than most DBMSs. In general, these systems are considered inexpensive to use relative to procedural languages, but in many cases the cost savings tend to be achieved through reduced labor rather than cost-effective methods of processing the data. Hence, if the data base is extremely large, as is true for some complex panel surveys, the cost savings might not be realized. Note that according to Doyle et al. (1987), the current public version of OSIRIS IV appears to be an exception to this rule, employing fairly cost-effective techniques for data manipulation and access.

Not all statistical packages are powerful enough to be used both to manage complex panel data and to serve as the basic tool for analysis. For example, TPL (Bureau of Labor Statistics, 1984) is strictly what its name implies—a table-producing language—and it is extremely efficient and effective for that purpose. While not being self-contained solutions to the access problems, systems like TPL can be effectively used in conjunction with other systems to analyze the data. More powerful systems such as SAS (SAS Institute, 1985) can be used either to fully manage the data (recognizing that there are likely to be limits on the available options for file organization) or in conjunction with another system.

Procedural languages: The user may also elect to develop special-purpose systems to access complex panel data using a procedural language. These are high-level languages that allow the user to achieve a specified goal by explicitly directing the operation of the machine. Given sufficient development funds, this may in fact prove to be the optimum approach since it can be geared specifically to the problems of accessing complex data and can be optimized for the system on which it will be used. Mathematica Policy Research (MPR) has used this approach in dealing with the cross-sectional public use files from SIPP (Mathematica Policy Research, 1986).

The user should be aware that the potential cost of development could be excessive for this approach. Development of programs for well-defined tasks such as the MPR application does not have to be expensive but it does require specialized staff. Development of a general-purpose system, on the other hand, can require an excessive amount of resources. Hence, it may be more cost effective to select an existing system and either correct its deficiencies or work within them.

For users who are considering the use of procedural languages, Doyle et al. (1987) contain some useful comparisons of performance between three different languages (COBOL, FORTRAN, and Pl/1) and OSIRIS IV. COBOL appears to be the most cost-effective language for applications on large data sets requiring extensive input/output, with PL/1 being the most expensive. OSIRIS IV proved to be more cost effective than two of the other three languages tested. Note, however, that COBOL is perhaps the least flexible of all of these languages in terms of statistical analysis. Its use is optimal if it is viewed primarily as a means to extract information from large master files to be used as input to other software packages for analysis.

5. SUMMARY

Complex panel data represent massive collections of information and have a myriad of uses. As a result there are numerous choices to make regarding data base design and access. All of these choices must be made within the physical and monetary constraints of the research and in light of the survey design and the specific applications planned. This paper addresses the issues but does not resolve them. It merely points out that there is no single solution considered optimal for all complex panel surveys. In fact, in some cases uses of the survey may be so varied that there is no single strategy that could be considered the best.

REFERENCES

Bureau of Labor Statistics (1984), *Table Producing Language System Version 6*, U.S. Government Printing Office, Washington, DC.

Carr, T. J., Doyle, P., & Lubitz, I. S. (1984), *Turnover in Food Stamp Participation A Preliminary Analysis* (Final report to the Food and Nutrition Service, USDA), Mathematica Policy Research, Inc., Washington, DC.

Center for Human Resource Research (1981), *The National Longitudinal Surveys Handbook*, College of Administrative Services, The Ohio State University, Worthington, OH.

Citro, C., Hernandez, D., & Herriot, R. (1986), "Longitudinal Household Concepts in SIPP: Preliminary Results," *Proceedings of the Census Bureau Second Annual Research Conference*, U.S. Bureau of the Census, Washington, DC.

Codd, E. F. (1985, Oct. 14, Oct. 21), "Is Your DBMS Really Relational?," *Computerworld*.

Curtis, R. M. (1986, Sept.), "The Relational Dilemma—A Management Perspective," *Newsletter on Data Management*, Arthur D. Little, Inc.

David, M. (1985), "Designing a Data Center for SIPP: An Observatory for the Social Sciences," *Proceedings of the Social Statistics Section*, American Statistical Association, Washington, DC.

Dicker, M., & Casady, R. J. (1982), "A Reciprocal Rule Model for Defining Longitudinal Families for the Analysis of Panel Survey Data," *Proceedings of the Social Statistics Section*, American Statistical Association, Washington, DC.

Doyle, P., Citro, C. F., & Cohen, R. L. (1987), *Feasibility Study of Long-Term Access to SIPP* (Final report to the Food and Nutrition Service, USDA), Mathematica Policy Research, Inc., Washington, DC.

Duncan, G. J., & Hill, M. S. (1985), "Conceptions of Longitudinal Households: Fertile or Futile?," *Journal of Economic and Social Measurement*, 13(3–4), 361–75.

Folsom, R. E. (1980), *Family Unit Analysis Weighting Methodology for the National Medical Care Expenditure Survey*, Research Triangle Institute, Research Triangle Park, NC.

Greene, W. H. (1986), *LIMDEP™ Users Manual* (Unpublished).

Goldstein, H. (1979), *The Design and Analysis of Panel Studies*, Academic Press, London.

Hammond, R. G. (1983), "RAPID: A Statistical Database Management System," in *Computer Science and Statistics: The Interface* (J. E. Gentle, ed.), North Holland, Amsterdam.

Kent, W. (1983), "A Simple Guide to Five Normal Forms in Relational Database Theory," *Communications of the ACM*, 26(2).

Kroenke, D. (1983), *Database Processing* (2nd ed.), Science Research Associates, Inc., Chicago.

Mathematica Policy Research (1986), *Survey of Income and Program Participation 1984 Panel Waves 3, 4, and 5: Technical Description of the Production of Person and Household Files* (Final report to the Food and Nutrition Service, USDA), Mathematica Policy Research, Inc., Washington, DC.

Nelson, D., McMillen, D. B., & Kasprzyk, D. (1985), *An Overview of the Survey of Income and Program Participation: Update 1* (SIPP Working Paper Series, No. 8401), U.S. Bureau of the Census, Washington, DC.

SAS Institute, Inc. (1985), *SAS Users' Guide: Basics* (5th ed.), SAS Institute, Cary, NC.

Survey Research Center (1984), *User Guide to the Panel Study of Income Dynamics*, ICPSR, University of Michigan, Ann Arbor.

Teitel, R. (1985), "Analysis of Computer Storage Structures and Access Methods from a Statistical Database Perspective," *Proceedings of the Statistical Computing Section*, American Statistical Association, Washington, DC.

Tuma, N. B. (1979), *Invoking Rate* (Unpublished program manual).

NOTES

[1]This classification of the uses of panel studies is similar to that proposed in the Solenberger et al. paper at this symposium except that analysis of individuals who reside together for part of the reference period is considered to be a special case of longitudinal aggregate studies. The author is grateful to Martha Hill, who created the term "longitudinal aggregate studies" in her review of an early draft of this paper.

[2]See, for example, Goldstein (1979).

[3]The two surveys were the Survey of Income and Program Participation (SIPP) and its predecessor, the Income Survey Development Research Panel (ISDP). In both

cases the personal identification was based on the address at first interview. When members of different initial addresses merged, the identifiers of one or more members were changed. Furthermore, when new sample members joined an original sample unit that had split into two, the new sample members received the same identifiers. In SIPP, the latter problem was intended to be corrected with the addition of a third component of the identification system (the entry address). However, this third component was not entirely incorporated into the processing system.

[4]This interpretation was derived from the description of longitudinal household concepts under consideration by the Census Bureau (McMillen & Herriot, 1985, p. 356).

[5]For an overview of SIPP see Nelson et al. (1985).

[6]The other relative was not interviewed in Wave 2. However, because he was in the sample early in the reference period, Wave 2 data were imputed.

[7]In a network system, data are stored on separate records and the association between records is accomplished through pointers or keys developed and "hard wired" when the data base is constructed. In a relational system, data are stored in a series of rectangular files called tables. The files are logically linked together by a series of keys, a variable or variables used to uniquely identify each record. Data are physically linked for each application according to the specification of the user.

[8]The term "program unit" refers to the grouping of participants into welfare or tax units according to the program regulations. These programs do not generally use households or families as defined in this paper as the method of grouping participants. Often the program units are subsets of households or families. However, they can occasionally span multiple households.

[9]One anonymous reviewer of this paper notes that you can model any time-unit structures as long as you have clear, unambiguous definitions. Furthermore, if these definitions are not clear, this is not a data base design problem. Unfortunately, the reality for complex panel studies like SIPP is that the time-unit structure *is* ambiguous and it *is* up to the data base designer to provide flexibility to researchers to choose the analytic unit that is most appropriate for the analysis at hand.

[10]There are a host of conceptual problems to deal with as well, such as weighting and imputation, which are outside the scope of this paper. Many are discussed elsewhere in this volume.

[11]For an overview of the PSID, see Survey Research Center (1984).

[12]A "spell" is the time period during which an observation maintains a specific characteristic. For example, a spell of employment is the period during which at least one job is held. This spell begins when a job is found after a period during which no job has been held and ends when another jobless period begins.

[13]As an illustration of the difference between address and household in a longitudinal context consider the hypothetical example. Address 11 contains one intact two parent family and another relative at the time of the first interview. At the time of the second interview address 11 consists of only the wife and child. From an analytic perspective it is unlikely that the members of address 11 at the second interview would be considered the same household as the members at the first interview because of the substantial disruption that occurred between the two waves.

[14]Sequential access refers to processing data one record at a time in sequence. Direct (or random) access refers to processing data in any sequence. Data files accessed sequentially can be stored on virtually any machine-readable device. However, direct access requires that the data be stored on conventional or floppy disks or mass storage devices.

[15]This statement applies to very large surveys for which the storage requirements well exceed the capacity of the largest reasonably priced hard disks available for microcomputers.

Data Base Management Approaches to Household Panel Studies*

Peter Solenberger, Marita Servais, and Greg J. Duncan

The purpose of this paper is to present an overview of the crucial design issues of managing data collected in longitudinal studies of households. How to organize the data collected in such studies is not obvious. Data are typically gathered at the level of the household, (e.g., household size) and at the level of persons living in a household (e.g., employment status). Some desirable household information is an aggregation of information from persons living in the household (e.g., total household income). Especially problematic are changes in the composition of households across time because they destroy the simple correspondence that exists at any point in time between persons and households.

There are a number of possible data base structures for longitudinal information about persons and households. A crucial concern is the need to balance a structure's flexibility against its costs, where the latter include the costs of organizing the data to fit the structure, updating the information as additional waves of data are collected, obtaining error-free linkages between persons and households, and the time analysts must spend to learn the structure. Solutions that minimize some costs but ignore others may be much costlier than seemingly less efficient approaches.

We begin in Section 1 with a generic description of a number of household panel surveys that have begun in several countries over the past two decades. The flexibility of these studies in providing representative information on the persons and households in their samples has led to a diverse set of demands on their data structures, some examples of which are listed. Section 2 presents an overview of the data management problem posed by these demands, followed by a listing of some possible solutions. A more detailed presentation of two approaches is given in Section 3, using the Panel Study of Income Dynamics (PSID) as an example.

1. DESIGN AND ANALYSIS FEATURES OF HOUSEHOLD PANELS

A number of household panel studies begun during the last two decades share a common set of design characteristics. The initial panel

190

households are selected from probability samples of dwellings or population lists. These samples provide representative samples of various subunits within dwellings, including households, families, subfamilies, and transfer program recipiency units and persons. With a properly specified set of rules regarding the definition of these subunits and the tracking of those units over time, panel studies of households can maintain representative samples of each of the various subunits over time. This requires that newly formed subunits of interest (families, AFDC recipiency units, person births, etc.) enter the sample with known selection probabilities in order to reflect the corresponding changes that are taking place in the population at large. These procedures produce data on a representative sample of persons living in stereotypic, stable nuclear families as well as persons involved in an assortment of interesting household composition situations. For example, household panels contain representative samples of couples who undergo a divorce or separation during the data collection period, children who leave home and establish independent households, and unmarried mothers who face the choice of enrolling in the AFDC program.

The need for sophisticated structures to handle the longitudinal data on persons and households stems from two sources. The first is the analysts, who have generated an exceedingly diverse set of demands for longitudinal analyses of various aggregations of persons and of the events they experience. The second is the panel households themselves. The movement of sample persons into and out of households and into nonresponse status over time leads to difficulties in organizing the longitudinal household and person data.

1.1. Examples of Problems Posed by Analytic Needs

We can illustrate the most common analytic demands through a set of examples based on combinations of household and person attributes. We use poverty status as the household attribute of interest. Official U.S. government definitions of poverty are based at the household level and consist of a comparison of a given household's total annual income with the annual income judged necessary to provide a minimally acceptable standard of living for the given household's composition. A household's poverty status is both an attribute of the household and of each person residing in the household (i.e., a child is deemed "poor" if he lives in a household with total income less than the official poverty threshold.) We use disability status as the person attribute in our examples, with the extent of ill health of each person determining his or her disability status. Note that disability status can also be used as an attribute of the household if one wants to know, for example, the fraction of children living in households in which the head is disabled, or the fraction of households containing one or more disabled preschool children. The following are examples of analytic needs involving longitudinal studies of poverty and/or disability status, ordered roughly by complexity.

(1) Cohort studies of persons: Many longitudinal analyses require only information on the attributes of persons over time. For example, descriptions of the disability "spells" of persons in later life—when they occur, the duration of the spells, the links between disability and retirement—would require longitudinal information on the health and, in the latter case, labor supply status of the person during the age span of interest. Another example of such studies would be of the labor market experiences of women in their child-bearing years, gathered from longitudinal information on a cohort of women, say age 21 to 35 at the beginning of the panel period. Data structures for these analyses could be quite straightforward, requiring only information at several points in time about the persons in the given cohort.

(2) Household attributes of persons: Data from the cross-sectional Current Population Survey are typically used to describe the characteristics of persons living in poor households in a given calendar year. Extensions to household panel studies might include a description of the multi-year poverty experiences of various cohorts of persons (e.g., under the age of 10 or over the age of 65 in the initial year of the panel period.) Data requirements for this type of analysis are somewhat more complicated, since persons must be linked to their appropriate household data in each year of the analysis period. A common extension of this kind of analysis is to link together longitudinal information about attributes of both persons and households, as in a study of the extent to which the onset of disability was accompanied by a transition into poverty.

(3) Aggregating person attributes to a household level: Measures of household income are sometimes obtained from responses to direct questions about the household total. More typical is the procedure of asking about the income of each individual in the household and then aggregating these income figures to a household total. Similarly, measures of household composition such as the number of minor children can be obtained by aggregating information from the demographic information about each individual in the household. Although some of this household-level information is important enough to be included in the aggregate form on the household record, it is generally desirable to be able to generate a range of household information by aggregating individual attributes to a household level.

(4) Households over time: Analogous to the example of poverty attributes of persons would be analyses of the poverty experiences of households over time, where the household replaces the person as the unit of analysis. There are problematic aspects to defining households across time, as persons move in and out of household units, creating in-

stances where several time two households may have descended from the same time 1 household. However, an analyst might sometimes think it desirable to follow aggregates of persons (e.g., households or recipiency units) over time. Obtaining units of analysis of groups of persons over time presents difficult data management problems, as detailed in subsequent sections.

(5) Persons who resided in the same household during a portion of the panel period: Suppose that an analyst wishes to compare the poverty status of ex-husbands and ex-wives following a divorce or separation that occurs during the panel period. Prior to the divorce they husbands and wives share the same household and poverty status, but subsequent to the divorce they will live in separate households with possibly different poverty statuses. Here the need is to be able to link information on the new household of the ex-husband with information on the new household of the ex-wife.

There are abundant examples of each of these types of analyses using longitudinal data on households and persons from the PSID. The first two are by far the most common, but the latter three are common enough to constitute legitimate tests for the suitability of any general purpose data structure that might be proposed to handle such data.

1.2. Examples of Problems Posed by Changes in Household Composition

Apart from the analytic challenges to data structures of household panels are the problems posed by the changing structure of households themselves. Most household panels follow so-called "split-off" persons— children leaving the parental nest, an ex-spouse after divorce—who leave original households and set up new ones. Prior to the split, these persons shared the same household and therefore the same household attributes as the persons with whom they had lived. After the split, of course, the household attributes of split-offs will differ. If they should happen to reunite with members of their former households, as when a couple separates but then reunites or a child returns to the parental home after a "false start" in an independent household, then they will constitute a household in which the family "histories" of the persons in the recombined families will differ. As shown below, properly accounting for the historical information of such persons is not a straightforward task for a data structure.

Two other examples of problems caused by sample composition dynamics will also be examined below. The first is the rare but especially perplexing set of problems caused by the movement of some persons from one wave one household into another. Clustered samples such as those in the PSID or the Survey of Income and Program Participation (SIPP) typically produce a handful of such cases. The second is nonresponse, which cannot be ignored as easily in panels as in cross-sectional surveys, since nonrespondents in later panel waves will have potentially valuable panel information during earlier waves.

2. DATA MANAGEMENT PROBLEMS AND POSSIBLE SOLUTIONS

2.1. Data Matrices

Suppose that a panel study has gathered information about persons and aggregates of persons at different points in time and wants to do the five types of analysis described in Section 1. For simplicity's sake, let us assume only one aggregate of persons (household, family, or food stamp unit, for example) and three points in time (years or months, for example).

Suppose further that the information gathered is somewhat different at each point in time and that the panel study staff, choosing the simplest possible organization, has built six data matrices to represent the aggregates and persons at the three points in time. This in fact is what the PSID has done with its now eighteen years of data.

The three person–data matrices labeled "Per1," "Per2," and "Per3" are groups of variables describing the persons at the three points in time. They are person–time groups. The three aggregate data matrices labeled "Agg1," "Agg2," and "Agg3" are groups of variables describing the aggregates at the three points in time. They are "aggregate–time groups."[1]

This is not the only possible organization of the data. For example, it would be possible to separate any variables with "timeless" information about persons and aggregates—birth date and date of formation of household, for example—into two additional groups. It would be possible to separate any variables with the same information about persons and aggregates at the different points in time—personal and household income, for example—into two more groups. And it would be possible to put the relationship of persons to aggregates into yet another group.

We are assuming the panel study staff has decided not to do these things because the proliferation of groups would complicate the structure too much for both humans and machines. In some cases, however, increasing the number of groups may improve logical clarity and increase machine efficiency. Such decisions need to be made before the data is collected in order to avoid unnecessary problems processing it.

2.2. Relationships among Groups

Logically, the relationship of the person–time groups to each other is one-to-one. (We leave aside the question of persons entering or leaving the panel.) For each occurrence of Per1 there is one and only one occurrence of Per2 and of Per3. For each occurrence of Per2 there is one and only one occurrence of Per1 and of Per3. And for each occurrence of Per3 there is one and only one occurrence of Per1 and of Per2.

The relationship of the aggregate–time groups to each other may be harder to determine, because the composition of the aggregates changes over time. When a person moves in or out, is the aggregate the same? For the moment we will assume that the panel study staff has decided the aggregates have no meaningful existence over time and thus the aggregate–time groups have no logical relationship to each other.

The aggregate–time groups do have a relationship to the corresponding person–time groups, however. This relationship is "one-to-many." For each occurrence of Agg1 there may be several occurrences of Per1. For each occurrence of Agg2 there may be several occurrences of Per2. And for each occurrence of Agg3 there may be several occurrences of Per3.

2.3. Identification and Link Variables

Each group has an identification variable whose values uniquely identify each occurrence of the group. Each group that must be related to another group has a link variable whose values correspond to the values of the identification variable of the other group.[2]

The person–time groups are identified by a person–ID variable. Note that there is no need for a time variable to identify the person–time groups, since the group itself identifies the time. The aggregate–time groups are identified by an aggregate–time ID variable. There is also no need for a time variable here, since the group identifies the time.

Corresponding occurrences of the person–time groups are linked to each other via their person–ID, which remains constant over time. Occurrences of the person–time groups are linked to corresponding occurrences of the aggregate–time groups via the current aggregate–time ID variable, which is included in the person–time groups.

2.4. Logical Structure

The logical structure described above is diagramed in Figure 1. One-to-one relationships are indicated by arrows that are single-headed in both directions. One-to-many relationships are indicated by arrows that are single-headed in one direction and double-headed in the other direction.[3]

(1) Cohort study of persons: The data management problem involved in doing a cohort study of persons with this structure is relatively simple. The example of such an analysis used in Section 1 is the disability spells of persons. The software must be able to locate the corresponding occurrences of the person–time groups Per1, Per2, and Per3

FIGURE 1. LOGICAL STRUCTURE OF THE
PERSON–AGGREGATE–TIME DATASET

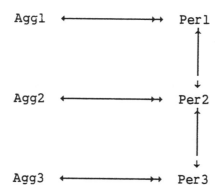

using the one-to-one relationships among them and obtain the desired variables from them. This procedure is diagramed in Figure 2.

FIGURE 2. COHORT STUDY OF PERSONS

(2) Aggregate attributes of persons: The data management problem involved in doing a study of the aggregate attributes of persons is a bit more complicated. The example of such an analysis used in Section 1 is the poverty experiences of persons. The software must first locate the corresponding occurrences of the person–time groups Per1, Per2, and Per3 and then locate the occurrences of the aggregate–time groups Agg1, Agg2, and Agg3 that correspond to them. This procedure is diagramed in Figure 3.

(3) Aggregating person attributes: The data management problem in-

FIGURE 3. AGGREGATE ATTRIBUTES OF PERSONS

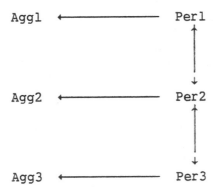

volved in aggregating person attributes is in a sense the reverse of the one involved in getting aggregate attributes of persons. The example of such an analysis used in Section 1 is determining the number of children less than six years old in a household. The software must first locate the occurrences of the aggregate-time groups Agg1, Agg2, and Agg3 and then locate the occurrences of the person-time groups Per1, Per2, and Per3 that correspond to them. This procedure is diagramed in Figure 4.

FIGURE 4. AGGREGATING PERSON ATTRIBUTES

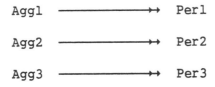

(4) Aggregates over time: The data management problem involved in doing a study of aggregates over time is a step more complicated, since it requires linking corresponding occurrences of the aggregate groups. The example of such an analysis used in Section 1 is the poverty experiences of households. The software must first locate the corresponding occurrences of the aggregate-time groups Agg1, Agg2, and Agg3 and then locate the occurrences of the person-time groups Per1, Per2, and Per3 that correspond to them.

Since the data structure itself does not imply a way to determine which occurrences of the aggregate-time groups should be regarded as corresponding to which other occurrences, the user will have to supply such a method. This procedure is diagramed in Figure 5.

FIGURE 5. AGGREGATES OVER TIME

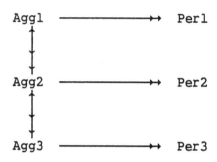

(5) Persons who resided in the same aggregate: The data management problem involved in doing a study of persons who resided in the same aggregate during a portion of the panel period is considerably more difficult than any of the previous ones.

The example of such an analysis used in Section 1 is the poverty status of ex-husbands and ex-wives following a divorce or separation. This requires a unit of analysis of persons who were married and then divorced at some time in the sample. The software must first locate the married occurrences of the person–time groups Per1 and Per2, then see if any of them correspond to divorced occurrences of the person–time groups Per2 and Per3. For appropriate pairs of the person–time groups it must then locate the corresponding aggregate–time groups Agg1, Agg2, and Agg3. This procedure is outlined in Figure 6.

FIGURE 6. PERSONS WHO RESIDED IN THE SAME AGGREGATE

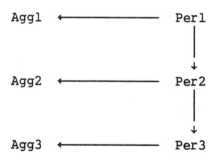

2.5. Adding Relationships among Aggregate–Time Groups

Up to now we have assumed that the aggregate–time groups have no relationship to each other worth building into the data structure. Suppose now that the aggregate–time groups do have a meaningful relationship to each other. Specifically, each aggregate–time group has a one-to-many relationship to *subsequent* aggregate–time groups. In other words, each aggregate unit has only one antecedent unit at each previous point in time but may have more than one subsequent unit at each succeeding point in time. This situation is diagramed in Figure 7, which would replace Figure 1.

FIGURE 7. ADDING RELATIONSHIPS AMONG AGGREGATE–TIME GROUPS

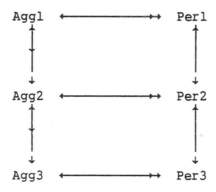

This change has no effect on any of the analyses described above except the aggregates-over–time analysis, which is made simpler by the inclusion in the data structure of an explicit link between the aggregate–time units.

2.6. Implementation

Let us now proceed to the problem of implementing the logical data structures we have been describing. We will discuss five possible approaches: (1) sorts and match-merges of the separate data matrices, (2) general-purpose data base management programs, (3) specialized data base management programs, (4) sequential access of hierarchical structures, and (5) sequential access of rectangular structures.

2.7. Sorts and Match-Merges

A first approach is the traditional method of sorts and match-merges, which has been revived in some implementations of relational data bases, for example, an INGRES data base with a HEAP or HEAPSORT structure. The data matrices are kept in separate files, sorted by their ID variables, and the requested units of analysis are pulled together with a series of sorts and match-merges.

To obtain data for a cohort study of persons—disability spells, for example—one would have to match-merge the person–time groups Per1, Per2, and Per3 on their person–ID variables. This would require two match-merges.

To obtain data for a study of aggregate attributes of persons— personal experiences of poverty, for example—one would have to sort Per1, Per2, and Per3 on their aggregate–time link variables; match-merge Per1 with Agg1, Per2 with Agg2, and Per3 with Agg3; sort the resulting groups on their person–ID's; and match-merge them. This would require six sorts and five match-merges.

To obtain data for a study of aggregated person attributes—number of children less than six years old, for example—one would have to sort Per1, Per2, and Per3 on their aggregate–time link variables; aggregate Per1, Per2, and Per3; and match-merge the aggregated Per1 with Agg1, the aggregated Per2 with Agg2, and the aggregated Per3 with Agg3. This would require three sorts and three match-merges.

To obtain data for a longitudinal study of aggregates—household experiences of poverty, for example—one would have to sort Per1, Per2, and Per3 on their aggregate–time link variables; match-merge all the corresponding occurrences of Per1 with Agg1, Per2 with Agg2, and Per3 with Agg3; sort the combined Agg3/Per3 group on a link variable with Agg2; match-merge Agg3/ Per3 with Agg2/Per2; sort the combined Agg2/Agg3/Per2/Per3 group on a link variable with Agg1; and match-merge it with Agg1/Per1. This would require five sorts and five match-merges.

To obtain data for a study of persons who resided in the same aggregate at the beginning of the study—the poverty experiences of exspouses, for example—one would have to sort the desired occurrences of Per1, Per2, and Per3 on their aggregate–time link variables; match-merge Per1 with Agg1, Per2 with Agg2, and Per3 with Agg3; sort the resulting groups on their person–ID variable; match-merge them; sort the combined Agg1/Agg2/Agg3/Per1/Per2/Per3 groups on their time 1 aggregate–ID variables; and match-merge them. This would require seven sorts and six match-merges.

Thinking about these sorts and match-merges may be useful for understanding the data management tasks involved in obtaining the units of analysis needed for the four types of studies, but actually doing them for each analysis would be quite tedious, error-prone, and expensive.

2.8. General-Purpose Data Base Management Programs

A second approach is to put the six groups into a general-purpose data base management program and let it pull together the necessary units of analysis using its internal pointers.

A good general-purpose data base management program will allow the user to access almost any data in almost any way, including many ways unforeseen by the writers of the program and the creators of the data base. This power has a price, however.

In order to access almost any data in almost any way, a general-purpose data base management program must go in one of two directions. Either it has a complicated initial setup that spells out every method of accessing the data that the applications programmer who designs the data base can imagine, or it has a relatively simple initial setup and requires the end user to do the programming. Examples of the former are IMS and IDMS; examples of the latter are DB2 and INGRES.

General-purpose data base management programs tend to be expensive, requiring relatively large amounts of CPU time, memory, and disk space. For very large datasets like the PSID dataset or the even more complicated datasets involved in the SIPP, the costs of unlimited access become prohibitive, unless the user has an in-house computer and staff which reduce marginal computing costs to essentially zero.

One way to limit the errors and costs is to limit the ways in which the data base can be accessed, the "views" available. But this means giving up one of the main reasons for turning to the general-purpose data base management program in the first place.

2.9. Specialized Data Base Management Programs

A third approach is to access the data through a specialized data base management program that takes advantage of the probable use of the data base to reduce costs.

Panel studies have quite definite data processing needs. They collect, enter, check, correct, recode, and aggregate the data at fixed, relatively infrequent intervals, such as quarters or years. Thereafter they read and perhaps further recode relatively large portions of the data for each analysis. The key data processing requirement is the ability to read large numbers of logically (if not physically) sequential records as fast as possible.

Libraries, banks, and department stores have quite different data processing needs. They collect data through transactions processed interactively minute-by-minute. They require frequent—at least daily—reports on the transactions and the status of the data base. The key data processing requirement is the ability to read and write single records randomly as fast as possible.

A general-purpose data base management program needs to be able to handle both kinds of data processing. But a specialized data base management program can focus on the known use of the data to achieve greater efficiency.

The University of Michigan's Survey Research Center has recently added a specialized "directly structured dataset" capability to its OSIRIS IV software package.[4] Like a general-purpose data base management system, OSIRIS uses pointers to locate the occurrences of the groups that need to be pulled together. The most common linkages can be made permanent at the time the data base is created. Other linkages can be improvised for less common analyses.

2.10. Hierarchical Structures

A fourth approach—the one actually used by the PSID for the last eight years—is based upon a hierarchical structure in which each group is immediately subordinate to one and only one other group. If a group is subordinate to another group, then any group subordinate to it is also subordinate to the other group. Such a structure can be described as a "tree" or "outline," and its records can be processed sequentially for greater speed.

Not every logical data structure fits easily into a hierarchy. For example, a dataset with neighborhood, household, and child data is a natural hierarchy, but a dataset with school, household, and child data is not. Children from the same household may go to different schools, and children from different households may go to the same school. The child group is immediately subordinate to both the school and household groups, neither of which is subordinate to the other.

Schools, households, and children can be reduced to a hierarchy by picking a structure and duplicating the occurrences of the nonhierarchical group to make it fit into the hierarchy. We could make it a hierarchy of children under households under schools by duplicating the household data. We could make it children under schools under households by duplicating the schools data. And we could make it schools and households under children by duplicating both the household and school data.

The person–aggregate–time dataset would be a natural hierarchy if every person started out in one aggregate unit and stayed there. If this were the case, Figure 1 and Figure 7 could be reduced to Figure 8.

However, the fact of the matter is that persons move in and out of aggregates. There is no simple hierarchy of aggregates over time or persons within aggregates. Different sets of persons will be together in aggregate units at different points in time. Duplicate occurrences of aggregate–time will be necessary to reduce the person–aggregate–time

FIGURE 8. SIMPLE HIERARCHY

dataset to a hierarchy.

The PSID has devised an effective way to organize the duplicate occurrences of aggregate–time groups. It requires the creation of two additional groups, labeled "Root" and "Hist."

Root is a group of all the time 1 (root) aggregates. It has one occurrence for each occurrence of the time 1 aggregate group. It contains no data, only the ID variable of the time 1 aggregate group.

Hist is a group of all the different aggregate histories of the persons in the dataset. Its variables are the ID variable of the root aggregate group with which the aggregate history is associated and the aggregate–time ID variables of those histories.[5]

The aggregate–history group can be linked to the root group, and the aggregate–time groups and person–time groups can be linked to the aggregate–history group. These linkages are diagramed in Figure 9. The groups are represented in outline form in Figure 10.

FIGURE 9. COMPLEX HIERARCHY

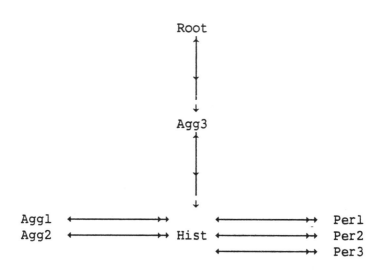

FIGURE 10. LEVELS IN COMPLEX HIERARCHY

Level 1	Level 2	Level 3	Level 4	Level 5
Root				
	Agg3			
		Hist		
			Agg1	
			Agg2	
				Per1
				Per2
				Per3

To obtain data for a cohort study of persons, one would enter the hierarchical structure at level 5 and access only Per1, Per3, and Per3.

To obtain data for a study of aggregate attributes of persons, one would enter the structure at level 5 and access all six data groups.

To obtain data for a study of aggregated person attributes, one would enter the structure at level 5 and access Per1, Per2, and Per3. These would need to be sorted by their aggregate–time link variables, aggregated, and merged back in with Agg1, Agg2, and Agg3.

To obtain data for a study of aggregates over time, looking backward from time 3, one would enter the structure at level 2 and access all data groups.

To obtain data for a study of persons who were in the same aggregate in the first time period, one would enter the structure at level 1 and access all data groups. This entry would wrongly associate persons who moved into a time 3 aggregate with a root aggregate different from their own, but there are very few of these.

2.11. Rectangular Structure

A fifth approach is based upon a rectangular structure. This would require putting the corresponding occurrences of all six data groups on a single record, as in Figure 11.

FIGURE 11. RECTANGULAR STRUCTURE

Agg1   Agg2   Agg3   Per1   Per2   Per3

A rectangular dataset would involve both duplication and padding of data. The aggregate–time data would have to be duplicated for the occurrences of persons sharing the aggregate units. The aggregate–time and person–time data would have to be padded whenever the data were missing because the person was not in the sample at a particular time. The duplication and padding would increase the storage required for the dataset and, more important, the cost of processing it.

Nonetheless, all the analyses that could be done with the hierarchically structured dataset could also be done with the rectangular dataset. Let us assume that the rectangular dataset is sorted from major to minor field by the time 1 aggregate–time ID variable, the most recent aggregate–time ID variable, and the person–ID variable.

To obtain data for a cohort study of persons, one need only read the cross–time aggregate–person records and extract the needed variables from Per1, Per2, and Per3.

To obtain data for a study of aggregate attributes of persons, one would read the records and extract the needed variables from all six groups.

To obtain data for a study of aggregated person attributes, one would read the records and extract the needed variables from all six groups. These would need to be sorted by their aggregate–time link variables and aggregated.

To obtain data for a study of aggregates over time, again looking backward from time 3, would be more difficult. One would have to read the cross–time aggregate–person records and accumulate those variables from all six groups that came from desired occurrences of the same time 3 aggregate unit. This involves an additional step of programming that would be done automatically by the data management routines for the hierarchically structured dataset.

To obtain data for a study of persons who were in the same aggregate in the first time period, one would read the cross-time aggregate–person records and accumulate the variables from all six groups that came from desired occurrences of the same time 1 aggregate unit. As in the case of the hierarchically structured dataset, this procedure would wrongly associate persons who moved into a time 3 aggregate with a different time 1 unit as a root.

We have been using a very simple hypothetical dataset to illustrate our points about data management problems and possible solutions. However, the methods we have described will work for more complicated datasets.

The PSID is currently handling 18 years of household and person data using rectangular, hierarchically structured, and now directly structured datasets. In principle, these methods could also be applied to more complicated data structures such as SIPP.

A SIPP dataset would start as a collection of monthly data matrices at the household, family, person, and income levels. These groups would then be accessed according to one or another of the methods described above. For other uses, the SIPP data might be organized into monthly groups at household, recipient-unit, person, and income levels.

3. EXAMPLES OF PSID DATA STRUCTURES

We shall now examine in some detail how two of the methods of arranging panel study data across time have been implemented with data from the PSID and how the five types of analysis outlined earlier can be accomplished using these structures. First we will examine four types of household situations encountered in collecting data about people in households across time that need to be represented in the data structure. Then we will examine how these four situations are accommodated using a hierarchical structure and using a specialized data base management approach.

3.1. Types of Household Situations

The PSID sampled 4,802 households in 1968 and has been collecting data each year since about persons living in these households and households that have derived from the original households. Panel studies that follow people living in households at several points in time are faced with different data management problems than are encountered in cross-sectional studies. At a single point in time each person can be associated with a single household. However with the passage of time households change, sometimes in quite complex ways. After 15 years only 12% of the PSID people lived in households that experienced no change in composition. While panel studies that cover a shorter time span will not encounter these situations as often, data managers must consider how the data structure will handle situations such as these.

(1) Split-off household: The most common type of change that households experience over time is that a child reaches maturity and leaves the parental household to form a household of his or her own. In the PSID members of original sample households who move away are interviewed as separate households in subsequent years. These are called "split-off" households. An example of this situation is illustrated in Figure 12. The 1968 household number 01 contained three members—father, mother, and son, designated respectively as 1-1, 1-2, and 1-3 in Figure 12; in 1969 they were also living together in the household number 59.[6] By 1986 the son had moved out, married, and formed a household of his own, which appears as 1986 household 45; mom and dad are interviewed as household 27 in 1986. In 1987 the son and his wife (1-70) have a child (1-30), and are interviewed as household 34; mom and dad are interviewed as household 53.

FIGURE 12. SPLIT-OFF HOUSEHOLD

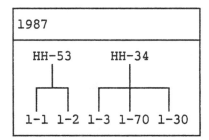

(2) Nonresponse household: While the annual response rate after the initial two years of the study has been high, 96% to 97%, by 1983 only

10,925 of the original 18,272 members of the 1968 panel households were still in the sample. People leave the panel because of death, refusal to participate further, or because they have moved and are not locatable. Figure 13 illustrates a couple that was interviewed as household 02 in 1968 and as household 54 in 1969. By 1986 the husband (2–1) had died and the wife (2–2) was interviewed as household 75; in 1987 the wife was institutionalized and an interview could not be obtained.

FIGURE 13. NONRESPONSE HOUSEHOLD

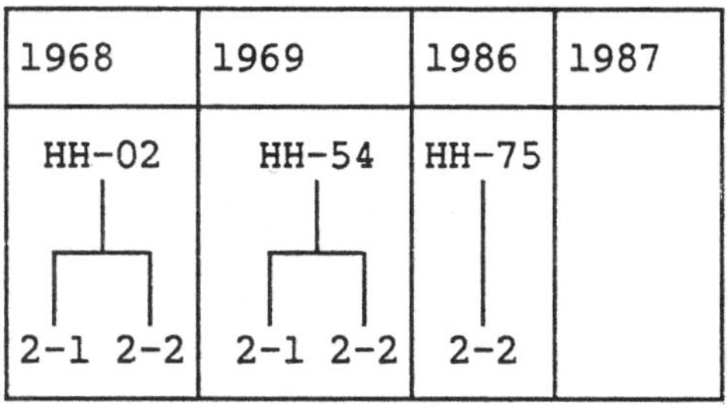

1968	1969	1986	1987
HH–02	HH–54	HH–75	
2–1 2–2	2–1 2–2	2–2	

(3) Reunited split-up household: People who were living together initially may later form separate households of their own and then recombine to form a single household in a later year. Figure 14 illustrates the situation of a couple and a child who were living together in 1968 as household 03 and in 1969 as household 17. In 1986 the husband (3–1) had left and was interviewed as household 51, while the wife (3–2) and child (3–3) were interviewed as household 24. By 1987 the family had reunited and was interviewed as household 32.

FIGURE 14. REUNITED SPLIT-UP HOUSEHOLD

1968	1969	1986	1987
HH–03	HH–17	HH–24 HH–51	HH–32
3–1 3–2 3–3	3–1 3–2 3–3	3–2 3–3 3–1	3–1 3–2 3–3

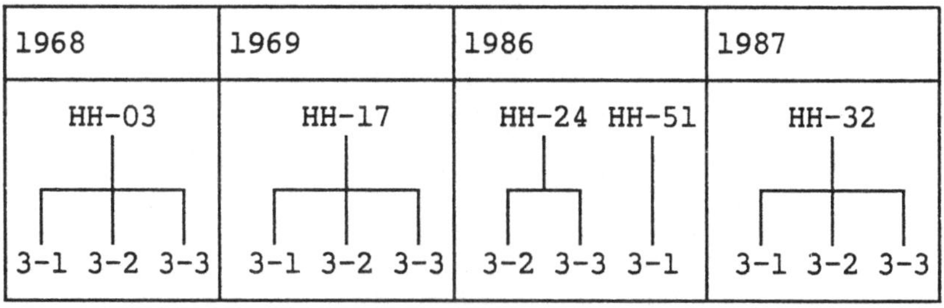

(4) Different original household: Very complex situations occur when people who are associated with different originating 1968 households are found in subsequent years in a single household.[7] Figure 15 illustrates a situation of this type. In 1968 two households were interviewed. Household 04 consisted of a husband (4–1) and a wife (4–2).

Household 05 consisted of a father (5-1), mother (5-2), and child (5-3). In 1969 the child had moved in with her aunt (4-2) and uncle (4-1), who were interviewed as household 82; the mother and father were interviewed as household 97. By 1986 the child moved back in with her mother, where they were interviewed as household 66, the aunt and uncle were interviewed as household 50, and the father had become nonresponse. In 1987 the aunt and uncle had split up, the uncle was interviewed as household 78, and the aunt had moved in with her sister and her niece where they were interviewed as household 25.

FIGURE 15. DIFFERENT ORIGINAL HOUSEHOLD

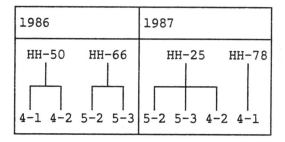

3.2. Structure of the PSID Hierarchical File

While data for people living in households at different points in time cannot be grouped into a natural hierarchy across time because people move between households, with the introduction of the concept of "household history"—grouping together all people who have shared the same set of household experiences during the years of the panel—the data, with modest duplication, can be arranged in a hierarchical format.

The structure of the PSID hierarchical dataset is illustrated in Figure 16, which has four levels. Level 1 is the household of origin. There is one group at this level, labeled G068 in the illustration, which contains just the 1968 household identifier. Level 2 is the current-year household. There is one group at this level, labeled G087 in the illustration, which contains the 1987 household information. Level 3 is the previous years' households. There are three groups at this level in the illustration,[8] labeled G168, G169, and G186, containing household information of the 1968, 1969, and 1986 households, respectively. The household history is used to build the hierarchical dataset but does not occur as an actual group.

FIGURE 16. 1987 PSID HIERARCHICAL STRUCTURE

Name	Level	Group Number	Identifiers	Data
Household of origin	Level 1		68 HH-ID	
		G068		—
Current year household	Level 2		68 HH-ID 87 HH-ID	
		G087		1987 Household
Previous years' households	Level 3		68 HH-ID 87 HH-ID History Num	
		G168		1968 Household
		G169		1969 Household
		.		.
		.		.
		G186		1986 Household
Person	Level 4		68 HH-ID 87 HH-ID History Num Person ID	
		G268		1968 Person
		G269		1969 Person
		.		.
		.		.
		G286		1986 Person
		G287		1987 Person

3.3. Identification Variables

The level 1 group, the household of origin, is identified by the 1968 household identifier. The level 2 group, the current-year household, is identified by the 1968 household identifier of the current-year head[9] and the 1987 household identifier. The level 3 groups, the previous years' households, are identified by the 1968 household identifier of the current-year head, the 1987 household identifier, and the "household history number." The history number is a sequential number that uniquely identifies sets of experiences that members of the current-year, 1987, household members have had. If all members of a 1987 household have lived in the same household during all of the panel, then the 1987 household will have only one household history.[10] Level 4 groups, person groups for each year of the panel, are identified by the 1968 household identifier of the current-year head, the 1987 household identifier, the "household history number," and the person identifier, the 1968 household of origin and person number.

(1) Split-off household in a hierarchical structure: The 1968 household 01 has generated two 1987 households—household 34, consisting of the son (01–03) and his wife (01–70) and child (01–30), and household 53, consisting of the father (01–01) and mother (01–02) from the 1968 household. All of the people in each of the 1987 households have had the same set of prior household experiences, so each 1987 household has only one "household history." There is duplication of the 1968 and 1969 household records (noted with * for the first occurrence of a record and ** for the second occurrence of a record in Figure 17) for the two years when the son lived in his parents' household. The 1968 household record 01 and the 1969 household record 59 are part of the household history of both 1987 households 34, consisting of the son, his wife and child, and household 53, consisting of the original father and mother. Person–year records appear only for the years the person was present in an interviewed household, 1986 and 1987 for the son's wife, and 1987 only for the child. The resulting hierarchical structure is shown in Figure 17.

(2) Nonresponse household in a hierarchical structure: There is no G087 because neither person present in 1968 household 02 was in a household that was interviewed in 1987. Zero is used as a level 2, current-year, identifier. By 1987 the two people in the 1968 household 02 had had two sets of previous years' household experiences—the husband (02–01) was present in a household interviewed in 1968 and 1969, households 02 and 54, respectively, and the wife (02–02) was present in a household interviewed in 1968, 1969, and 1686, households 02, 54, and 75, respectively. The household records for the years they lived in the same household, 1968 and 1969, appear in the household history of both the husband and the wife. No household records appear in the history for the years the person is not present. Person–year records appear only for the years the person was present in a household interviewed, 1968 and 1969 for the husband , and 1968, 1969 and 1986 for the wife. The resulting hierarchical data structure is shown in Figure 18.

(3) Reunited split-up household in a hierarchical structure: The 1968 household 03 has generated only one 1987 household, household 32, but members of this 1987 household have had two sets of household ex-

FIGURE 17. SPLIT-OFF HOUSEHOLD HIERARCHICAL STRUCTURE

Group Number	Level 1 Hhd. of Orgn.	Level 2 Curr. Year Hhd.	Level 3 Prev. Years Hhd.	Level 4 Person		Data
068	01					
087	01	34				87 Household 34
168	01	34	01			68 Household 01*
169	01	34	01			69 Household 59*
186	01	34	01			86 Household 45
268	01	34	01	01	03	68 Person 01–03
269	01	34	01	01	03	69 Person 01–03
286	01	34	01	01	03	86 Person 01–03
287	01	34	01	01	03	87 Person 01–03
287	01	34	01	01	30	87 Person 01–30
286	01	34	01	01	70	86 Person 01–70
287	01	34	01	01	70	87 Person 01–70
087	01	53				87 Household 53
168	01	53	01			68 Household 01**
169	01	53	01			69 Household 59**
186	01	53	01			86 Household 27
268	01	53	01	01	01	68 Person 01–01
269	01	53	01	01	01	69 Person 01–01
286	01	53	01	01	01	86 Person 01–01
287	01	53	01	01	01	87 Person 01–01
268	01	53	01	01	02	68 Person 01–02
269	01	53	01	01	02	69 Person 01–02
286	01	53	01	01	02	86 Person 01–02
287	01	53	01	01	02	87 Person 01–02

* First occurrence of record.
** Second occurrence of record.

periences. The husband (03–01) lived in households 03, 17, and 51 while the wife (03–02) and child (03–03) lived in households 03, 17, and 24. Thus the 1987 household 32 has two household histories. Duplicate household records appear for 1968 and 1969. The resulting hierarchical data structure is shown in Figure 19.

(4) *Different original household in a hierarchical structure:* The original 1968 household 04 generated one 1987 household, household 78, consisting of a single person, the uncle (04–01), and provided a member, the aunt (04–02) who lived with her sister in 1987, for the 1987 household 25, which was generated by the 1968 household 05. The 1968 household 05 had three members, one of whom—the father (05–01)—has become nonresponse by 1987. The remaining two members—the mother (05–02) and the child (05–03)—form the 1987

FIGURE 18. NONRESPONSE HOUSEHOLD HIERARCHICAL STRUCTURE

Group Number	Level 1 Hhd. of Orig.	Level 2 Curr. Year Hhd.	Level 3 Prev. Years Hhd.	Level 4 Person		Data
068	02					
168	02	00	01			68 Household 02*
169	02	00	01			69 Household 54*
268	02	00	01	02	01	68 Person 02–01
269	02	00	01	02	01	69 Person 02–01
168	02	00	02			68 Household 02**
169	02	00	02			69 Household 54**
186	02	00	02			86 Household 75
268	02	00	02	02	02	68 Person 02–02
269	02	00	02	02	02	69 Person 02–02
286	02	00	02	02	02	86 Person 02–02

* First occurrence of record.
** Second occurrence of record.

household 25 with the aunt. Each of the members of 1987 household 25 has had a unique set of household experiences—the mother lived in households 05, 97, and 66; the aunt (04–02) lived in households 04, 82, and 50; and the child lived in households 05, 82, and 66—and so 1987 household 25 has three histories. Duplicate household records appear when a household experience appears in more than one household history. The resulting hierarchical data structure is shown in Figure 20.

3.4. Implementation of Analysis Problems Using the Hierarchical File

We shall next review how the five data management problems described previously would be implemented with this hierarchical arrangement of the PSID data.

(1) Cohort study of persons: To obtain data for analysis of persons across time, one can enter the hierarchical structure at level 4, the person level, and access all the person–year groups (groups 268 through 287). All occurrences of relevant groups are identified by identical ID strings at level 4. If analysis is desired only of persons who resided in responding households in all years of the study, all groups will be required. Not all groups would be required if the analysis were of persons present at any time during the period; a person who entered the study in later years will be missing groups from the earlier years. Any missing groups could be filled with missing data or zeros.

(2) Household attributes of persons: To obtain data for analysis of persons across time with associated household data, one can again enter

FIGURE 19. REUNITED SPLIT-UP HOUSEHOLD HIERARCHICAL STRUCTURE

Group Number	Level 1 Hhd. of Orig.	Level 2 Curr. Year Hhd.	Level 3 Prev. Years Hhd.	Level 4 Person		Data
068	03					
087	03	32				87 Household 32
168	03	32	01			68 Household 03*
169	03	32	01			69 Household 17*
186	03	32	01			86 Household 51
268	03	32	01	03	01	68 Person 03–01
269	03	32	01	03	01	69 Person 03–01
286	03	32	01	03	01	86 Person 03–01
287	03	32	01	03	01	87 Person 03–01
168	03	32	02			68 Household 03**
169	03	32	02			69 Household 17**
186	03	32	02			86 Household 24
268	03	32	02	03	02	68 Person 03–02
269	03	32	02	03	02	69 Person 03–02
286	03	32	02	03	02	86 Person 03–02
287	03	32	02	03	02	87 Person 03–02
268	03	32	02	03	03	68 Person 03–03
269	03	32	02	03	03	69 Person 03–03
286	03	32	02	03	03	86 Person 03–03
287	03	32	02	03	03	87 Person 03–03

* First occurrence of record.
** Second occurrence of record.

the hierarchical structure at level 4, the person level, and access all the person-year groups (groups 268 through 287) and their associated household groups (groups 87 and 168 through 186). All occurrences of person–year groups needed to form a record are identified by identical ID strings at level 4, the person level; occurrences of previous years' household groups needed to form a record are identified by a subpart of this ID string at level 3; and the occurrence of the current-year household needed is identified by a subpart of the ID string at level 2. Not all of these groups need be required. A person who entered the study in later years will be missing person–year groups from the earlier years. A person who was not in an interviewed household for later years or for some intermediate years will be missing both household–year and person–year groups for the missing years. Any missing groups could be filled with missing data or zeros.

(3) Aggregating person attributes to a household level: The underlying assumption of the above procedure is that any required aggregation of person attributes to a household level, such as counting the number of children less than six years old in the household, has been done

FIGURE 20. DIFFERENT ORIGINAL HOUSEHOLD HIERARCHICAL STRUCTURE

Group Number	Level 1 Hhd. of Origin	Level 2 Curr. Year Hhd.	Level 3 Prev. Years Hhd.	Level 4 Person		Data
068	04					
087	04	78				87 Household 78
168	04	78	01			68 Household 04*
169	04	78	01			69 Household 82*
186	04	78	01			86 Household 50
268	04	78	01	04	01	68 Person 04–01
269	04	78	01	04	01	69 Person 04–01
286	04	78	01	04	01	86 Person 04–01
287	04	78	01	04	01	87 Person 04–01
068	05					
168	05	00	01			68 Household 05*
169	05	00	01			69 Household 97*
268	05	00	01	05	01	68 Person 05–01
269	05	00	01	05	01	69 Person 05–01
087	05	25				87 Household 25
168	05	25	01			68 Household 05**
169	05	25	01			69 Household 97**
186	05	25	01			86 Household 66*
268	05	25	01	05	02	68 Person 05–02
269	05	25	01	05	02	69 Person 05–02
286	05	25	01	05	02	86 Person 05–02
287	05	25	01	05	02	87 Person 05–02
168	05	25	02			68 Household 04**
169	05	25	02			69 Household 82**
186	05	25	02			86 Household 50
268	05	25	02	04	02	68 Person 04–02
269	05	25	02	04	02	69 Person 04–02
286	05	25	02	04	02	86 Person 04–02
287	05	25	02	04	02	87 Person 04–02
168	05	25	03			68 Household 05***
169	05	25	03			69 Household 82***
186	05	25	03			86 Household 66**
268	05	25	03	05	03	68 Person 05–03
269	05	25	03	05	03	69 Person 05–03
286	05	25	03	05	03	86 Person 05–03
287	05	25	03	05	03	87 Person 05–03

* First occurrence of record.
** Second occurrence of record.
*** Third occurrence of record.

previously. If this were not the case and new aggregations needed to be done, the hierarchical structure would not make this easy.

It would be reasonably straightforward to produce new aggregations for the current-year households or for the households of origin, since the person records for the current-year households and the households of origin are contiguous in the file. One would need to aggregate person attributes to the current-year household or to the household of origin and add the new aggregates back into the file—a two-step process—before analysis could proceed along the lines outlined in the previous section.

It would be more difficult to construct new aggregations of person attributes to the intermediate previous-year households. First, one would need to "reunite" members of the previous-year household (by sorting the person records on the previous year's household identifier) and to aggregate the person attributes to the previous-year household. Second, one would need to select all previous-year household records from the hierarchical file (including duplicated previous-year household records) and to merge the aggregated information onto these records. Third, it would be necessary to merge these records back into the hierarchical file. This three-step process, which would need to be repeated for each previous year before analysis, could proceed as described in the previous section.

(4) Households over time: There is not a generally agreed-upon definition of households across time. In the early years of the PSID a household headed by the same person at two points in time was defined as the "same" household. As years have passed and the study has gone on, many households have changed using this fairly static definition of household. As a result there has been less emphasis on analyzing households across time and more emphasis on analyzing persons across time, categorizing them as having lived in households that experienced different conditions; consequently, little work has been done with this study to develop alternative methods of linking households across time.

It is possible, as the data are currently structured, to analyze households across time linking current-year households with the previous households of the current-year head. One can then limit analysis to households that have had no change in head, include households where a wife at time 1 became a head at time 2, or include all current-year households. To obtain data for this type of analysis of households across time, one can enter the the hierarchical structure at level 2, the current-year household level, and access the current-year household group (group 087), the previous year's household groups (groups 168 through 186), and, if desired, the person–year groups (groups 268 through 287). If the person–year groups are included, multiple occurrences of each person–year group for a current-year

household must be allowed. All occurrences of relevant groups are identified by identical ID strings at level 2, the current-year household level.
Again, any missing groups could be filled with missing data or zeros.

(5) Persons who resided in the same household: A more difficult
problem is to link people who resided in the same household at time 1
and in different households at time 2. The hierarchical structure allows
only a limited amount of this type of analysis directly. It is possible to
obtain records for analysis of people who lived in the same household in
1968 and who subsequently live in different households. To obtain data
for analysis of people who lived in the same household in 1968 and in
separate households at later points in time, one can enter the hierarchy
at level 1, the household of origin, and access the current-year
household group (group 087), the previous years' household groups
(groups 168 through 186), and, if desired, the person–year groups
(groups 268 through 287). Multiple current-year household groups,
multiple intermediate-year households, and multiple person–year groups
for each household of origin may occur and accommodation for this
must be made.

3.5. Structure of the PSID Directly Structured File

The hierarchical file just described had an increasing amount of redundancy in it (especially after the nonresponse persons were retained in
the hierarchical dataset beginning in 1984) and was somewhat restricting in the type of analysis that was possible. To avoid these problems,
a specialized data base management approach has been developed using
the OSIRIS IV software.

There are 19 household–year groups, one for each year of the study,
1968 through 1987, labeled G168 through G187, respectively. A
household–year group contains one record for each household in the
study in year n. There are also 19 person–year groups, one for each
year of the study, 1968 through 1987, labeled G268 through G287,
respectively. A person–year group contains one record for each person
in the study in year n. In addition, there is a timeless person group
called the "master person" group and labeled G200. The master person
group contains one record for each person ever in the study. The
master person record contains the household number of each household
in which the person resided[11] for each year of the study. Representations of these data matrices are illustrated in Figures 21 and 22.

3.6. Identification and Link Variables

Each occurrence of a household–year group is uniquely identified by the
year n household number. Each occurrence of the master person group
and of the person–year groups is uniquely identified by the combination
of two variables, the 1968 household and person number. The master

FIGURE 21. DATA MATRICES FOR 1987 PSID DIRECTLY STRUC-
TURED DATASET, MASTER PERSON GROUP—GROUP 200

Person ID	1968 Link	1969 Link	1986 Link	1987 Link
Split-off Household				
Person 1–01	01	59	27	53
Person 1–02	01	59	27	53
Person 1–03	01	59	45	34
Person 1–30	01	59	45	34
Person 1–70	01	59	45	34
Nonresponse Household				
Person 2–01	02	54	00	00
Person 2–02	02	54	75	00
Reunited Split-up Household				
Person 3–01	03	17	51	32
Person 3–02	03	17	24	32
Person 3–03	03	17	24	32
Different Original Household				
Person 4–01	04	82	50	78
Person 4–02	04	82	50	25
Person 5–01	05	97	00	00
Person 5–02	05	97	66	25
Person 5–03	05	82	66	25

person group is linked to the household–year groups by year n household number. The master person group is also linked to the person–year groups by the 1968 household–person number. It would have been possible to create additional links, e.g., to link the year n person group to the year n household group and to link the year n person group to the year $(n - 1)$ household group, to the year $(n - 2)$ household group, etc. However, for considerations of economy and simplicity, this has not been done. The simpler structure can be used with the OSIRIS IV software for any analysis contemplated. An illustration of this arrangement is presented in Figure 22.

(1) Split-off household in the directly structured dataset: Each of the people who resided in the 1968 household 01—the father (01–01), mother (01–02), and son (01–03)—can be linked to 1968 household 01 data in group 168 by the 1968 household identifier in their master person record. PSID assigns earlier-year household identifiers to people who subsequently become associated with households that derive from the earlier-year households. Thus the son's wife (01–70) and child (01–

FIGURE 22. DATA MATRICES FOR 1987 PSID DIRECTLY STRUCTURED
DATASET, HOUSEHOLD–YEAR AND PERSON–YEAR GROUPS

Group 168	Group 169	Group 186	Group 187
HH 01 '68 data	HH 17 '69 data	HH 24 '86 data	HH 25 '87 data
HH 02 '68 data	HH 54 '69 data	HH 27 '86 data	HH 32 '87 data
HH 03 '68 data	HH 59 '69 data	HH 45 '86 data	HH 34 '87 data
HH 04 '68 data	HH 82 '69 data	HH 50 '86 data	HH 53 '87 data
HH 05 '68 data	HH 97 '69 data	HH 50 '86 data	HH 78 '87 data
		HH 51 '86 data	
		HH 66 '86 data	
		HH 75 '86 data	

Group 268	Group 269	Group 286	Group 287
Per1–01 '68 data	Per1–01 '69 data	Per1–01 '86 data	Per1–01 '87 data
Per1–02 '68 data	Per1–02 '69 data	Per1–02 '86 data	Per1–02 '87 data
Per1–03 '68 data	Per1–03 '69 data	Per1–03 '86 data	Per1–03 '87 data
		Per1–70 '86 data	Per1–30 '87 data
Per2–01 '68 data	Per2–01 '69 data		Per1–70 '87 data
Per2–02 '68 data	Per2–02 '69 data	Per2–02 '86 data	
			Per3–01 '87 data
Per3–01 '68 data	Per3–01 '69 data	Per3–01 '86 data	Per3–02 '87 data
Per3–02 '68 data	Per3–02 '69 data	Per3–02 '86 data	Per3–03 '87 data
Per3–03 '68 data	Per3–03 '69 data	Per3–03 '86 data	
			Per4–01 '87 data
Per4–01 '68 data	Per4–01 '69 data	Per4–01 '86 data	Per4–02 '87 data
Per4–02 '68 data	Per4–02 '69 data	Per4–02 '86 data	
			Per5–02 '87 data
Per5–01 '68 data	Per5–01 '69 data	Per5–02 '86 data	Per5–03 '87 data
Per5–02 '68 data	Per5–02 '69 data	Per5–03 '86 data	
Per5–03 '68 data	Per5–03 '69 data		

30) are also linked to 1968 household 01.[12] All five people are also
linked to the 1969 household 59 record in a similar way. In 1986 the
father and mother are linked to 1986 household 27 while the son and
his new wife (and their unborn child) are linked to 1986 household 45.
In 1987 the father and mother are linked to 1987 household 53, and
the son and his wife and child are linked to 1987 household 34.

(2) Nonresponse household in the directly structured dataset: The husband (02–01) was present in a responding household only in 1968 and
1969. His master person record is linked to records in household–year
and person–year groups only for these two years. His wife (02–02) was
in responding households 1968, 1969, and 1986. Her master person
record is linked to records in household–year and person–year records
for these three years.

(3) Reunited split-up household in the directly structured dataset: The
husband (03–01) is linked to household–year records 03, 17, 51, and 32
for years 1968, 1969, 1986, and 1987, respectively. The wife (03–02)

FIGURE 23. 1987 PSID DIRECTLY STRUCTURED DATASET

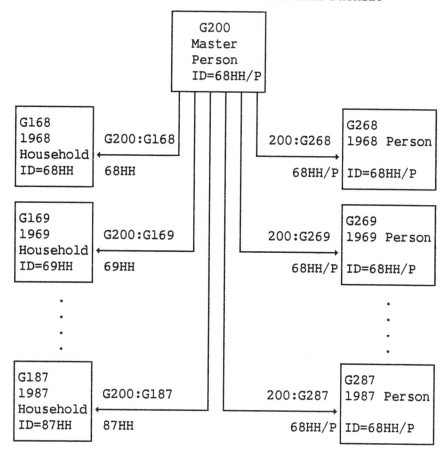

and child (03–03) are linked to household–year records 03, 17, 24, and 32 for 1968, 1969, 1986, and 1987. Each is linked to person–year records for each year.

(4) Different original household in the directly structured dataset: The uncle (04–01) is linked to household–year records 04, 82, 50, and 78, while the aunt (04–02), who moves into her sister's household in 1987, is linked to household–year records 04, 82, 50, and 25. The father (05–01) is linked to household–year records 05 and 97 for 1968 and 1969, the two years when he resided in an interviewed household. The mother (05–02) is linked to household–year records 05, 97, 66, and 25, while the child (05–03) who lived with the aunt and uncle in 1969 is linked to household–year records 05, 82, 66, and 25. Each person is also linked to appropriate person–year records by their 1968 household–person number.

3.7. Implementation of Analysis Problems Using the Directly Structured File

We shall next review how the five data management problems described previously would be implemented with the PSID directly structured file. Using directly structured files, one's entry point into the structure is a group rather than a level as in a hierarchically structured file. Since the only group in the PSID structure that is linked to more than one group is the master person group, entry at the master person group is the only entry that allows the linking of data from more than two groups.

(1) Cohort study of persons: To obtain data for analysis of persons across time, one would enter the directly structured dataset at the master person group and access all the person–year groups (groups 268 through 287) that are linked to it. If analysis is desired only of persons who resided in responding households in all years of study, all groups will be required. If analysis is desired of persons present at any time during the study period, all person–year groups may be optional. Any missing groups could be filled with missing data or zeros.

(2) Household attributes of persons: To obtain data for analysis of persons across time with associated household data, one can again enter the directly structured dataset at the master person group and access all person–year groups (groups 268 through 287) and all household–year groups (groups 168 through 187). Not all of these groups need be required. A person who entered the study in later years will be missing person–year groups from the earlier years. A person who was not in an interviewed household for later years or some intermediate years will be missing both household–year and person–year groups for the missing years. Any missing groups could be filled with missing data or zeros.

(3) Aggregating person attributes to a household level: The above approach will suffice when any required aggregation of person attributes to a household level, such as counting the number of children less than six years old in the household, has been done previously. If this were not the case and new aggregations needed to be done, the directly structured file would require modification. One would need to aggregate person attributes to the household group or all household groups and add the new aggregates back into the file before analysis could proceed as described in the previous section.

(4) Households over time: There has been lessening interest in the PSID on analyzing households across time. However, a limited amount of analysis of this type has been done. Generally the approach used has been to link household data from previous years of the current-year household heads, and then often to restrict analysis to some subset of

these current-year households, e.g., to limit analysis to current-year households who are headed by a person who was a head or wife in previous years. To obtain data for this type of analysis of households across time, one can enter the directly structured data set at the master person group taking only persons who were heads in 1987, and access the household–year groups (groups 168 through 187). Any missing groups could be filled with missing data or zeros.

Other definitions of households across time can also be handled with the directly structured file. For instance, a definition of a household across time might be that a time 2 household is to be considered the continuation of a time 1 household if the time 2 household contains the majority of the time 1 household members. To obtain data for this type of analysis of households across time, one accesses the directly structured dataset using a separate control dataset that contains the time 1 identifier (e.g., the 1968 household number) and the time 2 identifier (e.g., the 1987 household number) of households that are to be linked. Other definitions of households across time could be handled in a similar manner.

(5) Persons who resided in the same household: The directly structured dataset allows more flexibility to retrieve occurrences of records that are not explicitly linked within the structure. If the analyst wishes to examine the impact of divorce on poverty status, he/she may want to link data from the household of the couple predivorce with data from the postdivorce household of the ex-husband and data from the postdivorce household of the ex-wife. To obtain data for this type of analysis, one accesses the directly structured dataset using a separate control dataset that contains the predivorce household identifier of the couple, the postdivorce household identifier of the ex-husband and the postdivorce household identifier of the ex-wife.

4. SUMMARY

We have presented an overview of the issues involved in managing household- and person-level data collected in longitudinal studies of households. No simple structure can accommodate these data, because changes in household composition over time destroy the simple hierarchy between households and persons that exists at any one point in time.

We began with a generic description of household panel surveys and discussed their design and analysis features, introducing five examples of analyses requiring longitudinal data. It then discussed the data management problems involved in these analyses and five possible solutions. Using the analyses introduced in Section 1, we presented in some detail two solutions actually used by the Panel Study of Income Dynamics (PSID). The first was sequential access of a hierarchical structure created using the concept of "household histories." The second was a specialized data base management program, the directly structured dataset capabilities of the OSIRIS IV software package.

The preferred method for managing household panel study data will depend on the hardware, software, and support staff available to the user. The simplest approach is to force the data into a rectangular structure. This is probably adequate for 95% of the desired analyses. It requires more machine time for processing but is easier for the user to understand and thus for many users may be less expensive than other approaches. The PSID distributes its data in rectangular form, as does SIPP.

A more efficient approach is to build a hierarchical structure. With appropriate software the analyst can do the same things with a hierarchical structure as he or she can with a rectangular structure but more easily and more cheaply. The problem is that the analyst must abandon the mentality of rectangular structures, must have software that can handle hierarchical structures, and must have an adequate understanding of both the dataset and its hierarchical implementation.

The natural data structure for a panel study is not a hierarchy, however. This makes it tempting to turn to the traditional approach of sorts and match-merges or to data base management programs. This paper argues that sorts and match-merges are useful for an intellectual understanding of the data management problems involved in panel studies but too cumbersome and expensive to use in real life.

General-purpose data base management programs and specialized data base management programs are more appealing solutions to the data management problems discussed in this paper. The choice between the two depends mainly on the computing capability—hardware, software, and staff—available to the analyst. Generalized data base management programs are more available, but a specialized data base management program is likely to be cheaper and perhaps easier to use.

Many research facilities will not be able to implement either the approach of a general-purpose data base management program or of a specialized data base management program, because these programs require large amounts of disk space for large amounts of data. The disk space currently may not be available at a cost most researchers can afford. The processing costs may also be prohibitively expensive.

The PSID, for example, is already straining at the limits of what the very good computing facilities of The University of Michigan have to offer. Indeed, the PSID is in a kind of race with computer technology, in which the study is betting that the capabilities of computer technology will expand faster than its dataset.

The demands of much more complicated datasets, such as the SIPP dataset, exceed the capabilities of the computer technology currently available at most research facilities. At these facilities, rectangular and perhaps hierarchical structures may still be the most feasible approaches. In any case, they are still very much in the running.

NOTES

*This paper has benefited from comments by Peter Joftis, Dorothy Kempter, David McMillen, and Michael Nolte.

[1]The reader may wish to replace the terms "group," "occurrence," and "variable" with the terms "table," "row," and "column;" the terms "file," "record," and "field;" or the terms "entity," "tuple," and "attribute," if any of the latter are more familiar.

[2]The reader may wish to replace the terms "identification variable" and "link variable" with the terms "primary key" and "foreign key," if these are more familiar.

[3]The reader may wish to use the term "schema" to refer to the logical structure described here and diagramed in Figure 1 and the term "subschema" to refer to the various ways of using the schema.

[4]The OSIRIS IV directly structured dataset capability is currently available for IBM-like machines using the Michigan Terminal System (MTS) and IBM OS, MVS, and VM/CMS.

[5]The PSID uses the Hist group in building its hierarchical structure but then drops the group from the data base because it is never used as a unit of analysis. The hierarchical structure of the PSID data base is described in Section 3.

[6]Household numbers that uniquely identify a household in a particular year are assigned sequentially, each year starting with one, as the interviews are received in the field office. No attempt is made to link households across time with this number.

[7]Because of the clustering of the original 1968 sample, in several instances different portions of the same families residing at separate addresses but in the same block at the time of the 1968 interview were selected as sample households. One example would be marital partners who were separated in 1968, each living in households that were selected as sample households in 1968, who later reunited. Another example would be children moving back and forth between relatives' households when the households of both sets of relatives happened to be sampled in 1968. Of the 20,393 people in households that were interviewed in 1984, only 10 originated from a different 1968 household than the 1968 household of the head of the household in which they lived in 1984.

[8]Nineteen household–year groups, one for each year 1968 through 1986, would be present in the actual file. Level 4 is the person. There are four groups at this level in the illustration, G268, G269, G286, and G287, which contain person information for the years 1968, 1969, 1986, and 1987, respectively. Twenty person–year groups, one for each year 1968 through 1987, would be present in the actual file.

[9]In most instances this is also the 1968 household identifier of all the 1987 household members.

[10]Children who are born to sample members are assigned the "history" of their parent; nonsample members who reside with panel members in subsequent years of the panel are assigned the "history" of the panel member with whom they become associated. We shall examine in detail how this hierarchy works with data for people who live in households with different sets of experiences.

[11]Or became associated with at a later point in time through marriage or birth.

[12]This early information might be useful, for example, in a study of the poverty history of a household into which a child was born. If a study did not attribute household data to people who did not actually reside in the household in year n, then the linking variables would be zero for the years when no link was appropriate. Each person is also linked to appropriate person–year records by their 1968 household–person number. Person–year records are present only for years the per-

son was actually present in a responding household. Thus the son's wife is linked to person–year records only for 1986 and 1987, and the child is linked to a person–year record only for 1987.

Managing Panel Data for Scientific Analysis: The Role of Relational Data Base Management Systems*

Martin David

1. INTRODUCTION

This paper rests on two assumptions: (1) panel data are collected for later exploitation by a scientific community larger than the group engaged in the design and collection of the panel; and (2) the panel design, repeated observations on a sample of elements (individuals), creates complexities that will obscure the meaning of the observations unless the system of data management followed is more comprehensive than the system that is customarily applied to cross-sectional surveys.

The six guidelines proposed by Boruch and Pearson (1985, pp. 30–71; 1988, pp. 16–28) for the evaluation of panel data collection entail substantial improvements in the management of panel data:

1. Data should be made available as expeditiously and as widely as possible to the research community.
2. Simultaneous analyses of data by other researchers should be permitted and facilitated.
3. Greater resources should be devoted to assessing and improving the quality of measurement in contrast to sampling error.
4. Data should be collected in a manner that permits easy linkage to other data.
5. Researchers should be permitted to add on independently designed special-purpose studies to the longitudinal survey.
6. Longitudinal surveys should be enhanced by coupling them to experimental studies.

It follows from these guidelines that well-managed panel data are portable. Outsiders should be able to make the same interpretation of the observations as those who collected the data; they will not be misled into false inferences by lack of necessary information. This implies a management strategy for panel data that we discuss in Section 4. First, we demonstrate that panel complexity is conceptually different from the complexity that arises from proliferation of cross-sectional datasets.

2. COMPLEXITIES OF PANEL DATA

Let us begin by representing the information in a set of panel data as a set of arrays where C references the number of cases and K references the number of attributes in each array:

$$X(1), \qquad Y(2), \qquad Z(3), \qquad \qquad W(\Omega) \qquad \qquad (1)$$
$$(C_1 \times K_1) \quad (C_2 \times K_2) \quad (C_3 \times K_3) \quad \cdots \quad (C_\Omega \times K_\Omega)$$

where 1, 2, . . . Ω refer to the order (wave) of the measurement process. The arrays are named differently to reflect the fact that each wave of the panel contains different measures. Given the sampling principle in the original design, the number of cases in each wave varies. Changes in the sample universe (births and deaths), attrition, and noninterview assure varying numbers of observations (responses).

The challenge to longitudinal analysis of such a panel is to reconstruct a subset of the observations in array (1) as a time series of information:

$$[X(1), \quad X(2), \quad X(3), \ldots X(\Omega)] \qquad \qquad (2)$$
$$(C_{1-\Omega} \times \Omega K_{1-\Omega})$$

This task requires two major operations. First, corresponding units at each wave of measurements must be linked. Second, equivalent measures must be defined for the linked units. The original measurements can then be organized into array (2).

Linkage of observations across waves implies that an identity exists for each case in each wave. The identities must be matched across waves. Aliases and changes in identity must also be part of the information collected in the panel dataset. These information requirements are unique to panel designs.

Furthermore, the quality of the linkage must be evaluated (David, 1985b). Errors in linkage distort the covariance of measures taken at different points in time.

Because some cases exist for which there are observations at one time but not at another, linkage entails a task of reconciliation. Units that do not appear in each wave must be accounted for. The analysis of matched cases requires a conceptual perspective on the missing observations. One possibility is that the observation desired is impossible— the person died. The alternative is that the observation was censored. Appropriate analysis techniques to deal with these two possibilities are extremely different.

Equivalent measures derive from parallelism in the measurement process at different waves. Three conditions are sufficient (although they may not be necessary):

(1) The same question must be used to elicit the measure.

(2) The same coding scheme must be applied to capture the data.

(3) The same context must apply to the question asked. (Recall period and reference period are identical; the respondent is the same.)[1]

Information must be readily at hand to establish (1), (2), and (3); otherwise the meaning of measures of change is unclear.

3. RELATIONAL DATA BASE MANAGEMENT SYSTEMS

For the purposes of this discussion, relational data base management systems (RDBMS) have five essential characteristics:

(1) All information stored in an RDBMS is stored in arrays.

(2) Arrays are rectangular.

(3) An RDBMS can join any two arrays in a data base.

(4) The arrays in an RDBMS are stored and retrieved independently.

(5) One of the arrays in the RDBMS is a catalog of all the attributes and all the arrays.

3.1. Significance of These Characteristics

(1) *All information stored in an RDBMS is stored in arrays:* The array is a relation in the mathematical sense. (Below we shall refer to arrays as relations.) Each row, or case, is an ordered tuple. One element in the row is an identifier; the other elements in the row are attributes of *that* identifier. Looking at all rows together the array consists of one (or more) columns that identify the case and columns that contain the same attribute for all cases. The arrays can be thought of as tables of observations. (Insistence that attributes refer to the identifier precludes attributes that relate to other attributes and information that is an attribute of a different, but related identifier (Kent, 1983)).

For example, each relation always contains the identifier for an individual; one contains personal attributes, birthdate, gender, ethnicity; another relation contains the identifier and the address of the house in which that individual lives; and a third relation contains the address and attributes of that address, number of rooms in the dwelling, type of structure, etc.

(2) *Arrays are rectangular:* The number of columns corresponds to the number of attributes plus the components of the identifier. The number of rows corresponds to the number of cases for which the attributes have been measured and can be meaningfully assigned. There is no need to maintain a not-in-universe code for cases in which the attributes were not measured (by design) or do not apply.

(3) *An RDBMS can join any two arrays in a data base:* Arrays with the same unit will contain the same identifier. E.g., individuals are identified by their Social Security numbers. Joining two arrays pro-

duces the matched set of attributes from both arrays for all cases that
are present in both arrays. For two arrays that do not share a common
identifier the relationship between the identifiers must be established
from a third array. For example, the listing of persons living at a particular address (the relationship just described above) gives a relationship of individual identifiers to an address identifier.

(4) *The arrays in an RDBMS are stored and retrieved independently:*
Alterations in one array do not affect the validity or organization of
data in another.[2]

(5) *The RDBMS maintains a catalog of all the attributes and all the arrays:* The system records the location of each aspect of the dataset. It
associates each aspect with a unique name. Thus the dataset is named;
relations are named; attributes are named. The names can be searched
to locate specific observations. The user need not remember any details
of the physical organization of the data base.

3.2. Discussion and Examples

We provide some concrete examples. First, the arrays stored in a data
base might be array (1). The use of the RDBMS imposes no constraints
on the change in the arrays from one wave to the next. (This is the
result of characteristic (5); relations are stored and retrieved independently.)

How might the arrays corresponding to $X(1)$ appear in an RDBMS?
Refer to Figure 1. The observations are shown at the bottom of the
figure as a series of arrays XR_{A1}, XR_{I1}, ... XR_{S1}. Each array consists of a set of ordered tuples. Each tuple includes the identifier and a
number of attributes. For convenience we list the identifier as the first
attribute. The number of cases in the array is shown in the last
column. Note that the number of cases can be different for each relation. This reflects the fact that different parts of the data collected can
apply to different universes and can contain different numbers of identifiers. To illustrate the significance of this assume that the first relation contains attributes of addresses and that the second contains attributes of individuals. Over an entire panel these arrays will be:

$$\text{Addresses} \quad \begin{matrix} XR_{A1}, & YR_{A2}, & ZR_{A3}, & & WR_{A\Omega} \\ (C_{A1} \times K_{A1}) & (C_{A2} \times K_{A2}) & (C_{A3} \times K_{A3}) & \cdots & (C_{A\Omega} \times K_{A\Omega}) \end{matrix}$$

$$\text{Individuals} \quad \begin{matrix} XR_{I1}, & YR_{I2}, & ZR_{I3}, & & WR_{I\Omega} \\ (C_{I1} \times K_{I1}) & (C_{I2} \times K_{I2}) & (C_{I3} \times K_{I3}) & \cdots & (C_{I\Omega} \times K_{I\Omega}) \end{matrix}$$

The merit of separate arrays is that attributes of addresses need not be
repeated on every individual's record. It is unnecessary to keep redundant information. Address information can always be attributed to individuals, when needed, because the individual relation contains an attribute that identifies the address with which he or she is associated at

FIGURE 1. INFORMATION ARRAYED IN A DATASET

Range	Content	Information Type
	Nomenclature	
For $X(1)$:	Dataset name ND_1	[[1]]
For ND_1:	Relation names $NXR_A1, NXR_{I1}, \ldots NXR_{S1}$	[[2]]
For NXR_A1:	Attribute names $Nx_1, Nx_2, \ldots Nx_{KA}1$	[[3]]
NXR_{I1}:	Attribute names $Nx_1, Nx_2, \ldots Nx_{KI}1$	
.	.	
.	.	
.	.	
NXR_{S1}:	Attribute names $Nx_1, Nx_2, \ldots Nx_{KS}1$	
For Nx_j:	Domain	[[4]]
	Legal codes $\{(x, Mx)\}$	[[5]]
	Illegal codes	[[9]]
	Observations	[[10]]
Relation	*ID Attributes*	*Cases*
XR_A1:	$a, x_1, x_2, \ldots x_{KA}1$	C_A
XR_{I1}:	$i, x_1, x_2, \ldots x_{KI}1$	C_I
.	.	
.	.	
.	.	
XR_{S1}:	$i, x_1, x_2, \ldots x_{KS}1$	C_S

each wave of observation. As listed above, characteristic (3) will then let us join any characteristic of an address to any individual and conversely.

For example, using the array that relates individuals to addresses, the number of individuals at an address can be counted. The count, household size, is a derived attribute. It can be associated with an address and can then be distributed back to each member of the household. In this process information from individuals is aggregated to the address level and the aggregate is distributed back to individuals. The process is elementary for RDBMS and can be accomplished in three lines of computer instructions.

Finally, we note that at each point of time each individual is associated with a unique address. Over time, owing to the formation of new households, new jobs, etc., individuals may move from one address to another. This poses no difficulties for the organization of the data whatever. Analysis of moves and the housing characteristics in any residence implies that one may wish to extract from successive waves of the panel a relation that records episodes of tenure by each sample member at the residences that he or she occupies.

4. APPLICATION OF RELATIONAL DATA BASE MANAGEMENT SYSTEMS TO MANAGING PANEL DATA

Relational data base management systems can improve information management in panel datasets in at least six major ways:
(1) They enhance the storage of datasets.
(2) They automatically document data organization.
(3) They aid in extracting the longitudinal data structure suggested in equation (2).
(4) They can be driven by syntax that is more self-documenting than the programming code of most languages.
(5) The direct-addressing facilities of information retrieval lend themselves to detecting and correcting, rather than burying, errors.
(6) They can encompass metadata pertaining to the panel that are essential to the correct retrieval of measurements. (By metadata we refer to all of the material about observations and the process of collecting those observations.)
(See David, Robbin, & Flory, 1988a; Robbin & David, 1988.)

4.1. Enhanced Storage

Storage of datasets in relations achieves several advantages. The semantic principle underlying the relational organization clarifies the unit of analysis and the universe to which the observations apply. Thus it is clear that the value of an automobile, stored in a relation for which the automobile serial number is the identifier, is an attribute of the identifier. The difficulties when relational principles are not used can be illustrated by an example from the Income Survey Development Program. Values of automobiles were stored on a record identified by address; it was not clear to what universe the value of the car applied.

An obvious concomitant of the relational organization of the data is that since observations are accessible in arrays—i.e., rectangular blocks—relations that are not required for a particular investigation need not be retrieved. When data are stored in an alternative manner, some additional computation is required to retrieve the block of relevant attributes. Either pointers or the reading of irrelevant material is required.

A more subtle concomitant of relational organization is the reduction in the number of addressable items of information that must be carried in a dataset. In principle, it is possible to eliminate not-in-universe codes. It is possible to avoid pointers that organize records into a hierarchy or network. And it is possible to eliminate repeated storage of the same attribute (other than identifiers). In practice, redundant variable storage, pointers, and not-in-universe codes are not eliminated. However, they are sharply reduced and the attendant data management is correspondingly reduced.

4.2. Documentation of Data Organization

Refer again to Figure 1. Observations are made accessible by name in an RDBMS. This feature characterizes other data base management systems as well. First we discuss the importance of this feature. Then we indicate the unique contribution of RDBMS to facilitating nomenclature.

The nemesis of most data managers and analysts are procedures that store and retrieve information by the position occupied. Invariably there are errors in calculating the position, and totally irrelevant information is retrieved and not always detected (Coale & Stephan, 1962). In data base management systems this problem is reduced by addressing observations by names.[3] Information types 1–3 in Figure 1 are names for the corresponding dataset, relations, and attributes. Names are distinguished from the underlying observations by the initial N; thus NXR_{A1} is the name for the relation XR_{A1}. (We refer to information types by enclosing the number in double brackets, e.g., [[2]] is the name of a relation.) In RDBMS these names are stored in a catalog that is extended whenever new relations are constructed (by analysts) or additional information is added to the system.

The proliferation of nomenclature in a panel data base creates an obstacle to data management and retrieval. The nomenclature that is associated with replication of measures at Ω waves is potentially as large as the product of the nomenclature required for one wave multiplied by Ω.

How can this aggregate of information be reduced? The obvious device is to construct naming conventions that permit recognition of the identical information across waves. This requires that the data manager establish a root and infix (suffix) that name each measure, or each relation, and the wave in which that measure applies, respectively. This device can be applied to [[1–3]]. For example, the roster of persons living at an address is a relation. Call it the "HH–Person" relation. Using this name [[2]] as a root, the name for the relation at any wave of the panel can be generated as "HH–Person–#", where # is the number of the wave.

Unique names [[3]] for the identical, replicated attributes in several waves can then be generated by a constant attribute root to which the relation name is appended. For example, the attribute of being the householder can be named by the root "Householder". The unique name of the householder attribute in a particular wave can be generated by the root and suffix

Householder.HH–person–#

In this form only the # varies over waves. Note that this naming convention depends on the relational character of the data management system. A core of identical variables must first be established by applying the criteria (1), (2), and (3) listed in Section 2. When the identical attributes are stored in identical relations, the naming convention conveys important semantic information about longitudinal structure.

Because RDBMS can be used to reorganize observations into new relations, the process of extracting the parallelism implied by the design of the data collection can be retrieved even after data have been stored in the system under other names. New nomenclature, along the lines suggested above, can be substituted for other less informative names.

Recognition of particular measures and particular relations is heightened by the constancy of the roots across waves. The user can then be assured that identical roots convey identical information.

4.3. Extracting Longitudinal Data Structure

Establishing identical attributes is the beginning of panel data management. The next step is to provide tools for the construction of longitudinal analyses, casting the data into the form of equation (2). It is vital to create an array that establishes the sample reconciliation mentioned above. It is equally vital to facilitate the retrieval of information that is widely distributed in the dataset—for example, the information that certain persons terminated employment, retired, and began receiving old-age benefits. Each of these three items of information is characterized by high autocorrelation, so that one would typically need to search data for many persons in many waves of a panel before one identified a change in any one of these three statuses. In managing data from the Survey of Income and Program Participation survey we have found it convenient to generate longitudinal relations that describe:

- Response, nonresponse, and presence in the sample universe by wave
- Address—change of address
- Recipiency of transfer incomes—change of recipiency
- Marital status—change of marital status

These relations take advantage of the autocorrelation of status over time and the infrequency of transfer income recipiency to compress repetitive observations distributed over several waves into a compact form that is accessible without addressing numerous waves of data. The following discussion indicates the principle, one that is easily implemented in the RDBMS framework.

Finding dynamic episodes—spells: Consider the vectors that pertain to a single attribute, $X_k(t)$, that is periodically measured for $t_0 \leq t \leq t_0 + \Omega\theta \equiv T$, where θ is the reference period and Ω is the number of waves. For simplicity of exposition, assume that $X_k(t)$ is binary, e.g., recipiency of old-age benefits. We can generalize to discrete categorical variables later. For any unit, the time series of measurements will be a row vector:

$$\|x_{ki}(t)\| = (0, 0, 1, 0, \ldots 1, 1)$$

$$1 \times (T - t_0)$$

This vector can clearly be compacted by recoding the repeated observation of a given status in some manner. We will achieve the compaction by defining a new type of array that we call an event history array:

$$EH(NX_k) \equiv \| i, t_{ej}, x_{ki}(t_{ej}), t + d_{ej} \| \tag{3}$$

where e_j $(j = 0, 1, 2, \ldots, j^* < \Omega)$ is the index that records each date of observation when x_{ki} is observed as different from the preceding observation. Note that $e_0 = t_0$ when no information prior to the date of the first observation is elicited. That is, the first observed status is recorded in the first *EH* row for each unit of observation. Second, note that in this instance the t_0 entered is not the date of onset of the status reported by $x_{ki}(t_0)$, it is the earliest known or observed status. The true date of onset has been censored. Third, note that the terminal date recorded on the last record for each unit is also censored. $x_{ki}(t_{ej*})$ records the last known status and is right-censored.

We provide a simple example. Assume that the relation *NXR* is measured repeatedly, so that we have *NXR*[1], *NXR*[2], *NXR*[3], and *NXR*[4]. Correspondingly, we have the replicated arrays of attributes, $NX_k[1]$, $NX_k[2]$, $NX_k[3]$, and $NX_k[4]$. For concreteness, suppose that the attributes measure recipiency of unemployment benefits with the values shown in Table 1. It is easy to see that the data provide examples of four kinds of individual histories. The first individual continuously received benefits; his status is unchanged. The fourth individual changes status at each date of measurement. The event history relation, $EH(NX_4)$, will consist of 10 rows, illustrated in Table 2. The "?" show values that are censored; that is, they bound the time extent of the status and will be altered when observations for $NX_4[0]$ and $NX_4[5]$ are made.[4]

What the *EH* representation of the data achieves is to convert repetitive occurrences of static observations into a compact representation of "change" or transactions. Sorting the array by t makes it possible to retrieve all cases with a change by looking at only one field. This obviously facilitates study of change. Moreover, if the probabilities of

TABLE 1. WAVE-BY-WAVE PATTERNS OF
RECEIPT OF UNEMPLOYMENT BENEFITS

i	$NX_4[1]$	$NX_4[2]$	$NX_4[3]$	$NX_4[4]$
1	yes	yes	yes	yes
2	yes	no	no	no
3	yes	no	yes	yes
4	yes	no	yes	no

TABLE 2. EVENT HISTORY PATTERNS OF
RECEIPT OF UNEMPLOYMENT BENEFITS

i	t	$NX_4[t]$	$t + d$
1	1?	yes	4?
2	1?	yes	2
2	2	no	4?
3	1?	yes	2
3	2	no	3
3	3	yes	4?
4	1?	yes	2
4	2	no	3
4	3	yes	4
4	4	no	4?

change are small, the EH representation of the measurements can conserve storage positions. (Note that the original array—$NX_4[1]$, $NX_4[2]$, $NX_4[3]$, $NX_4[4]$— can be reconstructed from the EH representation.) Examples of retrievals using this type of relation are given in David, Robbin, and Flory (1988a) and Flory, Robbin, and David (1988).

Nothing in the foregoing discussion depends on the fact that $x_{ki}(t_{ej})$ is binary. It need only be discrete. It should have sufficiently few categories to limit the number of rows in the EH array.

4.4 Legibility and Portability

The RDBMS that encompass the five major characteristics listed above derive from a conceptual framework based in set theory. (Codd, 1985, discusses why a data base management system may not fulfill the characteristics above, and why it may not perform in a truly relational fashion.) This makes it possible to define a language for manipulating observations that is easily understood by analysts and capable of expressing major changes in the form of the data in a few statements. The corollary is that data management can be more completely and satisfactorily documented.

A second advantage of these languages is that they have a great deal of common structure and will be portable between computers. Thus the query language of SQL can easily be translated into the query language of INGRES. The converse is not true; see Date (1987).

4.5. Housekeeping and Documentation of Programs

Figure 2 indicates what is required to provide adequate documentation for data management. Each element is marked to indicate whether the documentation can be incorporated into the RDBMS or whether it resides externally in the computer system. The RDBMS will automatically catalog any new relations that are created by data management. It can also be used as a tool for housekeeping, particularly for keeping an inventory of the media on which data are stored. David et al. (1974) argue that no large data collection activity can survive without a catalog of media and programs.

At least four types of documentation lie outside the RDBMS. These are the help, the audit trail, the program files, and the journal or logs of execution.

One of the most valuable tools that has been generated from computer science is the help library, a hierarchical file whose records contain annotations of key words, including directions to the modifiers of those key words that are stored at lower levels in the hierarchy. It is convenient to think of the file as a graph whose nodes contain descriptive information on the name located at that node. The help library can be browsed. Its nodes can be addressed directly to locate an item of critical information. Additions can be made frequently to extend the information provided.

Every major data collection effort should develop a help library. It is probably most useful when the order of the graph does not extend to more than four levels; otherwise the novice needs to know a great deal about the cognitive structure within which the original annotations were written. Such structure is better supplied by using a few key words to isolate a reference to a structured piece of documentation that can be written as a self-supporting paper and cataloged within the RDBMS.

Clearly, one of the key words in a help library can refer to key data construction tasks, such as building the event history relation in the previous section. But the help library is far more general. Each of the names [[1–3]] recorded in the data base can be described. Relations can be described to give the user a clearer picture of the semantic principle on which they are generated, including the periodicity with which new cases are generated (monthly values for attributes that change each month, etc) and the universe to which the relation applies (all married couples).

FIGURE 2. DOCUMENTATION FOR THE CONSTRUCTION OF DATASETS

Help <Data construction or DC>   SYSTEM
AAreadme.<DC>

These files provide an abbreviated annotation of work done that can be described by the nouns and adjectives included in "< >" The advantage of Help is that it is a menu-driven browsing capability. The advantage of "AAreadme" is that it will be the first file alphabetically listed in a subdirectory of related files.

<DC>.audit   RDBMS   SYSTEM

The audit trail must provide information on where to find the data input, the programs, the execution log, and the output of the named activity.

<DC>.log   SYSTEM

The log of the program execution provides important information as to the environment in which the programs were executed, including the date of execution (which uniquely defines multiple versions). The log preserves any execution diagnostics that may later be required for interpretation.

<DC>.programs   RDBMS   SYSTEM

Programs should also be internally commented to provide motivation for each step of the algorithm and to annotate parameters that may be adjusted in future executions.

Data/output   RDBMS

Ideally data are managed in a system that provides internal dictionaries that describe the data structure for users, and assure that observations do not become separated from the information necessary to interpret them.

Media libraries   RDBMS

Entries need to be made in searchable directories of tape, document, and program holdings so that the physical location of each of the above forms of documentation can always be recovered, even after the information has been archived to a storage medium that is not loaded in a computer.

Figure 2 enumerates the elements of a scheme for documenting the processes by which data bases are created. The six elements are seldom available to archival users of the data. Yet if they are not preserved, future data managers and users have no assurance that intended observations were treated in the manner that is described in codebooks and external documentation. It is particularly important to provide the audit trail of how information comes to its permanent form, so that intermediate products can be recovered, and errors do not imply going back to the original data entry procedures.

4.6. Error Correction

The relational organization of a data base facilitates four operations that are essential to defining and correcting errors. (1) The attributes whose relationship to each other appears in question must be retrieved (projection). (2) The cases or subpopulation of cases on which problems arise must be isolated (restriction). (3) The information must be readily retrieved (query). (4) Inspection then determines whether they are to be replaced (update). These four operations are the operations at which RDBMS excel. RDBMS are designed to locate individual cases by direct addressing, rather than sequential processing. Their language provides integrity to the correction of errors, namely the system avoids situations in which the information contained in the data base includes both corrected and uncorrected values.

4.7. Documenting Metadata

Figure 3 outlines material that is needed to assess the quality of measurements (Bailar, 1984). To understand the adequacy of coverage, the intrusion of systematic errors into the data collection process, and the strengths of the dataset, a collection of material written about the design and the facts of the dataset is required. Much of the material on design is recorded, sometimes in user guides (Economic Behavior Program, 1984); but much is never recorded and exists only in the minds of the research team gathering the data.

Information type [[11]] deals with the description of the design. Some of this lends itself to narrative; some does not. Integrity rules describe consistencies demanded of the data by the logic of the collection process; they are seldom reported, because it requires the reporting of a set of logical formulae (that are often only implicitly defined in undocumented computer code). The failure to enforce consistency is usually left for others to discover, independently of the data collectors who have extensive tallies of such problems. Again, the difficulty in data management is that the tallies do not lend themselves to narrative expositions; they require extensive storage and indexing. It is often argued that the cost of indexing such information exceeds the cost of replicating the tally at another location.

Information type [[13]] records what is known about the data, including anomalies such as that just described. It is also useful to record counts of critical populations, such as the set of units for whom the longitudinal response is complete, membership in population subgroups that are of particular interest, etc. The additional information that is required to understand the conditioning of data collected in later waves on data collected in prior waves is highlighted in [[14]].

While it is not necessary to an analysis of a dataset, many research questions derive from or are informed by other analyses. This is the motive for counting completed work as a separate information type [[15]].

FIGURE 3. METADATA, INFORMATION ABOUT DESIGN, AND ANALYSIS

Metadata, Information about the Data Collection	Information Type
Design of ND$_t$	[[11]]
Sample	
Questionnaire	
Field procedures	
Coding	
Editing	
Linkage	
Treatment of missing data	
Integrity rules	
Design of the panel	[[12]]
Following rules	
Verification of linkage	
Periodicity	
Facts of ND$_t$	[[13]]
Consistencies (related to integrity rules)	
Sample reconciliation	
Control tabulations (demographic, design, and linkage)	
Facts of the panel—ND$_t$ × ND$_t$,	[[14]]
Intertemporal consistencies	
Sample reconciliation	
Control tabulations	
Analyses	[[15]]

RDBMS lend themselves to cataloging each of these types of information. Because such catalogs can be searched by the computer using the same syntax that is used to retrieve observations, data managers obtain powerful control over their data management and are able to do so with little investment in the programming or special-purpose activities otherwise required to handle these complex types of documentation. An excellent example of documentation systematically applied is David, Robbin, and Flory (1988b).

5. SUMMARY

This paper began with the assumption that panel data are portable for analysis by a large scientific community. It has outlined a number of ways in which RDBMS can contribute to completeness in the documentation of data management and simplification of the data management task. Exactly how to evaluate data management systems is not clear to the profession. The use of computer resources is probably the least important feature. I would measure the effectiveness of data management by other principles. In closing I ask readers how well their data management system handles the following major issues that lie at the heart of satisfying Boruch and Pearson's guidelines:

- The system catalogs metadata. Information of types [[11–15]] can be archived and retrieved through the system.
- The system is designed to encourage untutored outsiders to teach themselves.
- The system maintains an audit trail or journal of the transformation of observations from the time of collection to the routine rollover of information in an archival form of storage.
- The system has the flexibility to accommodate modifications in the panel design and follow-up data collection, without requiring major changes in the archival data that have been stored.
- The system will respond to analysts' desire for new units of analysis and new procedures to handle truncation.

RDBMS make a major step to each of these goals with a relatively simple language and an implementation that is highly portable between computers.

REFERENCES

Bailar, B. A. (1984), "The Quality of Survey Data," *Proceedings of the Survey Research Section*, American Statistical Association, 45–53.

Boruch, R. F., & Pearson, R. W. (1985), *The Comparative Evaluation of Longitudinal Surveys*, Social Science Research Council.

Boruch, R. F., & Pearson, R. W. (1988), "Assessing the Quality of Longitudinal Surveys," *Evaluation Review*, 12, 3–58.

Coale, A. J., & Stephan, F. F. (1962), "The Case of the Indians and the Teen-Age Widows," *Journal of the American Statistical Association*, 57, 338–47.

Codd, E. F. (1985a, October), "Is Your Database Really Relational?," *Computerworld*, ID1-ID9.

Codd, E. F. (1985b, October), "Does Your DBMS Run by the Rules?," *Computerworld*, 45–49ff.

Date, C. J. (1987), *A Guide to INGRES*, Addison-Wesley, Reading, MA.

David, M. H. (1985a), "Designing a Data Center for SIPP: An Observatory for the Social Sciences," *Proceedings of the Section on Survey Research Methods*, American Statistical Association.

David, M. H. (1985b), *The Language of Panel Data and Lacunae in Communication about Panel Data* (Center for Demography and Ecology Working Paper 85–20).

David, M. H. (1986a), *Data and Metadata: Exploiting Dualities in Observations and Nomenclature* (Unpublished manuscript), available from the author on request.

David, M. H. (1986b), *Didactics, Dynamics, Semantics, and System: Ingredients for a Comparative Panel Data Management System* (Unpublished manuscript), available from the author.

David, M. H., Gates, W. A., & Miller, R. F. (1974), *Linkage and Retrieval of Microeconomic Data*, Lexington Books.

David, M. H., & Robbin, A. (1981), "The Great Rift: Gaps between Administrative Records and Knowledge Created through Secondary Analysis," *Review of Public Data Use*, 9, 153–166.

David, M. H., Robbin, A., & Flory, T. (1988a), "Access to Data: Handling the 1984 SIPP," *Proceedings of the Social Statistics Section*, American Statistical Association, 205–210.

David, M. H., Robbin, A., & Flory, T. (1988b), *SIPP ACCESS Handbook*. (Prepared for workshop on SIPP, June 2-5, 1988, University of Wisconsin-Madison.) Available from the authors.

Economic Behavior Program, Survey Research Center (1984), *User Guide to the Panel Survey of Income Dynamics*, Institute for Social Research, The University of Michigan, Ann Arbor.

Flory, T., Robbin, A., & David, M. H. (1988), *Creating Longitudinal SIPP Files Using a Relational Data Base Management System*, University of Wisconsin, Institute for Research on Poverty, Madison. Available from the authors.

Kent, W. A. (1983), "A Simple Guide to Five Normal Forms in Relational Data Base Theory," *Communications of the Association of Computing Machinery (ACM)*, **26**(2),110–125.

Robbin, A., & David, M. H. (1988), "SIPP ACCESS: Information Tools to Improve Access to National Longitudinal Panel Surveys," *Review Quarterly*, **27**(4), 499–515.

Tsichritzis, D. C, & Lochovsky, F. H. (1982), *Data Models*, Prentice-Hall, Englewood Cliffs, NJ.

NOTES

*Financial support from the National Science Foundation SES #8411785 is gratefully acknowledged.

[1]It may be argued that these conditions are not sufficient; conditioning of the respondent may affect the answer. Alternatively, some longitudinal data can be constructed when these conditions do not apply, as when the coding scheme maps several categories at one point in time into a single category at another.

[2]Ideally, in an RDBMS the same attribute is stored in only one array. Updates to an attribute therefore affect only one array and complex procedures for eliminating inconsistent values of a particular attribute are avoided. In the less than ideal realization of RDBMS, attributes may be redundantly stored in more than one array. Then all instances of the attribute will need to be simultaneously updated. This problem is at least an order of magnitude smaller than in other data base management systems, where it is necessary to capture relationships through pointers that are widely distributed in the dataset. Alteration of an attribute that affects a pointer value entails updating of a large number of items in the dataset. Lest this be thought an arcane consideration, pointers will be modified by the storage of an additional wave of information (unless that information was fully anticipated in the design of the data base at the time that the first wave was stored).

[3]When the names are positions in a list, mnemonic devices offer little cognitive assistance to the analyst (David et al., 1974). When the nomenclature scheme offers linguistic content, error is reduced.

[4]The $EH(NX_4)$ array contains 40 positions. The original observations were stored in 20. This makes it clear that the advantage of the event history approach is not necessarily in compaction. The EH array contains a redundancy, namely, the $(t + d)$ array contains information that appears on the subsequent row in the t column. The "?" shown reflect the fact that some information is indeterminate. The date shown represents an episode in progress at the date of measurement.

Data Base Management: Discussion*

Halliman H. Winsborough

My background as a demographer and sociologist who has worked primarily on very large cross-sectional data sets gives me a very different perspective on the issues raised in these papers. It is clear that all the authors think that panel surveys are so unusual and complex that they need something special in the way of a computerized data management scheme. Two of the papers sound as though they are working their way toward a proposal to support a data base system design; the third is in the middle of using a fairly complex relational system. I sense an argument going on, but I am puzzled by some of the assumptions that seem to underlie the argument.

My quandary centers around three issues:

1. What is special about panel studies that data management issues arise and are discussed exclusively in this context? Aren't these issues that arise for all public data?
2. Why would any group of social scientists even think about writing their own data management programs? There is so much to know about data bases, and much skill and facility is available in packaged systems.
3. Where are the lines of responsibility drawn when it concerns data? What should we expect from the collectors and what from the analysts?

1. DO PANEL DATA REQUIRE SPECIAL MANAGEMENT?

A first assumption that winds through all of these papers is that panel studies present special data management problems. Two reasons are given. First, they are large and complex. Second, they involve a large number of entities and there is a problem of tracing some of these entities through time.

Although the first of these reasons, size and complexity, should make one think seriously about data management issues, it is not unique to panel studies. There are many large and complex data sets in regular use. A year's worth of Current Population Survey (CPS) files is many times larger than the Panel Study of Income Dynamics (PSID). Even ignoring CPS's panel aspects, a year's worth of data is very complex, especially if you keep careful record of the design elements. Some of

242

the "life history" type surveys are complex. Several surveys about to go in the field are, I think, similar in complexity, if not size, to the Survey of Income and Program Participation (SIPP). One example is the National Study of Families and Households.

This is not to say that size and complexity are bad reasons for panel analysts to worry about data management, but only to note that data management issues may be of increasing importance in the social sciences at large. Perhaps we should address the issue in general rather than just in the panel context.

The second reason given for the special data problems of panel studies is more interesting. It has to do with the number and kinds of entities that exist in the data and the problems associated with tracing these entities through time. I think that the number of entities is, like size and complexity, a good reason to worry about data management but it is not unique to panel studies. I will discuss this issue first and then the issue of tracing entities through time.

Some kinds of entities loom large in the design of nearly any survey: PSU's, housing units, households, persons. Everyone worries about them to some extent, with PSU's and housing units probably getting less attention in data base issues than is warranted by their importance in the design. Using a relational system, I like to set up a separate relation, i.e., a separate rectangular data file, about each of these entities. In each file the "variables" are information about that kind of entity. Thus, number of rooms goes in the housing unit file, number of cars in the household file, and years of school completed in the person file. Household number is also a variable in the person file, so you can aggregate over members of the same household and "join" information from the household file to that from the person file at will. A modern data base will complete the "join" operation slightly more slowly than it takes to read the record. You can join aggregated person information to a household file or you can join household characteristics to the appropriate records in a person file. Since a relational system is not hierarchical, it is possible for persons to belong to more than one household as well as for many people to belong to one household.

Other kinds of entities are obtained in the data collection schedule. For example, race and ethnic membership questions may be asked to identify membership in these entities. Members of some groups may be asked special questions not asked of others. Indeed, nearly every filter question involves the designation of an entity that may need to be recognized in the data management scheme. In Public Use Microdata Sample (PUMS) files, I divide the person file into two parts, one pertaining to the questions asked of all people and another for those questions asked only of persons over age 15. This saves space and helps keep track of the appropriate universe for variables. Of course, it is easy to join some or all of the questions asked of everyone to some or all of the questions asked of adults.

Other entities are formed not by common response to a question but by relationships to other entities. Household relationship questions identify families, subfamilies, secondary families, cohabiting couples and the like. Since these questions essentially ask for "pointers" to other records, they demand special consideration in designing a management scheme for any survey in which they appear. The traditional way to deal with these problems is through the rather restrictive convention of a household reference person. Household members' relation to that reference person is determined and from the various responses the network of relationships among household members is unraveled and new entities such as subfamilies and secondary families are created. The query language for most relational systems makes such work relatively easy. In addition, relational systems permit more flexible ways of indicating relationships between records. A field for the record number of the spouse, or mother, or father of the person is possible for one-to-one or many-to-one relations. For many-to-many relations—for example, siblings—there are a number of ways to proceed and the choice among them depends on the use one is likely to make of the sibling relationship.

Other entities are found through rosters in the schedule: autos, job transitions, residence changes, pregnancies, transfer payments, marriages. In relational systems a new file is created for each new entity where variables are strictly about that specific entity.

The foregoing kinds of entities are explicit in the data collection schedule. If you don't ask the questions, you don't get the entities. Lots of other entities are implicit rather than explicit in the design and the schedule. They can be made explicit by aggregation or disaggregation, e.g., birth cohorts or person-months. Birth cohort is an obvious aggregation on the person attribute, birth year. There are many aggregations like birth cohort; they are usually easy to accomplish even if the data are in pretty rough form.

Married couple is also an aggregation on person records, using the relationship questions to form the new entity. It can be a bit more difficult to produce. If your data base is not relational, you would be better off knowing you want this entity from the start; otherwise, you are likely to write a good deal of code to process the files.

Person-month, at its simplest, involves creating multiple records for each person in a wave, with the multiple depending on the months in the wave, then stacking the records over persons and then over waves. Attributes for such entities would be information about that person in that month. If you were not aware of the entity before structuring the data, you would need a good data base system to easily retrieve information about person-months.

These entities can exist in cross-sectional as well as panel studies, thus arguing for taking the data management issue seriously in general.

The uniqueness of entities to panel studies is persistence. Can you match entities over waves? Persons and housing units seem conceptually straightforward, although, as both David and Doyle point out, they may be operationally troublesome. Households and families are a mess to trace over time, because the survey household is a convention invented for cross-sectional surveys. It is a convention that holds the relationships and entities that exist among people sharing living quarters without having to specify each kind of relationship or entity individually.

The family concept used in survey work is little different in character. The idea was invented for the 1940 Census to specify some of the relationships and entities collected together in the household. Previously we assumed that members of a household shared expenses; now we assume that members of a family do. The concept should be called the cohabiting family to make clear that there is a larger and temporally more persistent entity that is different from this conventional idea. Since these conventional ideas of household and family are really convenient cross-sectional aggregations of entities and relations, it does not make much sense to try to follow them over time. The specific entity or relationship is usefully traceable over time rather than the construct it was in. Thus, one might wish to trace couples, or mothers and their children, or people who share income. Tracing each of those specific entities requires some thought but has the virtue of specificity.

From the foregoing I conclude that the puzzle about the persistence of entities in panel analysis is not something deep about the panel design but rather something shallow about the household and cohabiting family constructs.

2. SHOULD WE DEVELOP OUR OWN DATA BASE SYSTEM?

Several years ago I visited the Institute for Social Research and met the late Monica Blumenthal, a psychiatrist turned survey researcher doing wonderful studies of attitudes about violence. She remarked that social scientists are quite odd in their distrust of people with other speciality areas. If she had a patient with a heart problem, she said, she and most other physicians would call in a cardiologist. A social scientist would try to read up on the matter and do surgery himself. I thought of that remark in relationship to these papers, which seem like proposals for the design of a data base system for panel surveys.

There is much to know about designing data base systems. It is an area of concentration for the Ph.D. in Computer Science at the University of Wisconsin–Madison. Much research has accumulated; some of it is quite theoretical but much of it applied. The result of this work is many systems for machines ranging in size from PC's to Sierras. Nearly every major machine vender has several. The older CODASYL, or network, systems are being phased out. IBM, standby for years, for example, is officially on the way out. In these older systems you must pre-specify the joins you need. If you think of something new you want to do, it is trouble. Implementing these systems requires much effort and changing them requires much work. The directly structured file system for the PSID appears to be of this type.

Various implementations of the relational model appear to have sparked interest recently. Among other virtues, they support ad hoc queries. You do not need to know all the questions you are going to ask of the data, all the "joins" you will want to do, before you begin. IBM has two forms, SQL/DS and DB II. DEC has added its own relational system, RDB, to its longstanding CODASYL system. There are several systems not associated with manufacturers: Ingres, Oracle, ADABAS. Of course, systems for microcomputers abound; dBase III is well spoken of by the professionals I know.

Many of the available and affordable software systems are very powerful. Good systems have query optimizers that analyze what you want done in terms of what is known about your specific data base and which resources are scarce on your machine. The query optimizer determines the best way of satisfying your query given the constraints and environment. They are not perfect. Indeed, research is currently underway to try to make them better. But, as they stand, the query optimizers do a better job of deciding than many data base managers and they do it very rapidly.

Two specific reasons are given for not wanting to use one of the currently available systems. One seems to be a feeling that relational systems are only for businesses that process many transactions. Most implementations of the relational model try to make transaction processing go quickly. Other implementations optimize other applications. The Omega system designed for the Crystal Project optimizes the speed of the join and would be fine for social science work. The point is that it is not the relational model that is "for" transaction processing but rather the implementation. In a number of systems the analyst may tailor the implementation to make it work more efficiently for specific problems. You can choose the storage of the data and you can choose to create indexes. You can choose how much effort you wish to spend collecting marginals on the data for use by the optimizer. You can also choose to use or not use features. For example, the journaling David mentions in Ingres is quite resource-intensive and I avoid its use whenever possible. There is also great security control in many relational systems. For example, you can permit a given user access to only specific relations at specific times of the day from specific terminals.

Certainly the current packaged data base systems are not perfect.

They have developed over a number of years of intensive work by excellent computer scientists. Even when they are less than optimum, they are good compared to developing your own data base system.

A second reason mentioned in the papers for wanting to do it yourself is the issue of size and cost. I do not understand this point. Consider size first: 16 waves of PSID occupy about a half gigabyte of space on tape. I have just ordered an additional half gigabyte of disk storage for one of our Vaxes. It costs about $10,000. The PSID would probably take about half as much space if stored in a relational system, partly because of reduction in "padding" and partly through using integer format. Suppose it required a whole gigabyte to work effectively. This would cost about $20,000 for the drives, another $2,000 for cabinets, and say an additional $1,000 for installation. Add about $10,000 for the initial license for a data base system at the university price. The total cost thus far is $33,000, or a middle-level programer for about half a person-year. You do not get very far on a complex problem in half a person-year. Even if you spent another $30,000 on a Microvax or a Sun to get more cycles, I suspect you would still be better off.

Once machines were expensive and people comparatively cheap. Machines are now cheap compared to people. The cost of disk storage especially seems likely to continue its dramatic decline. An investment in efficiency must pay off pretty quickly.

3. WHOSE RESPONSIBILITY IS IT TO DO WHAT TO THE DATA?

This is the most implicit issue in these papers. The authors make little distinction between the responsibility of the analyst and that of the collector-disseminator. All the authors are really analysts at heart. As such, they are certain they know what information in the data is important and they are certain they know how to get it. This is natural. Often analysts have developed the project and participated in selling it to others. They know that their idea of what is important is widely held. They know that the design and content are focused on those issues. Thus it is not surprising that analysts want to structure the data in ways to make answering *their* questions most efficient. For their own work, that is exactly what they should do.

People in the data-providing role, I think, would be wise to remain a bit more agnostic about what will be the important information in a given survey. They know from experience that when the first round of questions are answered, there will be a second round. Questions in round 2 depend on what was found out in round 1, and the most important round 2 question is likely to come from the round 1 answers that were most surprising, i.e., the least predictable.

How should an agnosticism about what is important influence the provider's solutions to data management issues? The first thing it says is that the provider should provide *all* the information available in the survey. This often means information about the place of an entity in the design and also about the things that go wrong. Perhaps one reason there is less research on some methodological issues than would be desirable is that the necessary information isn't really on the publicly released file but in the provider's filing cabinets. The second thing an agnosticism about importance implies is that the data provider needs to be able to restructure the data set as new questions arise and as new entities become important. The relational model is the only one that provides that facility without a major retooling.

4. SUMMARY

First, data management problems are acute for all parts of social research, not just for analysis of panel data; the software and hardware environment is changing to provide helpful answers to our problems.

Second, custom programming or custom design of data management systems is likely not to be cost-effective except for the very largest agencies.

Third, analysts should do whatever it takes to the data to get good answers to the questions burning in their hearts and minds. But providers must conserve all the information in a survey and keep that information in a way that allows it to be easily reformed to answer a multitude of questions.

NOTE

*The preparation of this paper was carried out (in part) using facilities of the Center for Demography and Ecology, the University of Wisconsin–Madison, which receives core support for population research from the National Institute for Child Health and Human Development (Grant No. 05876).

Nonsampling Errors in Panel Surveys

Graham Kalton, Daniel Kasprzyk, and David B. McMillen

1. INTRODUCTION

As in any survey, nonsampling errors in panel surveys can arise from many sources. These sources may be divided into two broad classes: nonobservation errors and measurement errors. Nonobservation errors may be further subdivided into nonresponse errors, which occur when some of the required information is not collected for some sampled units, and noncoverage errors, which occur when some units in the target population have no chance of being selected for the sample. Measurement errors occur when incorrect data for some sampled units are used in the survey analysis. Measurement errors may be further subdivided into response and processing errors. The former occur when incorrect data are recorded on the survey questionnaires and the latter when errors are introduced in converting the survey responses into machine-readable form for the analysis. Response errors may arise because of a faulty questionnaire, the collection of data from inappropriate informants, memory errors, deliberate distortion of responses (e.g., prestige bias), interviewer effects, and misrecording of responses. Processing errors can arise from incorrect editing, coding, and data entry.

This paper cannot review all these sources of nonsampling error as they pertain to panel surveys. Instead, it concentrates on two sources of nonsampling error that are peculiar to panel surveys; it examines how certain aspects of a panel survey design can affect nonsampling errors; and it discusses the problems nonsampling errors cause for the measurement of gross change from a panel survey.

The two special nonsampling error concerns with panel surveys are nonresponse and conditioning. Nonresponse is a more serious problem with a panel survey than with a cross-sectional survey because nonresponse rates usually increase as a panel ages; it is also a more complex problem as more information about nonrespondents is available for use in nonresponse compensation procedures. Section 2 discusses the problem of panel nonresponse, and gives examples of the response rates achieved in several panel surveys. Panel conditioning refers to the situation when repeated questioning of panel members affects their survey responses, either by changing the behavior being reported or by changing the quality of the responses given. This topic, and the associated topic of rotation group bias, are discussed in Section 3.

A major consideration in the choice of methods to use in a particular survey is the control of nonsampling errors. In many cases an important reason for the choice of a panel design is to avoid or at least reduce one type of nonsampling error—that is, response errors caused by memory decay. Rather than rely on a cross-sectional survey with respondents' reports of events that may have occurred a long time ago, a panel survey may collect data on events shortly after their occurrence. It thus tries to minimize the serious problems of recall loss and telescoping that occur when respondents are asked to make longer-term recollections. Once a decision has been made to employ a panel design, there still remain a number of further design issues to be settled, and these can also have considerable effects on certain types of nonsampling errors. Section 4 discusses the effects on nonsampling errors of four of these issues: the length of time between waves of data collection; from whom to collect the survey data on each wave; which mode of data collection to employ for each wave; and which sample members are to be followed in a panel.

The final section discusses the problems that nonsampling errors cause for the measurement of gross change in a panel survey. A major potential analytic benefit of a panel design is the ability to measure gross change, that is change at the individual level. In practice, this benefit is hard to realize because nonsampling errors can seriously distort measures of gross change.

2. NONRESPONSE IN PANEL SURVEYS

A major concern with panel surveys is the increasing loss of sample cases that occurs as the panel ages. Like cross-sectional surveys, panel surveys are subject to nonresponse at the initial wave as a result of refusals, not-at-homes, inability to participate, untraced sampled units, and other reasons. In addition, they are subject to nonresponse at subsequent waves for the same reasons. It should be noted that the calculation of nonresponse rates for a panel survey needs to be performed with care; the nonresponse rate for a particular wave is generally not the simple proportion of the original sample that failed to respond at that wave, but rather allowances need to be made for sample members leaving the survey population (e.g., through death, emigration, or entering an institution) and, where applicable, for additions to the sample to reflect new entrants to the population (e.g., births, immigrants, and those leaving institutions). In particular, the practice of treating sampled units that leave the survey population as nonrespondents leads to an overstatement of the nonresponse rate.

As a rule the overall nonresponse rates in panel surveys increase with successive waves of data collection, thus increasing the risk of bias in the survey estimates. The following examples illustrate the non-response experiences of some panel surveys:

- The household survey component of the U.S. National Medical Care Utilization and Expenditure Survey (NMCUES) consisted of five interviews at approximately 3-month intervals during 1980 and 1981. The first two contacts were personal visits to sampled households, the next two were mainly made by telephone, and the last one was a personal visit. The nonresponse rate at the initial wave was 9.9%. By the fifth wave, the rate had increased to 22.1% (Wright, 1983).

- The U.S. Survey of Income and Program Participation (SIPP) follows members of initially sampled households for approximately two-and-a-half years. Interviews are conducted every four months to collect data on income and program participation in the preceding four months. The household sample loss rate (defined as the cumulative noninterview rate adjusted for the growth in the number of panel households over time) was 4.9% at the initial wave of the 1984 Panel. This rate had increased to 22.3% by the ninth wave (Kasprzyk & McMillen, 1987).

- The U.S. Panel Study of Income Dynamics (PSID) has followed sample members of 5,000 initially sampled families annually since its inception in 1968. The nonresponse in the first year was about 24%. In the second year, 86% of those interviewed in the first year were reinterviewed, and since then about 97% of those interviewed in one year have been interviewed again in the next year. After seventeen years, the cumulative nonresponse rate is a little over 50% (Survey Research Center, 1986).

- The National Child Development Study (1958 British cohort) started with about 16,000 births in a week in March 1958. Follow-ups of the NCDS cohort have taken place at ages 7, 11, 16, 20, and 23. After excluding deaths and emigrants, data have been collected on 91.2% of the sample at age 7, 90% at age 11, 87.3% at age 16 (Fogelman & Wedge, 1981), 85% at age 20, and 76% at age 23 (Shepherd, 1985).

- The Monitoring the Future surveys start with a nationwide sample of high school seniors each year. A subsample of 2,400 respondents is then selected each year for a ten-year panel. Members of the panel are surveyed biannually by mail questionnaire, with one half of them being surveyed in even-numbered calendar years and the other half in odd-numbered years. In the first follow-up after high school, on average over the years about 85% of the original panel members have returned questionnaires. The 1985 panel retention rate for the class of 1976 was 71% (Johnston et al., 1986).

The general picture that emerges from an examination of the non-response rates in these and other panel surveys is that the overall rates increase from one wave to the next, but the rate of increase declines over time. With well-conducted panel surveys, the increases in non-response rates are not as large as might be feared, but nevertheless the cumulative effect gives rise to substantial overall nonresponse rates at later waves. The main causes of panel nonresponse are refusals and failures to trace mobile sample members.

It seems likely that the rates of refusal increase with respondent burden, and hence that more refusals are likely when the intervals between waves are short and the demands placed on respondents are great. Minimizing respondent burden, the use of persuasion letters and feedback reports, assurances of anonymity, and the use of payments or other incentives may be effective in reducing losses from refusals.

The problems of tracing increase as the time interval between panel waves increases, both because the proportion of movers increases and because tracing becomes more difficult. On the other hand, a sizeable interval between waves provides the time needed to make extensive tracing efforts; movers are often not thoroughly traced in panel surveys in which interviews are conducted at short intervals, say every three or four months, because there is not sufficient time to fully investigate all the possible tracing sources. The use of telephone directories, mail forwarding, motor-vehicle registrations, employers, parents, neighbors, etc., can be extremely effective for tracing mobile population members over long periods of time, but a thorough tracing operation employing a combination of such sources can be extremely time-consuming. It is also very expensive. See the paper by Burgess in this volume for further discussion of tracing.

An issue that arises with nonresponse in panel surveys is whether attempts should be made to interview at a given wave persons who were nonrespondents at the previous wave. In some panel surveys nonrespondents at one wave are excluded from subsequent waves, whereas in other surveys attempts are made to collect data in subsequent waves for at least some types of wave nonrespondents. The case for attempting to recapture wave nonrespondents in later waves is stronger for panel surveys with short intervals between waves, since time pressures may result in a sizeable number of noncontacts and failures to trace in such surveys. The SIPP, for instance, does attempt to recapture wave nonrespondents; of those interviewed at the fifth wave, about 5% had missing interviews at one or more earlier waves, and consequently would have been nonrespondents at the fifth wave had wave nonrespondents not been recontacted (Kalton et al., 1986). A factor to be considered concerning the recontact of wave nonrespondents is how their incomplete data will be used in the survey analyses. For cross-sectional analyses of the data for individual waves, all the respondents for the given wave can be included, irrespective of whether they are nonrespondents on other waves, but often longitudinal analyses across waves are restricted to the subset of sampled units that responded on all waves. Thus, for this type of longitudinal analysis, it is not necessary to recapture sampled units after they have failed to respond at one wave. See Kalton (1986) and Lepkowski's paper in this volume on nonresponse compensation methods for missing interviews.

The main concern about panel nonresponse is that the nonrespondents may differ in systematic ways from the respondents. For nonresponse occurring after the first wave of a panel survey much data are available from which to check on the comparability of wave respondents and nonrespondents; the nonrespondents and respondents for a particular wave can be compared in terms of their responses to the first (and possibly other) waves. Comparative studies of this type have been conducted for a number of panel surveys (e.g., Kalton et al., 1986, for SIPP; Goldstein, 1976, for the NCDS; Sobol, 1959, for a panel survey of economic attitudes and plans for car purchase). These studies generally show close similarity between wave respondents and nonrespondents in terms of their response distributions for many first-wave variables, but there are also often a number of variables for which the distributions show appreciable differences.

Duncan et al. (1984) report the results of an interesting study in which they examined the effect of reducing the level of effort made in the PSID to retain sample members. With the intensive procedures employed in the PSID, few of the variables measured in the initial 1968 wave accounted for variation in the nonresponse in 1980. However, when a simulated less expensive PSID sample was created, the situation changed. The simulated sample was constructed by dropping families that would have been lost if interviewing had been terminated at a particular date each year or if interview attempts had been restricted to a maximum of four personal or seven telephone calls for waves from 1973 to 1980; the resultant sample included only about two-thirds of the actual 1980 PSID sample of individuals. With the simulated sample, the nonresponse after the first wave was associated to an appreciable extent with several first-wave characteristics, particularly race, income, and age.

Although these comparative studies of later-wave respondents and nonrespondents in terms of their first-wave responses have often found that the nonrespondents do differ from respondents in some respects, this does not necessarily imply that the panel estimates will be biased. When nonresponse biases in certain variables are identified, nonresponse adjustments can be made to counteract them. Duncan et al. (1984), for instance, found that when they reweighted their simulated sample for differential response rates across subgroups of the sample, there were minimal differences in outcome measures between the reweighted simulation sample and the actual sample. This finding is, of course, not conclusive and should not be taken to imply that relaxed efforts in tracing or nonresponse follow-up are acceptable. However, it does bring out the fact that the effects of panel nonresponse are properly assessed only after nonresponse adjustments have been made. Although the PSID response rate is near 50%, a detailed investigation of the PSID sample adjusted for nonresponse found little evidence of bias across a wide range of estimates (Becketti et al., 1983).

3. PANEL CONDITIONING

A second source of concern with regard to nonsampling error in panel surveys is panel conditioning. Conditioning refers to a change in response that occurs because the respondent has had one or more prior interviews. The effect may come about because the prior interviews change respondents' behaviors or because they change the way respondents answer questions. The prior interviews may, for example, provide respondents with information or cause them to reflect on an issue, thus changing their subsequent behavior or attitude. Alternatively, they may simply cause a change in the quality of respondents' reports. On the one hand, respondents may become better reporters because the prior interviews have taught them what is expected of them or have sensitized them to the events to be reported so that their recall is improved; on the other hand, they may become poorer reporters because they become unwilling to repeatedly make the effort required to report accurately.

Panel conditioning is thus the reactive effect of prior interviews on current responses. In practice, it is difficult to separate the effects of conditioning from those of other changes between waves, especially attrition. In experimental studies, Mooney (1962) found that older persons reported higher levels of illness on the first time in the panel than on subsequent times, and Neter and Waksberg (1964, 1965) found that resident owners reported fewer jobs and less expenditure on housing repairs and alterations on the second than on the third time in the panel. In a survey of British social attitudes, Lievesley and Waterton (1985) found that panel members gave fewer "don't know" and fewer "socially desirable" answers on the second round of the survey (one year after the first round) than did a fresh cross-sectional sample. Their analysis allowed the separation of such conditioning effects from the effect of panel attrition.

Cohen and Burt (1985) report on what they call a "data collection frequency" effect. In the National Medical Care Expenditure Survey, a departure from the original survey design allowed for an investigation of this effect on the reporting of health-care-related events. They found that individuals with four rounds of data collection were associated with a higher health care utilization and expenditure profile than individuals with five rounds of data collection during the same interval. When household-reported data were compared to record-check data they also found that individuals with four rounds of data collection were characterized by a significantly higher level of agreement between the data sources.

In addition, there are results from two validation studies that compared panel respondents' reports with record data. In one, Traugott and Katosh (1979) found that longer-term members of a panel survey of election behavior gave more accurate responses on voting behavior than the newer members, and, moreover, had higher levels of voting. These effects may be due to panel conditioning or, perhaps, to attrition. In the other validation study, Ferber (1964) found that the quality of reporting on savings holdings improved with length of time in the panel; this improvement was partly due to panel attrition, with the poorer reporters dropping out of the panel, and partly to an increase in accuracy of reporting for those that remained.

In theory, panel conditioning may be examined by means of a rotating panel design in which fresh replicate samples are added to the panel at each wave. Provided that all other survey conditions are the same for all the rotation groups at a particular wave, the comparison of the results for the various rotation groups for that wave provides a means of examining conditioning effects. In practice, however, it is virtually impossible to keep all other survey conditions constant for the different rotation groups. Sometimes the questionnaire is somewhat different for the various rotation groups (e.g., the incoming rotation group may be given a longer questionnaire to collect basic sociodemographic data that need not be collected again at reinterviews), the mode of data collection differs (e.g., all members of the incoming rotation group may be interviewed face-to-face, whereas many of the members of other rotation groups may be interviewed by telephone), or the type of informant differs (there may, for example, be more female household informants for the incoming rotation group than for the others). Even if factors such as these are controlled, there remains the problem that the rotation group samples are not exactly comparable because of the effects of panel nonresponse. It is consequently difficult to disentangle the effects of conditioning from the effects of other changes.

A number of studies have been conducted to compare the results obtained by different rotation groups, the variation between the results being termed the rotation group bias. The existence of rotation group bias has been established for various labor force characteristics in the Current Population Survey and several possible sources of the bias have been suggested (Bailar, 1975, 1979; McCarthy, 1980; U.S. Bureau of the Census, 1978). For instance, the estimate of the percentage unemployed from the households in the incoming rotation group is 10% larger than the average for all eight rotation groups. Rotation group bias has also been found in the Canadian Labour Force Survey (Ghangurde, 1982). An analysis of rotation group bias in the National Crime Survey found that victimization rates declined with length of time in panel, with the largest drop occurring between the second and third times in the panel (Woltman & Bushery, 1975). Lehnen and Reiss (1978) also found that respondent reports of incidents decrease monotonically with the number of prior interviews. This range of studies indicates that rotation group bias is a pervasive effect in rotating panel surveys, and the cause of the effect although not known may include nonresponse bias from attrition in a panel survey as well as a response bias from repeated interviews.

Unfortunately, documentation of the existence of rotation group bias is not sufficient; an issue that remains in a rotating panel design is the determination of which, if any, rotation group provides unbiased estimates. Unless the biases associated with each rotation group cancel, overall estimates of levels will also be biased. Moreover, the conventional assumption that rotation group bias does not affect estimates of change—that estimates of change are unbiased—relies on the assumption that rotation group biases are constant from one time to the next and that the biases are additive (Fienberg, 1980). The reasonableness of these assumptions needs to be established when discussing change estimates from panel surveys.

4. EFFECTS OF FOUR FEATURES OF PANEL DESIGN ON NONSAMPLING ERRORS

The extent of nonsampling errors in a panel survey depends on the design of the survey. This section discusses four design features that have considerable importance for nonsampling error. Three bear largely on the extent of response error—the interval between waves, respondent rules, and mode of data collection. The fourth feature, rules for following sample persons (following rules), is concerned with noncoverage errors.

4.1. Interval between Waves

A panel survey is generally employed to collect data on events occurring over a period of time. At any one wave the data collected usually relate to the events occurring since the previous wave. The longer the interval between waves, the lower is the cost of the survey. On the other hand, the longer the interval between waves, the longer is the reference period for which the respondent must recall the events.

Three factors are commonly considered in the choice of length of reference period: recall loss, telescoping effects, and respondent burden. Studies such as those by Cannell et al. (1965) on hospitalizations and

Turner (1972) on victimizations have shown that events that happened longer ago are less likely to be reported in survey interviews. Events that are less salient to the respondents appear to be the ones that are more likely to be missed. Therefore, the longer the reference period, the more likely it is that some events will not be reported. Bushery (1981) in conducting reference period research for the National Crime Survey found that reported victimization rates using a 3-month reference period were higher than those using a 6-month reference period; the 6-month reference period, in turn, obtained higher reported victimization rates than a 12-month reference period. This, again, reinforces the point of view that reporting tends to increase as the reference period is shortened. There appears to be less recall loss or memory decay with shorter reference periods.

In the development of the SIPP, a short (subannual) reference period for reporting sources and amounts of income received, especially monthly federal program participation, was viewed as desirable. The reference period issue was addressed in a pilot test preceding the SIPP. In the pilot test an experiment that varied the reference period was conducted; the experiment compared a single interview using a 6-month recall period with two consecutive interviews, both using 3-month reference periods. A summary of the analysis conducted in the pilot test is provided by Olson (1980). The proportion of respondents reporting some positive amount of income in the initial 3-month reference period was higher for the 3-month reference period group than for the 6-month reference period group; that is, the 6-month recall period appeared to understate the proportion of income reported in the earlier part of the period. This pattern seemed to hold for a number of specific income sources such as wages, Aid to Families with Dependent Children, general assistance, and unemployment compensation. The findings, though not definitive, support the presumption that longer recall periods increase chances of omission due to memory loss.

Telescoping refers to the frequently encountered phenomenon that respondents tend to draw into the reference period events that occurred before (or after) the period. Telescoping and recall loss are related to the interval between waves of a panel survey in opposite directions. As the reference period increases we have seen that reporting decreases; however, as the reference period decreases telescoping errors will increase as respondents tend to inflate the number of events by reporting earlier events as having occurred in the reference period. The extent of telescoping can be examined in a panel survey by determining those events reported on a given wave that had already been reported on a previous wave. Using this approach, Neter and Waksberg (1964) found that with a one-month reference period the reporting of the number of jobs of $100 or more for house alterations or repairs was about 55% higher with unbounded recall (i.e., before excluding those reported on a previous wave) than with bounded recall (i.e., after excluding those reported on the previous wave). With a 3-month reference period, the extent of telescoping was reduced to a 26% higher rate of reporting for the unbounded recall. Telescoping effects can be reduced after the first wave in a panel survey by using bounded recall, as is done in the National Crime Survey (Garofalo & Hindelang, 1977).

The longer the reference period, the greater the effort the respondent may need to recall the information being collected. Apart from this factor, however, the total respondent burden in reporting specified events in a panel survey of given duration is independent of the length of the reference period. Thus, for instance, collecting twelve months of income in two waves, with six months of income collected at each wave, places the same total burden on the respondent as collecting the twelve months of income in four waves, with three months of income collected at each wave. However, the respondent burden at a particular wave does depend directly on the length of the reference period. In their examination of respondent load effect, Neter and Waksberg (1965) concluded that it affected the number of jobs for house alterations and repairs under $10 reported; at least 20% fewer such jobs were reported for the preceding month when the reference period was lengthened from one to six months, and at least 11% fewer such jobs were reported for the preceding three months when the period was lengthened from three to six months. They detected no respondent load effect for these small jobs when the reference period was lengthened from one to three months, and they suggest that the effect is also likely to decline with size of job.

4.2. Respondent Rules

There are many ways of designating informants in a household survey that collects data on all members of the sample households. The "respondent rules" vary in the extent to which household members are required to be self-reporters, and in the range of other household members who are acceptable as proxy informants. For example, every adult in the household may be required to respond personally, one adult in the household may be the household informant providing data for all household members, or all adults present at the time of interview may be required to respond personally and a related adult may be interviewed for an absent adult. See Roshwalb (1982) for a review of this topic.

The main factors involved in the choice of appropriate respondent rules for a particular survey are the quality of the resultant data, the costs of data collection, and the level of nonresponse. The use of the more restrictive respondent rules is based on the premise that self-reporters will provide higher-quality responses than proxy informants, and that if proxy informants are used, informants related to the sample members will generally provide higher-quality responses than unrelated informants. While in general these premises seem reasonable, the amount of differences in data quality from using different respondent rules is likely to depend heavily on the data being collected; in some cases, proxy informants may be able to give almost as accurate responses as self-reporters, whereas in other cases their responses may be much less accurate. In the National Crime Survey, for example, there is some evidence that proxy reporting may produce underreporting of victimization for some types of crimes (Roebuck, 1984). Sometimes, it may even happen that proxy informants provide more accurate responses; proxy informants may, for instance, give more honest answers on some sensitive matters than self-reporters. There is also another way in which the respondent rules may affect the quality of the data, namely through the burden on the informant. In particular, when one household informant provides data for all household members, the heavy reporting load may lead to lower-quality responses.

The disadvantage of the more restrictive respondent rules is the greater difficulty they cause in data collection. They require more callbacks to collect the data—hence, the data collection costs are increased—and, in addition, they may lead to a higher nonresponse rate. This disadvantage has to be balanced against the potential advantage of better-quality data from the application of the more restrictive rules. The appropriate choice of rule for a particular survey thus must depend on the survey's subject matter.

The respondent rules adopted for the SIPP are that adults present at the time of interview report for themselves while proxy informants are accepted for absent adults. A hierarchy of proxy informants has been established for the SIPP so that a spouse is always the first choice as a proxy; the second-level proxy is the adult who was the proxy informant at the previous interview; the third-level proxy is an individual who was a proxy at any other interview; and, finally, a first-time proxy is accepted. In the event that a knowledgeable proxy is not available, a personal visit is scheduled to interview the uninterviewed persons. Using these rules, the self/proxy rates in the 1984 SIPP Panel were stable from wave to wave, with between 63% and 67% of the respondents at each interview reporting for themselves (Kasprzyk & McMillen, 1987).

A particular consideration involved in the specification of the respondent rules for a panel survey is the possible changes in informant between waves. About 40% of the individuals who participated in all eight interviews of the 1984 SIPP Panel were self-reporters at each interview. Another 19% of the individuals had only one or two proxy interviews conducted for them. At the other extreme, about 10.5% of the individuals participating in eight interviews never reported for themselves. A change in informant may give rise to changes in the measurement errors of the data being collected, and this can be particularly harmful for the measurement of gross change in a panel survey. Suppose, for example, that an individual's labor force status remains the same across two waves of a panel. On one wave the individual reports his status as "looking for work" while on the next wave a proxy informant reports that the individual is "not in the labor force". The survey data will thus erroneously show a change in labor force status. In this way, when the variable under study is highly stable across waves, changes in measurement error associated with changes in informants can lead to serious overestimates of the amount of gross change.

4.3. Mode of Data Collection

The main modes of survey data collection are face-to-face interviews, telephone interviews, and self-completion questionnaires. For a particular survey, the choice between modes involves an assessment for each mode of such factors as the feasibility of collecting the data required, the quality of the data that would be collected, the response rate that would be achieved, the availability of an adequate sampling frame, and the costs of data collection.

For a single-round survey of the general population, the likelihood of a very low response rate and the lack of a sampling frame nearly always rule out reliance on self-completion questionnaires. The choice between face-to-face and telephone interviewing is more open to debate. Face-to-face interviewing usually produces a higher response rate, and the noncoverage of the area sampling frame generally employed with face-to-face interviewing is probably less biased than the noncoverage resulting from households without telephones when telephone interviewing is used. On the other hand, telephone interviewing is less expensive than face-to-face interviewing. As a rule, face-to-face interviewing tends to be preferred when a long and complex questionnaire is employed.

A number of studies have been conducted to compare the quality of responses to face-to-face and telephone interviews. In practice it is difficult to disentangle the effect of mode on quality from its effects on other features of the survey, including nonresponse, different questionnaire forms, and different interviewers. Nevertheless, the finding that emerges from these studies is that responses to face-to-face and to telephone interviewing are generally quite comparable. A detailed study by Groves and Kahn (1979), for example, found few significant differences in univariate and bivariate response distributions by mode. The differences that did appear suggest that telephone respondents tend to give shorter responses to open questions and to give financial data in somewhat more rounded form than respondents to face-to-face interviewing. Cannell et al. (1981), on the other hand, found that the telephone respondents in their study reported somewhat more health events than the respondents to face-to-face interviewing, and they interpreted this as indicating higher-quality reporting of this information on the telephone. In general, the differences that have been found are not large, and thus do not indicate that one mode is appreciably superior to the other in terms of the quality of the data produced.

In panel surveys, the use of a combination of modes of data collection often proves effective. The Current Population Survey (CPS), for instance, requires interviewers to collect data by face-to-face interviewing for households entering the panel, but on certain other waves they may collect the data by telephone. At the end of 1976, about three out of five CPS interviews were completed by telephone (U.S. Bureau of the Census, 1978). A similar approach is used in the National Crime Survey, in which, nowadays, about four out of five interviews are conducted by telephone (Paez & Dodge, 1982). The National Medical Care Expenditure Survey (NMCES) and NMCUES have used a telephone mode in the third and fourth rounds of the panel (Horvitz & Folsom, 1980). This mixed-mode strategy capitalizes on the advantages of both modes, while also reducing their disadvantages. The use of face-to-face interviewing at the first wave avoids the noncoverage of nontelephone households and establishes a relationship with the sampled household. This relationship, together with the collection of the households' telephone numbers, helps with the response rate and the conduct of subsequent telephone interviews.

Another mixed-mode strategy that can be useful with panel surveys is to employ both face-to-face interviewing and self-completion questionnaires. Again, the face-to-face interviews establish a relationship with the sampled households that can help with the response rate for the self-completion questionnaires. If face-to-face interviewing is used for the main data collection, self-completion questionnaires mailed to respondents a short time before the next wave of the survey can be collected by the interviewers. Self-completion questionnaires have a particular use in panel surveys in which one household member is interviewed in each sampled household, providing data for all household members. When individual responses are wanted from each member of the household—for instance, responses to attitude items—they may be collected by means of a self-completion questionnaire. Self-administered questionnaires of this type are being employed with the National Medical Expenditure Survey to obtain information on health behavior, health opinions, and attitudes.

4.4. Rules for Following Sample Persons

Panel surveys often aim to provide data for a wide variety of analytic objectives. Some analyses are cross-sectional, producing estimates from individual waves of data, whereas others are longitudinal, focusing on temporal characteristics of individuals or other analytic units. Since populations are not static over time, these analyses often have different populations of inference. An important consideration in the design of a panel survey is then to ensure that the survey collects the data needed to provide valid estimates for the various cross-sectional and longitudinal populations of analytic interest. The sample design for the first wave of a panel survey is chosen to provide coverage of the defined survey population at the time the sample is selected. After that, "following rules" are needed to determine how the samples for subsequent waves are to be generated. These following rules specify which units of the sample remain and which drop out from one wave to the next, and, if necessary, how additional sample members are to be added to the panel, in order that the panel will provide valid samples of the desired populations of inference.

In panel surveys of individuals, where the individual is the unit of analysis, the sample for subsequent waves is easily defined as all individuals originally selected for the sample. Such studies are generally referred to as cohort studies and their results relate only to the particular cohort from which the sample was selected. The following rules are straightforward, namely, to follow all individuals in the originally selected sample. Examples of these studies abound—the National Longitudinal Surveys of Labor Market Experience, High School and Beyond, and the National Longitudinal Study of the High School Class of 1972 (Office of Management and Budget, 1986).

Other panel surveys identify a sample of dwelling units, and at each interview estimates of characteristics of households, families, and individuals are developed to represent these populations. Cox and Cohen (1985) refer to this type of design as a longitudinal dwelling unit design; under this design interviews are obtained from the current occupants of a sample dwelling unit. The basic sampling unit for the survey is the dwelling unit rather than an individual household or family. The implementation of this design is straightforward as the interviewer returns to the same addresses visited in previous rounds. Thus, following rules are trivial with this design, as the sample unit remains in the same location for each interview. The National Crime Survey, Current Population Survey, and American Housing Survey are examples of this type of design.

Other panel surveys adopt a design that Cox and Cohen (1985) label a longitudinal household design. In this design, the individual is the basic sampling unit, not the dwelling unit. Individuals living at the sample address at the time of the first interview are identified and interviewed in subsequent waves regardless of whether or not they reside at the originally sampled address. Typically, data are obtained on all individuals residing with the sample individuals in order to also provide household and family data for the given time period. Panel surveys designed to provide individual and family data for a specified period of time adopt following rules that try to locate and interview all sample individuals who have moved since the previous interview. Examples include the National Medical Care Expenditure Survey and National Medical Care Utilization and Expenditure Survey. Occasionally, geographic constraints on the following rules are imposed for cost reasons, as is the case with the SIPP. A further discussion with the SIPP as a case study follows.

The SIPP is designed to cover the U.S. civilian noninstitutionalized population aged 15 and over at each wave of the survey. The initial sample of households gives a sample of this population at the first wave. The rules for following persons selected into the sample then provide the procedures to update this sample for the subsequent waves. The ideal following rules for the SIPP would specify that all members of the original sample who remain in the U.S. civilian noninstitutionalized population should be followed, and that supplementary samples should be added at each wave to represent the new entrants to that population. In practice these rules have to be somewhat restricted because of the difficulties and high costs of implementing them. We will comment briefly on some of these restrictions; Kalton and Lepkowski (1985) provide a more detailed discussion of the SIPP following rules.

As noted, the SIPP rules aim to follow original sample persons throughout the life of the panel or, if they leave the U.S. civilian noninstitutionalized population (through death, emigration, joining the Armed Forces and living in barracks, or becoming an inmate of an institution), until the time they leave. For practical reasons two restrictions have been imposed on these rules. First, for processing reasons, persons in the 1984 SIPP Panel who were in the 4.9% of nonresponding households at the first wave were not followed in subsequent waves. Second, sample persons who moved to an address that was more than 100 miles from a SIPP primary sampling area within the United States were not followed for face-to-face interviews (but interviewers were instructed to conduct telephone interviews with them where possible). Since about 96.7% of the U.S. population lives within 100 miles of the SIPP primary sampling areas, this restriction is not a severe one.

In constructing following rules for panel surveys, it is necessary to consider not only the theoretical issues related to their coverage of the populations of inference, but also the practical feasibility of carrying them out. The SIPP rules specify, for example, that panel members are to be followed when they move, and those who move can be difficult to trace. Jean and McArthur (1984) report that only 80% of movers between the first and second waves of the 1984 SIPP Panel were traced.

Another issue relating to movers concerns those who temporarily move out of the U.S. civilian noninstitutionalized population and return during the life of the panel. A particular group of concern is those who move into institutions, such as nursing homes, for a period of time. If such persons return to a household containing another panel member, they are recaptured for the panel. If institutionalized persons do not return to a panel household, however, they are lost. An attempted remedy for this loss is to have interviewers check on whether a sample person is still institutionalized; if not, he or she is followed as any other mover.

One part of the following rules under the longitudinal household design deals with original panel members. The other part deals with the supplementary samples that may be needed to represent new entrants to the population. There are no effective economical ways of updating a panel sample for most types of new entrants to the population, and therefore the deficiencies in coverage have largely to be accepted. There are, however, remedies that can be used in certain cases.

The one type of entrant that can in general be readily handled comprises the "births"—in the case of the SIPP,·these are persons who reach the age of 15 during the life of the panel. Children under 15 in original sampled households are followed in a SIPP panel provided that they stay with one or more panel members. Those who reach 15 during the life of the panel become full panel members to be interviewed and followed for the rest of the panel. This procedure thus captures all the births from the original sample, except for the few that leave panel members before reaching their 15th birthday. Other entrants to the population are persons coming from abroad, persons returning from institutions, and persons returning from living in Armed Forces barracks. It would be a major undertaking to fully represent these entrants in the sample since no straightforward sampling frames are available for them.

Thus, following rules in a panel survey vary by the type of design adopted; in all cases, however, decisions concerning following rules need to be balanced by analytic and cost concerns.

5. MEASURING GROSS CHANGE

A major analytic advantage of panel surveys is that they provide the data needed to measure gross change, that is change at the individual level. The SIPP, for instance, collects data from which one can estimate not only the levels of use of different poverty programs, but also the flows into and out of those programs. While information on gross change is of great value, its measurement is bedeviled by serious response error and other problems. (See, for instance, Hogue & Flaim, 1986, on the problems with the labor force gross flow data from the

CPS, and U.S. Bureau of the Census and Bureau of Labor Statistics, 1985).

The critical difficulty in measuring gross change is that it is highly sensitive to changes in the measurement errors for sampled individuals between waves of the panel. Many aspects of a panel survey operation can give rise to changes in individual measurement errors, including:

1. *Simple response variability.* A respondent's answer may depend on his or her mood at the time of interview, the time and place of the interview, etc., irrespective of other changes.

2. *Changing respondents between waves.* In particular, information on a sampled person may be provided by a proxy informant on one wave, while the person may be a self-reporter on another.

3. *Changing the mode of data collection.* Although the discussion in Section 4.3 suggests that responses to face-to-face and telephone interviews are quite comparable, any differences that do exist may have a serious effect on measures of gross change.

4. *Changing interviewers.* If the interviewers affect the responses obtained, changes in the interviewers for a respondent from one wave to the next will affect changes in measurement error. This issue is particularly relevant when the data are collected on one wave by face-to-face interviews while they are collected on the next wave by centralized computer-assisted telephone interviewing. In this case, all those interviewed by the latter procedure will have different interviewers on the two waves.

5. *Changing questionnaires.* Several studies have shown that responses to a survey question can be sensitive to the context in which the question is asked, that is, to the other questions on the questionnaire (see, for instance, Kalton & Schuman, 1982; Shapiro, 1987). A change in the questionnaire from one wave to the next can thus give rise to a change in the measurement error for a given (unchanged) question.

6. *Changing interpretation of the meaning of a question.* During the life of a long-term panel, the meaning of words may change in subtle ways, so that over time respondents interpret the identical question differently.

7. *Panel conditioning.* As discussed in Section 3, membership in a panel may cause a change in responses. Panel members may, for instance, report fewer events on later waves because of a loss of motivation, or alternatively they may report more because they better understand the task. Panel membership may also alter their behavior.

8. *Changing coders.* If coders vary in the ways in which they code answers, a change in coders for a particular respondent will cause a change in measurement errors. Similarly, even if the coders remain the same, it is possible that as they become more

experienced with the survey, their coding practices may change.

9. *Imputation.* Unless care is taken, imputation for item non-responses can cause the amount of gross change to be over-stated (see Lepkowski's paper in this volume).

10. *Matching files.* If the sample is not properly accounted for, and the individual identification number assigned to a sample individual is not unique, mismatches of respondents across waves may generate a measure of change that is greater than is actually occurring.

11. *Keying errors.* Keying errors are a very small part of nonsampling errors in surveys. Nevertheless, they can give rise to incorrect changes in individual measurements.

Kalton et al. (1986) presented several examples of the problems of measuring gross change using SIPP data. One example concerns the sex and race items, which should remain constant across waves, and the age item, which should remain constant or increase by a single year between adjacent waves. Despite the fact that discrepancies in these demographic items were returned to the field for verification and correction, a small number of inconsistencies remained in the final data. The reasons for the inconsistencies are not known, but they are likely to be caused by a combination of interviewer, keying, and processing errors.

A second example relates to the occupation and industry classifications, where what appears to be an unduly large number of changes took place across waves. During the 1984 and 1985 SIPP Panels, occupation and industry data were collected and coded independently, even though an individual's employer and job may not have changed. Some evidence on the inconsistency of occupation and industry classifications is obtained by considering the 16,886 persons who reported the same employer for the first twelve months of the 1984 SIPP Panel. Of these persons, only about 50% had the same three-digit occupation code and only about 70% had the same three-digit industry code on all three waves. At first sight, this apparently excessive number of changes seems likely to be a coding problem, but this is not necessarily the case. Even though the job is the same, the description recorded by the interviewer may vary greatly from one wave to the next because of response variability, different interviewers, or different informants.

The procedures for collecting occupation and industry data were modified in the 1986 SIPP Panel to reduce the number of erroneous changes in occupation and industry codes, as well as to reduce interview time. The modification introduced a screener question that asked if activities or duties had changed in the past eight months. If a negative response was given to the screener question, the detailed occupation and industry questions were skipped, and the occupation and industry codes from the previous wave were brought forward to the current wave during data processing. There is, of course, the risk with this procedure that respondents may incorrectly give a negative response in order to avoid the more detailed questions, thus leading to an overstatement of the stability of occupations and industries across waves.

A third example of the problem of measuring gross change with SIPP data relates to month-to-month changes. The SIPP has an interval of four months between waves but for many income sources data are collected on a monthly basis. A common finding has been that more change has occurred between adjacent months when the data are collected in different waves than when they are collected in the same wave; that is, there is more change between months 4 and 5, 8 and 9, etc., than between months 1 and 2, 2 and 3, 3 and 4, 5 and 6, 6 and 7, 7 and 8, etc. This finding was first noted for transitions between recipiency and nonrecipiency for some social programs in the Income Survey Development Program (ISDP—a precursor to the SIPP) by Moore and Kasprzyk (1984). Kalton et al. (1985) observed a similar finding for monthly income amounts from various sources in the 1979 ISDP Research Panel; the correlations of amounts one month apart were higher when the months were within the same wave than when they were in different waves. Corresponding findings for recipiency/nonrecipiency of various income sources in the SIPP are reported by Burkhead and Coder (1985) and Weidman (1986). There are two possible explanations for these findings. One is that the greater amount of monthly change that occurs when the data are collected in different waves is due to changes in nonsampling errors between waves. The other is that the smaller amount of monthly change that occurs when the data are collected in the same wave is due to a false consistency in the responses given within a wave. Probably a combination of these two explanations is needed. Although the first explanation seems intuitively more appealing, there is some evidence that over-consistent reports within waves can occur (Kalton & Miller, 1986).

REFERENCES

Bailar, B. A. (1975), "The Effects of Rotation Group Bias on Estimates from Panel Surveys," *Journal of the American Statistical Association*, **70**, 23–30.

Bailar, B. A. (1979), "Rotation Sample Biases and Their Effects on Estimates of Change," *Bulletin of the International Statistical Institute*, **48**(2), 385–407.

Becketti, S., Gould, W., Lillard, L., & Welch, F. (1983), *Attrition from the PSID*, Unicon Research Corporation, Santa Monica, CA.

Burkhead, D., & Coder, J. (1985), "Gross Changes in Income Recipiency from the Survey of Income and Program Participation," *Proceedings of the Social Statistics Section*, American Statistical Association, 351–356.

Bushery, J. (1981), "Recall Biases for Different Reference Periods in the National Crime Survey," *Proceedings of the Section on Survey Research Methods*, American Statistical Association, 238–243.

Cannell, C. F., Fisher, G., & Bakker, T. (1965), *Reporting of Hospitalization in the Health Interview Survey* (Vital and Health Statistics, Series 2, No. 6), U.S. Government Printing Office, Washington, DC.

Cannell, C. F., Groves R., & Miller, P. V. (1981), "The Effects of Mode of Data Collection on Health Survey Data," *Proceedings of the Social Statistics Section*, American Statistical Association, 1–6.

Coder, J. F., & Feldman, A. M. (1984), "Early Indications of Item Nonresponse on the Survey of Income and Program Participation," *Proceedings of the Section on Survey Research Methods*, American Statistical Association, 693–697.

Cohen, S. B., & Burt, V. L. (1985), "Data Collection Frequency Effect in the National Medical Care Expenditure Survey," *Journal of Economic and Social Measurement*, **13**, 125–151.

Cox, B. G., & Cohen, S. B. (1985), *Methodological Issues for Health Care Surveys*, Marcel Dekker, New York.

Duncan, G., Juster, F. T., & Morgan, J. N. (1984), "The Role of Panel Studies in a World of Scarce Resources," in *The Collection and Analysis of Economic and Consumer Behavior Data* (S. Sudman & M. A. Spaeth, eds.), Bureau of Economic and Business Research, Champaign, IL, pp. 301–328.

Ferber, R. (1964), "Does a Panel Operation Increase the Reliability of Survey Data: The Case of Consumer Savings," *Proceedings of the Social Statistics Section*, American Statistical Association, 210–216.

Fienberg, S. F. (1980), "The Measurement of Crime Victimization: Prospects for Panel Analysis of a Panel Survey," *The Statistician*, **29**(4), 313–350.

Fogelman, K., & Wedge, P. (1981), "The National Child Development Study (1958 British Cohort)," in *Prospective Longitudinal Research: An Empirical Basis for the Primary Prevention of Psychosocial Disorders* (S.A. Mednick & A.E. Baert eds.), Oxford University Press, Oxford, pp. 30–43.

Garofalo, J., & Hindelang, M. J. (1977), *An Introduction to the National Crime Survey*, National Criminal Justice Information and Statistics Service, Law Enforcement Assistance Administration, U.S. Department of Justice, U.S. Government Printing Office, Washington, DC.

Ghangurde, P. D. (1982), "Rotation Group Bias in the LFS Estimates," *Survey Methodology*, **8**, 86–101.

Goldstein, H. (1976), "A Study of the Response Rates of Sixteen-Year-Olds in the National Child Development Study," in *Britain's Sixteen-Year-Olds: Preliminary Findings from the Third Follow-up of the National Child Development Study (1958 Cohort)* (K. Fogelman, ed.), National Children's Bureau, London, pp. 63–70.

Groves, R. M., & Kahn, R. L. (1979), *Surveys by Telephone: A National Comparison with Personal Interviews*, Academic Press, New York.

Hogue, C. R., & Flaim, P. O. (1986), "Measuring Gross Flows in the Labor Force: An Overview of a Special Conference," *Journal of Business and Economic Statistics*, **4**, 111–121.

Horvitz, D. G., & Folsom, R. (1980), "Methodological Issues in Medical Care Expenditure Surveys," *Proceedings of the Section on Survey Research Methods*, American Statistical Association, 21–29.

Jean, A. C., & McArthur, E. K. (1984), "Some Data Collection Issues for Panel Surveys with Application to the Survey of Income and Program Participation," *Proceedings of the Section on Survey Research Methods*, American Statistical Association, 745–750.

Johnston, L. D., O'Malley, P. M., & Bachman, J. G. (1986), *Drug Use among American High School Students, College Students, and Other Young Adults: National Trends through 1985* (DHSS Publication No. (ADM) 86–1450), U.S. Government Printing Office, Washington, DC.

Kalton, G. (1986), "Handling Wave Nonresponse in Panel Surveys," *Journal of Official Statistics*, **2**, 303–314.

Kalton, G., & Lepkowski, J. (1983), "Cross-Wave Item Imputation," in *Technical, Conceptual, and Administrative Lessons of the Income Survey Development Program (ISDP)* (M. H. David, ed.), Social Science Research Council, Washington, DC, pp. 171–198.

Kalton, G., & Lepkowski, J. (1985), "Following Rules in SIPP," *Journal of Economic and Social Measurement*, **13**, 319–329.

Kalton, G., Lepkowski, J., & Lin, T. (1985), "Compensating for Wave Nonresponse in the 1979 ISDP Research Panel," *Proceedings of the Section on Survey Research Methods*, American Statistical Association, 372–377.

Kalton, G., McMillen, D. B., & Kasprzyk, D. (1986), "Nonsampling Error Issues in the Survey of Income and Program Participation," *Proceedings of the U.S. Bureau of the Census Second Annual Research Conference*, 147–164.

Kalton, G., & Miller, M. E. (1986), "Effects of Adjustments for Wave Nonresponse on Panel Survey Estimates," *Proceedings of the Section on Survey Research Methods, American Statistical Association*, 194–199.

Kalton, G., & Schuman, H. (1982), "The Effect of the Question on Survey Responses: A Review," *Journal of the Royal Statistical Society, A*, 145, 42–73.

Kasprzyk, D., & McMillen, D. B. (1987), "SIPP: Characteristics of the 1984 Panel," *Proceedings of the Social Statistics Section, American Statistical Association*, 181–186.

Lehnen, R. G., & Reiss, A. J. (1978), "Response Effects in the National Crime Survey," *Victimology*, 3, 110–160.

Lievesley, D., & Waterton, J. (1985), "Measuring Individual Attitude Change," in *British Social Attitudes: The 1985 Report* (R. Jowell & S. Witherspoon, eds.), Gower Publishing Company, Aldershot, England, pp. 177–194.

McCarthy, P. J. (1980), "Some Sources of Error in Labor Force Estimates from the Current Population Survey," in *Data Collection, Processing, and Presentation: National and Local. Appendix Vol. 2: Counting the Labor Force*, National Commission on Employment and Unemployment Statistics, U.S. Government Printing Office, Washington, DC, pp. 2–39.

Mooney, H. W. (1962), *Methodology in Two California Health Surveys* (Public Health Monograph No. 70), U.S. Government Printing Office, Washington, DC.

Moore, J. C., & Kasprzyk, D. (1984), "Month-to-Month Recipiency Turnover in the ISDP," *Proceedings of the Section on Survey Research Methods, American Statistical Association*, 726–731.

Nelson, D., McMillen, D. B., & Kasprzyk, D. (1985), *An Overview of the Survey of Income and Program Participation: Update 1* (SIPP Working Paper Series No. 8401), U.S. Bureau of the Census, Washington, DC.

Neter, J., & Waksberg, J. (1964), "A Study of Response Errors in Expenditures Data from Household Surveys," *Journal of the American Statistical Association*, 59, 18–55.

Neter, J., & Waksberg, J. (1965), *Response Errors in Collection of Expenditures Data by Household Interviews: An Experimental Study* (Bureau of the Census Technical Paper No. 11), U.S. Government Printing Office, Washington, DC.

Office of Management and Budget (1986), *Federal Longitudinal Surveys* (Statistical Policy Working Paper No. 13), U.S. Government Printing Office, Washington, DC.

Olson, J., ed. (1980), *Reports from the Site Research Test*, Office of the Assistant Secretary for Planning and Evaluation, Department of Health and Human Services, Washington, DC.

Paez, A. D., & Dodge, R. N. (1982), *Criminal Victimization in the U.S., 1979–80 Changes, 1973–80 Trends* (Bureau of Justice Statistics Technical Report NCJ-80838), U.S. Department of Justice, Washington, DC.

Roebuck, J. (1984, Aug. 20), "Results of Investigation of Self vs. Proxy (Household) Respondent Interviewing in the National Crime Survey" (Unpublished memorandum for documentation), U.S. Bureau of the Census, Washington, DC.

Roshwalb, A. (1982), "Respondent Selection Procedures within Households," *Proceedings of the Section on Survey Research Methods, American Statistical Association*, 93–98.

Shapiro, G. M. (1987), "Interviewer-Respondent Bias Resulting from Adding Supplemental Questions," *Journal of Official Statistics*, 3, 155–168.

Shepherd, P. M. (1985), *The National Child Development Study* (Working Paper No. 1), Social Statistics Research Unit, City University, London.

Sobol, M. G. (1959), "Panel Mortality and Panel Bias," *Journal of the American Statistical Association*, 54, 52–68.

Survey Research Center (1986), *A Panel Study of Income Dynamics: Procedures and Tape Codes 1984 Interview Year*, Institute for Social Research, The University of Michigan, Ann Arbor.

Traugott, M. W., & Katosh, J. P. (1979), "Response Validity in Surveys of Voting Behavior," *Public Opinion Quarterly*, 43, 359–377.

Turner, A. G. (1972), *The San Jose Methods Test of Known Crime Victims*, National Criminal Justice Information and Statistics Service, Law Enforcement Assistance Administration, U.S. Department of Justice, Washington, DC.

U.S. Bureau of the Census (1978), *The Current Population Survey: Design and Methodology* (Bureau of the Census Technical Paper No. 40), U.S. Government Printing Office, Washington, DC.

U.S. Bureau of the Census and Bureau of Labor Statistics (1985), *Proceedings of the Conference on Gross Flows in Labor Force Statistics*, U.S. Bureau of the Census and Bureau of Labor Statistics, Washington, DC.

Weidman, L. (1986), "Investigation of Gross Changes in Income Recipiency from the Survey of Income and Program Participation," *Proceedings of the Section on Survey Research Methods*, American Statistical Association, 231–236.

Woltman, H., & Bushery, J. (1975), "A Panel Bias Study in the National Crime Survey," *Proceedings of the Social Statistics Section*, American Statistical Association, 159–167.

Wright, R. A. (1983), "Survey Design and Administration of the National Medical Care Utilization and Expenditure Survey," *Proceedings of the 19th National Meeting of the Public Health Conference on Records and Statistics* (DHHS-PHS84–1214), 177–180.

Sources of Nonsampling Error: Discussion

Colm O'Muircheartaigh

1. INTRODUCTION

I will address my remarks to the general issues of nonsampling errors in panel surveys and, rather than present a detailed discussion of the material in Kalton, Kasprzyk, and McMillen, I will present an overview of nonsampling errors, with some differences in emphasis from that presented by Kalton et al. In Section 2 I discuss the reasons why we are concerned about nonsampling errors. In Section 3 I describe some approaches to the measurement of quality. Section 4 deals with some specific topics in the area, and the final section gives my overall view of the way in which nonsampling errors should be tackled.

2. REASONS FOR INVESTIGATING NONSAMPLING ERRORS

The first reason for investigating nonsampling errors is to provide an estimate of the accuracy/precision of survey results. Two distinct factors affect this overall assessment: first, the *bias* or systematic error in our measurements; second, the *variability* of our measurements. Ideally we would like to estimate the mean square error of our survey results, incorporating both the bias and the variance. The formal boundary between bias and variance depends on the definition of the *essential survey conditions* for the survey. For example, if we allow the question form to vary from one iteration to another, then question form provides a component of variable error. On the other hand, if we define a single question form as the measurement procedure for the whole survey, then question form provides a possible source of bias in the survey. Even when we allow different question forms, there could still be an overall bias that is in some way an averaging of the gross biases of the different forms we include in the process.

I believe that it is irresponsible to conduct a survey and produce results in the form of estimators (either of means or of more complex statistics) without appending to these estimates an indication of their precision. Therefore, the design of surveys should, insofar as possible, include the possibility of obtaining quantified measures of the errors contained in the data. In this sense I suggest that all surveys should be *measurable;* i.e., it should be possible from within the survey data themselves to obtain an assessment of their probable error.

The second reason for investigating nonsampling errors is to identify special problems within the measurement process. In particular, it is desirable to identify procedures (for example, mode of data collection) that lead to particular types of error in the data. We should also be concerned with the possibility that certain variables, or certain indicators of underlying variables, are not adequately measured by our survey instruments. Finally, we also should consider the possibility that instruments that are otherwise satisfactory may lead to serious difficulties for particular subgroups of the population.

The identification of measurement problems is, in itself, a difficult issue. Unfortunately, merely identifying the sources of nonsampling error is not sufficient to redress data quality problems—steps must be taken to ameliorate these problems. We might, for example, develop new field procedures to overcome difficulties inherent in current data collection procedures. The definitions and implementation of variables in the study might be changed to take into account the difficulties identified. It is even possible that we will change radically the fieldwork strategy, both in terms of the mode of data collection and the allocation of work within that mode, in order to overcome some of the problems identified.

Even if we have succeeded in identifying and taking steps to eliminate the sources of nonsampling error, our task is not yet finished; we must evaluate the impact (if any) of the "improvements" that we have introduced into the survey. The procedures will not, in themselves, guarantee an improvement in the quality of data. It is therefore necessary to monitor the impact of these changes over time, using methods of evaluation similar to those used in identifying the problem in the first place.

Continuing surveys provide different opportunities from those available in one-time surveys. When dealing with one-time or ad hoc surveys, the *best* we can hope to achieve is the estimation of errors involved in the survey. We should, with appropriate design, be able to present at least the sampling error of the estimate, and some indication of the simple and correlated response variance. The identification of problem procedures, variables, or population subgroups would be more problematic. Moreover, by definition there is no possibility of incorporating changes or modifications of the procedures into a one-time survey. With continuing surveys, on the other hand, in particular repeated surveys or panel surveys, we have a much more promising field for inquiry and development. I believe that in such cases our strategy should be one of using the early phases or iterations of the survey as a time for identifying errors, incorporating modifications into later iterations or stages, and monitoring the impact of these modifications on still later iterations of the survey. Thus continuing surveys provide not only special problems in implementation, but also an enormous potential for the evolution of a systematic approach to the treatment and correction of errors.

3. APPROACHES TO QUALITY CONTROL

The most commonly used approach to quality control in social research is the pious hope that everything will be all right. This hope usually is presented in the guise of expert knowledge, the researcher believing that if s/he designs a program consistent with the manner in which such research has traditionally been carried out, no major problems will arise. The defense of the procedure is that the conventional wisdom in social research has approved the procedures being used.

A second approach, used by more conscientious researchers, is to consider explicitly alternative survey procedures. This includes, for example, the choice of mode of data collection—postal, telephone, or face-to-face. In these circumstances, a general knowledge of the literature, together with the prejudices and preconceived ideas of the researcher, guides the decision as to which mode to adopt. Neither this approach nor the preceding one involves any specific collection of information, nor any direct assessment of the appropriateness of the decisions taken.

A third approach to quality control involves direct examination of the data collected. Typically this entails making uncontrolled comparisons of procedures or subgroups. The basic intention is to find out what *is* happening in the survey, while recognizing that it is not possible to understand *why* things are happening (i.e., to attribute what happens to clearly defined causes). An example is the comparison of the quality of the data obtained from self-respondents and proxy-respondents. In every household for which self-reports are not required for every individual, there is an element of self-selection in the choice of respondent. If subsequently we compare the responses obtained from self-reports and proxy-reports, the comparison is contaminated by this nonrandom selection process. The contamination will be most severe if there is a strong relationship between the characteristics that determine the likelihood of self-response and the characteristics that are the object of the survey (i.e., the survey variables). The possibility of being found at home, and therefore of being a self-respondent, is often related to many of the variables with which survey research is concerned.

Even though an uncontrolled comparison of this kind is flawed, it still may provide an indication of an area in which more rigorous (and expensive) research would be justified. The advantage of this approach is that it provides an indication of possible problems and a basis for targeting methodological research at relatively low cost.

A fourth approach to quality control redresses the primary deficiencies of the third by employing controlled experiments to identify the root causes of nonsampling errors. The purpose of such experimentation is to determine not only what *does* happen in the survey, but what *could* happen if different procedures were adopted. It might, for example, be useful to test through experimentation whether differences observed by the direct examination of uncontrolled data persist when the experimental conditions are controlled.

Such experiments take many forms. One example is the use of a reinterview program in combination with the main survey to measure the reliability of the responses. Another example is the use of interpenetrated designs to examine the impact of the interviewer, or the coder, or the supervisor on the survey data. A third example, difficult to implement but potentially quite rewarding, is a field experiment that would select respondents based on a probability mechanism rather than on the whim of the interviewer and the transient characteristics of the situation.

These experiments are of two kinds. The first two examples involve the introduction of a modification to the survey procedures that permits the controlled evaluation of the impact of the existing procedures. This allows us to establish the reliability of responses through reinterview and the impact of the interviewer by randomization; thus, the first two examples essentially are concerned with estimation. The third example takes the process a step further by providing information that could form the basis for introducing changes into the survey procedures. Such a system of experimentation should be inductive, each experiment building sequentially on the results of earlier experiments.

The four approaches described above suffer from the flaw that they do not allow us to move outside the essential survey conditions under which the survey is conducted. They are therefore, in many ways, more appropriate for the examination of precision than they are for the examination of bias. If bias is also to be incorporated into our measures of measurement error, then it is necessary to have a source of information different from, and superior to, the survey data as a basis for comparison. It would therefore be necessary to carry out validation studies, such as record checks, in order to compare information obtained from the survey with the "true" values for respondents. Such validation studies are obviously desirable, always expensive, and sometimes impossible to conduct.

It often is not realized that validation studies also need to incorporate a formal experimental design. For example, if we were to find that proxy-reports were superior to self-reports, this would not necessarily mean that extending the use of proxy-reports would improve the quality of the data overall. The results would apply only to the comparison of the self- and proxy-reports generated by the current method of respondent selection, and would not necessarily carry over to a different set of procedures for selecting respondents.

A program of data quality evaluation ideally should progress from the least to the most of the controlled options outlined above. The program should use the results of data inspection to generate hypotheses for experimental evaluation, modify procedures on the basis of such an evaluation, and ultimately use experimental designs to monitor the impact of changes in procedure. Such studies could provide information not only on the precision but also on the accuracy of results.

4. SPECIFIC ISSUES

4.1. Nonresponse

Nonresponse has two effects on the results of sample surveys. First, and most obvious, is the reduction of sample size. This problem in itself is not particularly serious, since it could, in principle, be compensated for by selecting a larger sample in the first place. The second negative effect of nonresponse is the possible introduction of bias into the results. This occurs when nonrespondents systematically differ from respondents.

Bias due to nonresponse is potentially more serious for panel surveys than for one-time surveys, since in addition to first-wave nonresponse additional nonresponse arises during later waves in the panel. Panel surveys do, however, offer the possibility of obtaining considerable information about the nature of nonresponse at later waves. On this issue I would depart somewhat from the emphasis given by Kalton et al. The authors focus on a comparison of the characteristics of first-wave respondents with those of nonrespondents for later waves in panel surveys. Typically, when such comparisons are carried out, only the marginal distribution of the respondents and the nonrespondents for different waves are presented. Therefore, it appears that the respondents and nonrespondents are broadly similar. It is likely, however, that particular subgroups in the population are more seriously affected than it appears from these single variable comparisons. In the case of the Survey of Income and Program Participation (SIPP), for instance, the drop-out percentage is high for those who rented their accommodation for cash, for those between the ages of 15 and 24, for nonwhite individuals, and for married individuals whose spouse was absent. In each case, the drop-out percentage over five waves was of the order of 20%.

It is likely, though without a multivariate analysis it is impossible to be sure, that individuals who fall into a number of these categories simultaneously may be seriously underrepresented in later waves of the survey. Carrying out such an analysis usually is impossible because the number of cases in the cells of the multivariate table becomes too small. However, in the case of a rotating panel design, such as is found in the SIPP, it should be possible over a period of time to cumulate cases even for such small cells. There is also an opportunity to estimate the response propensities of different elements in the population more accurately than would be possible in a one-time survey or in a nonrotating panel. On the basis of such cumulation it should be possible to identify critical groups, i.e., groups where the attrition rate is particularly high, and then, having identified them, to consider the possibility of reweighting or alternatively to avoid making inferences about these population subgroups if the data are too sparse.

It also should be remembered that there should be a connection between the analysis of data during and after the survey and the implementation of changes in the fieldwork for the survey. If groups with a particular demographic profile can be identified, it may be possible to devise procedures in the data collection operation that take account of these data collection difficulties and may overcome them. Such procedures typically would be prohibitively expensive to use for the whole sample. However, if the analysis leads to the identification of subgroups that are at particularly high risk of nonresponse, it may be desirable when dealing with these groups to use such expensive procedures.

While not disagreeing with any of the specific points made by Kalton et al., I would add to them the suggestion that the overall thrust of the treatment of nonresponse, especially in rotating panel surveys, should be to carry out an analysis of the results obtained both in early waves and in early panels of such a survey, and to use the results in order to implement modifications of field procedures. The impact of such modifications similarly should be assessed over a set of waves and panels in the continuing survey. Basically, the approach that I favor is one of flexibility combined with feedback from the analysis of the data to the collection of later data.

4.2. Panel Conditioning

Conditioning refers to the change in response that occurs because the respondent has had one or more prior interviews. Panel conditioning is thus the reactive effect of prior interviews on current responses. The effect may come about because prior interviews change respondents' behaviors or because they change the way respondents answer questions. It also may come about because the interviewer's behavior changes at different waves in the panel study. This latter effect is most likely when the same interviewer carries out a sequence of interviews in a particular household.

For household surveys, it may be important to distinguish between the number of times the household has been in the sample and the number of times a particular respondent has been interviewed. If we take, for example, the respondent rules used for the SIPP, we can see that the procedure tries to maximize the frequency with which particular respondents are interviewed. The preferred choice of respondent in all cases is a self-report. If this is not possible, a hierarchy of proxy-respondents has been established for the SIPP so that a spouse is always the first choice as a proxy. The second-level proxy is the adult who was the proxy-informant at the previous interview; the third-level proxy is an individual who was a proxy at any other interview; and, finally, a first-time proxy is accepted.

Using these rules for eight waves of the 1984 SIPP panel, about two-thirds of the respondents were self-respondents for each of eight interviews. A similar pattern of respondents is said to be found in the Current Population Survey (CPS), although the proportion of self-respondents is lower in the CPS than in the SIPP. It would be desirable to adjust the analysis by "times in sample" of the results of both the CPS and SIPP using the number of times a particular individual has responded rather than the number of times the household has been in the sample.

In addition to the analysis suggested above, there is a second source of information available for analyzing the effect of conditioning and for looking at the differences between the waves of a panel survey, which is not referred to by Kalton et al. In some surveys, in particular in the CPS, a program of reinterviews has been carried out for some time, a subsample of respondents (or households) being reinterviewed shortly after the main interview. A comparison of the responses to the original interview and the reinterview provides information about the reliability, or simple response variance, of the responses. This provides a benchmark for the minimum expected variation between waves in the case where there has been no change in the respondent's real situation.

Research indicates that there is a strong between-interview correlation for the response errors themselves (O'Muircheartaigh, 1986). For CPS this correlation is of the order of 0.3. This suggests that researchers should be aware of the obverse of the conditioning effect usually observed—that is, that respondents may tend simply to repeat information given on an earlier occasion when a question is repeated on a later occasion. This is particularly important in the case of estimates of change between waves of a panel survey. Some evidence adduced by participants at this conference reinforces the view that the matching of interviewers and respondents over a long period for a panel survey may have its own dangers.

A further difficulty arises in panel surveys with regard to the accessibility of the previous information provided by the respondent. On the one hand it clearly is desirable that the interviewer should have access to this information in order to check that responses being given on a later occasion are not simply repetitions of information previously given. On the other hand, it is undesirable that the interviewer, and therefore the respondent, should be influenced by the answers given on the earlier occasion.

The overwhelming difficulty in assessing the extent of panel conditioning is that the same apparent result would be obtained if there were a change in the behavior of the interviewer, the behavior of the respondent, a change in the reporting quality of the respondent, or a change in the composition of the responding group. In order to disentangle these effects, it would be necessary to design a *set* of investigations in which some of these effects would be controlled.

The split-panel design (which selects an independent cross section of the population in parallel to each wave of the panel survey) is particularly useful in separating out the effects of panel conditioning. A comparison of the responses, both across waves of the panel and across independent replications of the cross-sectional sample, provides valuable information on this distinction. Waterton and Lievesley (1987) use this design in a survey of British social attitudes. In order to reach the ultimate goal of separating out the effects of interviewer conditioning, respondent conditioning, and interviewer–respondent interactions, it would be necessary to set up a controlled experiment in which different combinations of these are allocated randomly from within the available households. In the absence of such experiments, no real progress can be made.

4.3. Interval between Waves

The data collected at any wave in a panel survey typically relate to the events occurring since the previous wave. Practical considerations may influence the length of the interval between waves, for the longer the interval between waves, the lower is the cost of the survey. Undue spacing of waves may be indefensible, however, when the yardstick is not "cost per case" but rather "precision per dollar." The longer the interval between waves, the longer is the reference period for which the respondent must recall the events, and consequently the greater is the scope for errors of recall.

In a sense, the limiting positions are the cross-sectional retrospective survey, in which data are collected at one interview for the whole previous time period, and a diary survey, in which respondents are instructed to record events as they occur. Thus here, too, it is possible to view the cross-sectional survey as a special case of the panel survey situation. All of the problems that arise in cross-sectional surveys can, in principle, arise in panel surveys as well. The difference is that panel surveys provide opportunities for obtaining information about the *nature* of the response process—in particular, the accuracy with which the occurrence and timing of events are reported.

Panel surveys provide special examples of bounded survey interviews. Each wave after the first wave produces a bounded interview, with the boundary data provided by the responses on the previous wave. Much of the literature on panel surveys focuses on the difficulties that arise in linking responses from different waves and on the distribution of events during the period between waves. The principal problems identified are recall loss, telescoping effects, and respondent burden. Recall loss is known to increase as the reference period becomes longer. Data from the National Crime Survey and from SIPP support this view. Telescoping, on the other hand, tends to increase as the reference period decreases, as respondents tend to inflate the number of events by reporting earlier events into the reference period. The effect of telescoping on data quality can be examined by comparing events reported on a given wave with those reported in previous waves.

One major opportunity for estimating data quality in panel survey interviews (often neglected by survey researchers) is to collect information for an overlapping period in the two interviews. (The Consumer Expenditure Survey is one case in which this opportunity is partially exploited.) If interviews are conducted at three-month intervals, for example, it is possible to collect information at each interview for four months. This would give one full month's overlap between data collected for successive interviews, and the comparison of the two sets of reports for *that* month provides an excellent basis for the assessment of both telescoping and omission. An alternative would be to collect three months' data as before, but to have interviews at two-month rather than three-month intervals. The latter recommendation would require fairly drastic changes to current survey design; thus researchers and/or administrators may be reluctant to support it. It seems to me, however, that unless such an investigation is carried out, speculation and prejudgment will continue to provide the basis for the assessment of the magnitude and type of errors.

Despite the firmness of my position, I recognize that, in deference to practical considerations, it may well be desirable to execute a parallel experimental survey design in order to assess the impact of such changes before modifying the main survey in fundamental ways. Fortunately, panel survey designs provide an opportunity to build specific investigations of the patterns of responses and response errors into the structure of the survey procedures. It is important to remember that for this opportunity to be exploited fully, it is necessary to incorporate a succession of such investigations into the panel survey design. Otherwise, there can be no possibility of improvement and development in the survey.

In any event, the conclusions will depend on the specific topics being investigated, and on the saliency of events to the respondents. This means that the results of such an investigation will not necessarily be

portable from one survey to another, suggesting that it would be un-
wise, for example, to depend on the results of a study for household expenditure in planning a study of income or labor force status.

4.4. Respondent Rules

For household surveys there are many ways of designating informants.
The "respondent rules" vary in the extent to which household members
are required to be self-reporters, and in the range of other household
members who are acceptable as proxy informants. A voluminous
literature exists on the subject, although little is known definitely about
the impact of different respondent rules. A question of general interest
to survey researchers, and of specific interest to the designers of panel
surveys, is whether survey data suffer in quality when sampled individuals do not respond for themselves (i.e., to the extent that proxy-
response is accepted in lieu of self-response). Hill's (1987) examination
of selectivity and reporter bias in SIPP, for example, concluded that
there is strong evidence of the existence of substantial selectivity and
differential reporting biases between self- and proxy-reports. Moore
(1988) provides an interesting review of the literature in this area.

Most household surveys do not control rigorously the identity of the
respondent, and in particular do not control self/proxy assignment.
Therefore, it is not possible to distinguish response status effects from
effects due to other differences between the self- and proxy-samples.
The conventional wisdom is that the best information comes from self-
reports (see, for instance, Sudman & Bradburn, 1974). However, two
studies call this view into question. In a study of telephone interviewing for the Health Interview Survey, Mathiowetz and Groves (1985)
found "an overall tendency towards higher (better) proxy-reports
[which] runs directly counter to previous beliefs about self- versus
proxy-reports" (p. 96). The Mathiowetz and Groves results relate to
completeness of reporting rather than to the variability in the reports,
and higher reporting is assumed to represent more complete, and
therefore better, reporting. O'Muircheartaigh (1986) found in a study
of CPS reinterview data that the simple response variance was lower
for proxy-respondents than for self-respondents (when the data are controlled for some demographic characteristics of the responding individuals).

The fundamental problem with these results arises, of course, from
the nonexperimental nature of the selection of self- and proxy-
respondents. The factors that determine which individual is accepted as
the respondent are ill-defined, and there is no guarantee that these factors are not themselves related to the variables being investigated in
the survey. Any such relationship could invalidate conclusions based on
an uncontrolled comparison.

Three conditions must be met if the problem of determining the best "respondent rules" is to be solved. First, it is essential to define adequate measures of quality for the data. In some studies, the amount of reporting is used as a measure of quality, with the assumption that a greater amount of reporting indicates higher quality. In other studies, the reliability of the data (measured by simple response variance) is used as the criterion, with a high level of simple response variance being taken to indicate a low level of quality.

Second, data must be available (or collected) that permit the calculation of the measures of quality. The absence of validation data for most survey research makes it impossible to obtain comprehensive measures of quality. Moore and Marquis (1988) at the U.S. Census Bureau are conducting a record check study for SIPP that will overcome some of these problems, insofar as it will provide a measure of the accuracy of particular responses to the survey questions. However, this information still will be insufficient to determine the best set of respondent rules. The self-selection of respondents inherent in the way most household survey field work is carried out makes it impossible to decide whether in a particular case a different respondent would have given a more accurate answer. Thus, we have the third condition—the cases on which the comparison of different respondent rules is based must be generated by a (randomized) procedure that minimizes the probability of confounding the comparison of respondent rules with other possible effects.

Even in the absence of the perfect research design for the examination of the effects of self- versus proxy-reporting on data quality, however, it is still possible to derive from existing data a considerable amount of information about the way in which respondents behave. The patterns of response across waves of a panel survey can be tabulated and compared; the demographic composition of the set of respondent individuals can be examined for each wave and can be compared across waves; and imperfect measures of quality can be calculated in each case. Nevertheless, a final solution to the problem—or even a partial solution—can be obtained only through controlled experimentation.

4.5. Mode of Data Collection

The issues involved in deciding upon the mode of survey data collection are complex, involving the consideration of such factors as the availability of staff and facilities, coverage, response rates, data quality and costs. It is unlikely that the researcher will have complete information available on all of these factors for any given survey. Here, too, panel surveys are in a much stronger position than one-time surveys in terms of the amount of information that can be made available to the researcher.

A particularly important choice confronting the researcher these days is between face-to-face and telephone interviewing. It is generally believed that face-to-face interviewing produces a higher response rate and less noncoverage than telephone surveys. On the other hand, telephone interviewing is less expensive than face-to-face interviewing. It is less clear which mode of data collection yields higher quality data. A number of studies have been carried out in an attempt to answer this question, but many of them suffer from the disadvantage that they do not control important aspects of the survey process.

In the CPS, the simple response variance can be shown to be lower for telephone than for face-to-face interviews. However, the choice between face-to-face and telephone interviewing is not made randomly, but depends on the wave of the survey and on the discretion of the interviewer (O'Muircheartaigh, 1986). Groves and Magilavy (1986) show that the interviewer effect (measured by the intra-interviewer correlation coefficient) is lower, in general, for telephone surveys than for face-to-face surveys. Here again, though, the comparison is not made on the basis of a controlled experiment; Groves and Magilavy compare surveys where the decision to interview by telephone or face-to-face was made on other grounds.

It also is worth bearing in mind that workloads for the interviewers typically are considerably larger for telephone surveys than for face-to-face surveys. This inflates the impact of the interviewer effect on the precision of the survey estimates. Finally, there is disagreement in the literature on whether more reporting is obtained in telephone interviews than in face-to-face interviews. Cannell, Groves, and Miller (1981) found that telephone respondents reported somewhat more health events than the respondents to face-to-face interviewing, whereas Cantor in this volume describes a situation in which there was more reporting in face-to-face interviews.

These results, as is typical with empirical work, suggest that more research is necessary. The primary need is for controlled experiments in which the different modes of data collection are contrasted explicitly. Repeated surveys, including panel surveys, provide an opportunity to conduct such experiments. It is particularly important that experiments be designed that will obtain results applying to the specific variables being used in a particular study. For a one-time survey this typically is not possible, but for repeated surveys, the repetition of the data collection over different waves and panels provides an opportunity to learn sequentially about the effect of the mode of data collection on data quality for a specific subject area.

4.6. Characteristics of Respondents and of Subjects

Here I use the term "respondent" to denote the person actually providing the information to the interviewer, and the term "subject" to describe the individual on whom the data are being collected.

It is typically the case that in planning a survey the researcher wishes to devise a scheme whereby all individuals in the population are treated in the same way. Thus, the researcher designs a plan that is, on average, the appropriate plan for the population. It seems to me that this approach is too inflexible. When we consider the complexity of the information being requested from respondents in many panel surveys, and the wide variation in the quantity and detail of the information being demanded from different respondents, it seems clear that it is better to tailor procedures to the particular case with which the interviewer is dealing. Obviously, it is not possible to develop a scheme whereby each individual is treated differently. However, if it were possible to identify broad subgroups of the population that had either specific problems or unusual burdens in terms of the responses required, it would be appropriate to take this into account in planning the survey.

The research required to develop procedures of this kind is probably too basic to be included in an ongoing survey operation directly. But, once again, it should be possible by the analysis of existing data to identify particular groups in the population where difficulties often arise. One example of such an analysis would be to analyze the reliability of the data provided by different types of respondents for different items on the questionnaire and, subsequently, to devise a program of experiments to investigate the sources of these differences. The same approach could be used to investigate the problem of linking data obtained from different waves of the survey, or to identify a clear pattern of change in the level of response from one wave to another.

It seems obvious that if the procedures used by the interviewers differ for different respondents, then the circumstances under which each procedure might be used must be defined in advance. Clearly, it would be inappropriate to allow an interviewer *complete* discretion in approaching the questionnaire for each respondent; I suggest this probably would also be unacceptable to the interviewers. In principle, the interview situation should be "respondent-friendly." I believe this goal is best achieved when the collection of data is a *collaborative* exercise on the part of the interviewer and the respondent.

"Respondent-friendliness" could include enlisting the support of the respondent in determining the approach to be adopted. There is a regrettable tendency in survey questionnaires to mask the survey objectives from the respondent until the last possible moment. Two examples may illustrate this.

In the collection of data on labor force status, there are some people whose labor force status does not change over a period of decades, much less from month to month. On the other hand, there are groups in the population who clearly are error-prone because of the ambiguity of their position in the labor force. Thus an individual who has been working in the same full-time job for fifteen years is going to provide virtually error-free data, whereas a young person who recently left school and does occasional work clearly will be error-prone. By explaining to the respondents the purpose of the survey, it should be possible to undertake more detailed or more explicit questioning in cases where an ambiguous situation arises, and to do this with the active support of the respondent. In the analysis of CPS data, the particular group identified as being most subject to risk of response error was the group of young adults living at home with their parents or other relatives.

In the Consumer Expenditure Survey, data are collected for each individual in the household on a large number of variables. Some respondents may prefer to answer all the questions for one subject, and then answer all the questions for the next subject. Others may find it easier to consider each variable in turn for all the subjects. By explaining the structure of the survey to the respondent, the best strategy for each case could be agreed on at the outset, with (probably) beneficial effects on data quality.

The chief advantage of a panel survey design is that it is possible to cumulate information about individuals, and to modify survey procedures for obtaining information from particular groups of individuals over a period of time. Typically this advantage is *not* exploited, however; in general, data from earlier waves of the panel survey are not analyzed to identify difficulties of this kind. It is appropriate to mention that some of the preparatory research for this approach initially would need to be carried out experimentally, outside the range of the main survey. If useful results are obtained, these could be introduced gradually into the general survey procedures.

My basic recommendation is to begin *now* to investigate possible avenues of reducing the errors in the survey data. It might, for example, be useful to devise different questions, or more questions, for particular subgroups of the population, and to provide clearer definitions for cases where doubt can arise. One further recommendation in the case of long and demanding survey questionnaires is to define for the interviewer, and perhaps even for the respondent, the minimum information that is deemed necessary to make an interview useful. This would permit both the interviewer and the respondent to perform their tasks at an *acceptable* level, even if it were not possible for them to perform these tasks at the optimal level.

4.7. Measuring Gross Change

The major analytic advantage of panel surveys is that they provide the data needed to measure gross change—change at the individual level.

Unfortunately, many panel surveys do not use the full scope of the survey for such analysis. One of the reasons for this is that the measurement of gross change is seriously hampered by response error and other problems. Kalton et al. list eleven sources of such problems in a panel survey operation. They also present three particular examples where the problems have proved to be particularly severe for SIPP data.

The estimation of gross change brings together all of the problems previously described in this discussion. Nonresponse, conditioning, respondent rules, the time interval between waves, the characteristics of the respondents, and the mode of interview all contribute to the difficulty of measuring gross change. The issue also highlights the special nature of panel surveys, and demonstrates that the theoretical strengths of the panel survey design will be of no particular value unless some of these problems are overcome.

I believe that if we are to make substantial progress in the measurement of gross change, it will be necessary to adopt a more rigorous and quantified approach to research in this area. It is important not to lose sight of the fact that both the measurement of gross change and the individual observations themselves are affected by response errors. Ideally, the structure of the nonsampling errors should be embedded in the overall model being used to analyze the data. An example of such an approach is found in the work of Chua and Fuller (1987). In their work, they embed in the model to estimate gross change in labor force status (based on the CPS) a model for the response deviations in the data. They use the reinterview data from the CPS to estimate the inconsistencies in the data, and subsequently adjust the estimates of gross change to take these deviations into account. They expand their model to include separate treatment of different subgroups of respondents, in particular younger and older respondents.

5. CONCLUSION

A concern with quality is an integral part of any scientific measurement. In a real sense it is impossible to talk about data or results without some reference to the quality of these data or results. Therefore, all data collection programs should include a program of assessment and improvement of data quality. One-time surveys suffer from the disadvantage that changing populations, topics, or other characteristics of the survey make it difficult to build on the results of earlier work.

An enormous investment of time and resources is made in the preparation and development of repeated surveys. Panel surveys in particular provide a wealth of information about specific respondents, and indeed about the changing pattern of responses for the same respondents. They also provide a much longer time scale within which it is possible to modify, test and monitor the survey procedures. This work, however, will have relatively little value unless the errors are not only identified but quantified.

To some extent, the quantification of errors involves the modification, sometimes the drastic modification, of the survey procedures. This means that data may be collected that are not directly useful for the original purposes of the survey. It is entirely appropriate that in such cases we should collect data that we subsequently discard. This already happens in some surveys; for instance, in the Consumer Expenditure Survey the first of the five interviews is used purely for the purpose of bounding the second and subsequent interviews in the survey.

I believe that there is a strong case to be made for a general approach in which, parallel to the main survey, a supplementary survey is carried out in which experimental treatments are tested and the results of modifications are evaluated. It is frequently argued that such expense is not acceptable in the context of a particular survey. This contention is based on a false premise—that, in some way, data can be valuable where we have no knowledge of its quality.

Quality control of surveys is a particularly difficult problem because in many areas we do not have any readily available or acceptable criteria to determine quality. This shortcoming has been used in the past to justify the absence of effort, rather than as an inspiration for greater effort. The situation is exacerbated by the failure of different groups of people involved in survey work to communicate with each other. Field records and interviewer assessments have proved extremely valuable as an aid to the development of survey procedures. However, it is still rare that those involved in designing surveys participate actively in the field operation, or spend any time discussing the problems arising in the field with interviewers and their supervisors.

If the quality of survey research data is to improve, we must change the hidebound approach we take to the design and implementation of survey instruments. Conventional wisdom argues that it is essential to have exactly the same treatment for all individuals and households in the population. Such an approach ignores the obvious disparities both in the burden imposed on different respondents and in their capacity to cope with this burden. I am not proposing that each individual in the population be treated differently but, rather, that specific subgroups of the population (e.g., those who face a particular problem or for whom there are particular difficulties) be identified and dealt with in a more flexible way. The key idea here is flexibility—the researcher should recognize differences in the population and design survey instrument and procedures to take these into account.

Underlying all of these recommendations is the assumption that the findings can be made operational. In the past, a great deal of research in survey methodology has concentrated on identifying problem areas and specifying sources of error in survey data, rather than in formulating recommendations for dealing with those difficulties. The strength of research on nonsampling error associated with repeated surveys arises from the possibility of instituting a program of research in which each component of the program builds on the results of the previous components. In particular, given that the survey instruments and the survey population remain essentially stable, it is possible to carry out experiments parallel to the main survey operation and test operationally new procedures as they are developed.

A design like the SIPP design has tremendous potential for the measurement of survey errors. The emphasis in the paper is on the difficulties associated with panel surveys. The authors point out that panel surveys have, of course, the same nonsampling error problems that arise in cross-sectional surveys, but suggest that these problems are compounded in the case of repeated surveys and specifically for panel surveys. I would take a different view. The information collected in a panel survey and the possibility of using the panel as a benchmark for experimental procedures means that panel surveys can provide a much richer source of information on nonsampling errors in general. Methodological work in this context would contribute not only to the resolution of the specific problems of panel surveys but would also add to our understanding of the survey process in cross-sectional (over time) surveys in general. The impact of nonsampling errors on data quality in panel surveys not only imposes a responsibility on survey practitioners but also provides a challenge that should be welcomed for the opportunities for progress that it provides.

REFERENCES

Cannell, C. F., Groves, R., & Miller, P. V. (1981), "The Effects of Mode of Data Collection on Health Survey Data," *Proceedings of the Social Statistics Section*, American Statistical Association, 1–6.

Chua, T. C., & Fuller, W. A. (1987), "A Model for Multinomial Response Error Applied to Labor Flows," *Journal of the American Statistical Association*, 82, 46–51.

Groves, R. M., & Magilavy, L. J. (1986), "Measuring and Explaining Interviewer Effects in Centralized Telephone Surveys," *Public Opinion Quarterly*, 50, 751–766.

Hill, D. (1987), "Response Errors in Labor Surveys: Comparisons of Self and Proxy Reports in the SIPP," *Proceedings of the Third Annual Research Conference*, U.S. Bureau of the Census, 299–319.

Mathiowetz, N. A., & Groves, R. M. (1985), "The Effects of Respondent Rules on Health Survey Reports," *American Journal of Public Health*, 75, 639–644.

Moore, J. C. (1988), "Self/Proxy Response Status and Survey Response Quality: A Review of the Literature," *Journal of Official Statistics*, 4(2), 155–172.

Moore, J. C., & Marquis, K. H. (1988), "Using Administrative Record Data to Describe SIPP Response Errors," paper presented to the Section on Survey Research Methods at the Meetings of the American Statistical Association, August 1988.

O'Muircheartaigh, C. A. (1986), "Correlates of Reinterview Response Inconsistency in the Current Population Survey (CPS)," *Proceedings of the Second Annual Research Conference*, U.S. Bureau of the Census, 208–234.

Sudman, S., & Bradburn, N. M. (1974), *Response Effects in Surveys*, Aldine, Chicago.
Waterton, J., & Lievesley, D. (1987), "Attrition in a Panel Study of Attitudes," *Journal of Official Statistics*, **3**, 267–282.

Symptoms of Repeated Interview Effects in the Consumer Expenditure Interview Survey

Adriana R. Silberstein and Curtis A. Jacobs

The Consumer Expenditure Interview Survey is an ongoing panel survey of households interviewed five times at three-month intervals, and is one of two components of the Consumer Expenditure Survey Program, the other being the Diary Survey. This study examined the extent of time-in-sample and recall effects in the 1982/83 Interview Survey and identified systematic differences in the estimates symptomatic of reporting weakness, i.e., in the form of underreporting of expenditures. Comparisons to independent sources, such as estimates from the National Accounts, show much lower estimates for several published expenditure classes, and while these data are in themselves subject to error, the comparisons suggest underreporting in the survey (Gieseman, 1987). The results of this study indicate that recall bias is a major factor in this form of error and needs to be further examined. After a brief introduction of response errors in panel surveys, the paper describes important design features of the survey and some of the data adjustments. The method of analysis is then discussed followed by the presentation of findings and interpretations.

More factors affect the quality of responses in repeated interviews than in one-time surveys. Even the usual factors affecting survey responses, such as the type of data collected and the questionnaire design, play a different role at different stages of the panel survey as the respondent's interest and trust with the interviewer vary from one interview to the next. Panel respondents may report more selectively after the first or second interview in order to shorten the interview, and this tends to deflate the estimates. More cooperating respondents stay in the panel and this filtering and learning what is expected may improve reporting quality but, at the same time, may lead the respondents into consistent or set answers. These are manifestations of time-in-sample effects that condition response quality.

Recall effects are well known in retrospective surveys and typically tend to bias the estimates downward. Respondents remember less as the time of an event is further away from the interview time, and memory losses differ in degrees according to the frequency and importance of an event. The use of recall aids, such as checklists and screening questions, may help the recall of what is included in these materials, but may cause reporting less of what is not included. Fatigue due to a long interview is another factor affecting the quality of recall. Finally, recall errors may be in the form of "telescoping," or incorrectly reporting an event time of occurrence; telescoping may be external or internal to the reference period used in the survey.

Recall bias, telescoping, and the importance of bounding were systematically studied in a complex experiment of Residential Alterations and Repairs (Neter & Waksberg, 1965). Downward trends by time in sample were reported in the Current Population Survey (CPS) (Bailar, 1975, 1979), the National Crime Survey (Woltman & Bushery, 1975; Biderman & Cantor, 1984), and the Survey of Income and Program Participation (SIPP) (Kalton et al., 1985). Higher estimates for the most recent recall month were reported in the 1972 Consumer Expenditure Interview Survey (Pearl, 1979) and other surveys (Kemsley, 1979; Biderman & Lynch, 1981; Hill, 1986). Other studies have tried to model learning and memory; one such example is a model of the relationship of use of records and aided recall in various surveys (Sudman & Bradburn, 1973). Recently, cognitive laboratories are being used for introspective studies of respondent attitudes and memory effects (Jabine et al., 1984).

1. CONSUMER EXPENDITURE INTERVIEW SURVEY

1.1. Design Features

The sample of households is interviewed five times at three-month intervals. The first wave of interviews has one-month recall and is used mainly for bounding, a method that helps delimit the reference period of subsequent interviews. The second through fifth waves, with three-month recall, are used in the estimates; data from these waves are analyzed in this study. The reporting unit is the "consumer unit" (CU), i.e., those members of a household who are either related and/or who pool their income to make joint expenditure decisions.

The sample includes four rotation groups of households initiated over the course of a year; each rotation group includes three panels (one for each month of a calendar quarter). The panels are overlapping, i.e., when a panel is in the fifth wave, a replacement panel is in the first wave. Figure 1 shows the interviewing pattern for each of the four rotations by calendar quarter. The first wave for rotation group 1, for instance, is in the third calendar quarter and the three panels (shown collectively as 1) are first interviewed in July, August, or September, respectively. These panels replace panels in the same rotation at the fifth wave (5) and are interviewed for the second time in the fourth calendar quarter, in October, November, or December, respectively (see

FIGURE 1. SAMPLE WAVES BY REPORTING QUARTER

Quarter	Rotation Group 1	2	3	4
	Wave			
III	1 5	4	3	2
IV	2	1 5	4	3
I	3	2	1 5	4
II	4	3	2	1 5
III	1 5	4	3	2
IV	2	1 5	4	3
I	3	2	1 5	4
II	4	3	2	1 5

rotation 1, wave 2, quarter IV).

This design results in the reporting of expenses for any calendar month by respondents who are in the second, third, fourth, or fifth interview and reporting for the first, second, or third recall month. The "first recall month" refers to the month prior to the interview month, and "time in sample" is defined as the number of interviews a consumer unit has received.

Figure 2 shows how this design permits comparisons by time in sample and recall length that are controlled for month of expenditure. Denoted as "53" in the first line and first column is the mean of expenses that occurred in April as reported by panel 1, rotation 1 in the July interview; these respondents are at their fifth interview and report for the third recall month. The second panel in rotation 1 reports data for April in the May interview, fourth wave, first recall month ("41"); and the third panel in rotation 1 reports data for April in June, fourth wave, second recall month ("42"). Panels 2 and 3 are interviewed for the fifth time in August and September, respectively.

1.2. Data Collection

The same questionnaire is used in waves 2 to 5, with the variation that income and occupation data are collected in the second and fifth waves, but only updated in the third and fourth waves. Actual expenditure records are consulted during the interview, whenever possible. While no diaries are kept in this survey, about three-fourths of the respondents use some records of the expenses occurred in the period between two visits.

The reference period is three months preceding the interview month and the portion of the current month up to the interview day; three fourths of the interviews are conducted in the first two weeks of the month. There are four main types of expenditure questions in the in-

FIGURE 2. EXPENDITURE MEANS BY MONTH OF EXPENDITURE

	Rotation Group											
	1			**2**			**3**			**4**		
Panel:	1	2	3	1	2	3	1	2	3	1	2	3
Month				Wave and Recall Month								
April	53	41	42	43	31	32	33	21	22	23	51	52
May	52	53	41	42	43	31	32	33	21	22	23	51
June	51	52	53	41	42	43	31	32	33	21	22	23
→July	23	51	52	53	41	42	43	31	32	33	21	22
Aug.	22	23	51	52	53	41	42	43	31	32	33	21
Sept.	21	22	23	51	52	53	41	42	43	31	32	33

strument, but only two ask for month of expenditure; recall effects can be studied only for these "expenditure classes" (ECs). Section 8 (house furnishings) and section 9 (apparel) are examples of this kind of data collection (method 1). Section 4 (utilities) asks for bills received by the month they were received (method 2). This question is different from the recall month question in method 1, but the respondent is still asked to report on individual expenses, in most cases of a monthly nature.

Other parts of the questionnaire ask for the "usual" monthly expenditure (method 3). Examples of this kind are expenses for food and public transportation (section 20). A "global" figure is collected for the previous three months up to the interview day for other expenses (method 4); expenditures for the "current" month (the month of the interview) are collected separately in this case. Examples of this data collection method are books and entertainment expenses (section 17).

1.3. Data Adjustment

Numerous adjustments are made to the data after collection and some of them have a direct impact on data analyzed in this paper. Discussed briefly are the current-month adjustment, imputation, and aggregation.

The treatment of current-month expenditures influences the means by interview wave and recall month. A diagram of the relationship between month of interview, month of expenditure, and recall month is shown in Figure 3 for expenses collected with methods 1 and 2. Sample units interviewed for the second time in July, for instance, report expenses from April to July. These sample units are interviewed again in October and report expenses for the months of July to October.

Data reported for the same month as the interview month are transferred to the next interview data base during processing, and are added to expenses reported for the third recall month in the next interview. These expenditures are also transcribed onto the next interview questionnaire and made available to the interviewer to minimize duplication.

FIGURE 3. CURRENT MONTH EXPENDITURES

			Month of Interview:			
			2nd Wave			3rd Wave
Month of Expenditure:						
April	May	June	July	Aug.	Sept.	Oct.
Recall Month:						
3	2	1	Current 3	2	1	Current

The proportion of expenses reported for the current month depends on the timing of the interview during the month; on average, about one-third of the third recall month expenses are collected in the previous interview. Note that current-month expenditures reported in the fifth wave are deleted, since these data are out of scope.

A different technique is used for current-month data when global expenses are collected. Current-month data are subtracted from the total quoted by the respondent and not carried over in the data base; these data are not made available in the next interview.

Data imputation is used when the respondent reports an expense but is either unable or unwilling to provide cost information, and this process can affect the results. The imputation methods used are percentage distribution, weighted means, and hot-deck matching (Mopsik & Dippo, 1985).

The amount of item nonresponse is unknown, since purchases and other expenses not mentioned by the reporter cannot be inferred. Exceptions to this problem are obvious household expenses, such as mortgage and rent, which are imputed if not reported, once it is established whether the CU owns or rents. The weighted dollar amount imputed for these items in 1984 was 6% for mortgage interest paid on properties, and 9% for rent. The imputation of all other items occurs only if an item is mentioned but no dollar amount is stated; the percentage dollar amount imputed is about 1% for most items. Major exceptions in 1984 were: alcoholic beverages (6%), new car purchases (4%), used cars (13%), and "miscellaneous" (9%).

Imputation of the reference month occurs when the expenditure month is not stated. This imputation is less than 1% of data collected by month of expenditure.

The analysis of expenditure data is also affected by the level of detail and the classification of items. Expenditures aggregated in broad groups of items may be classified more correctly than specific ECs. At the same time, overall means include items collected and processed differently. One example is the adjustment made to mortgage interest data included in housing and shelter totals. These expenses are amortized at the processing level according to the reported interest rate. A decreasing trend by wave indicates that few new properties are reported during the course of the reporting cycle, but the trend itself is due to "computed" values; these expenses were not tested for time-in-sample effects, therefore.

2. METHOD OF ANALYSIS

Reporters that participated from the second through the fifth wave were selected for the study; this subset was 74% of all reporters in waves 2 to 5. Reporters with missing waves were excluded to allow meaningful comparisons. (Note the response rate for the survey is 85%.)

The analysis was done on a macro-level using reports of expenses occurred in calendar years 1982 and 1983, urban segment. Mean expenditures by time in sample, and, for selected expenditure classes, by recall month were compared for significant effects. The mean expenditure for interview $i = 2, 3, 4, 5$ and expenditure class $J = 1, \ldots, T$ is:

$$\hat{\bar{E}}(i, J) = \sum_{m=1}^{L} W_{im} E(i, J, m) \Big/ \sum_{m=1}^{L} W_{im} \tag{1}$$

where:

$E(i, J, m)$ = expense reported by the mth reporter for interview i and class J, and

W_{im} = weight for mth reporter at interview i (principal person weights).

Differences between means by time in sample for all recall months combined were tested with large sample chi-square statistics X^2 (the sample was approximately $L = 6,600$ for each of the four interviews):

$$X_J^2 = \hat{\Delta}_J^2 / s_{\Delta_J}^2 \sim \chi_1^2 \tag{2}$$

where:

$\hat{\Delta} = \hat{\bar{E}}_i - \hat{\bar{E}}_{i+k}, k = 1, 2, 3, (i + k \leq 5)$, with estimated variance $s_{\Delta_J}^2$.

Computations of the X^2 statistic were done with a method suggested by Koch and Lemeshow (1972) for complex surveys.

For each expenditure class:

$$X^2 = \bar{E}'C'[CVC']^{-1}C\bar{E} \tag{3}$$

where: $\bar{E}$ is a 4 × 1 vector of means by interview,

V is the covariance matrix computed with the method of balanced repeated replication (20 half-samples), and

C is either a 4 × 1 vector or a 3 × 4 matrix testing the following contrasts:

	Means compared	DF
$c' = [1 \quad 0 \quad 0 \quad -1]$	2nd vs. 5th	1
$c' = [0 \quad 1 \quad -1 \quad 0]$	3rd vs. 4th	1
$c' = [1 \quad -1 \quad -1 \quad 1]$	2nd vs. 3rd 4th vs. 5th	1

$$C = \begin{bmatrix} 1 & 0 & 0 & -1 \\ 0 & 1 & -1 & 0 \\ 1 & -1 & -1 & 1 \end{bmatrix}$$ Combined test   3

	Means compared	DF
$c' = [1 \quad -1 \quad 0 \quad 0]$	2nd vs. 3rd	1
$c' = [0 \quad 0 \quad 1 \quad -1]$	4th vs. 5th	1
$c' = [1 \quad 0 \quad -1 \quad 0]$	2nd vs. 4th	1

The analysis of time in sample and recall length used a linear model to test significant effects of both variables and their interaction; concerns over the stability of the covariance matrix led to the choice of an alternative technique for this part of the analysis. The model used logs of weighted mean expenditures for each calendar month of the two years 1982 and 1983 derived from a matrix as shown in Figure 2 (12 × 2 × 4 × 3 = 288 means). Main effects were tested for calendar month (seasonality), recall month, and interview (time in sample). Interaction was tested for recall month and interview. The variable "year" was not explicitly tested for effects since inflation changes were minimal in the years studied. Model estimation was performed with SAS General Linear Models (GLM), a program that does not take into account design effects.

3. FINDINGS AND INTERPRETATION

Although significance testing was carried out for differences in mean expenditures, detailed data are displayed in the form of percentage ratios of the means by wave and recall month to the overall mean for each expenditure class. The overall means, shown in column 1 of Table 1, are methodologically comparable with estimates published in the 1982/83 *Bulletin* (U.S. Bureau of Labor Statistics, 1986); they are somewhat higher than the estimates since the subset of reporters selected for the study had 10% more homeowners and the median family income was about 10% higher than the total sample.

3.1. Time-in-Sample Effects

The findings suggest that time in sample has limited conditioning effects judging from the comparisons of aggregate survey data; when variation is exhibited, time in sample appears to discourage reporting certain types of expenditures but to improve reporting certain other types. Decreased reporting from interview 2 to interview 5 was noted for house furnishing and apparel categories. The percentage ratios for these two interviews, respectively, were 112 and 92 for house furnishings and 103 and 97 for apparel. Increased reporting after the second interview was noted for vehicles, public transportation, and some of the utilities categories (Table 1).

The analysis based on the X^2 test statistic indicates no evidence of response conditioning in over half (10/17) of the EC totals tested; the comparisons for specific categories were generally in the same direction as those for the respective totals, even though not always significant. Food, alcoholic beverages, home maintenance, rent, household operations, entertainment, personal care, reading, education, and personal insurance had no significant comparisons. Utilities, house furnishings, apparel, transportation, health care, tobacco, and miscellaneous had one or more significant comparisons ($\alpha = .05$).

Table 2 shows the X^2 values for selected ECs illustrating particular points of interest in the comparisons. The combined test with df = 3 (column 5) includes the first three tests but is not a mere sum of the individual test values. Critical values for each test were derived from the χ_3^2 distribution, in accordance with simultaneous comparisons allowing any number of contrasts to be tested (Johnson & Wichern, 1982).

A significantly higher fifth interview mean compared to all other means, especially the fourth, was found for the "miscellaneous" section of the questionnaire. This finding points out an aspect of panel survey reporting known as the "last effort" exhibited by the interviewer and the respondent at the last interview. It may indicate a type of conditioning effect confined to certain parts of the questionnaire; notably, "miscellaneous" is one of the last sections (section 19).

A case that lends itself to two concurrent interpretations is the reporting of health care expenses. A significant difference was noted when the second and fifth interviews were compared, the fifth being higher (Table 2, column 2). This class includes outlays (positive values in the data base) and "reimbursements" (negative values in the data base). The upward trend, therefore, may indicate that a smaller number of reimbursements was reported during the fifth interview when

TABLE 1. CONSUMER EXPENDITURE INTERVIEW SURVEY: PERCENTAGE RATIOS BY TIME IN SAMPLE, 1982/83
(Subset of Respondents to Waves 2 to 5) (Urban)

Expenditure Class	Mean Exp. (Annual)	Percentage Ratio[a] Interview				Standard Error of Ratio Interview			
		2	3	4	5	2	3	4	5
Total Expenditures[b]	17,507	100	101	100	100	0.6	0.9	0.8	0.6
Food	3,274	99	101	100	99	0.7	0.8	0.7	0.6
Food at home	2,348	100	100	99	101	0.6	0.6	0.5	0.6
Food away from home	927	100	102	101	96	1.4	2.1	1.8	1.5
Alcoholic beverages	277	100	100	101	100	1.7	1.6	1.3	1.2
Housing	5,922	99	101	99	99	0.9	0.8	0.7	0.8
Shelter	3,261	101	101	99	100	1.1	1.0	1.0	1.4
Owned dwellings	2,141	100	102	99	98	1.6	1.3	1.2	1.9
Mortgage interest	1,274	101	101	99	106	1.6	1.0	1.2	1.2
Property taxes	437	91	99	103	101	2.9	3.9	3.6	4.4
Maintenance, repairs, insurance	430	107	101	93	101	4.4	3.9	4.3	4.8
Rented dwellings	823	100	100	98	100	1.3	1.3	1.0	2.0
Other lodging	297	94	100	105	101	4.1	3.6	3.8	3.2
Utilities, fuels, public services	1,620	99	102	99	102	0.5	0.6	0.5	0.6
Natural gas	349	99	101	100	97	1.1	1.7	1.2	1.7
Electricity	582	97	102	100	99	1.1	0.6	0.7	0.9
Fuel oil and other fuels	126	107	101	95	100	3.7	4.3	3.5	4.2
Telephone	407	100	102	99	101	0.9	0.7	0.7	0.7
Water and other public services	157	99	100	100	92	1.4	2.0	0.8	1.8
Household operations	278	100	101	98	93	3.9	3.1	2.4	4.8
House furnishings and equipment	763	112	98	98	88	3.5	2.7	3.6	2.7
Household textiles	80	119	94	93	127	7.0	4.0	5.6	3.6
Furniture	235	118	104	90	86	8.2	5.9	8.1	7.3
Floor coverings	43	107	77	90	103	14.9	10.2	17.6	16.8
Major appliances	126	120	102	92	112	6.5	5.1	5.2	4.3
Small appliances, housewares	65	100	96	103	101	6.4	2.6	5.8	5.6
Miscellaneous house equipment	214	102	96	112	91	4.8	3.0	7.6	4.8

continues

TABLE 1, continued

Expenditure Class	Mean Exp. (Annual)	Percentage Ratio Interview 2	3	4	5	Standard Error of Ratio Interview 2	3	4	5
Apparel	1,069	103	101	98	97	1.4	1.4	1.4	1.2
Men and boys	272	105	95	101	98	2.0	2.6	2.6	1.9
Women and girls	437	102	103	97	97	2.2	2.7	1.6	1.8
Children under 2	36	100	106	95	99	3.7	2.8	2.7	3.7
Footwear	125	104	102	98	97	2.1	1.9	2.3	2.2
Other apparel products, services	198	104	103	96	97	4.3	3.5	3.4	2.7
Transportation	3,849	94	103	104	98	1.9	2.9	2.1	2.4
Cars and trucks, new (net outlay)	832	81	116	113	89	7.1	10.1	11.1	7.7
Cars and trucks, used (net outlay)	561	91	95	107	107	6.5	12.1	8.5	8.7
Vehicle finance charges	178	96	100	101	103	1.7	1.3	1.3	1.6
Gasoline and motor oil	1,117	100	101	99	101	1.2	0.8	1.0	0.8
Maintenance and repairs	452	105	100	97	99	2.3	2.4	1.8	1.9
Vehicle insurance	340	99	97	103	100	2.1	2.5	2.0	2.7
Public transportation	229	92	111	106	90	3.6	5.2	4.7	4.0
Vehicle rental, licenses, charges	113	103	105	95	96	3.4	4.5	4.2	4.0
Health care	896	95	98	99	109	2.1	2.2	2.2	2.0
Health insurance	259	89	97	103	111	2.6	2.3	3.1	2.0
Medical services	479	97	98	95	110	3.4	3.2	3.3	3.6
Prescription drugs, medical supplies	158	97	102	101	100	2.4	1.7	2.3	1.4
Entertainment	918	100	105	98	97	2.7	2.2	2.9	2.7
Personal care	188	100	100	99	101	1.3	1.1	1.4	1.5
Reading	135	101	99	100	100	1.6	1.4	2.1	1.5
Education	263	95	99	105	101	5.1	5.9	4.3	7.7
Tobacco and smoking supplies	212	101	99	102	98	1.4	1.0	0.9	1.2
Miscellaneous	207	89	85	76	152	6.6	8.5	7.7	8.7
Personal insurance	297	94	97	109	100	6.2	3.0	9.2	4.2

aBase of percentage is the mean expenditure shown in column 1.

bTotal excludes "cash contributions" and "retirement, pensions, social security" expenses.

TABLE 2. TIME-IN-SAMPLE RESULTS FOR SELECTED EXPENDITURE CLASSES (X^2 Values)

| Contrast: | $2\leftrightarrow5$ | | $3\leftrightarrow4$ | $2\leftrightarrow3$ $4\leftrightarrow5$ | Comb. | $2\leftrightarrow3$ | $4\leftrightarrow5$ | $2\leftrightarrow4$ |
| | Decrease[b] | Increase[b] | | | | | | |
EC:[a]	(1)	(2)	(3)	(4)	(5)	(6)	(7)	(8)
Utilities	15.8*	3.2	4.9	0.5	17.2*	7.3	2.9	0.1
House furnishings	8.2*	.	—[c]	1.3	15.9*	7.2	1.0	5.6
Apparel	.	0.6	2.1	0.4	14.2*	0.9	—	8.0*
New cars	.	0.1	1.7	9.2*	10.4*	5.7	1.9	5.7
Gasoline	.	.	0.8	0.9	4.7	0.1	4.3	0.6
Car maintenance	3.0	.	.	1.7	5.1	1.5	0.7	4.7
Health care	.	15.0*	—	2.0	18.9*	1.2	10.3*	1.1
Tobacco	2.6	.	2.0	1.3	10.6*	0.9	9.9*	0.1
Miscellaneous	.	26.8*	0.4	17.8*	35.1*	0.1	25.5*	1.6

*Significant, $\alpha = .05$.

[a]Expenditure classes as defined in Table 1.

[b]Decrease: fifth mean lower than second mean. Increase: vice versa.

[c]Less than 0.05.

compared to the second. Another factor influencing this "upward" pattern could be that more expense records were made available at later stages of panel reporting.

3.2. Time-in-Sample and Recall Effects

Mean expenditures by recall month for the month most distant in time from the interview (recall month 3) were between 15% and 40% lower than the means for the closest month (recall month 1). This relationship was found in the means for all waves combined and in the means by time in sample. The percentage ratios of these means to the overall mean for class totals analyzed by recall length are displayed in Table 3. The ratios for apparel for the four waves combined were: 124 for the first recall month, 100 for the second recall month, and 76 for the third recall month; for house furnishings, these ratios were 116, 96, and 88, respectively. These results were consistent for detailed categories not shown in the table.

Summary results from the significance testing performed are displayed in Table 4. Significant recall effects were found for all the expenditure classes that could be tested, with the exception only of new and used cars, not surprisingly. The test for utilities was marginally significant and it did not indicate the downward pattern by recall month; this was probably due to the way the question is asked: "bills received" rather than expenditures, i.e., a "time gap" effect.

The calendar month was found to be significantly related to nearly all expenditures, i.e., seasonal variation; an exception was medical services. The interaction between time in sample and recall effects was not significant, and this indicates that recall effects are present in all waves of the survey. Time in sample was found to have significant effects only for house furnishings; the significance was borderline for apparel, considering design effects were not taken into account in the test procedure.

The recall effect findings point out the need to know the extent of telescoping. Internal forward telescoping is considered the most likely kind of telescoping in this survey, since each interview bounds the next. This form of telescoping would alter the means in the same direction as recall length bias, so that what is apparent is a combined effect; these two factors tend to affect the various expenses differently. There is no way to separate these two effects using the survey data, but in most cases it can be easily argued that recall effects are evident. Expenditures reported for a given month were actually made earlier in the reporting period, when internal telescoping is present; in this case, current-month expenditures transferred to recall month 3 in the next quarter would be overstated. A steeper decline by recall month would be observed if no adjustments were made for current-month expendi-

TABLE 3. CONSUMER EXPENDITURE INTERVIEW SURVEY: PERCENTAGE RATIOS BY WAVE AND RECALL MONTH, 1982/83
(Subset of Respondents to Waves 2 to 5) (Urban)

| Expenditure Class | Recall Month | Percentage Ratio[a] | | | | | Standard Error of Ratio | | | | |
| | | Interview | | | | | Interview | | | | |
		2–5	2	3	4	5	2–5	2	3	4	5
Home maintenance, repairs	1	112	119	112	109	108	3.4	5.5	6.8	6.6	6.8
	2	109	118	104	95	118	4.0	11.5	6.4	5.1	10.2
	3	79	85	80	77	76	3.4	6.0	5.3	5.7	4.5
Utilities, fuels, services	1	99	98	100	97	100	0.3	0.8	0.5	0.6	0.8
	2	102	101	101	102	102	0.3	0.7	0.9	0.6	0.8
	3	100	98	102	99	99	0.4	0.8	1.3	0.6	1.0
House furnishings	1	116	122	115	111	115	4.2	8.9	5.6	8.4	5.3
	2	96	110	87	100	85	4.2	9.2	3.9	8.1	4.6
	3	88	103	92	81	77	2.8	6.3	6.7	6.2	5.0
Apparel	1	124	126	127	123	120	1.3	2.5	3.1	3.3	2.5
	2	100	105	99	98	100	1.5	2.9	2.7	2.6	2.3
	3	76	80	78	73	73	1.2	2.4	2.6	2.1	2.1
Cars and trucks, new (net)	1	108	84	121	123	105	6.5	12.8	15.4	19.7	18.8
	2	95	93	93	98	97	9.5	10.8	18.2	18.2	17.2
	3	96	66	136	117	66	9.9	20.3	27.7	29.0	12.5
Cars and trucks, used (net)	1	118	121	98	126	129	8.7	17.8	16.8	20.7	13.9
	2	86	87	88	84	87	5.3	10.0	17.5	12.8	13.0
	3	95	65	100	112	104	8.6	7.1	17.7	16.0	20.5
Car maintenance and repairs	1	122	128	125	121	113	2.8	5.1	5.1	4.7	5.7
	2	98	103	97	91	103	1.9	4.3	3.4	5.1	5.3
	3	80	83	76	78	81	2.8	5.9	3.9	3.9	4.1
Medical services	1	112	114	92	111	131	3.8	7.3	7.3	7.3	7.4
	2	94	92	95	83	105	3.9	3.6	5.8	14.0	7.7
	3	94	85	106	93	95	4.2	8.1	9.8	13.0	7.3
Education	1	112	116	116	106	111	6.1	15.5	17.9	6.5	13.5
	2	93	85	80	97	112	4.0	7.0	8.6	8.2	13.5
	3	95	85	101	112	81	5.0	9.2	13.3	11.1	8.4

[a]Base of percentage is the mean expenditure shown in column 1 of Table 1.

TABLE 4. RESULTS FROM GLM: PR > F

EC:[a]	REF-MO[b]	RECMO[c]	INTV-N[d]	RECMO*INTV-N
Home maintenance	.0001	.0001	.28	.78
Utilities	.0001	.05	.27	.72
House furnishings	.0001	.0001	.0009	.64
Apparel	.0001	.0001	.05	.88
New cars	.01	.17	.20	.49
Used cars	.005	.08	.43	.28
Car maintenance	.0001	.0001	.40	.35
Medical services	.23	.0004	.17	.75
Education	.001	.0015	.84	.63

[a]Expenditure classes as defined in Table 1.

[b]REF-MO = 1, . . . , 12 calendar month.

[c]RECMO = 1, 2, 3 recall month.

[d]INTV-N = 2, 3, 4, 5 interview wave.

tures. The present findings from the recall month analysis would be reinforced rather than weakened in the presence of telescoping.

Similar relationships were found in the means from all sample reporters, especially the trends for house furnishings and apparel. More variations were noted in these means, due to the fact that one-quarter of these reporters are at different stages of the panel cycle.

4. CONCLUSIONS

Significant variation in the means by time in sample was found in less than half the published means; these conditioning effects were moderate and, in some cases, minimal. Recall-length effects were widespread among the expenditure classes for which expenditure month is collected; these effects were substantial. Typical time-in-sample and recall effects (i.e., downward trends) were found for a number of expenditures; house furnishings and apparel sections appear the most affected. Different patterns are more typical for other expenditures, and these may also suggest incomplete reporting; shifting the reporting of certain expenditures to later interviews due to better availability of records may cause certain expenditures never to be reported; higher reports for the last interview may point out the previous interview reports are low.

The recall-month findings raise questions that extend to parts of the survey not analyzed in this study, i.e., those expenses collected as usual or global amounts. Since a bias could be observed for sections collected by month of expenditure, it is possible that this is symptomatic of the rest of the survey sections. Conversely, the question of bias could indicate a selective phenomenon due to the sequence of sections, the frequency and saliency of purchases, or other factors related to certain expenditures. Research should be conducted that would identify which expenditures tend to be forgotten, omitted, and/or telescoped and test the extent to which questionnaire length influences these response errors.

REFERENCES

Bailar, B. A. (1975), "The Effects of Rotation Group Bias on Estimates from Panel Surveys," *Journal of the American Statistical Association*, **70**, 23–30.

Bailar, B. A. (1979), "Rotation Sample Biases and Their Effects on Estimates of Change," *Bulletin of the International Statistical Institute*, **48**(2), 385–407.

Biderman, A. D., & Cantor D. (1984), "A Longitudinal Analysis of Bounding, Respondent Conditioning and Mobility as Sources of Panel Bias," *Proceedings of the Section on Survey Research Methods*, American Statistical Association, 708–713.

Biderman, A. D., & Lynch, J. C. (1981), "Recency Bias in Data on Self-Reported Victimization," *Proceedings of the Social Statistics Section*, American Statistical Association, 31–40.

Gieseman, R. W. (1987), "The Consumer Expenditure Survey: Quality Control by Comparative Analysis," *Monthly Labor Review*, March 8–14.

Hill, D. H. (1986), "An Additive Model of Recall Error: Analysis of SIPP Data," *Proceedings of the Section on Survey Research Methods*, American Statistical Association, 226–230.

Jabine, T. B., Straf, M. L., Tanur, J. M., & Tourangeau, R. (1984), *Cognitive Aspects of Survey Methodology: Building a Bridge between Disciplines*, National Academy Press, Washington, DC.

Johnson, R. A., & Wichern, W. (1982), *Applied Multivariate Statistical Analysis*, Prentice-Hall, Englewood Cliffs, NJ, p. 203.

Kalton, G., McMillen, D. B., & Kasprzyk, D. (1986), *Nonsampling Error Issues in the Survey of Income and Program Participation* (SIPP Working Paper Series No. 8602), U.S. Bureau of the Census.

Kemsley, W. F. (1979), "Collecting Data on Economic Flow Variables Using Interviews and Record Keeping," in *The Recall Method in Social Surveys* (L. Moss & H. Goldstein, eds.), London University Institute of Education, London, pp. 115–133.

Koch, G. G., & Lemeshow, S. (1972), "An Application of Multivariate Analysis in Complex Sample Surveys," *Journal of the American Statistical Association*, **67**, 780–783.

Mopsik, J. A., & Dippo, C. S. (1985), "The Data Adjustment for the Consumer Expenditure Interview Survey," *Proceedings of the Statistical Computing Section*, American Statistical Association, 134–139.

Neter, J., & Waksberg, J. (1965), *Response Errors in Collection of Expenditure Data by Household Interviews: An Experimental Study* (Technical Paper No. 11), U.S. Bureau of the Census.

Pearl, R. B. (1979), *Reevaluation of the 1972–73 U.S. Consumer Expenditure Survey* (Technical Paper No. 46), U.S. Bureau of the Census.

Sudman, S., & Bradburn, N. M. (1973), "Effects of Time and Memory Factors on Response in Surveys," *Journal of the American Statistical Association*, **68**, 805–815.

U.S. Bureau of Economic Analysis (1986), "Business Conditions Digest," *Bulletin*, **26**(7).

U.S. Bureau of Labor Statistics (1986), *Consumer Expenditure Survey, Interview Survey, 1982–83* (Bulletin No. 2246), U.S. Government Printing Office, Washington, DC.

Woltman, H., & Bushery, J. (1975), "A Panel Bias Study in the National Crime Survey," *Proceedings of the Social Statistics Section*, American Statistical Association, 159–167.

Panel Effects in the National Medical Care Utilization and Expenditure Survey*

Larry S. Corder and Daniel G. Horvitz

The National Medical Care Utilization and Expenditure Survey (NMCUES) and the Health Interview Survey (HIS) are unusual in that both collected similar information for the same calendar year (1980) while employing different survey designs. The NMCUES was a five-round (or five-wave) longitudinal survey that followed a cohort of individuals throughout the calendar year (Bonham, 1983) while the HIS employs a 26 two-week accumulating sample design (Jack, 1981). From an analytical viewpoint, the principal difference between the two studies lies in their respective abilities to estimate the incidence and prevalence of medical conditions (HIS) and medical care expenditures and their sources of payment (NMCUES). The structure of the HIS, a continuous survey since 1958, directly informed and guided the construction of the NMCUES data collection instrument in several important analytical areas. These included estimation of reported medical conditions, disability days, and perceived health status. Collection of medical care utilization data was conducted at a much more detailed and extensive level in the NMCUES than in the HIS. Estimates of physician utilization, hospital discharges, length of stay, and dental utilization were made in each study.

This unusual combination of two study designs using similar data collection instruments to collect data in the same calendar year presents an opportunity to examine the differences in reporting basic social statistics when those statistics are generated with a cross-sectional design (HIS) and a panel design (NMCUES). We concentrate on medical conditions and medical care utilization and examine the differences between and among quarterly and annual estimates in the HIS relative to the NMCUES. We also examine quarter-by-quarter trends in the NMCUES data for conditioning, e.g., time in panel (the effect, positive or negative, of being interviewed before), telescoping (the misplacement of events in time by the respondent, such as moving a hospital stay that occurred six months ago to three months ago when interviewed), and omission effects (the behaviors of respondents that cause the omission of events from the survey report based on the "learning" that occurs during time in panel). The results we obtain are preliminary.

304

1. LIMITATIONS

The approach of empirical comparisons that we use has merit, since many of the conditions for a natural experiment to determine the effect of a panel design on reporting of basic social statistics in the health area are satisfied. These conditions are common time frames, use of similar data collection procedures in some cases and identical ones in others, and an accumulating cross-sectional design versus a five-wave panel design. Despite some common conditions, it is important to recognize that design differences could result in empirical differences. For example, the first two NMCUES contacts were personal interviews, the next two were mainly conducted by telephone, and the last was a personal interview. With this design, it is possible that NMCUES estimates for the third and fourth quarters of the year could be less than HIS estimates if, in fact, telephone interviews tend to produce fewer reports of medical care events than do personal interviews. In addition, the last NMCUES interview, conducted in 1981, collected data on medical care events occurring only in 1980. Thus, with any significant telescoping of 1980 events into 1981, interviews taken in the last NMCUES round would show fewer medical care events than the HIS for the last quarter of 1980. Finally, if the number of proxy interviews increased by round in the NMCUES and if fewer events are reported by proxy respondents, then this, too, could be a possible source of explanation for differences between the two surveys. These design differences demonstrate that we can only speculate on the causes of any observed differences.

Thus, of the two main elements we wish to test for in this paper—conditioning within a panel and differences between panel and cross-sectional reporting—the latter is confounded to some unknown degree by differences in reference period and necessarily associated differences in telescoping of reporting and errors of omission. These concerns should be kept in mind when examining the comparisons between the estimates from the two surveys.

2. THE STUDY

Quarterly and annual estimates for 1980 are produced by the HIS for medical conditions (acute), disability days, and physician and hospital utilization. Counts of events and rates of reporting across a wide variety of variables were made available by National Center for Health Statistics (NCHS) staff in the form of unpublished tables in order to make this study possible. For NMCUES, quarterly estimates of the incidence of acute medical conditions and physician and hospital utilization were generated from the NMCUES public use files by employing visit date and condition date information where available. The pattern of comparisons presented in the next section is determined by: (1) the unpublished HIS tables, which are available for all the comparisons, (2) the extent to which dates could be assigned to events in the NMCUES so that quarterly estimates could be generated, and (3) the extent to which data collection instruments and strategies were comparable, apart from differences in reference period.

These requirements led to the comparison of a limited number of quarterly tables for each study. For example, dates could not be assigned to NMCUES disability days, thus excluding the possibility of quarterly comparisons; intervals since last physician and dental visits were not asked in the NMCUES; and the limitation-of-activity questionnaire battery asking about functional limitations was included in the first NMCUES round of interviewing only. Thus, only a limited number of the 25 unpublished HIS quarterly tables for 1980 were reproduced for the 1980 NMCUES.

There are three components to the following analysis. First, the available and appropriate quarter-based estimates from HIS and NMCUES are presented and tests of significance of the differences between them are performed. However, two observations should be made. Certain HIS tables, while based on quarterly data, present annual estimates in each quarter. In this situation, quarterly estimates from NMCUES are annualized to ensure comparability for comparison purposes. The HIS population estimates represent the 1980 mid-year civilian, noninstitutionalized population while the NMCUES represents, in this case, the 1980 civilian, noninstitutionalized population that remained in the panel throughout the calendar year. Certain NMCUES tables were run both with all deceased persons included and with all deceased persons excluded, and comparison of rates is conducted in both cases.

The second component comprises a variety of comparisons of annual data and reports the results of tests of the significance of the differences between the two independent samples. Results may be interpreted in the context of known differences in study design. The third component examines the phenomenon of the time-in-panel effect.

3. THE COMPARISONS

Three basic sets of comparisons are made in this study. First, quarterly estimates are presented, with annual estimates being included for acute conditions and medical care utilization variables. Second, these estimates are examined and tested for a "conditioning" or "panel" effect. That is, observed significant within-quarter differences between HIS and NMCUES estimates for these variables could be attributed to a combination of a time-in-panel effect and a recency effect (that is, an effect due to the length of the recall period). Specifically, quarter 1 comparisons in levels of reporting reflect differences in study design, primarily telescoping and recall period, while the reporting levels for quarters 2–4 in NMCUES reflect conditioning and omissions as well as seasonal effects. Trends in HIS quarterly reporting reflect seasonal differences only, while NMCUES between-quarter comparisons in quarters 2–4 reflect differences, if any, that may be attributed to recall period and conditioning. Note that conditioning may have positive or negative effects on reporting. Third, differences in both direct quarterly estimates and annual estimates based on quarterly data are tested to determine if they are significant. In general, a pattern of increasing difference over time between NMCUES and HIS would indicate a panel effect. In practice, both the quarter-by-quarter change in each survey and differences between surveys must be examined (Figure 1).

Different trends across quarters clearly can be a reflection of conditioning in NMCUES and not in HIS. That is, a pattern of increasing differences across quarters would indicate conditioning in NMCUES. Both within-quarter comparisons and across-quarter differences are important for understanding the effect of design on reporting levels in this less-than-perfect natural experiment. Furthermore, it is arguable that extended recall periods could offset any positive gain from conditioning present in the NMCUES design. Indeed, given the studies' designs, it is difficult to disentangle these two effects.

4. FINDINGS

As we noted above, the number of tables available from the NMCUES that correspond to HIS quarterly detail is limited. They include acute medical conditions per 100 persons by condition group, physician visits per 100 persons, hospital discharges per 100 persons, and the distribution of the number of hospital episodes for each person. This limited set of tables is presented in Tables 1–5. Since NMCUES arguably captures more medical utilization associated with deaths than HIS, several NMCUES tables are presented with and without deaths (Horvitz, 1966). Further, we note that we have conducted numerous separate significance tests, each with a significance level of 0.05. However, we present the Z scores to allow for more conservative tests of significance to be conducted with the same data in light of the probability that some of the multiple comparisons conducted would achieve significance by chance alone.

Table 1 illustrates a pattern of differences in the reporting of a basic utilization variable. The most dramatic difference between HIS and NMCUES occurs in quarter 1, indicating NMCUES telescoping in an unbounded interview. The pattern across quarters is much smoother in HIS than NMCUES. The differences between the NMCUES and HIS estimates are not significant for quarters 2–4. The pattern of differen-

FIGURE 1. EXAMPLES OF A NMCUES PANEL EFFECT BENCHMARKED BY HIS

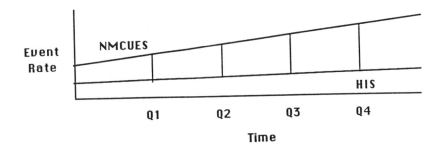

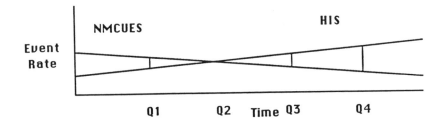

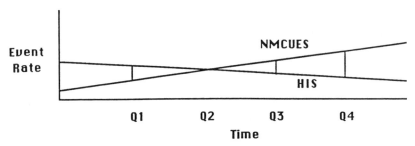

ces between quarters is more pronounced in NMCUES than in HIS—
due to a substantial drop in the rate between quarters 1 and 2 in
NMCUES—but no significant differences are observed. No time-in-
panel effect is apparent from these differences; that is, the pattern of
differences between quarter 2 and quarter 3 is relatively flat. Recency
of interview may account for these findings along with telescoping (forward in quarter 1, and backward in quarter 4) and errors of omission,
as NMCUES used twelve-week recall versus six months in HIS.

Table 2 illustrates that, across quarters, NMCUES captured a
higher level of initial hospitalizations than HIS. NMCUES hospital
data have shorter average recall compared to HIS and therefore more
get reported. These are likely to be reports of very short-stay, first

TABLE 1. HOSPITAL DISCHARGES PER 100 PERSONS:
NMCUES AND HIS BY QUARTER AND ANNUALLY[a]

| | NMCUES | | | | HIS | | Significance Tests | |
| | Without Deaths | | With Deaths | | | | Without Deaths | With Deaths |
	Mean	S.E.	Mean	S.E.	Mean	S.E.	Z_1	Z_2
Quarter								
1	16.4	.89	17.6	.94	13.7	.89	−2.15	−3.01
2	14.3	.77	15.3	.83	13.8	.88	−0.43	−1.24
3	14.4	.75	15.3	.77	14.5	.87	0.09	−0.77
4	14.8	.85	15.6	.87	13.7	.89	−0.89	−1.53
Annual	15.00	.45	19.00	.52	13.9	.90	−1.09	−2.02

Significance Tests of Quarter–Quarter Differences−Z

1–2	1.78		1.83		0.08			
1–3	1.70		1.81		−0.64			
1–4	1.30		1.56		0.00			

[a]Annualized estimates based on quarterly data for NMCUES and six-month data for HIS.

hospitalizations (Givens & Massey, 1985). We also suspect that the counting rules for hospitalizations were implemented slightly differently in the two surveys. The main finding from this table is the consistency of HIS estimates across quarters versus the initial hospitalization spike that is apparent in the NMCUES data, further illustrating the recency of interview effect.

Twelve-week versus two-week recall dominates the interpretation of Table 3. NMCUES also collected ambulatory visits in far more detail than the HIS, employing several questionnaire batteries where HIS employed one basic set of questions. Significant differences exist between the two surveys, annually and in quarters 1, 3, and 4. Indeed, the comparison is not affected by the inclusion of deaths and the pattern of differences resembles hospitalizations except that NMCUES shows consistently lower rates than does HIS. The trend across quarters does reveal some potential for a conditioning effect in NMCUES in quarter 2 in that the NMCUES rate does not fall as does the HIS. Patterns of difference between quarters are flat except in quarter 2. No appreciable difference in magnitude of reporting is present between quarter 1 and quarter 2. One interpretation of the findings from Table 3 may be that NMCUES respondents learn to anticipate the effect of responding affirmatively to probe questions that re-

TABLE 2. DISTRIBUTION OF PERSONS' HOSPITAL EPISODES BY QUARTER: NMCUES AND HIS ANNUALIZED ESTIMATES

Hospital Episodes	Quarter	NMCUES Without Deaths %	NMCUES Without Deaths S.E.	NMCUES With Deaths %	NMCUES With Deaths S.E.	HIS %	HIS S.E.	Significance Tests Without Deaths Z_1	Significance Tests With Deaths Z_2
0	1	85.4		84.9		89.4			
	2	87.5		86.7		89.9			
	3	86.8		86.3		89.6			
	4	87.1		86.4		89.5			
1	1	12.8	.7	13.1	.8	8.6	.4	−5.21	−5.03
	2	11.1	.6	11.7	.6	8.2	.4	−4.02	−4.85
	3	12.0	.6	12.4	.6	8.5	.4	−4.85	−5.41
	4	11.5	.6	11.9	.6	8.6	.4	−4.02	−4.58
2	1	1.6	.2	1.2	.2	1.5	.4	−0.22	0.67
	2	1.1	.2	1.2	.2	1.4	.4	0.67	0.45
	3	.8	.1	.9	.2	1.4	.4	1.46	1.12
	4	1.2	.2	1.4	.2	1.3	.4	0.22	−0.22
3+	1	.2	.1	.3	.1	.5	.4	0.73	0.49
	2	.3	.1	.4	.1	.5	.4	0.49	0.24
	3	.3	.1	.4	.1	.6	.4	0.73	0.49
	4	.2	.1	.3	.1	.5	.4	0.73	0.49

Significance Tests of Quarter–Quarter Differences—Z

1	1–2	1.84	1.40	0.71
	1–3	0.87	0.70	0.18
	1–4	1.41	1.20	0.00
2	1–2	1.77	0.00	0.18
	1–3	3.58	1.06	0.18
	1–4	1.41	−0.71	0.35
3+	1–2	−0.71	−0.71	0.00
	1–3	−0.71	−0.71	−0.18
	1–4	0.00	0.00	0.00

quire extensive responses later in the questionnaire. Significance tests of quarterly differences may be interpreted in this way, indicating growing respondent fatigue over time.

In Table 4, NMCUES reports significantly lower numbers of acute conditions annually and in each quarter of comparison than does HIS. Both HIS and NMCUES link condition reporting to medical events. Both surveys identify medical conditions using similar survey methods. Trends in quarterly reporting of conditions in each survey show the highest levels in quarter 1 and quarter 4. The largest significant differences between surveys also occur in these quarters. These estimates

TABLE 3. PHYSICIAN VISITS PER PERSON: NMCUES
AND HIS BY QUARTER AND ANNUALLY

	NMCUES				HIS		Significance Tests	
	Without Deaths		With Deaths				Without Deaths	With Deaths
	Mean	S.E.	Mean	S.E.	Mean	S.E.	Z_1	Z_2
Quarter								
1	1.07	.02	1.09	.02	1.30	.02	8.13	7.14
2	1.05	.02	1.05	.02	1.10	.02	1.77	1.77
3	.97	.03	.98	.03	1.20	.02	6.38	6.10
4	.96	.02	.96	.02	1.20	.02	8.49	8.49
Annual	4.05	.08	4.07	.08	4.80	.05	7.95	7.74

Significance Tests of Quarter–Quarter Differences—Z

1–2	6.71	1.41	7.07
1–3	2.77	3.05	3.54
1–4	3.89	4.60	3.54

exhibit a pattern of differences that could be accounted for by telescoping, present to some extent in the NMCUES hospital discharge data. Further, comparison of NMCUES data with deaths included makes no apparent difference in the findings. Assignment of acute condition codes was conducted in a comparable manner in both studies; either a disability or a medical consultation triggered the completion of a condition section for coding. One hypothesis is that two-week versus twelve-week recall directly affected reporting levels (Sudman & Bradburn, 1973). One might also speculate that certain conditions that were initially reported as acute became chronic over the study period and were coded as such. That is, NMCUES offers the survey respondent the opportunity to redefine a condition over the the study period while HIS does not. Nevertheless, acute conditions are reported in HIS at more than double the NMCUES overall rate, which allows speculation that the study design itself is the cause of the difference (Bonham, 1981). The quarter-by-quarter patterns of reporting and differences between quarters are similar in the two studies. Again, no time-in-panel effect is evident. That is, the pattern of quarter-to-quarter differences between the two surveys does not exhibit increases over time. Indeed the differences go up and down by quarter. However, both forward and backward telescoping may be present in NMCUES quarters 1 and 4.

The detail presented in Table 5 adds the observation that the NMCUES data on acute conditions are consistently reported at lower

TABLE 4. ACUTE CONDITIONS PER 100 PERSONS
BY QUARTER AND ANNUALLY[a]

	NMCUES		HIS		Significance Tests
	Mean	S.E.	Mean	S.E.	Z
Quarter					
1	33.4	.8	73.4	1.8	20.31
2	17.4	.4	72.4	1.3	18.38
3	18.1	.5	43.4	1.2	19.46
4	28.6	.6	63.0	1.6	20.13
Annual	97.5	2.8	222.2	4.00	25.54

Significance Tests of Quarter–Quarter Differences – Z

1–2	17.89	13.96
1–3	16.22	13.87
1–4	4.80	4.32

[a]NMCUES estimates include deaths. The annual estimate without deaths is 97.5 in NMCUES.

levels than HIS, but that time effects are not present. Further, injuries are not apparently reported at as high a level in NMCUES as in HIS. Other detailed conditions exhibit the same general pattern of significant differences observed in the aggregated data.

5. DISCUSSION

The findings presented in Tables 1–5 illustrate the complex nature of the relationship between estimates generated by the two study designs. They further illustrate the difficulty inherent in attempting to determine the conditioning effect of a panel design on reporting by the method of this paper, which attempts to take advantage of the similarities (questionnaire design) and differences (cross-sectional control on panel quarterly estimates) in the two survey designs conducted in the same year. Formally disentangling panel and recency effects cannot be fully addressed with the available data. Thus our results are certainly tentative. We show a clear number of differences between studies and trends over time within studies across a small number of health indicators. Our clearest evidence indicates that accurately accounting for deaths increases hospital utilization reporting in the NMCUES versus the HIS while it is apparent that the unbounded reference period in NMCUES in quarters 1 and 4 led to some telescoping. Also, differences in levels of condition reports are clear. Further, some evidence exists that suggests higher levels of reporting of short-term

TABLE 5. DETAILED ACUTE CONDITIONS PER 100
PERSONS BY QUARTER AND ANNUALLY[a]

	NMCUES		HIS		Significance Tests
	Mean	S.E.	Mean	S.E.	Z
Infective and Parasitic					
Quarter					
1	3.4	.2	7.4	.1	19.43
2	2.9	.2	5.6	.4	6.04
3	2.7	.2	5.8	.4	6.93
4	2.8	.2	5.8	.4	6.71
Annual	11.8	.5	24.6	.9	12.43
Significance Tests of Quarter–Quarter Differences—Z					
1–2	1.86		4.37		
1–3	2.60		3.88		
1–4	2.23		3.88		
Respiratory					
Quarter					
1	24.0	.6	45.5	1.3	15.02
2	8.0	.3	16.8	.6	13.12
3	8.0	.3	16.2	.6	12.22
4	18.8	.5	37.7	1.1	15.64
Annual	58.0	1.2	116.2	2.4	21.69
Significance Tests of Quarter–Quarter Differences—Z					
1–2	23.85		20.04		
1–3	23.85		20.46		
1–4	6.66		4.58		
Digestive					
Quarter					
1	1.7	.1	3.2	.3	4.74
2	1.6	.1	2.8	.2	5.37
3	1.9	.1	2.7	.3	2.53
4	1.5	.1	2.8	.3	4.11
Annual	6.9	.3	11.4	.6	5.71
Significance Tests of Quarter–Quarter Differences—Z					
1–2	0.71		1.11		
1–3	− 1.41		1.18		
1–4	1.41		0.94		

continues

TABLE 5, continued

	NMCUES		HIS		Significance Tests
	Mean	S.E.	Mean	S.E.	Z
Injuries					
Quarter					
1	.7	.1	7.7	.7	9.90
2	1.0	.1	8.9	.5	15.49
3	1.1	.1	9.3	.5	16.08
4	1.2	.1	7.5	.5	12.36
Annual	4.0	.2	33.4	1.0	28.83

Significance Tests of Quarter–Quarter Differences — Z

1–2	−2.12		−1.39		
1–3	−2.83		−1.86		
1–4	−3.54		0.23		

All Other Acute					
Quarter					
1	3.3	.2	9.6	.5	11.70
2	3.8	.2	8.4	.5	8.54
3	4.3	.2	9.4	.5	9.47
4	4.3	.2	9.2	.5	9.10
Annual	15.9	.5	36.6	1.1	17.13

Significance Tests of Quarter–Quarter Differences — Z

1–2	−1.77		1.70		
1–3	−3.54		0.28		
1–4	−3.54		0.57		

[a]NMCUES estimates include all deaths.

hospitalization in the NMCUES due to a recency effect just as lower levels of reporting of physician visits may be attributed to that cause or respondent omissions/fatigue. Last, some evidence of conditioning is apparent in the NMCUES physician visit data, which may have its source in errors of omission as well as recency.

No consistent results for level of reporting were found across the variables of interest. Physician visits showed a clear pattern of differences across quarters and annually between the two studies. HIS rates were higher. Last, a pattern of lower levels of reporting of acute conditions was noted in NMCUES.

All of the three measures of interest were different, to some degree, between the two surveys, where quarterly data were available for a detailed examination. However, no obvious conditioning effect could be observed in the reported data. Rather, the observed differences could be attributed to treatment of deaths, differences in reference period, telescoping, or errors of omission with some degree of assurance in most comparison cases.

Therefore, these initial results, with all their limitations, do not strongly favor the collection of medical care data on a national basis with either a cross-sectional or a panel design. Rather, they illustrate the importance of design considerations and their effects on survey results. However, our principal finding is that quarterly estimates from NMCUES, benchmarked by HIS, show no clear pattern of panel conditioning by our method of comparing patterns of change in quarterly results between studies for a five-round, one-year panel of continuously enrolled individuals.

It should be noted that differences in study design between NMCUES and HIS, guided by the need to collect expenditure data, are clearly as important as understanding the extent of any response effects. Our limited results do not, in our opinion, demonstrate negative aspects of the panel approach, but rather should be viewed as a limited confirmation that a panel approach dictated by data collection requirements may be designed to produce statistics unaffected by conditioning.

6. FURTHER RESEARCH

Accurately concluding the presence or absence of time-in-panel effects by the comparison of estimates as presented in this paper is very difficult. What methods are open to the serious survey practitioner concerned with such effects in a panel survey? Two alternatives come to mind. The first is to design the panel survey in such a way that the data collected can be used to estimate one or more measurement error components. The second is to use multivariate model-based estimation, again using the data collected to estimate the parameters in the model. Both methods require what Horvitz (1985) and Horvitz et al. (1987) have referred to as "standard measures" of accuracy for systematic errors in the measurement process.

The notion of acceptable standards of accuracy for systematic measurement errors deserves attention, in view of the fact that most surveys, and panel surveys in particular, need to consider the mode of measurement, respondent rules, reference periods, and potential panel effects. The survey research community has not been as concerned with bias in the measurement process as it should be. It is quite clear that surveys that require respondents to report past events are subject to rather significant systematic measurement errors, oftentimes more so if proxy respondents are permitted. The level of error (underreporting) is correlated, as we might expect, with the presence or absence of a temporal bound and with the recency of the event, that is, the length of time between the date of interview and the date of the event. It is also associated with whether respondents are reporting events that occurred to themselves or events that occurred to someone else in their family or household.

The use of a "standard measure" of accuracy implies deciding, for example, that personal interviews, self-responses, two-week reference periods, the second interview in a panel survey, etc., are suitable standards against which to compare alternatives to these design features. The bias in the alternatives can be measured relative to the standards even though in exact terms the standards may themselves not be unbiased. The use of such standards opens up considerable potential for either design-based (or model-based) adjustments for response biases arising from the use of proxy respondents, unbounded interviews, time in panel, lengthy reference periods, etc. Estimates of survey measurement errors that are based on "standard measures" of accuracy for the measurement process are designated as "standard unbiased" estimates in the sense that such estimates are accurate relative to a standard measure (or a set of standard measures) that, in absolute terms, could still be biased.

The Bureau of the Census already uses survey designs that permit "standard unbiased" estimates of some measurement error components. The most effective is the rotating panel design with overlapping reference periods, such as is used in the National Crime Survey (NCS) and in the Survey of Income and Program Participation (SIPP). It is possible with these designs to use the data collected to derive "standard unbiased" estimates of time-in-panel bias, bounding errors, and recency effects. For example, the "standard unbiased" estimate of time-in-panel bias is made possible by the fact that, with a rotating panel design, comparable samples in every respect, except for time in panel, provide data for each time period. Thus, one sample may be in the panel for the sixth time, a second sample for the fifth time, a third sample for the fourth time, etc., when reporting data for the same time period. Using this design feature, plus the assumption of a "standard measure" such as "the second interview has the least time-in-panel bias" will produce useful "standard unbiased" estimates of time-in-panel bias that are unconfounded with the period effects.

Design-based "standard unbiased" estimates of measurement error components require judicious use of randomization, such as in the above illustration of the use of rotated (or staggered) samples to remove the possible confounding of time-in-panel effects with period effects. Multivariate regression estimation of measurement error components requires, at a minimum, sufficient balancing of the sample with respect to length of reference period for data collected about a specific time period, self versus proxy interviews, bounded versus unbounded interviews covering a specific time period, etc., in order to minimize the level of confounding in the resulting estimates. Design-based estimates are also possible for each of these components, and would be preferred, since the likelihood of confounding is essentially removed. Thus, it is preferable to designate a random subsample to be interviewed with both a proxy respondent and a self-respondent to estimate the bias in proxy respondent data.

Multivariate model-based comparisons of measurement error components in the National Crime Survey have been made by LaVange and Folsom (1985). These estimates had the advantage of the rotation design for estimating lack of bounding, recency, and time-in-panel effects, but relied on sample balancing for the estimates of proxy-respondent and mode-of-interview effects. The resulting measurement error components were then used to adjust the survey estimates using a set of "standard measures," including self-response interviews.

Multivariate regression can be used to estimate "standard unbiased" time-in-panel effects in the NMCUES data, along with proxy-respondent, mode-of-interview, and length-of-reference-period effects, but not without admitting the potential for confounding, particularly with period effects. This, however, is another stage of research.

REFERENCES

Bonham, G. S. (1981), "Comparison of National Estimates: National Medical Care Utilization and Expenditure Survey and National Health Interview Survey" (Unpublished paper), University of Louisville, Louisville, KY.

Bonham, G. S. (1983), *Procedures and Questionnaires of the National Medical Care Utilization and Expenditure Survey* (Series A, Methodological Report No. 1, DHHS Pub. No. 83-20001). U.S. Public Health Service, U.S. Government Printing Office, Washington, DC.

Dicker, M., & Sunshine, J. (1987), *Family Use of Health Care, 1980 National Medical Care Utilization and Expenditure Survey* (Series B, Descriptive Report No. 10, DHHS Pub. No. 87-20210), U.S. Public Health Service, U.S. Government Printing Office, Washington, DC.

Givens, S. D., & Massey, J. T. (1985), "Comparison of Statistics from a Cross-Sectional and a Panel Study," *Proceedings of the Section on Survey Research Methods*, American Statistical Association, 640-645.

Horvitz, D. G. (1966), *Computer Simulation of Hospital Discharges* (Series 2, No. 13, PHS Pub. No. 1000), U.S. Public Health Service, U.S. Government Printing Office, Washington, DC.

Horvitz, D. G. (1985), "Discussion," *First Annual Research Conference*, U.S. Bureau of the Census, 372-375.

Horvitz, D. G., Folsom, R. E., & LaVange, L. M. (1987), "The Use of Standards in Survey Estimation," *Proceedings of the Section on Survey Research Methods*, American Statistical Association, 546-551.

Jack, S. S. (1981), *Current Estimates from the National Health Interview Survey* (Series 10, No. 139, DHHS Pub. No. 82-1567), U.S. Public Health Service, U.S. Government Printing Office, Washington, DC.

LaVange, L. M., & Folsom, R. E. (1985), "Regression Estimates of National Crime Survey Operations Effects: Adjustments for Nonsampling Bias," *Proceedings of the Social Statistics Section*, American Statistical Association, 109-114.

Sudman, S., & Bradburn, N. M. (1973), "Effects of Time and Memory Factors on Response in Surveys," *Journal of the American Statistical Association*, **68**, 805-815.

NOTE

*The authors acknowledge the support of the Division of Health Interview Statistics of the National Center for Health Statistics for providing 1980 quarterly estimates from the National Health Interview Survey for the purpose of preparing this paper.

Evidence of Conditioning Effects in the British Social Attitudes Panel*

Jennifer Waterton and Denise Lievesley

---

1. INTRODUCTION

The analytic advantages of a panel design over repeated cross-sectional studies are well-known: Change is measured with higher precision and can be charted not only at an aggregate but also at an individual level. Despite these advantages the panel approach is often rejected because of the fears of severe attrition and conditioning. Respondents may be "contaminated" by having been interviewed previously, making the panel unrepresentative. Certainly most ·published research reinforces this view of panels by highlighting the existence of conditioning effects under particular circumstances. Few papers put these results in perspective by discussing the many occasions when conditioning might have occurred but did not. This study has been designed specifically to allow a systematic review of conditioning—its nature and its scale—and to put conditioning effects in the context of the other advantages and disadvantages of panel surveys. The paper begins by considering the nature of conditioning: its definition, how it can be measured, and its merits and demerits.

In order to explain how the British Social Attitudes Panel fits into the framework provided by existing research, some background information on the survey is given in Section 3.

Section 4 discusses our approach to the research, beginning with a review of relevant literature and going on to consider how we explored the evidence of conditioning within the Social Attitudes Panel.

Section 5 outlines the various hypotheses about the way in which conditioning might affect survey responses. For each hypothesis in turn the research findings that give rise to it are described and then a report on the relevant evidence from the Social Attitudes Panel is given. Our conclusions are presented in Section 6.

2. THE NATURE OF CONDITIONING

2.1. Definition of Conditioning

Panel conditioning, panel bias, reinterview effects, time-in-sample bias, and rotation group bias are synonyms used to describe the phenomenon whereby the very act of being interviewed changes attitudes or behavior or—more likely—changes the *reporting* of attitudes or behavior.

Although conditioning is most frequently discussed in relation to repeated interviewing, it may occur *within* an interview.

2.2. Measurement of Conditioning

The lack of evidence on conditioning is probably due to the requirement for a control sample in order that the reinterview effects may be disentangled from attrition and real change in the population. This control sample could be in the form of a separate fresh sample alongside the panel but, in practice, more often consists of a rotating design in which some respondents are interviewed for the first time while others are interviewed for the second or subsequent time.

Examining conditioning effects *within* an interview requires a split design (i.e., where a control sample is asked only parts of the interview) or rotation or repetition of some questions.

2.3. Real Change vs. Change in Reporting

As indicated in the definition of conditioning, there are two categories of conditioning effects. First, the interview may bring about real changes in the behavior and/or the attitudes of the respondents. The second possibility is that respondents may change the way they *report* their behavior and/or attitudes in response to the interview. These two types of conditioning effects, although clearly distinguishable in the abstract, are impossible to separate in practice in the absence of external validation data.

When researchers have attributed conditioning effects to "real changes" they have often done so because this view is supported by respondents' own claims. For example, Kemsley (1961) reports that 20% of women and 5% of men claimed that the act of record keeping influenced their expenditure (see also Quackenbush & Shaffer, 1960). However, one should probably treat respondents' retrospective reports of their own changes with some skepticism (see, for example, Lievesley & Waterton, 1985). We return to the distinction between real change and change in reporting behavior in Section 4.3.

2.4. The Desirability of Conditioning

The common assumption is that conditioning is automatically detrimental to the quality of the responses. Thus conditioning, in the sense of changes in reporting behavior, is represented as a disadvantage of panel surveys. The view that the prior experience of being interviewed might *improve* the quality of reporting is not widely held.

Of course the argument about whether conditioning is a "good" or a "bad" thing is a difficult one to resolve. *Criterion* validity (the degree to which the measurement is predictive of the properties it purports to measure) is usually impossible to assess in surveys, since external data for validation are rarely available. So, to measure improvements (or otherwise) in data quality, researchers have to employ three rather weaker criteria: (1) *construct* validity (the extent to which answers relate to each other in a way that is consistent with theory); (2) *face* validity (the extent to which answers appeal to common sense); and (3) *process* validity (the extent to which the survey procedures employed meet specified criteria of quality—often measured by examining the "errors" displayed by the survey data). For a fuller discussion of validity see Bateson (1984) and Turner and Martin (1985).

According to these criteria, conditioning does have some beneficial effects. These are discussed later in relation to the Social Attitudes Panel survey. But first it is necessary to give some background information on the survey.

3. THE SOCIAL ATTITUDES PANEL

3.1. The Design of the Survey

In Spring 1983, the first of an annual series of surveys about social attitudes in Britain was launched by Social and Community Planning Research (SCPR).

The core survey, the Social Attitudes Survey, relies on a cross-sectional design employing face-to-face interviews that concentrate on attitudes, values, and mores relating to a wide range of social issues. Key topic areas (including attitudes to government spending, the economy, defense issues, and sexual morality) are included each year, together with demographic details. Other topics, such as attitudes to the environment, the National Health Service, crime, sexual equality, new technology, and so on, are asked less frequently. The target population for the survey series is all adults aged 18 years or over living in private households in Great Britain. The sample size for the first three rounds was around 1,700, now increased to 3,000 per annum for the fourth year and beyond.

The primary motivation for the series was the paucity of data relating to the British public's attitudes, values, and beliefs—in marked contrast to the depth of coverage afforded to surveys of social conditions and behavior patterns. The primary objective of the Social Attitudes Survey is to provide a series of profiles of public attitudes in the 1980s. Partly because of worries about the possible effects of conditioning and attrition, the study was designed as a cross-sectional series. However, since we felt that the survey had an important function in monitoring trends and changes in attitudes over time, we also obtained additional funding from the Economic and Social Research Council to explore the methodological and analytical implications of a simultaneous panel approach to the same data. Thus for the four years 1983 to 1986 the Social Attitudes Data Base consists of a larger cross-sectional survey and a smaller panel survey.

The content of the two questionnaires in any given year is substantially the same, although not identical. Any questions included in both surveys are asked in identical form. The fieldwork periods for the two components coincide approximately, taking place between March and May each year.

3.2. Response and Attrition

Each year a response rate of approximately 70% has been achieved on the cross-sectional survey. This rate is not atypical in Britain for an academic survey with no single "policy" purpose and nongovernment funders.

In 1984, a random sample of 769 individuals was drawn from the pool of respondents to the 1983 survey. These individuals constitute the panel sample. Table 1 traces them from their responses in 1983 through all subsequent stages to 1986.

Broadly speaking, the panel in 1984 suffered attrition rates similar to the nonresponse recorded for the cross-sections (just under 30%). At the next round the position was somewhat improved (20% nonresponse) and at the last round the losses were very small (about 5%). The panel members who still remain seem to be highly motivated and committed to the survey. We would expect response rates for subsequent waves to be maintained at about 95%. The response rates recorded here, which may at first sight seem rather disappointing, do not compare unfavorably with other longitudinal studies, especially when the nature of the present survey is taken into account (for a fuller discussion of this issue see Waterton and Lievesley, 1986). Erosion of a panel over time might not be a problem if it were evenly spread over subgroups. Unfortunately, the pattern of attrition is not like this: Differential attrition was experienced for half of the demographic questions (12 out of 22) and half of the attitudinal questions (101 out of 211).

The effects of attrition on the Social Attitudes Panel data are discussed more fully in a separate paper that also describes in detail the weighting procedures adopted to compensate for the differential attrition (Waterton & Lievesley, 1986). However, even though differential attrition was experienced in the panel survey this does not prevent an examination of the effects of conditioning. The procedure used is described in Section 4.4.

TABLE 1. ATTRITION IN THE SOCIAL ATTITUDES PANEL SURVEY

	1984	1985	1986
A. Selected for panel	769	602	470
B. Refused interview in advance	43	22	6
C. Issued sample (A–B)	726	580	464
D. Attrition—inevitable (dead, emigrated, etc.)	11	11	7
E. Attrition—noncontacts	27	12	1
F. Attrition—refusals	85	55	12
G. Attrition—untraceable	33	22	4
H. Attrition—other nonresponse	17	14	8
J. Successful interviews	553	466	432
Response rate [J/(A–D)]	73%	79%	93%
Cumulative response rate	73%	62%	58%

4. THE RESEARCH METHOD

4.1. Introduction

The first stage in our research on conditioning was to review the relevant literature. This search led to the identification of six different types of conditioning effects that we then explored using the Social Attitudes Panel. Each of the steps in our research procedure is described below. The hypotheses and the findings are presented in Section 5.

4.2. Literature Review

It was suggested, in Section 2.2, that the absence of a large body of data on conditioning effects is due to the need for split-panel or rotating designs. However, even when such designs have been used, a systematic appraisal of conditioning effects has rarely been undertaken. This means that the literature is rather sparse, fragmentary, and unsystematic in its treatment of the subject. Such evidence as has been adduced seems to owe a lot to serendipity! Many of the documented effects have been identified by chance because, for example, the rotation groups are seen to differ systematically in a way not accounted for by differential attrition. As mentioned before, few reviews have been undertaken that might lead to conclusions concerning the lack of conditioning effects.

With regard to subject matter, a substantial proportion of the literature on conditioning relates to surveys of consumer behavior, labor force participation, and voting behavior. The concentration on surveys of consumer behavior is probably due to the preponderance of panel designs for this type of research (see Ehrenberg, 1960; Quackenbush & Shaffer, 1960; Buck et al., 1977). Similarly the overrepresentation of labor force variables in the conditioning literature is probably due to the large and influential U.S. Current Population Survey and the Canadian Labour Force Survey, both of which employ rotating designs (see Bailar, 1975; Ghangurde, 1982). It is more illuminating, perhaps, to examine why conditioning effects have been of particular interest to political researchers. In part this is probably due to the seminal paper by Clausen (1968), which reported that people interviewed before an election voted in greater proportions than those who were not interviewed. This research stimulated others to try to replicate and build on Clausen's finding (Kraut & McConahay, 1973; Yalch, 1976; Traugott & Katosh, 1979). The "stimulus effect" (using Clausen's terminology) engendered by the interview is not merely a methodological curiosity for academic political researchers: It is also of substantive interest. The stability of political attitudes and voting behavior, their interrelationship, and the ways in which voting behavior can be affected by, for example, canvassing or the "bandwagon effect," have long been a focus of political research (see, for instance, Bochel & Denver, 1977; Marsh, 1985). The interest in conditioning effects in voting behavior research is probably also due to the availability of voting turnout statistics that permit aggregate validation of the survey results. Occasionally, individual validation is also possible (Kraut & McConahay, 1973) so that conditioning effects on the behavior itself can be separated from effects on the *reporting* of the behavior.

4.3. Investigating Conditioning Effects

Having reviewed the literature on conditioning, we generated six hypotheses on its effects. Repeated interviewing:

1. May bring about changes in attitudes or behavior by raising consciousness.
2. Freezes attitudes.
3. Leads to more honest reporting by reducing social desirability bias.
4. Leads to improved understanding of the rules that govern the interview process.
5. Increases motivation and improves the quality of reporting.
6. Increases respondent burden and reduces data quality.

We then turned to the Social Attitudes Panel to examine the evidence for these effects. In the absence of well-established theories of how conditioning effects might apply in a survey of attitudes our approach has been exploratory. Indeed, our allocation of research findings to categories is necessarily subjective and a different categorization might easily be made—particularly as the six types deal with overlapping rather than discrete mechanisms.

Of the six hypotheses, the first two are concerned with real changes while the latter four are concerned with changes in reporting behavior. We have already mentioned the difficulty of distinguishing between "real changes" and changes in reporting behavior, but, a priori, in a survey of attitudes we assume that any differences we find are attributable to changes in reporting, since this this seems more credible.

The survey literature suggests that, where conditioning effects do occur, they tend to manifest themselves early in the interview sequence rather than after many exposures. In most of the findings we have presented in this paper the search for evidence of conditioning effects begins at the *second* interview. We have occasionally looked for effects *within* the initial interview but are constrained by the lack of control within the design (see the discussion in Section 2.2).

Given that the panel survey includes mainly attitudinal variables, which are inherently difficult to validate, we can only speculate about whether conditioning has had harmful or beneficial effects on the quality of the responses. Our ability to apply criterion validity is, of course, extremely limited. But the wide range of topics covered in the Social Attitudes Survey and the relatively small number of questions on any one topic makes the assessment of construct validity problematic too.

4.4. Estimating Conditioning Effects

The design of the Social Attitudes Survey—with parallel cross-sectional and panel elements—allows change in response due to conditioning to be separated from other possible explanations of change—such as attrition or "real change"—with which they might otherwise be confounded. The following example will serve to demonstrate the method.

Respondents were asked their attitudes to the level benefits for the unemployed. The proportion who answered "don't know" (DK) fell from 6.8% in the 1983 panel to 5.1% in the 1984 panel. We found that those who dropped out were more likely to reply "don't know" at the first interview than those who continued. Having excluded those who dropped out from the analysis—thus confining our comparison to the *same* people in 1983 and 1984—we still find a fall in "don't know" responses from 5.5% in 1983 to 5.1% in 1984. This fall must be due to real change and/or conditioning, or variable measurement errors. (The fall from 6.8% to 5.5%, on the other hand, is due to attrition.) We have tried as far as possible to keep the measurement errors constant both across the different rounds of the panel and between the panel and cross-sectional surveys. However, we cannot guarantee that the conditioning effect does not incorporate a small amount of variable measurement errors.

In order to separate the effects of *conditioning* from those of real change, information from the cross-sectional surveys is necessary: In both 1983 and 1984 an identical proportion of respondents to the cross-sectional surveys replied "don't know," implying that there was in fact no "real" change in the population. Hence we concluded that the small drop from 5.5% to 5.1% was due entirely to respondent conditioning in the panel. This example illustrates the method by which the components of change may be disentangled. The Appendix describes the method in a little more detail, extends it to the situation where the cross-sectional surveys do show a change, and outlines calculation of the variance of conditioning effects.

4.5. Testing the Significance of Conditioning Effects

Throughout the analyses we have conducted formal tests of significance. However, given that this exercise is intended to be exploratory and that we hope other researchers will test our ideas in their own data sets, we have not restricted ourselves to presenting only significant results, although we do distinguish between significant and insignificant findings.

4.6. Conditioning Effects on Subgroups

We examined each of the hypothesized conditioning effects not only in relation to the sample as a whole but also to various subgroups:
  (1) Age of respondent.
  (2) The age at which the respondent completed full-time education.
  (3) Their general knowledge (the answers to several knowledge questions have been combined to provide a knowledge score that takes the values 0, 1, 2, or 3 and represents low knowledge levels (0) through high knowledge levels (3)).
  (4) Their political party identification.
  (5) The strength of their party identification.
  (6) Whether the same or different interviewers conducted the interview.

These subgroups were chosen in the light of published research on the correlates of response variability such as interviewer and question-wording effects (see Jackman, 1973; Sudman & Bradburn, 1974; Schuman & Presser, 1977; Collins, 1982). If the results relating to these sources of response variability carry across to conditioning, we might expect older people, those with weak or no party identification, and those with lower education and knowledge levels to be more affected by conditioning.

Although we have adopted an exhaustive approach to the subgroup analysis (in that we have examined each of the subgroups for each of the findings on conditioning), it is not necessary to present all of these results in this paper. We have given information only on these detailed analyses where the conditioning effect acted differently on subgroups.

5. THE FINDINGS

Hypothesis I: Repeated interviewing may bring about changes in attitudes or behavior by raising consciousness.

The idea that repeated exposure to the same stimulus can influence behavior lies at the heart of consumer advertising. When considered in the context of an interview about attitudes this suggests that asking questions may prompt a respondent to gather information and either to form opinions where none existed before or to alter pre-existing opinions.

Studies of the effects of pre-election interviews on political attitudes and voting turnout have provided a focus for research in this area (Glock, 1952; Clausen, 1968; Kraut & McConahay, 1973; Yalch, 1976; Bochel & Denver, 1977; Traugott & Katosh, 1979). All these studies conclude that interviews do bring about changes, although they do not agree about the magnitude of such effects, the extent to which they may cumulate with additional interviews, or the persistence of the conditioning effects over time.

Two panel studies of attitudes are particularly relevant to this discussion. Sobol (1959) uses a consumer buying intentions survey with a rotating design to examine the hypothesis of changing attitudes. The conclusion from this five-wave, three-year study is that there is no evidence of any reinterview effects. On the other hand, Bridge et al. (1977) find that respondents interviewed about attitudes to cancer are more likely to rate good health as important than those not interviewed about cancer. In the same study, however, no effect was produced on a rating about the importance of safety from crime among those interviewed earlier about burglary prevention. They conclude:

> Interviews may change respondents' attitudes about a topic *if* they come to see themselves as having insufficient information or opinions about something that they otherwise perceive to be socially important. But if a topic is seen as unimportant or if the salience of the topic is already very high, question asking will not shift attitudes and information levels. The interview situation fosters the perception that an issue is socially important and something about which respondents should be concerned. (p. 63)

If the Social Attitudes interview is an event of little importance to respondents, then it would be unlikely to affect their attitudes. In fact, in-depth follow-up interviews conducted the day after the initial interviews revealed that the survey generated a considerable amount of discussion between respondents and their friends and relatives.

Our investigation of changes in attitudes indicated that panel respondents become "politicized" as a result of the interview. The proportions classified as partisans—strong supporters of a political party (Young, 1984)—increase over time for the panel respondents, but remain stable in the cross-sectional survey. Table 2 presents estimates of the conditioning effect in one-year intervals from 1983 to 1986 and their associated errors. In all cases, the estimates are in the expected direction but do not reach statistical significance until 1986.

Examining the change in partisanship by the composite indicator of knowledge (described earlier in Section 4.4) we find, as expected, that the conditioning effect is more pronounced for the subgroups with lower knowledge scores (Table 3).

If panel membership does indeed heighten awareness, it might be reasonable to assume that respondents' views on a topic become more well-defined and firmer with repeated interviews. Although we have no direct way of measuring this, a reasonable proxy for strength of feeling might be the selection of extreme points on a range of attitude scales. To examine this hypothesis we identified 18 items asked in both 1983 and 1984 that used four- or five-point attitude scales. The points at either end were designated "extreme." The analysis was confined to those answers that the cross-sectional survey found to display no real change. Twenty-nine categories of answer were included.

Counter to the hypothesis that extreme answers increase with exposure, there was actually a drop in the endorsement of extreme answers among the 1984 panel respondents. However, the effects are very small and they are not unidirectional for all subgroups (see, for example, the breakdown by educational subgroup, included in Table 4.)

Hypothesis II: Repeated interviewing freezes attitudes.

Hypothesis I considered the evidence that interviewing may stimulate attitude formation and change. In contrast, this second hypothesis concerns itself with the opposing view that interviewing may actually inhibit a change that might otherwise have occurred. As Bridge et al. (1977) explain, "This freezing may occur when the respondent perceives the interview response to be a sort of public commitment to a position, or when the respondent in a panel survey feels some pressure to be consistent across interviews" (p. 57).

Although freezing of attitudes is often cited as a disadvantage of repeated interviewing, there is virtually no evidence that it takes place. In an unpublished doctoral thesis, Glock (1952) concluded that frequent pre-election interviewing caused respondents to form voting intentions rapidly that then remained stable. Although Bridge et al. interpret this as possible evidence of a freezing effect, another explanation is that interviewing simply helps to speed up the process of attitude formation.

Reeder et al. (1977) report a study in which public attitudes to population growth remain stable over waves of a panel survey, while

TABLE 2. POLITICAL PARTISANSHIP ESTIMATES OF CONDITIONING

| Answers in . . . | Political Partisanship | | | |
	Cross-Sectional Samples	1984 Panel Respondents	1985 Panel Respondents	1986 Panel Respondents
1983	47.2%	49.4%	50.9%	50.7%
1984	47.0	50.5	51.7	51.4
1985	46.9	—	54.3	53.8
1986	46.8	—	—	58.2
Base	1,700 (3,100 in 1986)	550	465	430
		1983–84	1984–85	1985–86
Estimates of conditioning:		1.3%	2.7%	4.5%
Standard error of this estimate:		2.2	2.3	2.2

TABLE 3. POLITICAL PARTISANSHIP FOR 1986 PANEL
RESPONDENTS BY KNOWLEDGE LEVELS

| Knowledge Level | Political Partisanship | | | |
	1983	1984	1985	1986
Very low	30.4%	33.9%	39.3%	48.2%
Low	41.4	45.4	45.0	48.6
High	52.5	52.0	58.1	58.4
Very high	70.4	67.1	62.9	75.3

TABLE 4. AVERAGE NUMBER OF "EXTREME" ANSWERS GIVEN
BY 1984 PANEL RESPONDENTS IN 1983 AND 1984

| | Age at Completing Full-Time Education | | | | |
	≤ 15 yrs.	16 yrs.	17 or 18 yrs.	≥ 19 yrs.	All Ages
1983	4.21	3.81	3.84	3.57	4.01
1984	3.90	3.55	3.96	3.76	3.81

significant changes take place in repeated cross-sections.

We uncovered no findings in the Social Attitudes Panel survey that fit this model of conditioning. This is hardly surprising, since if freezing of attitudes takes place at all it is more likely to occur in a single-issue survey that covers one topic in depth. Our survey, on the other hand, covers a wide range of topics in the course of any one interview. Also it might be expected that attitudes—if frozen—would remain so only for a relatively short period, not necessarily as long as a year.

Hypothesis III: Repeated interviewing leads to more honest reporting by reducing social desirability bias.

Social desirability bias has frequently been identified in survey responses. (For a review of the research see Sudman & Bradburn, 1982, or Turner & Martin, 1985.) It may arise when questions deal with issues of a private or personal nature or when some responses are clearly perceived to be more socially desirable. There is some evidence that the degree of such bias depends on the mode of data collection (see Waterton & Duffy, 1984, and Sykes & Hoinville, 1985).

The expectation underlying this hypothesis is that respondents will relax and feel more at ease with each successive wave of interviewing and, as a consequence, not feel the need to give socially appropriate responses.

Probably the main evidence in support of this hypothesis is from research on employment. The labor force surveys in Canada, the United States, and France have all experienced lower reports of unemployment in second and subsequent interviews (see Bailar, 1975, and Ghangurde, 1982). It has been suggested that these lower reports are due to increased willingness to admit to a socially undesirable status, including both working in the underground economy and not working at all (see Collins, 1986).

The follow-up in-depth interviews conducted after the initial Social Attitudes interview indicated that, even at that stage, respondents had begun to relax and to feel more comfortable about revealing their opinions to the interviewer. We identified six questions asked on both the 1983 and 1984 surveys that we believed might be subject to social desirability effects. These questions are listed in Table 5 alongside the direction of the effect we predicted *if* social desirability bias were reduced.

Table 6 shows the proportions of the 1984 panel respondents who endorsed each of the six categories outlined in Table 5 in 1983 and in 1984, together with the cross-sectional results, an estimate of the conditioning effect, and the standard error of this estimate.

Four of the six items show results in the expected direction. For only one of these—relating to racial prejudice—is the effect statistically sig-

TABLE 5. QUESTIONS USED IN TESTING HYPOTHESES
OF REDUCED SOCIAL DESIRABILITY BIAS

Questions	Predictions
	If social desirability bias were reduced, then expect . . .
1. Racial prejudice	Increase in percentage admitting to being very racially prejudiced.
2. Protest at unjust law	Increase in percentage answering that they would do nothing to stop an unjust law being passed.
3. Income disclosure	Decrease in percentage refusing to answer income question.
4. Advocate law-breaking	Increase in percentage answering that people should follow their conscience even if it means breaking the law.
5. Personal propensity to break law	Increase in percentage saying they would break the law on occasion
6. Newspaper readership	Increase in percentage who report they do not regularly read a newspaper.

TABLE 6. CONDITIONING EFFECTS ON SENSITIVE QUESTIONS

	1 Very racially prejudiced	2 Would do nothing	3 Refused to give income	4 Would follow conscience	5 Would break law	6 Does not read newspaper
Panel						
1983	4.9%	8.0%	6.9%	49.2%	35.3%	18.3%
1984	6.1	4.7	5.9	47.0	32.8	21.3
Cross-Section						
1983	4.4	13.6	11.6	46.4	30.4	22.9
1984	2.9	8.3	12.2	42.7	29.4	27.7
Expected direction of effort	+	+	−	+	+	+
Estimate of effect (E)	+2.7	+2.0	−1.6	+1.5	−1.5	−1.8
Standard error (E)	0.9	1.5	1.5	2.5	2.4	1.9

nificant. The subgroup analyses reveal no consistent differences for items 1-4 in Table 6 by party identification, strength of identification, or education. However, there is again some evidence that those with low knowledge scores are more susceptible to conditioning. (For example, the percentage admitting to being very racially prejudiced increases from 1.6% in 1983 to 11.5% in 1984 for the low-knowledge group and the estimate of conditioning is found to be statistically significant.) Also, for two of the four items (racial prejudice and taking no action) the social desirability bias is smaller when respondents are interviewed by the same interviewer. However, for the item about incomes the reverse is true (there are fewer refusals for the group interviewed by a different interviewer).

Two of the questions (racial prejudice and income) were also asked in 1985 and 1986. Any conditioning effects on the racial prejudice question seem to disappear in subsequent reinterviews. However, for the 1986 panel respondents, the percentage refusing to answer the question about income continues to fall—from 7.3% to 4.4% to 4.3% and finally, in 1986, to 2.4%.

Another aspect of social desirability bias is the putative pressure on respondents to appear consistent in their answers. Although people may well hold mutually contradictory views, it is suggested that they may feel embarrassed about revealing inconsistency in an unfamiliar situation such as an interview. If this is so, then as people become more familiar with being interviewed, we would expect them to be increasingly less on their guard and thus more willing to express inconsistent views. We would thus expect to find reductions at each successive wave in the inter-item correlations of related questions asked in the *same* year. Our exploration of this has focused upon three groups of questions—those relating to defense (five questions), economic affairs (four questions), and welfare issues (three questions). The average inter-item correlations for each of the four waves of the panel are given in Table 7. As is apparent from the table, there is no evidence of systematic changes in the inter-item correlations over time, so the hypothesis finds no support.

Hypotheses IV, V, and VI, which follow, deal with highly related aspects of conditioning, with the result that the literature is difficult to apportion among the different hypotheses. We therefore depart from the current style and review here the literature relevant to all three remaining hypotheses.

Recognition of the complexity of the interview rules and of the skills that are necessary in order to become a "good" respondent has led some researchers (particularly those in the qualitative field) to advocate the use of "expert" panels (see, for example, Gordon & Robson, 1982). In consumer panels, data from the first few interviews are routinely rejected because they are believed to be of low quality. It is assumed

TABLE 7. AVERAGE WITHIN-GROUP CORRELATIONS
FOR EACH OF THE PANEL WAVES

Questions on ...	1983	1984	1985	1986
Defense	.33	.31	.35	.32
Economic affairs	.19	.17	.21	.19
Welfare issues	.11	.27	.18	.17

that this lack of quality stems from a lack of expertise on the part of panel members on how to report their purchases. Unfortunately this issue is rarely discussed in the market research literature. Most studies of conditioning in both purchasing behavior and in attitudes examine effects *after* the running-in period and have concluded that the effects are very small (Ehrenberg, 1960; Barnard et al., 1986). Only one study (Buck et al., 1977) examines the conditioning effects as they apply in these early interviews, and it concludes that the first reports do not differ significantly from later ones. A study reported by Collins (1986) took a rather different approach by examining the conditioning or "learning" effects during a single interview. The experimental results he reports relate to a survey in which a series of six split-ballot question-wording experiments were repeated, with the stimuli reversed within the *same* interview (so each respondent answered *both* forms of each question, only the sequence being reversed between the two systematic subsamples). Question form effects, which emerged in five of the six split ballots in the first run, were all missing in the second run. Collins suggests that this "form-resistance," which emerges at the second time of asking, is an indicator of improved quality of response.

The above examples assume that increased respondent knowledge and understanding lead to an improvement in data quality. In some circumstances it can be detrimental and a naive respondent is to be preferred. The most common example of this occurs when asking a series of filter questions all of which have a battery of questions dependent upon them. Respondents may learn that by giving, say, a negative response to the filters, they can thereby avoid the dependent questions. Researchers are thus often advised to ask all such filter questions first. Although this problem is well-known, there are few documented examples of it in practice. One exception is the U.S. National Crime Survey, in which respondents apparently learned that by saying they had not been victims they avoided answering questions on the nature and circumstances of a crime (Turner, 1984).

Hypothesis IV: Repeated interviewing leads to improved understanding of the rules that govern the interview process.

For most people, participating in an interview is an unfamiliar activity. This means they need to learn about their roles as respondents.

The major learning effect in the Social Attitudes Panel relates to the likelihood of answering "don't know" (DK): Respondents seemed to learn that their attitudinal *orientation* is of interest even if they did not hold a well-formulated or precise position on an issue. This knowledge seemed to reduce their propensity to answer "don't know" in favor of one of the precodes.

Using 25 attitude questions common to all four years, we examined the rate of answering "don't know" for the 1986 respondents. Our concentration upon the 1986 panel members means that differential attrition can be disregarded as an explanation of change. There was, incidentally, differential attrition on the rate of DK answering—(see Waterton & Lievesley, 1986). Although the absolute level of DK answering is very low, it shows a marked fall: from 0.54% in 1983 to 0.42% in 1984 to 0.35% in 1985, where it stabilized (0.34%) in 1986. On the other hand the level of DK answering on the cross-section was stable over this whole period.

If respondents are learning to give an answer other than "don't know" one might expect this effect to manifest itself *within* the first interview. It is not possible to examine this in a rigorous way but we plotted the proportion of DK answers by question sequence, and it did not display a pattern that supports the hypothesis that DK answers decrease during the interview.

Hypothesis V: Repeated interviewing increases motivation and improves the quality of reporting.

This hypothesis is likely to be of particular importance in panel surveys that require respondents to work hard. For example, they may be required to recall or articulate difficult or detailed information, or to perform a task such as keeping an accurate diary of their behavior. It is perhaps less likely to be relevant in a study such as the Social Attitudes Survey, which requires immediate answers to opinion questions.

The Social Attitudes Panel does not collect any direct evidence on respondent motivation or data quality. However, two proxy measures have been examined to see whether they can shed any light on this hypothesis. First, it may be reasonable to expect that increased motivation would be associated with increased enjoyment of the interview. A question about enjoyment was first asked in 1984 and then repeated in the following two years. The proportion saying they "enjoyed the interview a great deal" increased substantially over time. For the 1986 panel respondents, the percentage increased from 34% (1984) to 43% (1985) to 48% (1986), so if motivation is positively correlated with enjoyment, motivation should also have increased over time.

Another indication of respondent involvement might be the frequency of answers he or she gives that are outside the specified precodes. This implies effort on the part of respondents, which is a form of commitment. Ten questions were identified that were asked in both 1983 and 1984 and for which at least ten respondents gave answers outside the precodes. Each panel respondent was assigned a score for 1983 and one for 1984 according to the number of occasions he/she gave an "other answer." The z-statistic for assessing the difference between the two average scores is 0.09, indicating no change in the propensity of respondents to give unspecified answers.

A third indication of respondent involvement might be the sheer number of different answers given in response to open-ended questions. Since, however, the Social Attitudes Survey includes only a few open-ended questions, not all of which are repeated from year to year, an examination along these lines has not been possible.

Hypothesis VI: Repeated interviewing increases respondent burden and reduces data quality.

This hypothesis is, of course, based on the opposite premise to that of the previous hypothesis. Once again the effects are more likely to be apparent in surveys where the task is demanding. However, since respondents who tire of participating are more likely to drop out of the panel altogether than to alter their reporting behavior, we should ideally look for evidence of burden or fatigue during the course of a single interview. What we are able to get of any effect comes from self-reports: A number of respondents to the follow-up survey volunteered that they had indeed found the experience extremely tiring. Whether or not this necessarily leads to poorer reporting, however, is by no means certain.

6. CONCLUSIONS

SCPR's British Social Attitudes Panel study was set up as a methodological project specifically to examine the effects of attrition and conditioning on a panel study of attitudes. This was made possible by being able to juxtapose the panel findings against findings from the national cross-sectional survey, which takes place annually. In a systematic attempt to review the evidence for the presence (or absence) of any panel conditioning we have referred to six hypotheses about the expected effects that reinterviewing might have.

We find some suggestion that respondents become politicized by the interview, that they report more honestly, and that they become less likely to answer "don't know." We also find evidence that the effects were more pronounced for respondents with low knowledge scores. However, the magnitude of such effects is small and would not prevent comparisons with the fresh cross-sectional surveys, particularly because they do not cumulate with later reinterviews.

The encouraging but tentative conclusion from this survey is therefore that conditioning is not a major hazard. Indeed, dangers for representativeness appear much more likely to arise from differential attrition than from conditioning. However, we do not know whether or not these findings will prove to be generalizable. For instance, the elapsed time between interviews on this survey was one year. If this had been reduced, respondents may have been correspondingly more likely to remember their answers at the previous round and perhaps to have altered their responses accordingly. In any event, it is reassuring to note that on such a diverse survey of attitudes—some of which were elicited on subjects that must have been only marginal to most respondents' interests—the conditioning effects of repeated interviews incorporating many of the same questions seem at most to have been only minor.

The Social Attitudes Panel is small, so our conclusions have to be tentative. But our research design provides a rare analytic opportunity to separate possible conditioning effects from other sources of change, e.g., temporal change or panel attrition. It yields results that, at the least, should focus the efforts of other researchers.

APPENDIX

In this presentation, the concern is with categorical data (and hence with proportions). The estimate (E) of conditioning and the estimate of its variance are given below.

Let

c_1 be the estimate from the cross-sectional survey at t_1,

c_2 be the estimate from the cross-sectional survey at t_2,

p_1 be the estimate from the panel survey at t_1 based on respondents at t_2,

p_2 be the estimate from the panel survey at t_2 based on respondents at t_2,

r be the correlation between these two panel estimates, and

v be the proportion of the initial cross-section who are recruited to the panel.

Then

$p_2 - p_1$ is the estimate of conditioning + real change, and

$c_2 - c_1$ is the estimate of real change.

And, assuming that real change on the panel is equal to real change on the cross-section (i.e., that those respondents who drop out of the panel are not systematically different, in respect of their characteristics of change, from those who remain in the panel) then an estimate of the conditioning effect is given by

$$E = (p_2 - p_1) - (c_2 - c_1).$$

If the panel and cross-sections were independent, then

$$\text{var } (E) = \text{var } (p_1) + \text{var } (p_2) - 2 \text{ cov } (p_1, p_2) + \text{var } (c_1) + \text{var } (c_2).$$

However, because of the nonindependence of the panel and the original cross-section, an extra term is introduced and we have

$$\text{var } (E) = \text{var } (p_1) + \text{var } (p_2) - 2 \text{ cov } (p_1, p_2) + \text{var } (c_1) + \text{var } (c_2)$$
$$- 2v[\text{var } (p_1) - \text{cov } (p_1, p_2)].$$

Given that $\text{var } (p_1) \simeq \text{var } (p_2)$ and $\text{var } (c_1) \simeq \text{var } (c_2)$ (except for 1985 and 1986 when $\text{var } (c_2) = 1/2 \text{ var } (c_1)$ due to the increased sample size), this expression may be simplified considerably to give

$$\text{var } (E) \simeq 2 \text{ var } (c_1) + 2(1 - r)(1 - v) \text{ var } (p_1).$$

The calculation of the error on the estimate of conditioning assumes that simple random sampling has been employed. We have calculated complex sampling errors for the cross-sectional surveys (see Jowell & Airey, 1984). The design factors for attitudinal questions were mainly in the range 1.1 to 1.4. They will almost certainly be even smaller than this for the panel survey, which is less clustered than the cross-sectional surveys. Thus it is likely that the error on the conditioning effect would have been at most marginally larger had we taken account of the complex nature of the sample.

REFERENCES

Bailar, B. A. (1975), "The Effects cf Rotation Group Bias in Estimates from Panel Surveys," *Journal of the American Statistical Association*, 70, 23–30.

Barnard, N. R., Barwise, T. P., & Ehrenberg, A. S. C. (1986), "Reinterviews in Attitude Research: Early Results," *Proceedings of the Market Research Society Conference*, Brighton.

Bateson, N. B. (1984), *Date Construction in Social Surveys*, George Allen and Unwin, Contemporary Social Research Series, London.

Bochel, J. M., & Denver, D. T. (1977), "Canvassing, Turnout, and Party Support: An Experiment," *British Journal of Political Science*, 1, 257–269.

Bridge, R. G., Reeder, L. G., Kanouse, D., Kinder, D. R., Nagy, V. T., & Judd, C. M. (1977), "Interviewing Changes Attitudes—Sometimes," *Public Opinion*

Quarterly, **41**, 56–64.

Buck, S. F., Fairclough, E. H., Jephcott, J. St. G., & Ringer, D. W. C. (1977), "Conditioning and Bias in Consumer Panels—Some New Results," *Journal of the Market Research Society,* **19**(2), 59–75.

Clausen, A. (1968), "Response Validity: Vote Report," *Public Opinion Quarterly,* **32**, 588–606.

Collins, M. C. (1986), *Might Conditioning Be a Good Thing?* (Unpublished paper), Survey Methods Centre, SCPR, 35 Northampton Square, London EC1V OAX.

Collins, M. C. (1982), *Some Statistical Problems in Interviews with Older Respondents* (Paper presented to the Social Statistics Section, Royal Statistical Society), Survey Methods Centre, SCPR, 35, Northampton Square, London ECIV OAX.

Ehrenberg, A. (1960), "A Study of Some Potential Biases in the Operation of a Consumer Panel," *Applied Statistics,* **9**, 20–27.

Ghangurde, P. D. (1982), "Rotation Group Bias in the LFS Estimates," *Survey Methodology,* **8**, 86–101.

Glock, C. (1952), *Participation Bias and Reinterview Effects in Panel Studies* (Unpublished Ph.D. dissertation), Columbia University.

Gordon, W., & Robson, S. (1982), "Respondent through the Looking Glass: Towards a Better Understanding of the Qualitative Interviewing Process," *Proceedings of the Annual Conference of the Market Research Society,* pp. 455–474.

Jackman, M. (1973), "Education and Prejudice or Education and Response Set?," *American Sociological Review,* **38**(3), 327–339.

Jowell, R., & Airey, C. (1984), *British Social Attitudes—the 1984 Report* (Appendix I), Gower.

Kemsley, W. F. F. (1961), "The Household Expenditure Enquiry of the Ministry of Labor," *Applied Statistics,* **10**(3), 117–135.

Kraut, R. E., & McConahay, J. B. (1973), "How Being Interviewed Affects Voting: An Experiment," *Public Opinion Quarterly,* **37**, 398–406.

Lievesley, D. A., & Waterton, J. J. (1985), "Measuring Individual Attitude Changes," in *British Social Attitudes—the 1985 Report* (R. Jowell & Witherspoon, eds.), Gower.

Marsh, C. (1985), "Do Polls Affect What People Think?," in *Surveying Subjective Phenomena* (C. Turner & E. Martin, eds.), Russell Sage Foundation, New York, Vol. 2, pp. 565–591.

Quackenbush, G., & Shaffer, J. (1960), *Collecting Food Purchase Data by Consumer Panel* (Technical Bulletin No. 279), Michigan State University.

Reeder, L. G., Bridge, R. G., & Ong, K. P. (1977), *The Attitudinal Effects of Interviewing* (Unpublished).

Schuman, H., & Presser, S. (1977), "Question Wording as an Independent Variable in Survey Analysis," *Sociological Methods and Research Journal,* **6**, 151–170.

Sobol, M. G. (1959), "Panel Mortality and Panel Bias," *Journal of the American Statistical Association,* **54**, 52–68.

Sudman, S., & Bradburn, N. (1974), *Response Effects in Surveys,* Aldine, Chicago.

Sudman, S., & Bradburn, N. (1982), *Asking Questions: A Practical Guide to Questionnaire Design,* Jossey-Bass, San Francisco.

Sykes, W., & Hoinville, G. (1985), *Telephone Interviewing on a Survey of Social Attitudes: A Comparison with Face-to-Face Procedures,* Information and Publications, SCPR, 35, Northampton Square, London EC1V OAX.

Turner, A. G. (1984), "The Effect of Memory Bias on the Design of the National Crime Survey," in *The National Crime Survey: Workshop Papers, Vol. 2. Methodological Studies,* U.S. Department of Justice, Bureau of Justice Studies.

Turner, C., & Martin, E., eds. (1985), *Surveying Subjective Phenomena,* Russell Sage Foundation, New York.

Traugott, M. W., & Katosh, J. P. (1979), "Response Validity in Surveys of Voting Behavior," *Public Opinion Quarterly,* **43**, 359–377.

Waterton, J. J., & Duffy, J. C. (1984), "A Comparison of Computer Interviewing Techniques and Traditional Methods in the Collection of Self-Report Alcohol Consumption Data in a Field Survey," *International Statistical Review,* **52**(2), 173–182.

Waterton, J. J., & Lievesley, D. A. (1986), *Attrition in a Panel Study of Attitudes* (Unpublished paper), Survey Methods Centre, SCPR, 35, Northampton Square London EC1V OAX.

Yalch, R. F. (1976), "Pre-election Interview Effects on Voter Turnout," *Public Opinion Quarterly*, **40**, 331–336.

Young, K. (1984), "Political Attitudes," in *British Social Attitudes—the 1984 Report* (R. Jowell & C. Airey, eds.), Gower.

NOTE

*The authors would like to thank their colleagues in the Survey Methods Centre and the research team for the Social Attitudes Survey. In particular they are grateful to Kevin McGrath for computing assistance and Penny Young for help with the design of the panel survey.

Panel Conditioning: Discussion

D. Holt

1. INTRODUCTION

The papers by Silberstein and Jacobs, Corder and Horvitz, and Waterton and Lievesley investigate conditioning of respondents as a direct result of repeated interviewing. The issue is whether respondents faced with the same questionnaire under the same conditions and with the same definitions will respond differently depending upon the number of times they have previously been interviewed in earlier rounds of the survey. Numerical investigations of this issue have been made before, but in this session each paper represents an attempt to use surveys of different design to advance our understanding further.

The basic question is: How do our methods (repeated interviewing) affect our results? I shall review the findings on this basic question, but I will pose an additional question: How do our methods of investigating our methods affect our results?

The three studies have different designs and these have a strong impact on the power of the analyses presented by each set of authors. None of the studies was specially designed to investigate the basic question.

The essential features of each design are as follows:

Silberstein and Jacobs (S–J) deal with a standard repeated survey design with partial overlap, since 20% of the sample are replaced each quarter. The first interview is used only for bounding so as to overcome the problem of telescoping of purchases into the reference period. The other point of interest is that in each quarter, for each rotation group, the sample is divided into those interviewed in the first, second, and third month of the quarter. When estimates are required for purchases in a particular month, some members of each rotation group have to recall purchases up to one month previously, others must recall for up to two months previously, and the remainder for up to three months previously.

Corder and Horvitz (C–H) use the National Medical Care Utilization and Expenditure Survey (NMCUES), which is longitudinal and has no rotation. From this survey alone it is impossible to separate genuine changes in reporting from rotation group effects since in each quarter all of the respondents have been interviewed the same number of times.

To disentangle these two effects they need a separate estimate of genuine changes from one quarter to the next, and this is obtained from a separate survey, the Health Interview Survey (HIS), which purports to collect the same data on the same basis. It is clear from their paper that important differences exist between NMCUES and HIS, and these contribute to the difficulties of their analysis task.

Waterton and Lievesley (W–L) use the same design as C–H, and the same basic comments apply. However, the important advantage of W–L over C–H is that for W–L the separate estimates with which the longitudinal data are to be compared are derived from a cross-sectional survey that uses precisely the same concepts, definitions, and methods. Thus the major differences between surveys (NMCUES and HIS) that cause problems for C–H are not such a difficulty for W–L. There remain, however, some differences between the longitudinal and cross-sectional surveys because of differences in attrition due to nonresponse in each case.

We also note that W–L use a survey of attitudes rather than of previous behavior. There is therefore no question of respondents being required to recall previous behavior. Also, since the waves of the survey are one year apart we might suspect that conditioning effects would be slight.

2. A FRAMEWORK

There are various reasons, apart from sampling error, for the fact that estimates of a population mean may differ from one period to the next. Some of these are listed below:

- *Period Effect:* There may be genuine changes in the population mean.
- *Conditioning Effect:* Respondents who have been interviewed previously may be conditioned, and this affects their response.
- *Measurement Bias:* If all measurements are equally biased, this should not be a problem, since all the analyses presented rely on comparisons between means and constant bias will not affect the results. However, if measurement bias is different for different subgroups of the sample, then these measurement biases need not be eliminated when making comparisons between means. For S–J, the recall effect is of this kind, since the measurement bias seems to vary with one-month, two-month, or three-month recall. For C–H, the two surveys (NMCUES and HIS) are not precisely the same in important respects such as reference periods and there are even differences within NMCUES between the first, last, and other quarters. I will term this the recall/ reference-period effect, although for C–H there may also be other differences between NMCUES and HIS that contribute to this.

- *Age Effect:* Each rotation group can only represent the population at the time the sample was drawn. Over the period that the
  rotation group remains in the survey, the population will change
  in composition and later rotation groups will not be precisely
  replicated samples from the identical population. It is, in principle, possible to remove this problem by selecting all rotation
  groups at the same time or by eliminating new population members from later rotation groups. However, the effect is probably
  slight in many situations with partial replacement when rotation
  groups are regularly renewed.
- *Attrition—Unit Nonresponse:* In all surveys a nonresponse
  problem exists, but in repeated surveys the sample attrition
  tends to accumulate through the successive waves of the survey,
  making the rotation group less representative and therefore leading to a different mean for each rotation group.
- *Attrition—Item Nonresponse:* In addition to unit nonresponse
  there is item nonresponse. S–J refer to this as a potential source
  of conditioning.
- *Other Sources:* There may be other systematic differences between waves of the survey or rotation groups. In particular, the
  use of proxy responses may vary between the first interview and
  subsequent occasions. Also, C–H refer to the fact that telephone
  interviewing was used for some interviews during the survey.

We need to understand the factors that influence the quality of the
survey data. In particular, to investigate conditioning effects we must
be able to separate these from the effects of the other factors. Each of
the three papers attempts to do this within the constraints imposed by
the sample designs used. Since the surveys were not designed to isolate
conditioning effects, the authors must struggle to use their data to disentangle these effects and they meet with varying degrees of success.

3. A MODEL FOR EFFECTS

Consider a simple model for the mth observation in the ith period, the
jth rotation group, and the kth recall/reference-period group.

$$Y_{ijkm} = \mu + \alpha_1 + \beta_j + \xi_k + \epsilon_{ijkm} \tag{1}$$

For S–J the subscript k defines the groups who recall for one, two, or
three months, whereas for C–H and W–L the subscript defines the survey (NMCUES or HIS for C–H, and longitudinal or cross-sectional for
W–L). This model is almost certainly too simplistic in the following
ways:

(1) A linear model may be inappropriate. It is plausible that the
model should be multiplicative and if this were so the lack of fit
could show itself in the form of interactions between the main

effects.

(2) Even if a linear model is appropriate, some interaction terms may be needed. For example, there may be recall × period interactions δ_{ik} since when the period effect is high, respondents have more to report and this may interact with recall effects. Similarly, there may be time-in-sample × recall interactions η_{jk} if respondents' ability/willingness to recall changes with the number of previous interviews. Note that δ_{ik} is not a conditioning effect whereas η_{jk} is. If $\eta_{jk} = 0$, then the conditioning effects are described by $\{\beta_j\}$, which is then the main objective of the analysis.

(3) It may be necessary to introduce stratum effects v_h either as simple additive terms or as interaction terms between strata and the other effects.

The main point to note is that while model (1) may be too simple, each of the papers present analyses as if the model held.

Whatever the exact model formulation the impact of the various survey designs should be noted. S–J are in the strongest position, since in every period i all rotation groups j and recall effect groups k occur and no effects are confounded. For C–H and W–L, the longitudinal survey ($k = 1$, say) has $i = j$ at all periods, so that from this survey alone, period and rotation group are confounded. For the cross-sectional survey ($k = 2$), all respondents are in their first interview ($j = 1$). The period, conditioning, and recall effects may be disentangled so long as model (1) applies, but additional interaction terms (δ_{ik}, η_{ik}) will not be estimable from the data since they are confounded with other effects.

Each paper uses comparisons between a set of estimated means

$$\bar{Y}_{ijk} = \Sigma \, W_{ijkm} \, Y_{ijkm} \, . \tag{2}$$

The weights $\{W_{ijkm}\}$ are used to allow for adjusting to population totals (poststratification) and as an attempt to allow for age effects. The weights also make allowance for differential nonresponse and so are an attempt to allow for sample attrition. In addition, each set of authors attempts to eliminate the differential effects of attrition over time in sample by using only those respondents who respond throughout. Thus data from a sample member who fails to respond in the last quarter will not be used in any of the preceding quarters so that time in sample is not confounded with sample attrition. I support this approach, although it could be argued that the cumulative pattern of nonresponse is one aspect of a conditioning effect, but this is removed from the analyses presented. For S–J, each rotation group is restricted to those who respond throughout and the whole question of sample attrition is removed. For C–H and W–L, the cross-sectional survey has a response rate and pattern that is different from that of the longitudinal survey. The cross-sectional survey respondents have only one opportunity to be nonrespondents, whereas the longitudinal sample members have several. This contributes to the effect ξ_k, since k is used in both these investigations to identify the longitudinal and cross-sectional survey respondents.

It is valid to ask the question, What is a conditioning effect?

In Table 1, quarter 1 has a different length of reference period and, unlike the subsequent quarters, there is no bounding by a previous interview. Thus the conditions under which data are collected are not precisely the same for all quarters and this is completely confounded with the time in sample. Whether the difference between quarter 1 and the remainder is a conditioning effect or a recall effect is impossible to answer from the data.

The use of weights $\{W_{ijkm}\}$ is probably based on the premise that when estimating means it is better to adjust for nonresponse and to weight up to population counts. The objective in these papers is not to estimate population means but to estimate the conditioning effect. If model (1) is accepted, the parameters $\{\beta_j\}$ are the objective. Since we can only observe respondents, a separate question is whether the parameters $\{\beta_j\}$ would change if the response rate could be improved, and if nonresponse is a form of conditioning this could well be the case.

Assuming model (1) to be true

$$E\{\hat{\bar{Y}}_{j'..} - \hat{\bar{Y}}_{j..}\} = \frac{1}{IK}\sum_i\sum_k\sum_m (W_{ij'km} - W_{ijkm})(\mu + \alpha_i + \xi_k)$$

$$+ \frac{1}{IK}\sum_i\sum_k\sum_m (W_{ij'km}\beta_{j'} - W_{ijkm}\beta_j).$$

Thus in general the difference of means between rotation groups j and j' does not estimate $\beta_{j'} - \beta_j$ unless $W_{ij'km} = W_{ijkm}$ for all sampled units. Thus even under the simplest of models the weighted analysis may be misleading and it could be argued that since we are trying to estimate contrasts between the $\{\beta_j\}$ an unweighted analysis would be preferred. A further point suggests that if attrition is the same for all rotation groups, the number of observations for each i, j, and k would be approximately equal. In this case, the estimates for α, β, and ξ would be orthogonal and this would benefit the analysis. It is not certain but it is possible that using unequal weights could add to the lack of orthogonality.

A separate and perhaps more important point is the need for agreement on the target of inference. Whether the authors estimate a contrast between rotation group means or the corresponding function of $\{\beta_j\}$, it is very important to identify the target. With four quarters there are only three degrees of freedom and possible contrasts are $\beta_1 - \beta_2$, $\beta_1 - \beta_3$, $\beta_1 - \beta_4$, a linear trend $\beta_i = \beta i$, and possible quad-

TABLE 1. HOSPITAL DISCHARGES PER 1,000
PERSONS (FROM CORDER–HORVITZ)

Quarter	NMCUES
1	17.6
2	15.3
3	15.3
4	15.6

ratic effects, also. S–J estimate a wide range of contrasts and then interpret those that are most significant. I cannot see why some of their variables display linear conditioning effects, others U-shaped, and others none. Given the range of choices for possible contrasts, I am strongly supportive of the approach of W–L. They identify various kinds of conditioning effects from previous theory and then choose variables that might be expected to follow a prespecified pattern if conditioning exists. This formulates a hypothesis for each variable of a particular kind of conditioning effect and helps to overcome the problem of calculating too many contrasts. By embedding their analysis in existing theory they also advance our understanding rather than simply presenting additional evidence from a new survey that we, the survey community, must integrate with previous results. To some extent this approach also overcomes a problem they refer to, the tendency to analyze a large number of variables and to draw attention to "statistically significant" results.

4. THE INTERPRETATION OF RESULTS

The issues referred to above influence the way in which we should interpret the results presented by the three sets of authors.

S–J place particular emphasis on the analysis of the marginal means for each interview number (71.0, 62.4, 62.1, 58.7; $\chi^2_3 = 15.85$) (see Table 2). However, it should be observed that the recall effects are much larger than any conditioning effects, and also that there is interaction between the two. It is likely that the χ^2 value on the marginal means is actually capturing part of the recall effect and the marginal analysis of means may be inappropriate. In their final section S–J approach this problem in a better way by using a linear model to separate the two effects, although further analysis is needed to fully investigate the existence of conditioning effects.

S–J also refer to the problem of item nonresponse, particularly where skip questions have been used. Failure to respond to a skip question causes the subsequent expenditure values to be treated as zeros and the

TABLE 2. MEANS BY INTERVIEW NUMBER AND
RECALL MONTH (FROM SILBERSTEIN–JACOBS)

Recall Month	Interview Number				
	1	2	3	4	
1	77.8	73.4	70.7	73.2	
2	69.7	55.5	63.8	53.9	
3	65.6	58.2	51.7	49.0	
	71.0	62.4	62.1	58.7	$\chi_3^2 = 15.85$

potential impact of item nonresponse of this kind may be substantial. They should explore whether item nonresponse changes with interview number—especially for skip questions.

In the second quarter, NMCUES does not drop in a way that mirrors the HIS and, given the very small standard errors, C–H conclude that this is evidence of a conditioning effect. The essential assumption is that there is no interaction between survey and period effects ($\delta_{ik} = 0$). The survey effects ξ_k are very large (i.e., the NMCUES and HIS estimates for each quarter are quite different). Also, reference periods vary between quarters and telephone interviews have been used for some quarters. In these circumstances the standard errors give a spurious sense of precision since the underlying model assumptions are likely to have a much bigger impact on the comparison made. To be precise (see Table 3), for quarters 1 and 2

$$(\bar{Y}_{1,NMCUES} - \bar{Y}_{1,HIS}) - (\bar{Y}_{2,NMCUES} - \bar{Y}_{2,HIS})$$

is a valid estimator of the conditioning effects only if there is no interaction. There is no way of investigating this question from the data because of the confounding of effects, but it seems optimistic to assume that no interactions exist.

We contrast Table 4 (from W–L) with Table 3 (from C–H). In Table 4 there is a "survey effect" ($\xi_k \neq 0$) since the cross-sectional survey shows a systematic difference to the panel. This is mainly due to the extra attrition the panel experiences. However, this main effect is very small and the period effects (1983 to 1985) are also very small. In these circumstances we have much more confidence that no sizeable interaction between period and survey (δ_{ik}) exists and we give much greater credence to the estimate of the conditioning effect. Thus although W–L and C–H use data with the same design, W–L's data seem more suited to the implicit assumption of no interaction that underpins their choice of a contrast in means.

TABLE 3. PHYSICIAN VISITS PER PERSON (FROM CORDER–HORVITZ)

Quarter	NMCUES		HIS	
	$\bar{x}$	SE	$\bar{x}$	SE
1	1.07	.02	1.30	.02
2	1.05	.02	1.10	.02
3	.97	.03	1.20	.02
4	.96	.02	1.20	.02

TABLE 4. POLITICAL IDENTIFICATION BETWEEN 1983
AND 1985 (FROM WATERTON–LIEVESLEY)

	Panel		Cross-Sectional Survey	
	Partisans	Sympathizers	Partisans	Sympathizers
1983	50.7	27.2	47.2	25.6
1984	51.4	28.7	47.0	25.7
1985	53.8	28.1	46.9	25.5

5. CONCLUSION

The overall impression from the three papers is that the case for the existence of strong conditioning effects is not really proven. For both S–J and C–H the clear message is that even if conditioning effects exist, they are much smaller than the recall effects, which are a much more serious influence on the quality of the data. In some senses the analysis in W–L is more convincing, although they would readily accept that most of their conclusions are negative. Where conditioning effects appear to exist they are small but this is perhaps to be expected given the frequency of the survey and the type of data collected.

The other conclusion to be drawn from the studies is that survey designs that are suited to the primary purposes of the survey are not necessarily ideal when we try to disentangle the various factors that affect the quality of the survey data. There may be a need to carefully design and analyze special studies, including fieldwork, if the various effects are to be properly estimated.

Treatment of Wave Nonresponse
in Panel Surveys*

James M. Lepkowski

1. INTRODUCTION

Missing data in surveys are generally considered to be of two types. Unit nonresponse occurs when no data are obtained for a sampled unit, while item nonresponse occurs when one or more items are not completed for a unit that otherwise provides responses. The amount of missing data and the amount of information available about the non-responding unit influences the strategy employed to compensate for the missing data. For unit nonrespondents, all survey data are missing and hence limited sample design information may be the only available data for them. Compensating for unit nonresponse is typically made by weighting, a process in which appropriate respondent records are assigned increased importance in order to compensate for the nonrespondents. With item nonresponse, more extensive data are available about the nonresponding units (e.g., sample design data plus responses to other survey items). These data can be used to improve the quality of the compensation. Imputation, a process in which values are assigned for missing responses, is typically employed to compensate for item missing data.

In addition to unit and item nonresponse, panel surveys experience another type of nonresponse that may be termed wave nonresponse. Panel surveys collect information from the same sample elements at several different points of time, or waves. Each wave of data collection may cover either the same reference period or different reference periods. Wave nonresponse occurs when one or more waves of panel data are missing for a unit that has provided data for at least one wave. The amount of missing data for a record with wave nonresponse is typically greater than that encountered for item nonresponses, but data available from completed waves provide more detailed information about the wave nonrespondent than is available for unit nonrespondents. Weighting, imputation, or a combination of weighting and imputation may be considered for compensating for missing data due to wave nonresponse.

This paper reviews the nature of wave nonresponse in panel surveys and examines missing data compensation strategies for it. The review

348

is not intended to cover every conceivable wave nonresponse compensation strategy, but rather it describes the characteristics, strengths, and weaknesses of several general strategies. After a discussion of general panel surveys and wave nonresponse issues in Section 2, Sections 3 and 4 consider characteristics of weighting and imputation as compensation methods for wave missing data. Section 5 reviews an empirical comparison of weighting and imputation compensations in a data set with simulated missing data. Section 6 examines combined weighting and imputation strategies, and the paper concludes in Section 7 with a discussion of criteria that might be applied in the selection of a suitable compensation strategy for wave nonresponse.

2. PANEL SURVEYS AND WAVE NONRESPONSE

In preparation for the review of wave nonresponse compensation strategies, it is useful first to consider three issues in panel survey design: (1) the forms of analysis of panel survey data, (2) the types of data collected in surveys, and (3) wave nonresponse patterns.

For the present discussion, it is important to distinguish three forms of analysis for panel survey data: longitudinal comparisons, cross-sectional estimation, and longitudinal cumulation. Although panel surveys are typically considered most useful for collecting longitudinal information, particularly information on changes over time, panel survey data may also be analyzed for fixed points in time to provide cross-sectional estimates for a population at these times. Panel surveys are also used to cumulate information over time. Panel data may be cumulated over several short reference periods to provide information about a longer reference period. Cumulation may involve summing data across reference periods, the cumulation of rare events, or the construction of event histories over extended time periods.

These analytic uses of panel survey data must be taken into account when considering compensation strategies for wave nonresponse. A compensation method that provides reasonably accurate results for one form of analysis may provide poor results for another. For example, an imputation suitable for cross-sectional analyses of panel data may introduce an incorrect change in status because an actual response is recorded on one wave and an imputed response for the same item is recorded on another wave.

The type of data elements may also be an important factor when choosing a wave nonresponse compensation strategy. Three types of elements may be distinguished: continuous, categorical, and conditional. Continuous items are interval-scaled measures, such as income. Categorical items are characteristics of an event or condition that are recorded in discrete units. They may include indicator items denoting presence or absence of a condition, discrete scale choices, or even counts. Conditional items are measures for which a value of a continuous or categorical item is present only if the sample element meets certain eligibility criteria defining a subgroup of the population. Conditional items arise in surveys through the use of branching or conditional sequencing in a questionnaire. For sample elements not in the subgroup the value of the item may be a code denoting ineligibility for this item, while for subgroup members the value is a continuous or categorical response for the subject. For example, wage and salary income is available only for individuals with wage earnings or salaried jobs. Alternatively, a categorical item such as type of employer or receipt of income during a given period will be zero for unemployed persons.

Compensation for wave nonresponse will be concerned with several of these types of measures simultaneously. As will be discussed in more detail later, global (as opposed to item) weighting strategies handle all data types at the same time, and hence type of measure is not important. On the other hand, the type of measure is important for many imputation strategies. For instance, imputation methods that obtain an imputed value through a prediction equation must consider the type of measure being imputed. Categorical items may need a different type of procedure than continuous (although imputation for both types of items may be summarized in terms of a similar model). Conditional items pose particular problems since both eligibility and the value of the item must be handled. A two-stage or dependent imputation procedure may be used to impute eligibility first, followed by imputation of the value of the item for imputed eligible units.

Finally, the patterns of wave nonresponse must also be considered in developing an appropriate compensation strategy. The complexity of compensation for wave nonresponse depends on the patterns of wave nonresponse that occur. For example, the schematic in Figure 1 indicates several patterns of wave nonresponse that may occur in a three-wave panel survey. The patterns of wave response are represented for this three-wave panel as a pattern of X's (representing a wave response) and O's (representing wave nonresponse). (This representation is used even though the waves may vary in length from one individual to the next.) There are $2^3 = 8$ wave response/nonresponse patterns for the three-wave panel; all but the three-wave nonresponse pattern OOO are represented in the figure.

The frequency distribution of the wave response/nonresponse patterns will vary across surveys depending on the survey topic, survey organization, and other factors. Table 1 presents the frequency distribution of respondents to two panel surveys with similar topics and designs. The Income Survey Development Program 1979 Research Panel (ISDP) consisted of residents of approximately 7,500 households who were interviewed every three months for a total of six interviews

FIGURE 1. WAVE NONRESPONSE PATTERNS
FOR A THREE-WAVE PANEL SURVEY

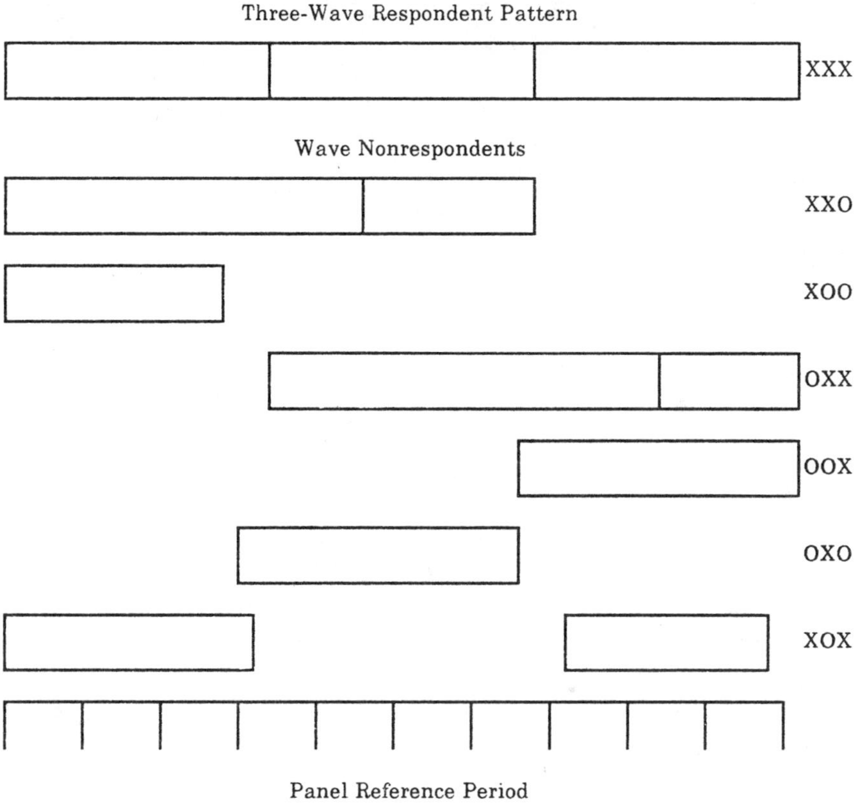

Three-Wave Respondent Pattern

Panel Reference Period

concerning income and participation in government. programs (Ycas &
Lininger, 1981). The 1984 Panel of the Survey of Income and Program
Participation (SIPP) consisted of residents of approximately 20,000
households interviewed every 4 months for a total of up to nine interviews (Herriot & Kasprzyk, 1984). The data in Table 1 are for the first
three waves of each panel; they relate to persons aged 16 and over in
the ISDP panel and persons aged 15 and over in the SIPP panel.

In both surveys, the largest percentage of persons are three-wave
respondents. Among wave nonresponse patterns, the attrition patterns,
in which the respondent appears in an early wave and then fails to
respond at later waves, are the next most frequent patterns. The nonattrition patterns are the least frequent in both surveys, but three of
these patterns (OXX, OXO, and OOX) do not appear in the SIPP 1984
Panel since wave 1 nonrespondents were not followed at later waves.
Thus, the frequencies of the patterns must be examined, keeping in

TABLE 1. PERCENTAGE DISTRIBUTION OF PERSON RESPONSE PATTERNS
FOR THE FIRST THREE WAVES OF THE ISDP 1979 AND SIPP 1984 PANELS

Pattern[a]	ISDP		SIPP
	All Persons	Wave 1 Respondents	
Respondents:			
XXX	80.2%	83.3%	90.0%
Attrition Wave Nonrespondents:			
XXO	7.2	7.5	4.9
XOO	6.7	7.0	4.2
Nonattrition Wave Nonrespondents:			
XOX	2.3	2.4	1.0
OXX	2.2	–	–[b]
OXO	0.6	–	–
OOX	0.9	–	–

[a]Response = X, nonresponse = O.

[b]Persons not responding at wave 1 in the SIPP were not followed for subsequent interviews.

mind that the SIPP 1984 Panel did not follow wave 1 nonrespondents. For the sake of comparison, the percentage distribution for ISDP wave 1 respondents is provided removing the nonattrition patterns OXX, OXO, and OOX .

As more waves are considered, the relative frequency of the "complete respondent" pattern decreases (usually slowly), while the number of patterns increases rapidly. For example, for the full 9 waves of the SIPP 1984 Panel there will be $2^9 = 512$ wave response/nonresponse patterns. The complete wave respondent pattern (i.e., XXXXXXXXX) is likely to be the most frequent pattern, followed by the patterns with one and two missing waves. The other patterns are each likely to have small relative frequencies, but cumulatively their frequency will not be negligible. In the subsequent discussions of compensation strategies, the number of patterns and their relative frequencies will be important in considering different methods.

3. WEIGHTING TO COMPENSATE FOR WAVE NONRESPONSE

One popular method of weighting for nonresponse divides the sample into a number of adjustment cells based on auxiliary information available for both respondents and nonrespondents. Within the adjustment cells, the weights, which may be the inverse of the probability of selection for the individual unit, are summed for all eligible units and for responding units. The nonresponse adjustment is computed as the ratio of the sum of weights for all eligible units to the sum for responding units. This ratio is applied to the weights of each of the responding units in the adjustment cell in order to compensate for the nonrespondents in that cell.

There are many other weighting methods to compensate for nonresponse (see, for instance, reviews by Chapman, 1976; Kalton & Kasprzyk, 1986; Oh & Schueren, 1983; and Sande, 1982). They include matching nonrespondents to respondents and adding the nonrespondent's weight to that of the matched respondent; duplicating respondent records instead of weighting; obtaining predicted probabilities of response for respondents through probit or logistic regression and weighting by the inverse of predicted probabilities; and raking ratio procedures to develop weights based on known marginal distributions for population characteristics. For the ease of presentation, the subsequent discussion uses the simple adjustment cell weighting procedure as a reference method.

Several features of adjustment cell weighting deserve comment. To the extent that the auxiliary variables used to form adjustment cells are correlated with other survey variables with missing values, responding and nonresponding units in each cell will tend to have similar values for the missing survey items. If the response rates vary across the adjustment cells as well, the nonresponse adjustment weights will reduce the bias due to nonresponse (Rubin, 1976). In addition, the effects of nonresponse adjustment are spread across many respondents in the same adjustment cell. There will be some increase in the variance of estimates due to the increased dispersion of weights, but it will generally be less than weighting methods that generate different weights for each respondent (e.g., matching nonrespondents to a single respondent).

For unit nonresponse, weighting is a global strategy in the sense that a single adjustment is used to compensate for all nonresponses in a survey. While this may be a practical approach to implement, a single set of auxiliary variables, mostly limited to sample design information, will seldom be strong correlates for all, or even many, survey items. For wave nonresponse, there are many more auxiliary variables available for creating weighting classes including the same variable collected on other waves, information that may be highly correlated with the missing value. Weighting to compensate for wave nonresponse can use this expanded set of auxiliary variables to improve the accuracy of the adjustment.

Another consideration in weighting for wave nonresponse is the multiple analytic uses of the data. Consider the three-wave panel survey illustrated in Figure 1, and suppose that three cross-sectional estimates are to be calculated from the panel data, one for each of waves

1, 2, and 3. For the wave 1 cross-sectional estimate, each of the response patterns XXX, XXO, XOO, and XOX provide wave 1 data, and they would have to be weighted to account for the nonresponding patterns OOO, OOX, OXX, and OXO. The response patterns for the wave 2 cross-sectional estimate are XXX, XXO, OXX, and OXO, while for wave 3 the responding patterns are XXX, OXX, XOX, and OOX. Since different sets of records are involved in providing cross-sectional estimates for waves 1, 2, and 3, three different sets of weights could be used for cross-sectional estimation.

When longitudinal comparisons are of interest and one wishes to use as much of the data as possible, other sets of weights are required. For example, only the XXX and XXO patterns provide data for comparing waves 1 and 2; compensation is needed for five wave nonresponse patterns and the complete nonrespondents (i.e., pattern OOO) in making wave 1 and 2 comparisons. Longitudinal comparisons for waves 1 and 2, 2 and 3, 1 and 3, and 1, 2, and 3 require four more sets of wave nonresponse compensation weights. Thus, for a three-wave panel survey designed to meet both cross-sectional and all varieties of longitudinal estimation objectives, seven sets of wave nonresponse weights are needed. The number of weights rapidly increases with the number of waves: the nine-wave SIPP 1984 Panel could require $2^9 - 1 = 511$ weights for both cross-sectional and longitudinal purposes.

The number of weights needed for longitudinal analysis objectives can be reduced if fewer patterns are used. At a minimum, only the three-wave respondents (i.e., pattern XXX) can be used to meet all three objectives mentioned earlier with weights to account for the complete wave nonrespondents (i.e., pattern OOO) and the six wave nonresponse patterns. This approach is attractive if the frequency of the wave nonresponse patterns is small relative to the frequency of the three-wave response pattern. However, even for a three-wave panel this approach may discard sizable amounts of data, causing substantial losses of precision for cross-sectional estimates and longitudinal comparisons.

Alternatively, some of the wave nonresponse patterns may be used in combination with the complete wave respondents. For example, using the attrition patterns and the three-wave respondents only, one weight per wave is needed to meet cross-sectional and longitudinal analysis objectives. Wave 1 cross-sectional analysis could use the XXX, XXO, and XOO patterns compensated for the remaining wave nonresponse patterns (including the OOO pattern). For wave 2 cross-sectional or wave 1 and 2 longitudinal analyses, patterns XXX and XXO are compensated for the remaining patterns. For all remaining analyses (e.g., wave 3 cross-sectional estimates, wave 2 and 3 longitudinal analysis), the XXX pattern is compensated for all the other wave nonresponse patterns.

The amount of data discarded by this approach will be limited to data from the less frequent nonattrition patterns. The amount of discarded data could be further limited by including "re-entry" patterns as attrition patterns, discarding data obtained after re-entry. For example, the third wave of data in the XOX pattern could be discarded to create additional attrition nonrespondents. Although this is the only re-entry pattern in a three-wave panel, panels with more waves will have a number of other re-entry patterns that can be converted to attrition patterns by discarding later waves of data.

Little and David (1983) have proposed a method for incorporating more auxiliary variables into the development of weights for attrition patterns of wave nonresponse. The only auxiliary information available for both first-wave respondents (patterns XXX, XXO, XOX, and XOO) and nonrespondents (patterns OXX, OXO, OOX, and OOO) are the sample design variables, denoted by the vector z, such as strata, sampling units, and characteristics of those units. As described previously, adjustment cells can be created using the design variables z, and the inverse of the response rate within each cell can be used as the weight for the first-wave respondents. Alternatively, a wave 1 response indicator r_1, equal to 1 for wave 1 respondents and 0 otherwise, can be regressed on the design variables z using probit or logistic regression. The wave 1 weights w_1 are then the inverses of the predicted probabilities of response for the wave 1 respondents conditional on their values of z. Only this latter regression approach is considered in the subsequent discussion.

Both the design variables z and the responses obtained for the wave 1 respondents, say x_1, are available for the wave 1 respondents. A wave 2 response indicator r_2 can be regressed on z and x_1 for the wave 1 respondents. From this regression, weights $w_{2.1}$ can be computed for the wave 2 respondents to compensate for the lost responses from wave 1 to 2. The overall weights for the wave 2 respondents are then computed as adjustments to the wave 1 weights to compensate for the additional losses incurred at wave 2; that is, $w_2 = w_1 \cdot w_{2.1}$

For a later wave, say the t th, the auxiliary data include the sample design variables z and the responses at all the previous waves for the respondents at the $(t-1)$th wave: $x_1, x_2, \ldots, x_{t-1}$. The attrition compensation weight $w_{t.1,2,\ldots,t-1}$ can again be computed as the inverse of the predicted probability of response at wave t obtained from the regression of the response indicator r_t on $z, x_1, x_2, \ldots, x_{t-1}$. The t th wave weight is computed as $w_t = w_{t-1} \cdot w_{t.1,2,\ldots,t-1}$.

Attrition patterns of wave nonresponse can again be created by eliminating waves of data from some patterns (e.g., treating XOX as XOO) or by adopting a data collection strategy of following only those units that responded on the previous wave. The elimination of waves of data from nonattrition patterns may cause a substantial loss of data for panels with a large number of waves. Thus, weighting strategies for nonattrition patterns of wave nonresponse are also of interest.

Unfortunately, the simplicity of this attrition weighting scheme is largely lost when nonattrition patterns are retained. The wave 1 weight, w_1, is again the inverse of predicted response probabilities obtained from the regression of a response indicator r_1 on the sample design variables z. In the nonattrition situation, however, the wave 2 weight requires two separate regressions. The first regresses the wave 2 response indicator r_2 on z and the wave 1 auxiliary variables x_1 for wave 1 respondents (i.e., patterns XXX, XXO, XOO, and XOX). The second regresses r_2 on just the sample design variables z for the remaining patterns that do not have a wave 1 response (i.e., OXX, OXO, OOX, and OOO). The wave 2 weight, w_2, is the inverse of the predicted response probability obtained from the appropriate regression.

Computation of the wave 3 weight, w_3, involves four regressions of the wave 3 response indicator r_3 on various combinations of sample design and previous wave auxiliary data x_1 and x_2:

(1) r_3 regressed on z, x_1, and x_2 for patterns XXX and XXO.

(2) r_3 regressed on z and x_1 for patterns XOX and XOO.

(3) r_3 regressed on z and x_2 for patterns OXX and OXO.

(4) r_3 regressed on z for patterns OOX and OOO .

Products of the weights created at each wave are needed for cross-sectional and longitudinal analyses of the data. Cross-sectional analysis of wave 2, for instance, would involve wave nonresponse patterns XXX, XXO, OXX, and OXO with the weight $w_1 \cdot w_2$. Longitudinal analysis of three waves would use the XXX pattern respondents and the product $w_1 \cdot w_2 \cdot w_3$.

Obviously, the nonattrition weighting scheme that attempts to use as much of the auxiliary information as possible requires substantially more computation than other schemes. The number of computations increases geometrically as the number of waves increases. In addition, the number of respondents for a set of matched patterns (e.g., treating pattern OOX as a complete three-wave nonresponse pattern) may be small, while the number of nonrespondents is large. For example, there are likely to be few OOX respondents but many OOO nonrespondents for developing the wave 3 weights. In consequence, the resulting weights for the OOX respondents could vary markedly, adversely affecting the precision of survey estimates. Collapsing of patterns can reduce this problem, but the price paid is not to use all the auxiliary information available for each pattern.

If cumulation of information across waves is of interest, an item-level weighting strategy may be employed. Each record may be viewed as data available for the portion of the entire reference period during which the respondent was eligible for the survey. When data are missing for a portion of the respondent's period of eligibility, the nonmissing portion may be inflated to account for the missing portion. For example, records with wave nonresponse patterns XXO, XOX, and OXX receive a weight of $3/2 = 1.5$ to account for the one missing wave. Records with patterns XOO, OXO, and OOX are weighted by a factor of 3/1. This longitudinal weighting strategy can be generalized from waves to other time units for which data may be recorded, such as months, weeks, or days. For example, if a wave nonrespondent pattern XXO represented three months of missing data during a one-year reference period, the nonresponse adjustment weight would be $12/9 = 1.33$. This longitudinal weighting strategy does not handle total unit nonresponse (i.e., pattern OOO); a separate weighting adjustment will be needed to account for them. Cox and Cohen (1985) illustrate a similar approach for the National Medical Care Utilization and Expenditure Survey (NMCUES).

This within-record inflation assumes that the respondent's available data are more strongly (and positively) correlated with the respondent's own missing data than are data from other respondents. It is a useful strategy for compensating for wave nonresponse when one is interested in cumulating information over waves, but it does not handle cross-sectional estimation or longitudinal comparisons.

Cross-sectional weights can be used to cumulate information over waves, but the cumulation can only be completed at an aggregate level (i.e., cumulating wave totals). The longitudinal weights also cumulate information over waves by inflating cumulated values of selected patterns of wave nonresponse. For example, cumulation over the first two waves of a three-wave panel could be accomplished by cumulating data from the records with both waves available, patterns XXX and XXO. The longitudinal weights can then be applied to the cumulated values to obtain cumulated totals adjusted for patterns that are missing one or both of the first two waves.

A final consideration in weighting is consistency between cross-sectional and longitudinal findings. Under some wave nonresponse weighting schemes, cross-sectional and longitudinal analyses may be based on different sets of cases and may be inconsistent. For example, consider the comparison of wave 1 and 2 totals for a three-wave panel. Wave 1 and 2 totals computed from a longitudinal file containing three-wave respondents will differ from the totals obtained by considering wave 1 respondents and wave 2 respondents separately.

Weighting for wave nonresponse is complicated by the need to handle multiple objectives and by the desire to use the available data as fully as possible. Simplified weighting schemes may be preferred even though they involve losses in precision of estimates due to the discarding of collected data. However, the complexity of wave nonresponse weighting should be considered relative to the potentially more extensive (and expensive) tasks involved with imputation for wave nonresponse.

4. IMPUTATION TO COMPENSATE FOR WAVE NONRESPONSE

By filling in the missing values in a data set, imputation makes analysis appear to be easy to conduct and results easier to present. A data set in which waves of data that are missing have been imputed is easier to handle analytically than one in which weights have been used to compensate for the missing waves. For example, when weighting compensates for wave nonresponse, the purpose of the analysis may have to be considered when deciding which set of weights to use. In contrast, a data set completed by imputation can be used for any analytic purpose. Imputation also assures that results from analyses that employ cross-sectional methods are consistent with those using longitudinal methods.

While imputation has a number of attractive features, it has major drawbacks as well. Imputation fabricates data to give a data set the appearance that it is complete. This appearance is misleading and can lead to a false sense of security in the reliability and accuracy of the findings. In addition, imputed values tend to attenuate the covariance among survey items when the observed covariation among the responses is biased toward zero by the presence of imputed values (Santos, 1981). For analysts concerned with examining relationships among a number of variables, some of which have imputed values, the amount of imputation may be an important issue to consider in determining appropriate analytic methods. For instance, imputation may be so extensive as to completely obscure a relationship between two survey measures. Analyses may be conducted without imputed values to determine when imputation is seriously attenuating the strength of relationships in the data (Lepkowski et al., 1987).

Imputation to compensate for wave nonresponse has an additional disadvantage relative to weighting. While weighting typically involves computation of a factor or a set of factors to handle all missing data, imputation must insert a data value for each item on a missing wave. Thus, imputation must consider every data item across a multiple-wave panel survey, a substantial task for most panel surveys. Some of the complexity can be alleviated by imputing blocks of information, such as an entire wave, for an individual.

4.1. General Imputation Methods

Following Kalton and Lepkowski (1983), imputation can be represented in a general form by the model $y_i = f(x_{1i}, x_{2i}, \ldots, x_{pi}) + e_i$ where y_i is the value imputed for the ith respondent, $f(x)$ is a function of p auxiliary variables in x, and e_i is an estimated residual. The function $f(x)$ is often expressed as a linear function $\beta_0 + \sum_j \beta_j x_{ij}$ where the β_j's are estimated from data for respondents. If the e_i are set at 0, the imputation method is a deterministic one, the distribution of the imputed variable is distorted, and the variance is attenuated. Stochastic imputation procedures add an estimated residual to the function $f(x)$. Those procedures are generally preferred to deterministic methods because they avoid the distortion of the distribution and attenuation of the variance. However, they can also increase the variance of survey estimates.

In addition to linear regression, this general imputation model also covers other types of imputation procedures. For example, imputation class methods are represented by a set of x_j's that are indicator variables jointly defining the imputation classes. Similarly, the familiar hot-deck imputation can be viewed as a function of auxiliary variables defining imputation classes for which a mean value plus a residual is to be assigned. The imputation consists of the imputation class mean plus a residual estimated from a similar donor.

Kalton and Lepkowski (1983) describe a variety of methods for imputing across waves of a panel survey. The hot-deck method can impute missing values for a variable on one wave by using the value of the same variable on another wave to determine imputation classes. For example, consider a wave nonrespondent whose income is known for wave 1 but is missing for wave 2. The wave 1 income values can be categorized and classes formed. The imputed value for the wave nonrespondent's missing wave 2 value would then be taken as the wave 2 value from a two-wave respondent who comes from the same wave 1 income class.

Similarly, regression and other item imputation methods can use auxiliary data from another wave in the imputation process. For instance, imputation of wave 2 income y_i given the individual's wave 1 income x_i can be represented by the regression model $y_i = a + bx_i + e_i$. Cross-wave regression imputation constructs a new variable $\hat{y} = a + bx_i$ for all individuals, imputes the e_i's for wave 2 nonrespondents, and imputes y_i as $\hat{y}_i + e_i$. The e_i's may be imputed by a hot-deck method using other auxiliary variables or by a stochastic mechanism that generates them from a known distribution.

The constant and slope terms, a and b, can be obtained from least squares estimates for the regression of y_i on x_i for two-wave respondents. Alternatively, assigning $a = 0$ and $b = 1$ without a residual term is a "carry-over" imputation, perhaps the simplest cross-wave imputation procedure to implement. Setting $a = 0$ and estimating b from the respondent data is a model of proportionate change; setting $b = 0$ and estimating a from the respondent data is a model of additive change. The quality of these cross-wave imputations depends on the strength of the correlation for the same item over time.

4.2. Cross-Wave Imputation Methods

When responses to an item are highly correlated over time, responses on one wave will be powerful auxiliary variables for imputing the missing response for the same item on another wave. Table 2, adapted from Kalton, Lepkowski, and Lin (1985), demonstrates the consistency of several quarterly categorical labor force and income recipiency items across the first two waves of the ISDP 1979 Research Panel. For instance, 18.3% of the 13,151 original sample respondents with data on both waves reported receipt of Social Security income on both waves, and 80.3% reported receipt on neither wave. A total of 98.7% of the respondents were thus consistent for the first two waves of the panel (i.e., gave the same answer on both waves).

If wave nonrespondents have the same cross-wave consistency as the wave respondents demonstrated in Table 2, a simple carry-over imputation would assign a high proportion of missing-wave income recipiency and labor force items correctly. The direction of the carry-over imputation could be forward or backward depending on which waves are missing. The deterministic carry-over imputation has, however, two disadvantages. First, the distribution of the carry-over imputations may differ from that of the nonmissing responses on the wave. Suppose, for example, that the wave nonrespondents have the distribution of responses shown in Table 2. The percentage of "Yes" responses on wave 1 for Social Security income is $18.4 + 0.4 = 18.8$, which would be the carry-over imputed percentage as well. But the actual wave 2 percentage "Yes" is $18.4 + 0.9 = 19.3$. Second, the carry-over imputation forces stability of responses for wave nonresponses, and thus causes an understatement of the extent of changes between waves.

A stochastic cross-wave imputation method can avoid these deficiencies. Instead of carrying over the wave 1 Social Security response, a stochastic mechanism could assign an expected $18.4/(18.4 + 0.4) = 97.9\%$ of the wave 1 Social Security income recipients with missing data on wave 2 a "Yes" response at wave 2 and an expected 2.1% a wave 2 "No" response. Similarly, an expected 99.5% of those with a "No" response on wave 1 and missing data on wave 2 would be assigned a "No" response on wave 2. The wave 2 stochastically imputed responses will have, in expectation, the same distribution on wave 2 as the wave 2 responses, and they will have imputed changes in Social Security income recipiency across the waves.

Consistency for continuous items can be examined in terms of cross-wave correlations. Table 3 presents cross-wave correlations from the ISDP 1979 Panel for three income items summarized by computing average correlations when the income items were one, two, or three or more months apart. One- and two-month differences could be between

TABLE 2. PERCENTAGE DISTRIBUTION OF RESPONSES FOR
SIX ITEMS ACROSS WAVES 1 AND 2 FOR THE ISDP 1979
RESEARCH PANEL RESPONDENTS AGES 16 AND OLDER

Item	Wave 1/Wave 2 Response				Consis-tency	Sample Size
	Yes/Yes	Yes/No	No/Yes	No/No		
Received Social Security	18.4%	0.4%	0.9%	80.3%	98.7%	13,151
Received federal SSI[a]	3.2	0.3	0.3	96.2	99.5	13,151
Worked in quarter	58.2	3.5	3.8	34.5	92.8	13,119
Reasons for Not Working:						
Going to school	11.0	0.9	0.7	87.4	98.4	4,520
Didn't want to work	4.9	6.5	8.5	80.1	84.9	4,520
Retired	15.3	5.0	6.5	73.2	88.5	4,520

Source: Kalton et al., 1985.

[a]Supplemental Security Income.

two monthly responses from the same wave or between different waves, while three-or-more-month differences were necessarily between different waves.

The average cross-month correlations are high for the earnings and Social Security items, and somewhat smaller for unemployment compensation. The correlations within a wave are higher than those between waves, probably because of response errors. As the number of months between values increases for the earnings and income items, the correlations decrease. However, for unemployment compensation, there is a decrease followed by an increase in correlation. It appears that there is substantial short term change in the amount of unemployment compensation received, but once the compensation has been received for a longer period, the amount of compensation begins to stabilize.

If wave nonrespondents have the same cross-wave correlation as the wave respondents, the carry-over imputation would assign a high proportion of missing-wave income recipiency and labor force items correctly. But, of course, this deterministic imputation has disadvantages that stochastic cross-wave imputation can avoid. A stochastic mechanism could add an estimated or an imputed residual to the carry-over value. The wave 2 stochastically imputed responses will, in the presence of high cross-wave correlations, have a similar distribution on wave 2 as the wave 2 responses, and they will have imputed changes in income across the waves.

The data in Tables 2 and 3 primarily address the problem of wave nonresponse on one of two consecutive waves where imputation can be

TABLE 3. AVERAGE CROSS-MONTH CORRELATIONS FOR
THREE AMOUNT ITEMS FOR ISDP 1979 RESEARCH PANEL
RESPONDENTS AGES 16 AND OLDER

Item	Wage and Salary Earnings	Social Security Income	Unemployment Compensation
One-Month Difference:			
Within wave	0.993	0.983	0.651
Between wave	0.842	0.921	0.408
Within and between	0.910	0.968	0.590
Two-Month Difference:			
Within wave	0.890	0.978	0.645
Between wave	0.839	0.924	0.448
Within and between	0.861	0.946	0.532
Three-or-More-Month Difference:	0.813	0.903	0.599

Source: Kalton et al., 1985.

made forward or backward, depending on which wave is missing. If both waves are wave nonresponses, the cross-wave imputation procedures are not applicable, and a back-up within-wave imputation or wave nonresponse weights may be needed.

Alternatively, imputation between nonconsecutive waves may be used. For example, in a three-wave panel the pattern OOX may appear to be inapplicable for cross-wave imputation for waves 1 and 2, when in fact the imputations could be backcast from wave 3. Of course, the longer the period across which the imputations must be made, the lower the quality of the imputations that may be expected because of decreasing cross-wave consistencies and correlations.

Other three-wave patterns offer additional alternatives for cross-wave imputations. The XXO and OXX patterns could use data from the nearest wave to forecast or backcast, respectively, for the missing wave. The XOX pattern can provide either a forecast or a backcast for the missing wave. The availability of two-wave responses can be used to improve the quality of imputations by examining three-wave consistency patterns or correlation matrices. Finally, data can be both forecast and backcast for the OXO pattern.

Of course, the complete nonrespondent pattern OOO cannot be imputed by cross-wave methods. These nonrespondents can be handled as unit nonresponses and weighted in the usual way, or some within-wave imputation procedure such as a hot deck can be employed. Weighting is easier and is generally the preferred method when compensating for unit nonresponse (see, for example, Kalton, 1983).

Not all panel survey items will be highly positively correlated over time and thus good candidates for cross-wave imputation. For example, health care utilization or expenditures during a one-year period are not likely to be highly correlated over time for most persons. The NMCUES, designed to collect such data from a panel of approximately 17,000 persons interviewed four or five times during a one-year period (Bonham, 1983), did not consider cross-wave imputation appropriate for wave nonresponse compensation because of low correlation over time for the principal survey items (Cox & Cohen, 1985).

A few items may actually be negatively correlated over time, such as expenditures for an automobile. Once such a major purchase of a durable good has been made, the expenditure is not likely to be made again for some time. Similarly, events such as births will be negatively correlated over limited time periods. These negative correlations may be useful for developing cross-wave imputation methods by indicating when an expenditure or event will not occur in a preceding or succeeding wave.

This discussion, and the functional imputation model described previously, has been concerned with one-at-a-time item imputations. Each variable is considered separately, and different imputation procedures may be used for each item depending on correlations with other items and types of variable (i.e., categorical, continuous, or conditional). Most item imputation systems do not develop separate procedures for each item to be imputed, but often use the same procedure to impute groups of similar items at the same time.

Wave nonresponse presents a set of missing items that can be compensated simultaneously through the imputation of an entire wave of data. The functional form for imputation can be generalized by considering the imputed value to be a vector of values, and hence, item imputation methods can be generalized to wave imputation methods in a straightforward way. For example, consider imputation of class means for missing items. The wave imputation alternative would assign a vector of class means for the missing items on an entire wave. Similarly, in a hot-deck wave imputation, a donor is matched to a recipient using auxiliary information from the sample design or from other nonmissing waves; a vector of the donor's wave data is imputed to the recipient's missing wave. Regression imputation in its various forms can also be generalized through multivariate regression methods (i.e., regression with multiple dependent variables) to handle wave imputation. A simple form of wave regression imputation is the wave carry-over method: all data items that are repeated on two or more waves are carried over to the missing wave. The carry-over imputations can be adjusted for known changes that might occur from one wave to the next, such as a change in the length of the reference period or a cost of living increase in Social Security payments.

An important advantage for wave imputation is its ability to handle several types of measures simultaneously. Item imputation becomes cumbersome for moderately complicated surveys because the same form of imputation is not suitable for all types of items. Continuous and categorical items may require different imputation models, while conditional measures may require imputing eligibility first followed by an imputation of values for imputed eligibles. Such imputation must be designed on an item-by-item basis.

5. EMPIRICAL COMPARISONS OF WEIGHTING AND IMPUTATION

Weighting and imputation have been described as two separate methods for compensating for missing data, but they are in fact closely related (Kalton, 1983; Kalton & Kasprzyk, 1986; Oh & Scheuren, 1983). The relationship is illustrated by considering a simplification of the traditional hot-deck imputation procedure: divide a sample of respondents and nonrespondents into imputation classes, and assign a nonrespondent within a class the value for the missing item from a respondent in the same cell. For analysis of this single item, this imputation scheme is equivalent to a weighting scheme in which the weight of the nonrespondent is added to the weight of the matched respondent who donated the imputed value. The mean and variance of the item are the same under either the simple hot-deck imputation or the weighting compensation scheme.

Nonrespondent weights are typically not added to individual respondent weights. This strategy avoids decreases in the precision of estimates from increased variability in the weights. Instead, all respondent weights within a class are increased proportionately to spread the adjustment across class members. Thus, imputation can increase the variance of estimates more than weighting class adjustments, although the increase in variance from imputation can be reduced by appropriate selection of donors (Kalton & Kish, 1984) or multiple imputations (Rubin, 1987).

Practical considerations also exist in choosing between weighting or imputation for compensating for wave nonresponse. For instance, weighting tends to be easier to implement, but imputed data are easier to use than weighted data. Imputation reduces the ability to detect important relationships among survey variables through attenuation of the strength of observed covariances, while weighting preserves the relationships in the observed data. Several sets of weights may be needed to serve multiple analytic objectives, while an imputed data set can be used to meet all analytic objectives.

An important criteria for choosing between weighting and imputation is the quality of the prediction provided by each. Weighting can seldom use the auxiliary variable with the strongest correlation for a given item to form weighting classes, and weighting must often categorize the auxiliary variables to form classes. Imputation can use auxiliary variables with the strongest correlations and does not have to categorize the auxiliary variables. These considerations suggest that weighting may provide poorer quality nonresponse compensation than imputation. However, imputation has a number of features that detract from its quality also. Many forms of imputation distort distributions, and longitudinal comparisons can be seriously distorted as well. For instance, Heeringa and Lepkowski (1986) report that cross-wave and within-wave hot-deck imputation both seriously distorted measures of change in income and employment from the SIPP 1984 Panel.

The relative quality of weighting and imputation for compensating for missing data in panel surveys has been investigated empirically (Cox & Cohen, 1985; Kalton & Miller, 1986; Mulvihill & Lawes, 1980). The Kalton and Miller investigation specifically addresses the use of imputation and weighting for wave nonresponse and hence offers a number of insights into the quality of the two strategies for the problem of wave nonresponse. Kalton and Miller conducted a simulation study among three-wave respondents from the first three waves of the SIPP 1984 Panel, simulating missing waves of data by a two-step process. First, an AID analysis (Sonquist et al., 1973) was used to identify a complex prediction model for four nonresponse patterns in the data, namely, XXX, XXO, XOO, and XOX. In the second step, the simulation data set was constructed by randomly selecting within the groups formed by the AID analysis three-wave respondents who were randomly assigned to a simulated wave nonresponse pattern. The assigned nonresponse patterns corresponded in distribution to the overall sample nonresponse patterns for the AID group from which the three-wave respondent was selected.

Two methods for compensating for wave nonresponse were applied to the simulated wave nonrespondents: (1) a simple carry-over imputation and (2) wave nonresponse weighting. Imputations were made for a small number of items, several categorical and one continuous. The simulated missing value for the item of interest was obtained from the closest nonmissing value for that item on that unit's record. The continuous measure, Social Security income, was reported on a monthly basis (i.e., four times per wave rather than once per wave), and the closest value available for imputation was usually the income reported in the last month of the most recent wave with nonmissing values. The wave nonresponse weighting procedure deleted all simulated wave non-response patterns, and weights were assigned to the simulated three-wave respondents within weighting classes formed by cross-classifying age group, sex, household income group, race, education level, whether receiving certain types of welfare, whether in the labor force, and whether unemployed.

Only 10% of the sample had a wave nonresponse on at least one wave. Consequently, comparison of survey estimates computed using imputed values and computed using weights would not have demonstrated particularly large differences between methods. The imputed data could have been compared directly to the actual data for the simulated wave nonrespondents to assess the quality of the imputations. But, since weighted estimates do not use the wave nonrespondents, a direct comparison of weighted and actual values for wave nonrespondents was not possible. However, for the weighting compensation method used by Kalton and Miller, the factor $(w_i - 1)$ is the increase in weight assigned to compensate for wave nonresponse, where w_i is the weight for the ith three-wave respondent. Therefore, weighted estimates computed using the weighting factor $(w_i - 1)$ could be compared for simulated wave nonrespondents to estimates computed using the imputed data for those same individuals. Those two estimates compared to the actual data for simulated wave nonrespondents provide a more sensitive comparison of the two methods than a comparison of survey estimates.

The results of the comparisons indicate the quality of the compensation provided by these two procedures depends on the type of analysis of interest. For example, for two categorical items the carry-over imputation was slightly better than the weighting procedure at maintaining the marginal distribution of responses for each wave. However, when the wave-to-wave change across these marginal distributions was examined, the imputed data for the wave nonrespondents indicated less change (i.e., attenuated the strength of the net change) than actually occurred. The weighted results also attenuated change, but less so than the imputed data.

When gross change was investigated for these same items, the attenuation among the imputed data was much more severe. From wave 1 to 2 the imputed data understated gross change by more than 50%, while from wave 2 to 3 it understated the gross change by 90%. In contrast, the weighted results had virtually the same two-wave gross change as actually occurred among the wave nonrespondents. Across all three waves, the imputed results also understated gross change by over 50% for each categorical item examined. The weighted data also understated three-wave gross change, but by less than 10% for one item and by 46% for the other.

Some of the attenuation problem for the carry-over imputations could be attributed to several structural limitations of the deterministic carry-over procedure. Two patterns of three-wave gross change could not appear among the simulated wave nonrespondents receiving imputations because they were impossible to impute with the deterministic carry-over procedure. Other patterns that involved gross change occurred infrequently among the imputed data because they could only be generated from infrequently appearing wave nonresponse patterns. Presumably a stochastic carry-over method could overcome these structural deficiencies, but then the simplicity of the deterministic procedure would be lost.

Comparison of the imputed and weighted results to the actual but simulated missing results for the continuous item indicated another deficiency of the carry-over imputation procedure. The weighted monthly Social Security income means were, for the most part, similar to the actual means for the wave nonrespondents, seldom departing from the actual by more than 1%. However, monthly means based on imputed data, although also seldom departing from the actual by more than 1%, were consistently smaller than the actual means after the second month of wave 2. It was discovered that Social Security recipients had received a cost of living increase in January, 1984. The carry-over imputation was forecasting for many, but not all, wave nonrespondents missing Social Security income for months after January using incomes reported prior to January. The result was a consistent understatement of the actual Social Security income. Kalton and Miller modified the carry-over imputation procedure by adding a cost of living increment to Social Security incomes before January, 1984, that were used as donors for later waves. The correction brought the imputed results much more closely into line with the actual means.

Kalton and Miller found it difficult to draw general conclusions from their limited investigation of a few variables from one survey. Considering that they compared the procedures for simulated wave nonrespondents, the impact of the differences they did observe is likely to be small and inconsequential when survey estimates are examined. They did find the quality of the weighting adjustment for wave nonresponse to be comparable to the simple carry-over imputation; weighting was better with respect to estimation of gross change. On the other hand, the weighting adjustment did discard 10% of the records, decreasing sample sizes and the precision of weighted estimates. The loss of precision is small with potentially just six wave nonresponse patterns discarded. However, if the full 8 or 9 waves of the SIPP 1984 Panel had been used in their investigation, the amount of discarded data would have been much larger.

6. COMBINED WEIGHTING AND IMPUTATION METHODS

Given the complementary strengths and weaknesses of the weighting and imputation strategies, it is natural to consider combinations of the two approaches. For example, imputation could be used to complete wave nonresponses for those patterns in which only one wave is missing (e.g., in a three-wave panel, patterns XXO, XOX, and OXX), and the remaining wave nonresponse patterns deleted and compensated by weighting. In the three-wave panel this "imputation for completing wave nonrespondents" procedure retains records with two waves of data, and cross-wave imputations are made between "neighboring" waves where the higher correlations of data closer in time can be used. In addition, only less frequent patterns are deleted, and there should not be serious losses in precision due to smaller sample sizes.

When there are more than three waves, the choice of patterns to impute and delete is more complex. For example, in a nine-wave panel, patterns with one, two, or perhaps three missing waves are candidates for imputation; four or more missing-wave patterns are perhaps most appropriately weighted. But the patterns with three missing waves might be imputed or deleted. For instance, the wave pattern XXXXOXOXO might be a good candidate for imputation from one neighboring wave to the next, while the pattern XXXXXOOOX, which has one missing wave that is at least one wave away from the most recent non-missing wave, might be better compensated by weighting. Decisions on how to handle each pattern depend on a variety of factors, and many of the choices are likely to be ad hoc.

Variable length reference periods will also influence the amount of missing data due to wave nonresponse. For example, because of field operations scheduling, some panel respondents may have a shorter period between interviews than others. In such a case, the decision about whether to handle wave nonresponse with imputation or weighting might be based on the proportion of the cumulative reference period that is missing. In the NMCUES, for instance, records with missing data for more than a fixed fraction (i.e., one-third) of the reference period were handled by weighting, and the others were compensated by imputation (Cox & Cohen, 1985). Again, the choice of a suitable strategy depends on many factors and is likely to be ad hoc.

Another combined imputation-weighting approach first uses imputation to convert nonattrition patterns into attrition patterns and then employs attrition weighting methods to compensate for the remaining wave nonresponse. For a three-wave panel the nonattrition patterns OXX and XOX could be completed by imputation, the attrition patterns XXO and XOO remain unimputed, and the nonattrition patterns OXO and OOX deleted. Alternatively, the OXO pattern could be completed as an attrition pattern by backcasting from wave 2 to wave 1 leaving only the OOX pattern and the OOO unit nonrespondent pattern. This "imputation for completing attrition nonresponse" approach retains more of the data than does the "imputation for completing wave nonrespondents" approach, since the pattern XOO (and possibly OXO) that was deleted for the previous combination is retained. In addition, less data are fabricated since the XXO pattern, which is likely to be a frequent wave nonresponse pattern, is not subject to imputation. On the other hand, imputation for "completing attrition nonresponse" requires more complicated weighting than the imputation for "completing wave nonrespondents" combination, especially for panel surveys with a large numbers of waves.

Another combination could use imputation for those data that can be imputed well (e.g., the same item across waves with high correlation across time) and weighting for other items (e.g., the same item across waves with low correlation; topical items asked only one time). However, this approach is problematic when there is analytic interest in examining relationships between the items receiving imputations and the other items that do not. It is not clear how the analysis of items that have different weights within the same record should be handled.

The treatment of wave nonresponse does not need to be limited to the production of a single data set. The need for different weights for different analytic needs can be avoided to some extent by providing data sets suitable for different types of analysis. For example, the SIPP produces individual data sets for each wave with weights suitable for conducting cross-sectional analysis. Wave nonrespondents for a given cross-sectional file are deleted, and their missing data are compensated by the cross-sectional weighting scheme. For longitudinal purposes, SIPP also produces a file composed of complete three-wave respondents with a single weight to compensate for all deleted units and wave nonrespondents. Of course, this approach has several disadvantages, especially the deletion of a large amount of nonmissing data for wave nonrespondents in the longitudinal file. The sample sizes, and the precision of estimates obtained from the longitudinal file, are greatly reduced by the deletion of wave nonresponse records. Further, cross-sectional results obtained from the longitudinal file will not be consistent with those from the cross-sectional data sets, nor will they be comparable to longitudinal analyses based on linked cross-sectional data sets.

7. SELECTING A WAVE NONRESPONSE COMPENSATION STRATEGY

The complexities of panel survey design and analysis preclude the recommendation of any single procedure for compensating for wave nonresponse. Survey specific procedures must be developed with a consideration of many factors, and it is difficult, and potentially misleading, to recommend a procedure or set of procedures for the diverse designs and settings encountered in panel surveys. Nonetheless, for the purpose of guiding the choice of a wave nonresponse adjustment procedure, and to attempt to summarize the preceding discussion, the relative strengths of several of the compensation procedures are compared across several criteria in Table 4.

Three weighting, three imputation, and two combined strategies are

TABLE 4. RELATIVE STRENGTHS OF WAVE NONRESPONSE
COMPENSATION STRATEGIES FOR FIVE GENERAL CRITERIA

Compensation Strategy	Criteria				
	Practicality	Flexibility	Quality	Precision	Preservation of Relationships
Weighting:					
Complete wave respondent	+ +	+ +	+	− −	+
Attrition patterns	o	+	+	o	+
Nonattrition patterns	o	+	+	+	+ +
Wave Imputation:					
Carry-over	+	o	+/−	+	−
Cross-wave hot deck	+	+	o	+	− −
Item Imputation:					
Item or multivariate regression	− −	− −	+ +/−	+	−
Combined Strategies:					
Imputation for completing wave nonrespondents	o	o	o	o	+
Imputation for completing attrition patterns	o	o	o	+	+

Key: + indicates strength; + + indicates greater strength; − indicates weakness; − − indicates greater weakness; o indicates neither strength nor weakness; +/− indicates both strength and weakness.

displayed:

(1) *Complete wave respondent weighting:* Units responding on all panel waves are weighted to account for all wave and unit nonrespondents.

(2) *Attrition pattern weighting:* Attrition nonresponse and complete wave respondent patterns are compensated for all other units.

(3) *Nonattrition pattern weighting:* All patterns of wave nonresponse are used to employ as much of the auxiliary information as possible to create weights.

(4) *Carry-over wave imputation:* Data from a responding wave are imputed to a nonresponding wave for the same unit.

(5) *Cross-wave hot-deck imputation:* Donor and recipient are matched using data from a responding wave for each, and donor data on the nonresponding wave are transferred to the recipient.

(6) *Regression imputation:* A regression model with the response to the same item on another wave as a predictor is used to estimate the value of a missing item.

(7) *Imputation for completing wave nonrespondents:* Wave non-response patterns with limited numbers of missing waves are completed by imputation; the other nonresponse patterns are compensated by weighting.

(8) *Imputation for completing attrition patterns:* Selected wave non-response patterns are imputed to achieve an attrition pattern; multiple weights are produced under an attrition weighting procedure to compensate for remaining nonattrition patterns.

Many criteria could be used to compare these various procedures; five broad categories of criteria are shown in Table 4. The first, *practicality*, refers to the ease of implementation of the procedure and to the ease of use of subsequent data. Since there are no widely available computer programs to implement most wave nonresponse compensation procedures, special purpose software must be developed for each survey. For a one-time panel survey, extensive software development for one of these approaches may be operationally infeasible; continuing survey operations can more readily accommodate the added cost and maintenance of such software. In general, imputation procedures that complete all missing items will be the easiest for analysts to use, although weighted data does not present serious difficulties for most survey analysts. Data sets with multiple sets of weights can present problems for the less sophisticated analyst.

The complete wave respondent weighting is likely to be the easiest to implement and use for one-time survey operations. Attrition and nonattrition pattern weighting require a series of probit or logistic regression models to determine weights (although weighting cell adjustments might also be employed); users must choose appropriate weights each time an analysis is conducted. The wave imputation strategies may be more difficult to implement than the weighting strategies, depending on how much each wave repeats items from a previous wave. The regression imputation strategies are likely the most difficult of all the strategies to implement because of the need to develop regression models for each item. On the other hand, the data sets produced under an imputation strategy will be easier to use than those with attrition or nonattrition pattern derived weights since, at most, only one set of weights need be considered in an analysis of imputed data. The combined strategy of imputation for completing nonrespondents has the analytic advantage of providing only one set of weights. However, both combined strategies may require substantial effort to implement since both imputation and weighting must be done for the same data set.

The *flexibility* criterion refers to the ability of the procedure to handle multiple data types (i.e., continuous, categorical, and conditional measures) in a data record. Clearly, the weighting procedures are the most flexible. Wave imputation procedures can handle multiple data types as well, although to the extent that there are items on a wave not appearing on another wave, wave imputation may not cover all data items. Regression imputation is clearly at a disadvantage when there are multiple data types since even multivariate regression models will be difficult to develop when continuous, categorical, and conditional measures must be simultaneously imputed. The combined strategies have adequate flexibility provided that some type of wave imputation can be used prior to weighting.

Quality refers to the ability of the compensation procedure to predict the missing value correctly. Kalton and Miller (1986) and others (Cox & Cohen, 1985; Mulvihill & Lawes, 1980) have found little difference in the quality of weighting and imputation approaches when cross-sectional estimation is of interest. But some forms of imputation are clearly inferior to weighting when longitudinal analysis is considered. Regression imputation can be highly accurate provided the model is correctly specified. The combined strategies should be no more accurate than the weighting strategy overall.

The *precision* of estimates depends on two factors: the sample size available and variability introduced by the procedure. Weighting strategies that delete records will necessarily produce less precise estimates than the other approaches. The combined strategies will be less affected by deletion of wave nonrespondent records since some wave nonresponses are completed by imputation. Weighting can also increase the variance of sample estimates through variability of the weights themselves. For instance, nonattrition pattern weighting can be subject to large variation in weights when the responding pattern in a regression has only a few cases.

The effect of imputation on precision have also been noted. Imputed results are subject to greater variability than weighted results because imputation operates on sample elements while weighting averages its effects over respondents within a weighting class. More troublesome for imputation is the ability to estimate the precision of estimates computed from imputed data. If the imputed values are used in standard statistical estimation software, the precision of survey estimates can be greatly overestimated (depending on the amount of imputation that occurs) since imputed values are treated as additional sample elements

The final criterion, *preservation of relationships*, is clearly a strength of weighting strategies. Among the combined strategies, the nonattrition pattern weighting preserves more of the relationships since more of the data is retained. The combined strategies will also be strong on this criteria since the imputation that is made will be limited and unlikely to attenuate the strength of relationships substantially.

Given these considerations, no clear favorite is apparent. When the amount of wave nonresponse is limited (e.g., when there are only a few waves of data collection), the weighting strategies appear preferable, at least on these limited criteria. When the amount of wave nonresponse is substantial, the imputation strategies have several clear advantages over the weighting methods. But the attenuation of the strength of relationships due to data fabrication remains a serious concern. The combined strategies retain some of the strengths, and the weaknesses, of both the weighting and imputation strategies. The combined procedures appear to be a useful compromise when there are a large number of waves and weighting procedures have serious deficiencies.

REFERENCES

Bonham, G. (1983), *Procedures and Questionnaires of the National Medical Care Utilization and Expenditure Survey* (National Medical Care Utilization and Expenditure Survey, Series A, Methodology Report No. 1, DHHS Publication No. 83-20001, Public Health Service), U.S. Government Printing Office, Washington, DC.

Chapman, D. W. (1976), "A Survey of Nonresponse Adjustment Procedures," *Proceedings of the Social Statistics Section*, American Statistical Association, 245-251.

Cox, B. G., & Cohen, S. B. (1985), *Methodological Issues for Health Care Surveys*, Marcel Dekker, New York.

Heeringa, S. H., & Lepkowski, J. M. (1986), "Longitudinal Imputation for the SIPP," *Proceedings of the Section on Survey Research Methods*, American Statistical Association, 206-210.

Herriot, R., & Kasprzyk, D. (1984), "The Survey of Income and Program Participation," *Proceedings of the Social Statistics Section*, American Statistical Association, 107-116.

Kalton, G. (1983), *Compensating for Missing Survey Data*, Survey Research Center, The University of Michigan, Ann Arbor.

Kalton, G. (1986), "Handling Wave Nonresponse in Panel Surveys," *Journal of Official Statistics*, 2, 303-314.

Kalton, G., & Kasprzyk, D. (1982), "Imputing for Missing Survey Responses," *Proceedings of the Section on Survey Research Methods*, American Statistical Association, 22-31.

Kalton, G., & Kasprzyk, D. (1986), "The Treatment of Missing Survey Data," *Survey Methodology*, 12(1), 1-16.

Kalton, G., & Kish, L. (1984), "Some Efficient Random Imputation Methods," *Communications in Statistics, Theory and Methods*, 13, 1919-1939.

Kalton, G., & Lepkowski, J. M. (1983), "Cross-Wave Item Imputation," in *Lessons of the Income Survey Development Program (ISDP)* (M. David, ed.), Social Science Research Council, New York, pp. 171-198.

Kalton, G., Lepkowski, J. M., & Lin, T. (1985), "Compensating for Wave Nonresponse in the 1979 ISDP Research Panel," *Proceedings of the Section on Survey Research Methods*, American Statistical Association, 372-377.

Kalton, G., & Miller, M. (1986), "Effects of Adjustments for Wave Nonresponse on Panel Survey Estimates," *Proceedings of the Section on Survey Research Methods*, American Statistical Association, 194-199.

Lepkowski, J. M., Landis, J. R., & Stehouwer, S. A. (1987), "Strategies for the Analysis of Imputed Data in a Sample Survey: The National Medical Care Utilization and Expenditure Survey," *Medical Care*, 25, 705-715.

Little, R. (1984), "Survey Nonresponse Adjustments," *Proceedings of the Section on Survey Research Methods*, American Statistical Association, 1-10.

Little, R., & David, M. (1983), *Weighting Adjustments for Nonresponse in Panel Surveys* (U.S. Bureau of the Census working paper).

Mulvihill, J., & Lawes, M. (1980), *Imputation Procedures for LFS Longitudinal Files* (Statistics Canada internal memorandum).

Oh, H. L., & Scheuren, F. (1983), "Weighting Adjustment for Unit Nonresponse," in *Incomplete Data in Sample Surveys, Vol. 2: Theory and Bibliographies* (W. Madow, I. Olkin, & D. Rubin, eds.), Academic Press, New York, pp. 143-184.

Rubin, D. (1976), "Inference and Missing Data," *Biometrika*, 63, 581-592.

Rubin, D. (1987), *Multiple Imputation for Nonresponse in Surveys*, John Wiley & Sons, New York.

Sande, I. G. (1982), "Imputation in Sample Surveys: Coping with Reality," *The American Statistician*, **36**, 145–150.

Santos, R. (1981), "Effects of Imputation on Regression Coefficients," *Proceedings of the Section on Survey Research Methods*, American Statistical Association, 140–145.

Sonquist, J., Baker, E., & Morgan, J. (1973), *Searching for Structure*, Institute for Social Research, The University of Michigan, Ann Arbor.

Ycas, M., & Lininger, C. (1981), "The Income Survey Development Program: Design Features and Initial Findings," *Social Security Bulletin*, **44**(11), 13–19.

NOTE

*The author thanks Graham Kalton for encouragement and valuable suggestions during the preparation of the manuscript, and also thanks Graham Kalton, two referees, and the discussant for useful comments and suggestions that improved the clarity of the presentation.

Estimating Nonignorable Nonresponse in Longitudinal Surveys through Causal Modeling*

Robert E. Fay

1. INTRODUCTION

In a recent paper comparing survey practice to theoretical models for missing data, Little (1986) enumerated six desirable properties for imputation procedures for government surveys:

(1) Imputations should be based on the predictive distribution of the missing values, given the observed values for a case.

(2) All observed items for a case should be taken into account in developing the imputations.

(3) Imputations should be selected from the predictive distribution in (1), not means.

(4) Multiple imputation (Rubin, 1987) should be used to provide measures of the added variance from imputation.

(5) Nonrandom (i.e., nonignorable in the sense defined later in this paper) nonresponse models should not be used to supply imputations, unless they are founded on objective data such as follow-ups of nonrespondents. Nonrandom nonresponse models may be useful, however, for the investigation of alternative assumptions about the missing data mechanism.

(6) Models for prediction should take into account substantive knowledge about the variables being imputed.

Although these comments were originally directed at government surveys, most of his arguments appear to extend readily to other large sample surveys confronted with significant problems of missing data, including longitudinal surveys. In general, Little's recommendations are well-taken and serve to clarify the purposes and properties of adjustments for missing data. The central themes of this paper stand in apparent conflict with some of the preceding points as stated. Although they contain technical concepts defined later in this paper, four points (7–10) serve to express the principal messages of this paper:

(7) It is both helpful and important to consider the process of nonresponse in place of an exclusive focus on the problem of predicting the missing data given the observed data.

375

(8) The variety, features, and potential of available nonignorable nonresponse models suggest the possibility of their use, if only on an illustrative basis initially, for a number of missing data problems in the near future.

(9) Although multiple imputation to represent additional variance from the imputation models is a desirable goal, equal attention needs to be focused on the effect of assumptions and models for nonresponse on missing data adjustments. Nonignorable nonresponse models have an important role in the investigation of alternative assumptions. As Rubin (1978, 1987) noted, multiple imputation may be used to investigate alternative assumptions.

(10) Some imputation procedures may appear to comply with the spirit of points (1)–(6), yet fall short of the ignorable nonresponse model. The ignorable nonresponse model represents the most common assumption in the theoretical literature on missing data and may be seen as part of the underlying motivation for these six points. Causal modeling, in particular, and increased sensitivity to possible nonignorable effects, more generally, may help to identify large discrepancies between survey practice and the theory of ignorable nonresponse in some instances. Thus, knowledge of nonignorable models provides additional insight into the implications of ignorable nonresponse.

The last four points appear to take exception to many of the first six. Point (7), in partial contrast to points (1) and (3) of Little, shifts the focus from prediction to representation of the process of nonresponse. Point (8) takes a bit more optimistic view of the potential application of nonignorable models than does Little's point (5). Point (9) adds choice of nonresponse model to imputation variance in (4) as a source of uncertainty in the interpretation of the survey data. Point (10) calls for greater attention to modeling the process of nonresponse in order to recognize parallels or disagreements between theoretical models for missing data and current practice.

This paper addresses these issues in more detail. On closer examination, much of the apparent contrast between the first six points and the last four is simply a matter of emphasis. Nonetheless, the evident differences between the central themes of this paper and the first six points, which I cite as a thoughtful summary of current views held by many researchers, are nonetheless striking and indicate the degree to which the viewpoint expressed here differs from current practice. The last section of the paper includes further discussion of the relation between the principal themes of this paper and other views.

A recent paper (Fay, 1986) identified a class of models for missing data in categorical situations by applying ideas of Goodman (1972, 1973a, 1973b) for analysis of causal models in categorical situations to indicators of response. Some specific nonresponse models from this class had been investigated earlier by others. Most models of this class assume nonignorable nonresponse. Roughly speaking, nonignorable nonresponse occurs when the probability of response, given the observed data, depends upon the unobserved data; a more complete definition of this concept appears in Section 3. Section 2 introduces these notions by discussing a simple hypothetical example from a cross-sectional survey. Three alternative treatments of the missing data are considered there: a predictive model based only on entirely complete cases, the ignorable response model, and a simple causal model for nonresponse. The effect of each set of assumptions is illustrated in Section 2. Section 3 provides a more formal statement of each model, including a general description of causal models for nonresponse and some discussion of alternative models for nonignorable nonresponse.

The remaining sections outline different situations in which causal modeling of nonresponse could be particularly important to the analysis of longitudinal surveys. Section 4 examines possible circumstances in which causal models may be used to describe unit nonresponse. Section 5 treats the problem of item nonresponse. Section 6 suggests practical strategies that may be used to evaluate the implications of nonignorable nonresponse or to implement such a model in the estimation. The concluding section returns to the summary presented in this first section and adds final comments.

2. AN EXAMPLE OF NONRESPONSE FOLLOWING A CAUSAL MODEL

The March Supplement of the Current Population Survey (CPS) provides estimates of income during the previous year for persons, families and households. The official annual estimates of median family income and of the numbers of persons and families in poverty in the United States derive from this source.

Information on income is collected from respondents by asking for each household member 14 years old or older whether specific types of income—for example, wages and salaries, self-employment income, interest income, Social Security payments, etc.—were received during the previous year. The interviewer asks the total amount for each type of income reported. For several of the most important items, such as wages and salaries, nonfarm and farm self-employment income, interest, and dividends, the recipiency and amount of income are noted individually in the final data, while for other items, such as pensions and annuities, a combined amount from a group of related sources is shown.

The CPS data are subject to a considerable degree of nonresponse. No information on recipiency or amounts is obtained for approximately 10% of the sample persons. Partial information is collected for another 20%. (The exact rates vary somewhat from year to year.) Because of the importance of the income statistics and the severity of nonresponse, a number of research efforts both within and outside the Census Bureau have focused on this problem. (For example, see Chapman et al., 1986; David et al., 1986; Greenlees et al., 1982; Oh et al., 1980; Rubin, 1982; Welniak & Coder, 1980.)

The Census Bureau currently imputes missing income information in the CPS through a statistical matching procedure. Each incomplete case is matched to a similar complete case, and missing items on income recipiency and amounts are substituted from the complete case. Because this strategy is employed separately for earnings and for other income, two different donors may supply values for different income items for the same incomplete case.

The problem of missing data in the CPS is a useful point of departure for the construction of an example that illustrates differences among current survey practices, the full use of the assumption of ignorable nonresponse, and the effect of model-based analyses of alternative assumptions of nonignorable response. A more rigorous expression of each model follows in Section 3. The following example is not derived directly from CPS data but instead illustrates in a simplified form some of the issues that a more complete analysis of missing data problems in the CPS might encounter.

Table 1 shows an artificially constructed cross-classification of 10,000 hypothetical survey respondents according to their total earnings income and their recipiency of interest income. The amount of interest income for recipients is classified into only two categories: below $1,000 or $1,000 and above. This simplification facilitates the illustration; clearly, however, any serious application would require a more detailed classification. To further simplify matters, only items on interest income will be subject to nonresponse; responses on total earnings will be assumed complete for all respondents. This assumption is contrary to the actual patterns of nonresponse in the CPS.

Table 1 was constructed to show a high statistical relation between total earnings and interest income. Thus, in a statistical sense, each can be used to make a prediction of the value of the other, although the relation is far from deterministic.

Table 2 shows hypothetical responses as they might actually be reported in an income survey such as the CPS in which a question on recipiency of a type of income precedes a question on amount. Like the CPS, Table 2 shows a substantial number of respondents who are able to report recipiency of interest income but who decline or are unable to supply the amount. The figures in Table 2 were computed by assuming that respondents fail to answer the recipiency question for interest income with probabilities .04, .10, and .28 for persons with no interest income, $1−$999, and $1,000 or over, respectively. Conditional on having reported recipiency, respondents are assumed to provide the

TABLE 1. RECIPIENCY AND AMOUNT OF INTEREST INCOME BY
TOTAL EARNINGS FOR HYPOTHETICAL SURVEY SAMPLE

Total Earnings	Interest Income		
	None	$1 – $999	$1,000 or More
Loss, $0 – $9,999	2400	1400	200
$10,000 – $24,999	1600	1800	600
$25,000 or more	400	800	800
Total	4400	4000	1600

amounts with probability .85 for those with interest income $1 – $999
and .70 for those with $1,000 or more. The values in Table 2 are ex-
pected values rounded to the nearest person. Note that these values
have been computed under an assumption that reporting of interest income depends upon the actual value of interest income: this assumption
is an instance of nonignorable nonresponse and is discussed further in
Section 3.

To illustrate the consequences of nonignorable nonresponse, three
sets of assumptions about missing data will be considered. The first of
these is similar to the assumptions implicitly underlying the current
CPS imputation procedure but will be termed here "allocation based on
the distribution of complete cases." Table 3 shows the consequences of
this assumption. To compute the values in Table 3, cases missing information on recipiency were allocated according to the distribution of complete cases with the same values for observed variables. For example,
for those with total earnings income under $10,000, the 292 cases with
missing response on recipiency are distributed in proportion to the complete cases; that is, 292 × [2304]/(2304 + 1071 + 101)] are allocated
to no interest income, etc. Similarly, the 232 cases with reported
recipiency but unknown amount for this category of earnings income
are distributed in the relative proportions of 1071/(1071 + 101) and
101/(1071 + 101) to the two categories of interest amounts. This
procedure resembles the current imputation procedure in the CPS, in
the sense that partial information is employed to the extent possible,
but the distribution used to assign missing values is derived only from
complete cases. Table 3 shows a considerable underestimation of the
number of recipients of $1,000 or more compared to the true values in
Table 1.

Table 4 shows the estimated cross-classification under an assumption
of ignorable nonresponse. The ignorable nonresponse assumption is further described in the next section. To derive the estimates in Table 4,
recipients without reported amounts are distributed in proportion to the

TABLE 2. RECIPIENCY AND AMOUNT OF INTEREST INCOME BY TOTAL
EARNINGS REPORTED BY THE HYPOTHETICAL SURVEY POPULATION

	Interest Income				
Earnings Income	Recipiency Not Reported	None	Recipient, Amount Not Reported	$1−$999	$1,000 or More
Loss, $0−$9,999	292	2304	232	1071	101
$10,000−$24,999	412	1536	373	1377	302
$25,900 or more	320	384	281	612	403
Total	1024	4224	886	3060	806

complete cases, in the same manner as the construction of Table 3. For example, the 232 cases without reported amounts for total earnings less than $10,000 are again distributed according to 1071/ (1071 + 101) and 101/(1071 + 101), increasing the first category by 212.0 and the second by 20.0. The method of distribution of cases without reported recipiency differs from that used to create Table 3, however; these cases are allocated according to the distribution arising from including cases with amounts allocated according to the previous step. For example, the 292 cases with unknown recipiency for those with total earnings less than $10,000 are allocated in proportion to 2304, 1283.0 (= 1071 + 212.0), and 121.0 (= 101 + 20.0).

Differences between Tables 3 and 4 are not dramatic but are nonetheless notable. The differences arise from the presumed distribution of the 292 persons with unknown recipiency: Instead of using only the distribution of complete cases, the ignorable model incorporates the partial information from those reporting recipiency but no amount in estimating the overall distribution. Consequently, Table 4 shows modestly higher estimates, by slightly less than 1%, of persons receiving interest income. In this and many other situations, the ignorable nonresponse model reaches different conclusions than predictive models based only on entirely complete cases, even when the observed covariates, such as total earnings in this example, have been taken into account. Further comments on this point follow in the next section.

Note that the estimates in Table 3 and Table 4 both attempt to take into account the partial information represented by incomplete cases. Neither set of estimates follows the frequent statistical practice of analyzing only the complete cases without any consideration of the incomplete cases. The methods underlying Tables 3 and 4 differ only in the manner in which the information from incomplete cases is incorporated in the estimates.

TABLE 3. ESTIMATED RECIPIENCY AND AMOUNT OF INTEREST
INCOME BY TOTAL EARNINGS, WITH MISSING DATA ESTIMATED
ACCORDING TO THE DISTRIBUTION OF COMPLETE CASES

| | Interest Income | | |
Total Earnings	None	$1 – $999	$1,000 or more
Loss, $0 – $9,999	2497.5	1373.0	129.5
$10,000 – $24,999	1732.8	1859.4	407.8
$25,000 or more	471.8	921.4	606.7
Total	4702.1	4153.8	1144.0

The third model described in this section is a specific instance of a
causal model for nonresponse in the form considered by Fay (1986).
This model, specified in more detail in Section 3, assumes that the
probability of nonresponse to the recipiency question depends upon the
actual recipiency and amount of interest income, and that the
probability of response to the question on amount depends upon the
amount of interest income. These assumptions correspond to the
process by which Table 2 was created from Table 1. Without independent knowledge of the probabilities of nonresponse, it is possible to estimate these probabilities directly from the data in Table 2. The model
yields the results shown in Table 5, which, except for the effects of
rounding, exactly reproduces Table 1.

Differences between Table 5 and Table 4 or Table 3 are substantial.
The estimated number of recipients of interest income of $1,000 or
more increases by almost 50%. If Table 2 represented the results from
an actual income survey, the choice of appropriate model for nonresponse would be of critical importance.

The ignorable response model given in Table 4 allows response to depend upon total earnings, but not on interest income. The third model
permits a dependence between response and interest income, but assumes that response and earnings are conditionally independent, given
interest income. Both models may be directly estimated from the data
in Table 2 without recourse to outside information. A model that permitted response to depend on both earnings and interest income without
any restriction on the form of this dependence could not be estimated
directly from the data. This situation is typical in estimable nonignorable nonresponse models, where relationships that are not directly
observed can only be inferred from the data by restricting the possible
relationships between response and one or more observed variables.
Section 3 discusses this point further.

TABLE 4. ESTIMATED RECIPIENCY AND AMOUNT OF INTEREST
INCOME BY TOTAL EARNINGS, WITH MISSING DATA ESTIMATED
ACCORDING TO AN ASSUMPTION OF IGNORABLE NONRESPONSE

	Interest Income		
Total Earnings	None	$1 – $999	$1,000 or more
Loss, $0 – $9,999	2485.4	1384.0	130.5
$10,000 – $24,999	1712.4	1876.2	411.5
$25,000 or more	457.1	930.3	612.6
Total	4654.9	4190.5	1154.6

3. IGNORABLE AND CAUSAL MODELS FOR NONRESPONSE

3.1. Ignorable Nonresponse

This section states in formal terms the concepts introduced in the preceding section. These terms form the basis of the discussion in subsequent sections on problems of nonresponse in longitudinal surveys. For each survey characteristic v_i subject to nonresponse, let r_i represent a variable with the value 1 if the respondent provides the information for v_i, and 2 otherwise. For each respondent h let v_{hi} be the value of v_i, and r_{hi} the corresponding indicator of response. A set of outcomes for the response indicators r_i is termed a *pattern* of response (or nonresponse) in this paper.

For purposes of this paper, and in a manner essentially consistent with the use of this term in the statistical literature (e.g., Rubin, 1976; Little, 1982; Little & Rubin, 1987; Fuchs, 1982), nonresponse is *ignorable* if, for each observed pattern of nonresponse, the probability of the pattern does not depend upon values of the unobserved characteristics (i.e., the values of variables v_i with $r_i = 2$), given the values of the observed characteristics (i.e., the values of variables v_i with $r_i = 1$ and any variables u observed for all cases).

When only one variable, v_1, is subject to nonresponse, the assumption of ignorable nonresponse asserts that nonresponse, r_1, is conditionally independent of v_1, given the values of the observed variables. Note that ignorable nonresponse neither requires that response be independent of all of the variables nor asserts that r_1 is unconditionally independent of v_i, unless v_1 is the only variable.

The implications of the definition of ignorable response are also easily understood for a *nested* or *monotone* pattern of response (e.g., Little, 1982). Figure 1 illustrates the concept of a nested pattern of response. For nested patterns, it is possible to order the variables in such a way

TABLE 5. ESTIMATED RECIPIENCY AND AMOUNT OF INTEREST
INCOME BY TOTAL EARNINGS, WITH MISSING DATA ESTIMATED
ACCORDING TO AN ALTERNATIVE CAUSAL MODEL

Total Earnings	Interest Income		
	None	$1–$999	$1,000 or more
Loss, $0–$9,999	2399.7	1399.8	200.5
$10,000–$24,999	1599.8	1800.3	599.9
$25,000 or more	400.0	800.0	800.1
Total	4399.5	4000.1	1600.5

that nonresponse on one variable implies nonresponse on all subsequent variables. In Figure 1, u is reported for each respondent; v_1 is subject to some nonresponse; v_2 is subject to possibly more nonresponse, including all subjects who were nonrespondents to v_1, etc.

The hypothetical example on income reporting provides an instance of a nested pattern of response. In the example, total earnings corresponds to a variable, u, observed for all respondents; recipiency, v_1, is observed for some but not all respondents; and v_2 is subject to additional nonresponse, including all cases with nonresponse on v_1.

Several aspects of the definition of ignorable nonresponse are directly related to the six basic points made by Little (1986) and cited in Section 1. The (1) construction of the predictive distribution for each case, (2) taking into account all observed items for the case, is consistent with the definition that ignorable nonresponse allows the probability of a pattern of nonresponse to depend upon the observed variables but not the unobserved variables given the values of the observed variables. In addition, (3) imputation viewed as draws from the predictive distribution given the observed data, instead of predictive means, allows the imputed values to serve as representations of the missing data in a manner consistent with ignorable nonresponse.

In categorical situations, the ignorable nonresponse in nested situations corresponds to a specific causal model for nonresponse (Fay, 1986), in the sense of Goodman (1972, 1973a, 1973b). To illustrate this connection in terms of the example in Section 2, let F_{hijkm}, for $h = 1,2,3$, $i = 1,2$, $j = 1,2,3$, $k = 1,2$, and $m = 1,2$, denote the expected values under ignorable nonresponse for the cross-classification of total earnings (h), recipiency of interest income ($i = 1$/yes, $i = 2$/no), income amount ($j = 1$/none, $j = 2$/$1–$999, $j = 3$/$1,000 or more), response on recipiency ($k = 1$/yes, $k = 2$/no), and response on income amount ($m = 1$/yes, $m = 2$/no). Because of the nested pattern of response, $k = 2$ does not occur with

FIGURE 1. A NESTED OR MONOTONE RESPONSE PATTERN[a]

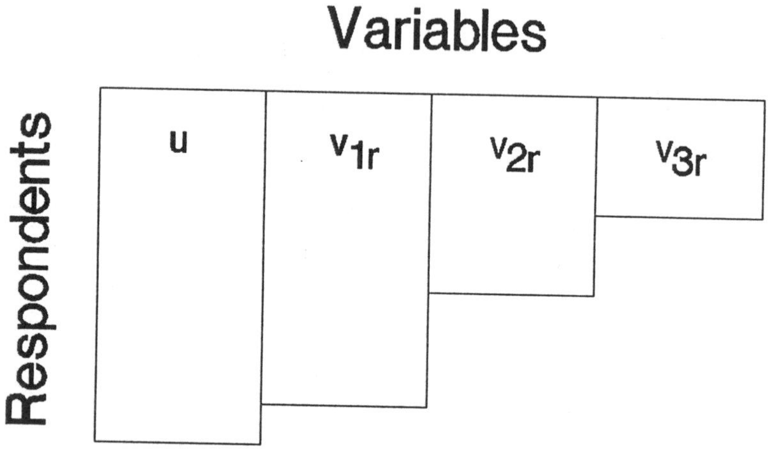

Variables

[a]The boxed area represents observed data; v_{1r} represents the observed values of v_1.

$m = 1$, implying $F_{hij21} = 0$ for all h, i, and j. The interrelatedness of the income questions implies additional cells containing zeros: F_{h11km}, F_{h22km}, and F_{h23km} for all h, k, and m, since these cells would represent impossible combinations for income recipiency and amounts; and F_{h2112} for all h, since persons without interest income who answer the recipiency question are necessarily considered respondents to the question on amount. Let f_{hijkm} denote the sample observations whose expected values are F_{hijkm}. Not all the f_{hijkm} are directly observed; instead, the available data are the distinct f_{hij11}, and the marginal totals $f_{hi.12}$ and $f_{h..22}$, where "." indicates summation over the subscript.

The ignorable nonresponse model places restrictions on the probabilities of total nonresponse, F_{hij22}, and partial nonresponse, F_{hij12}. Under ignorable nonresponse, the probability of total nonresponse given values of h, i, and j, $F_{hij22}/(F_{hij11} + F_{hij12} + F_{hij22})$, must depend solely on the observed value of h, but not on the unobserved values of i and j. This requirement follows from the general definition of ignorable response, which states that the probability of the given pattern of response may depend upon observed data h but not upon unobserved data i and j, given the observed data. Equivalently, the odds of this pattern, $F_{hij22}/(F_{hij11} + F_{hij12}) = F_{hij2.}/F_{hij1.}$, must de-the odds of this pattern, $F_{hij22}/(F_{hij11} + F_{hij12}) = F_{hij2.}/F_{hij1.}$, must depend only on h. Similarly, the odds of the partial pattern, $F_{hij12}/(F_{hij11} + F_{hij22})$, must depend only upon h and i but not j.

The first condition on odds may be represented by

$$\Omega^K hij = F_{hij1.}/F_{hij2.} = \gamma^{\bar{K}}\gamma^{H\bar{K}}h \tag{1}$$

where $\gamma^{\bar{K}}$ is an overall term and $\gamma^{H\bar{K}}h$ depends only on h. Consequently, this equation shows that the inverse of the odds of total nonresponse is a function only of h. The overbar notation over K in equation (1) serves to identify the coefficients as pertaining to the odds of K.

By using this relation and the condition that the odds $F_{hij12}/(F_{hij11} + F_{hij22})$ of the partial pattern must depend only upon h and i but not j, it is possible to show that the conditional odds of response to amount, given a response to recipiency, may be expressed as

$$\Omega^M hij1 = F_{hij11}/F_{hij12} = \gamma^{\bar{M}}\gamma^{H\bar{M}}h\gamma^{I\bar{M}}i\gamma^{HI\bar{M}}hi \tag{2}$$

In this application, the second equation is meaningful only for $i = 1$ but is shown in a more general form useful for applications to other nested situations. Restrictions, such as $\gamma^{HK}1\gamma^{HK}2\gamma^{HK}3 = 1$ and $\gamma^{HIM}11\gamma^{HIM}21\gamma^{HIM}31 = 1$, are required for the sake of identifiability.

Equation (1) specifies odds for levels of k for the marginal table collapsed across m, while equation (2) specifies odds for levels of m for the complete table. This aspect of the model, combined with the multiplicative form of the odds equations, are characteristic features of the system of causal modeling for cross-classified data developed by Goodman (1972, 1973a, 1973b).

Figure 2 shows a diagram representing these two odds equations. Single-headed arrows in such causal diagrams correspond to direct effects with the head of the arrow indicating the outcome variable. Double-headed arrows denote correlations without causal interpretation. Figure 2 includes the substantive associations between the survey characteristics, u, v_1, and v_2. Variables u, v_1, and v_2 are shown as causally antecedent to response indicators r_1 and r_2, since all arrows between survey variables and response indicators point to the response indicators.

As stated earlier, the ignorable response model for nested patterns of nonresponse to categorical variables can be represented as a causal model, as this example illustrates. The odds equation for each response indicator may be specified in terms of the complete table, as in equation (2) or a marginal table, as in equation (1). Causal models do not provide a similar exact representation of the ignorable model for non-nested patterns of nonresponse, but often a specific causal model will approximate the implications of the ignorable model in these cases (Fay, 1986).

FIGURE 2. A CAUSAL MODEL FOR NONRESPONSE
FOR THE INTEREST INCOME EXAMPLE[a]

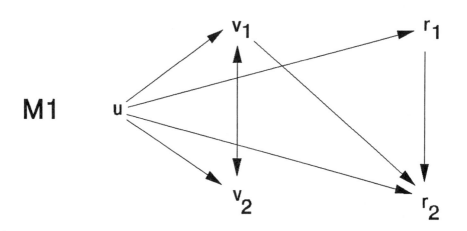

[a]This model represents the assumption of ignorable nonresponse.

3.2. Causal Models for Nonresponse

Besides providing a representation for the assumption of ignorable non-response for nested patterns, causal modeling offers a number of alternative interpretations of nonresponse for both nested and nonnested patterns. Model $M2$ of Figure 3 depicts the causal model discussed in Section 2. In place of arrows from u to r_1 and r_2, the arrows now originate from v_1 and v_2. This model is represented by the equations

$$\Omega^K h_{ij} = \gamma^{\bar{K}} \gamma^{I\bar{K}_i} \gamma^{J\bar{K}_j}, \tag{3}$$

and

$$\Omega^M h_{ij1} = \gamma^{\bar{M}} \gamma^{I\bar{M}_i} \gamma^{J\bar{M}_j}. \tag{4}$$

Figure 3 displays two other possible causal models for this example. Model $M3$ combines the assumption that nonresponse to recipiency depends upon total earnings income (equation 1) with an assumption that nonresponse to amount depends upon interest income (equation 4). (Actually, the arrow shown between v_1 and r_2 is redundant in this case but would be meaningful in other applications.) Similarly, model $M4$ rever-

FIGURE 3. ADDITIONAL CAUSAL MODELS FOR NONRESPONSE
FOR THE INTEREST INCOME EXAMPLE[a]

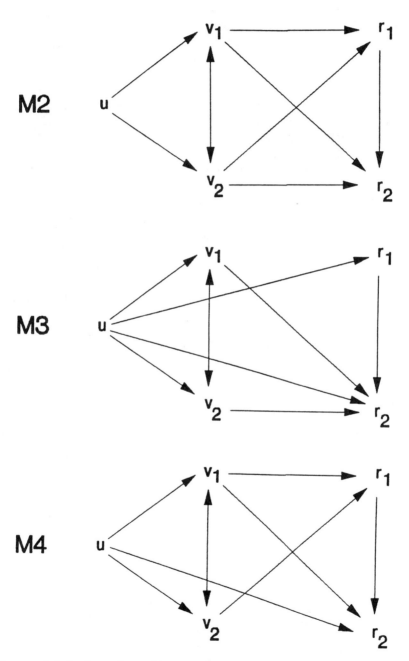

[a]These models imply nonignorable nonresponse.

ses the assumptions by pairing equation (2) with equation (3). Models $M3$ and $M4$ represent two compromise positions between $M1$ and $M2$ on the issue of whether response depends on interest income or on total earnings. Models $M3$ and $M4$ are also estimable from the data. These models do not exhaust the possible alternatives, and illustration of other causal models may be found in Fay (1986).

The assumption of ignorable nonresponse occupies a unique position among nonresponse models because of its properties for inference based on likelihoods (Little, 1982; Rubin, 1976). The correspondence between the ignorable model and a specific causal model for nonresponse for nested patterns raises interesting questions about the merits of each approach. Should we regard causal models as worthwhile because they represent a possible route to generalize the assumption of ignorable nonresponse? Should we instead regard causal models as a natural representation of nonresponse processes, and conclude that the frequent reasonableness of the ignorable model derives from its ability to either be or to approximate the appropriate causal model in many situations? An affirmative answer to the second question suggests missing data situations should be confronted as a problem of determining the appropriate mechanisms governing nonresponse, and that the model of ignorable nonresponse be regarded simply as one option among several. The last section returns to these questions.

3.3. Estimating Causal Models for Nonresponse

In some cases, closed-form estimators are available for causal models corresponding to nonignorable nonresponse. The example presented by Fay and Cowan (1983) provides one such instance of a closed-form nonignorable model. Usually, however, estimates may be obtained only through iterative algorithms. Application of the EM algorithm (Dempster et al., 1977) is particularly easy, and the details of this approach have been presented elsewhere (Fay, 1986). More general optimization procedures, such as those described by Leaver (1984), may be helpful for applications to some problems.

3.4. Prediction Based on the Distribution of Complete Cases

Table 3 displays estimates for the example in Section 2 under a predictive model based only on entirely complete cases. Since this procedure most closely resembles survey practice and is easily described, it is instructive to examine the assumptions on response required to rationalize this approach. For each h, the odds of complete nonresponse, compared to complete response, must be a constant depending only on h,

$$F_{hij22}/F_{hij11} = c_h ,$$

and the odds of partial nonresponse, compared again to complete response, must be a constant depending only on h,

$$F_{hij12}/F_{hij11} = d_h ,$$

for $j = 2,3$. In other words, the use of the distribution of complete cases to estimate the distribution for cases missing data requires for its correctness that the odds of a specific nonresponse pattern relative to complete response may depend only upon observed characteristics. These two odds imply, however, that the odds of nonresponse to the first question must satisfy

$$F_{hij22}/(F_{hij11} + F_{hij12}) = c_h \qquad \text{for } j = 1$$
$$= c_h (1 + d_h)^{-1} \qquad \text{for } j = 2,3.$$

In other words, the probabilities of nonresponse to the first of the two questions depend upon the probability of nonresponse to the second question in a complex way. The form of this model is dubious, even though in the example in Section 2, the results were reasonably close to those for the ignorable nonresponse model.

Except for special situations—for instance, if only one variable is subject to nonresponse—the preceding example illustrates a general phenomenon, namely, that prediction based on complete cases only, even when all observed items are taken into account as in Little's recommendation (2), typically imposes unusual conditions on the probabilities of response to individual items. This suggests that it is important to examine the implications of prediction based on only complete cases to assess how closely such a procedure approximates in practice the conclusions of a more correctly structured model, such as the assumption of ignorable nonresponse.

4. CAUSAL MODELING OF UNIT NONRESPONSE

The distinction between unit and item nonresponse is well known in sample surveys. In cross-sectional surveys, unit nonresponse almost always implies that no interview data are obtained from the sample unit. Longitudinal or panel surveys involve collection of data at more than one point in time, so that unit nonresponse may include instances of response for some periods and not others. Item nonresponse denotes failure to obtain individual items in otherwise complete interviews.

The distinction between unit and item nonresponse is typically reflected in the choice of methods to handle missing data. For multi-item surveys, the strategy of weighting adjustments for unit nonresponse and imputations for item nonresponse often represents the most effective solution. Kalton (1986) discussed the merits of this approach relative to imputation of missing unit responses.

DeMaio et al. (1986) recently reviewed a number of studies of non-response in the 1980 census. They found the distinction between motivational factors and difficulty of task to provide an effective summary of these studies. They observed that motivational factors were the more frequent determinant of mail response rate, whereas missing data on individual questionnaire items more often stemmed from the difficulty of the task. Although additional factors—difficulty of reaching respondents, concerns of privacy, etc.—may affect both unit and item nonresponse in longitudinal surveys, the distinction suggests that unit and item nonresponse are best treated as separate processes. By extension, discussion of possible causal models for nonresponse in longitudinal surveys is also facilitated by a division between unit and item nonresponse. The focus here is restricted to nonignorable models, since other researchers have discussed the implications of ignorable response for longitudinal surveys.

The Survey of Income and Program Participation (SIPP), conducted by the U.S. Bureau of the Census, has experienced a significant degree of unit nonresponse. The most frequent pattern of unit nonresponse is attrition, that is, a pattern of consistent response followed by consistent nonresponse. When combined with sample units not affected by unit nonresponse, the units subject to attrition form a nested pattern of response, as illustrated by Figure 1. A considerable fraction of unit nonresponse does not conform to the nested pattern, however. Aspects of the SIPP are described here as illustrations of general features common to many longitudinal surveys.

Some demographic characteristics in the SIPP—year of birth, sex, race, and ethnicity—are fixed for the sample persons. Once this information is collected, any dependence of nonresponse on such variables is a question of ignorable nonresponse. Nonignorable causal models for nonresponse are restricted to variables that are possibly unobserved.

Effects of survey variables on unit nonresponse are likely to involve a combination of two situations: one in which specific values of the survey variables are associated with higher than average unit nonresponse, and the second in which changes in a variable over time, such as those representing transitions connected with residence, employment, family composition, etc., may affect the probability of unit response.

The application of causal notions to nonignorable unit nonresponse depending upon the current levels of variables, as opposed to changes in the levels of variables, may find limited use in the SIPP and longitudinal surveys in general. Usually, such models would give results that differ little from the ignorable model when characteristics that are relatively stable over time are considered. Nonetheless, nonignorable nonresponse models, such as those described in the next section, may possibly prove useful in some instances.

A more likely application of causal models would be to represent the effects of transitions on nonresponse. The ability to implement such models depends largely upon the presence of nonattrition patterns in addition to attrition patterns. For longitudinal surveys in which nonresponse is restricted to attrition, so that the nonresponse pattern is nested, causal models depending upon transitions will generally be difficult to estimate, since transitions contributing to unit nonresponse will not be directly observed for that very reason. The most realistic possibility to model such effects in situations of entirely nested nonresponse rests on covariates that serve as precursors of the transitions, so that the association between the precursors and the unit nonresponse on the subsequent interview can be employed to infer the unobserved effect of the transition on response. Such precursors will often be unavailable or of limited utility, however.

The most promising application of causal models to the problem of unit nonresponse in panel surveys is to situations in which both attrition and nonattrition patterns of nonresponse occur. In the case of the SIPP, for example, attrition is the dominant form of unit nonresponse but other forms appear as well. Let a series of the interviews be denoted by a sequence of 1's and 0's, where 1 denotes a completed interview and 0 a unit nonresponse. Through causal models, it may be possible to infer the effect of transitions on unit nonresponse for attrition patterns, such as 11000 and 11110, from observed transitions persisting over a period of nonresponse in nonattrition patterns, such as 11011 or 10100.

To illustrate, the pattern 11011 does not provide direct information on transitions between the events measured by the second and third interview, or the third and fourth. Changes in status between the second and fourth interviews, however, are a direct indication that at least one transition occurred during this period. Adjustments for nonresponse using all important observed items, as Little suggests, would necessarily represent the effect of this transition, whether through a weighting adjustment of complete cases with similar transitions between the second and fourth interviews, or imputation that would reflect the observed data.

Thus, any reasonable adjustment for nonresponse, including specifically the ignorable nonresponse model, that is able to reflect the significant information in the observed data will represent an observed transition between the second and fourth interview in 11011 as taking place in that interval. Quite possibly, evidence may accumulate from such patterns that transitions are a significant contributing factor to nonresponse.

In situations in which evidence accumulates from nonnested patterns of nonresponse that transitions may be an important determinant of nonresponse, causal modeling and related forms of nonignorable models are a means to generalize from these patterns to attrition patterns. In other words, if transitions appear to be a significant contributing factor to unit nonresponse in instances of nonattrition patterns, then transitions may similarly contribute to attrition patterns. The ignorable non-response model and models using the distribution of complete cases only do not reflect observed relationships between transitions and response in adjusting for attrition. Causal models for nonignorable nonresponse can be used to infer such effects of transitions on attrition patterns as well. Since attrition patterns are the principal form of unit nonresponse in the SIPP, such models could have important implications on the estimation of the number of transitions occurring in the survey population.

5. CAUSAL MODELING OF ITEM NONRESPONSE

The example in Section 2 represented an instance of item nonresponse in a cross-sectional survey. Similar applications arise in interviews in longitudinal surveys, and methods such as those described in Section 3 or by Fay (1986) may be applied.

An interesting possible application of causal modeling to longitudinal surveys arises for survey questions asked at different points of time for the same characteristic. For example, such questions might concern labor force status or receipt of a specific type of income during a four-month period. For simplicity, we consider a characteristic v_1 measured at the first interview and v_2 measured at the second. Four patterns of response are possible: response or nonresponse on both interviews, and response on one interview but not the other.

The model of ignorable nonresponse is approximately but not exactly equivalent to a specific causal model, shown as model $M5$ in Figure 4 (Fay, 1986). When v_1 is observed but not v_2, ignorable nonresponse allows this pattern of response to depend on v_1, but not on v_2 given v_1 and the other observed variables. The arrow between v_1 and r_2 denotes this relationship. Similarly, nonresponse on v_1 but not v_2 may depend on v_2 but not v_1 given the value of v_2.

The model of ignorable nonresponse is disappointing as a plausible accounting of human behavior for such patterns. The assumptions of the model generate paradoxical relationships across time. For example, nonresponse r_1 is presumed independent of v_1, a characteristic that probably is available to the respondent, given the value of v_2, a future state that is certainly unavailable to the respondent or anyone else. Similarly, r_2 is allowed to depend on the past, v_1, but not the current state, v_2. Although some possible arguments may be offered for this second situation, the author's opinion is that current state generally would be a more important factor than past state. Finally, non-

FIGURE 4. TWO CAUSAL MODELS FOR ITEM NONRESPONSE
IN A LONGITUDINAL SURVEY[a]

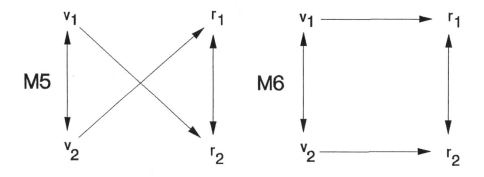

[a]$M5$ approximates the assumption of ignorable nonresponse.

response, r_1 is allowed to depend on the future state, v_2, only if response is obtained at the future date, i.e., $r_2 = 1$. If, instead, response is not obtained at the future date (i.e., $r_2 = 2$), r_1 is allowed to depend on neither v_1 nor v_2. Partial justification for this last assumption may be sought in attempting to categorize different classes of respondents, but generally this last requirement of the ignorable nonresponse is the most implausible of the three, in assuming the influence of behavior in future interviews on the effect of other variables on current response.

Model $M6$ of Figure 4 offers an intuitively acceptable alternative to the ignorable response model. This model, studied independently by Stasny (1985) and Fay (1986), shows effects of current status on response. These dependencies, unlike ignorable nonresponse, do not propose paradoxical time reversals, and for that reason alone, $M6$ appears a more acceptable hypothesis about nonresponse.

Causal models such as $M6$, alternatives presented by Fay (1986), and their generalizations to more than two interviews offer a variety of nonignorable nonresponse analyses that should provide considerable insight into the possible effects of nonresponse on the analysis of item nonresponse in longitudinal surveys. Possibly, such models may also be useful for the analysis of unit nonresponse, although the previous section suggested that models focusing on the effect of transitions may usually prove more appropriate. Conversely, models for the effects of transitions may be examined for item nonresponse, using techniques suggested in the previous section for unit nonresponse.

6. IMPLEMENTING CAUSAL MODELS FOR NONRESPONSE

6.1. Causal Models to Evaluate Assumptions

Imputation for nonresponse or weighting adjustments to survey data typically demand extensive professional involvement. Consequently, application of causal models or other nonignorable models generally would profit from estimating the gains from full-scale implementations of these approaches to the survey data by first attempting approximate, but simpler, alternatives. If the simpler alternatives show that response assumptions appear to have little impact on the final results, then more elaborate implementations can be avoided.

As a first step in a simplified analysis, a limited number of critical variables subject to considerable nonresponse or related to the process of nonresponse may be tallied in a cross-classification of perhaps four to ten variables. Alternative causal models could be investigated for the extracted variables and compared to results from the model of ignorable nonresponse applied to the same cross-classification. If the original imputation procedure was so designed, the results of allocating the missing data according to the distribution for complete cases could also be derived for the table. If alternative assumptions about missing data have limited substantive impact on the interpretation of the data, then there would be little or no incentive to proceed further with a more formal implementation.

If the simplified analysis shows that alternative assumptions have a large impact on the analysis, or if a specific causal model appears preferable on the basis of logic or external evidence, then a full implementation of the model or a set of models may be in order. Both weighting and imputation methods are available and could be selected according to the same principles that usually govern the choice between these two alternatives. In other words, a weighting adjustment may provide the most effective implementation of a causal model in those situations, particularly unit nonresponse, in which weighting would represent the typical choice for the standard nonresponse adjustments. Similarly, an implementation of a causal model through imputation would generally be possible for problems, particularly item nonresponse, where imputation would usually represent the preferred method.

6.2. Implementation of Causal Models through Weighting

Weighting adjustments for nonresponse are typically implemented by classifying both the response and nonresponse cases into a number of cells according to a number of characteristics observed for both response and nonresponse cases. Weights of response cases within each cell are proportionately adjusted to represent nonresponse cases in the same cell. A causal model implying nonignorable nonresponse could model a dependence between nonresponse and characteristics observed only for responding cases. The fitted causal model would predict, for each completed case, the probability of response.

One approach is to employ a response propensity weighting (Rosenbaum & Rubin, 1983; David et al., 1983; Little, 1986; Scheuren, 1986). Generally, this approach weights each complete case by the inverse of its probability to respond under the nonresponse model. This propensity (probability) to respond is estimated for complete cases under the causal model.

A second approach would follow a response propensity weighting under the causal model by a proportional readjustment within the usual weighting cells to the original sum of weights for observed and unobserved cells. In other words, within each of the usual weighting adjustment cells, the reweighted complete cases would be further adjusted by the factor f_1,

$$f_1 = \frac{\text{(Sum of original weights over response and nonresponse cases)}}{\text{(Sum of response cases after propensity adjustment)}}.$$

Application of the factor f_1 results in reweighted cases under the propensity model with weights that sum to the same total in each adjustment cell as the usual weighting adjustment. Nonetheless, the nonignorable features of the propensity weighting would still be reflected in the differential reweighting of complete cases within each cell. If the response model were correctly specified, the factors f_1 within each weighting cell should vary randomly about 1. The adjustment in this case would simply serve the purpose of maintaining an appearance of consistency with earlier procedures. If the response model failed to capture part of the information reflected in the design of the adjustment cells, this adjustment might reduce bias.

A third alternative provides a variation on the second. After the propensity adjustment of each completed case by p^{-1}, where p is the estimated propensity to respond, a factor f_2 could be derived as

$$f_2 = \frac{\text{(Sum of original weights for nonresponse cases)}}{\text{(Sum of propensity minus original weights for response cases)}},$$

and each completed case could be weighted by $[1 + (p^{-1} - 1)f_2]$. This alternative, like the simple propensity weighting but unlike the second alternative, insures that no complete case is reduced in weight, compared to its original weight, as a consequence of the adjustment.

The choice among the three alternatives or some other variant of them may rest on grounds of variance, robustness under model failure, or other considerations relevant to the problem at hand. The important point is that these alternatives do provide means to implement a causal model for nonresponse through weighting adjustments.

6.3. Implementation of Causal Models through Imputation

Imputation assigns a value or set of values in place of a missing value or set of missing values. When imputation is entirely model-based, the causal model for nonresponse may be incorporated into the predictive model. Imputations for many surveys, however, including most government surveys, are carried out through a "hot-deck" or statistical matching approach. Thus, it is important in practice to be able to implement a causal model through a form resembling the typical hot deck.

Generally, hot decks or statistical matching procedures provide a means to link a case with missing data with complete cases on the basis of matching characteristics. Conceptually, these approaches provide a sequence of complete cases ordered by preference. In the standard hot deck, the preference is usually determined according to how recently the complete case within the same adjustment cell was processed relative to the case requiring imputation. In a statistical matching approach, such as the imputation of income in the CPS, the ordering is determined by the degree of match on a number of key characteristics, while the ordering of complete cases matching the case with missing data to the same degree is arbitrary.

The usual hot-deck or statistical matching approach considers only the order of the preference sequence of matching complete cases in assigning the missing value. Often, the procedure assigns the missing value from the first complete case in the preference sequence; nonetheless, it is essential to envision the preference sequence implicitly defined by the procedure in order to define an alternative procedure reflecting a causal model.

In order to reflect a causal model of nonresponse, cases within the preference sequence must be given differential weight as possible choices for assigning a missing value. For a missing case with response pattern z, the causal model estimates for each complete case in the preference sequence with characteristics x the probability of response pattern z, $p(z, x)$, and the probability of complete response, $p(c, x)$. In the same spirit as the response propensity weighting, a complete case from the preference sequence should be weighted in proportion to

$$f = p(z, x)/p(c, x).$$

By assuming that the first several cases in the preference sequence are in a random order with respect to response probabilities, this reweighting of the imputation probabilities may be accomplished by estimating a maximum value, f_{max}, for the value f over all probable elements, retaining each complete case in the sequence with probability f/f_{max}, and selecting the first retained element of the sequence. A reasonable overestimate of f_{max} does not alter the probability properties of this procedure, but larger values of f_{max} relative to typical values of f will result in longer average searches down the preference sequence until a complete case is not rejected. If appropriate, design weights could also be reflected in this procedure as well. Other methods to implement a weighted hot-deck or statistical matching procedure could be used instead.

7. CONCLUDING REMARKS

The introductory section summarized the principal themes of this paper. In short, the purpose here has been to illustrate a class of models that may prove useful in the analysis of longitudinal surveys. Clearly, inappropriate models for nonresponse may produce extreme results in some situations. Nonetheless, when appropriate care is reflected in the choice of plausible models, nonignorable models may serve to represent uncertainty in the analysis of the data arising from lack of firm knowledge of the nonresponse process.

When knowledge on nonresponse becomes available, for example, from studies involving administrative records, evidence of nonignorable nonresponse effects deserves to be modeled as well as presented in the original form. If a causal model or other nonignorable model can be shown to fit the observed patterns considerably better than the ignorable model, then a nonresponse adjustment based on such a model appears as a prudent and reasonable choice.

As experience with nonignorable models grows, the requirement that a nonignorable model be validated with an outside standard may be relaxed to a degree. The ignorable model is now applied to many problems without external validation; researchers may come to recognize situations in which a specific nonignorable model should be used instead. In this respect, this paper questions the preference for the ignorable model implied by Little's remark (5) in the first section. Although some care is required in application, nonignorable models in longitudinal studies deserve consideration.

REFERENCES

Chapman, D. W., Bailey, L., & Kasprzyk, D. (1986), "Nonresponse Adjustment Procedures at the U.S. Bureau of the Census," *Survey Methodology*, **12**(2), 161–180.

David, M., Little, R. J. A., Samuhel, M. E., & Triest, R. K. (1983), "Nonrandom Nonresponse Models Based on the Propensity to Respond," *Proceedings of the Business and Economic Statistics Section*, American Statistical Association, 168–173.

David, M., Little, R J. A., Samuhel, M. E., & Triest, R. K. (1986), "Alternative Methods of CPS Income Imputation," *Journal of the American Statistical Association*, **81**, 29–41.

DeMaio, T., Marquis, K., McDonald, S.-K., Moore, J., Sedlacek, D., & Bush-Straf, C. (1986), "Cognitive and Motivational Bases of Census and Survey Response," *Proceedings of the Second Annual Research Conference*, U.S. Bureau of the Census, 271–295.

Dempster, A. P., Laird, N. M., & Rubin, D. B. (1977), "Maximum Likelihood from Incomplete Data via the EM Algorithm" (with discussion), *Journal of the Royal Statistical Society*, Series B, **39**, 1–38.

Fay, R. E. (1986), "Causal Models for Patterns of Nonresponse," *Journal of the American Statistical Association*, **81**, 354–-365.

Fay, R., & Cowan, C. (1983), "Missing Data Problems in Coverage Evaluation Studies," *Proceedings of the Section on Survey Research Methods*, American Statistical Association, 158–163.

Fuchs, C. (1982), "Maximum Likelihood Estimation and Model Selection in Contingency Tables with Missing Data," *Journal of the American Statistical Association*, **77**, 270–278.

Goodman, L. A. (1972), "A General Model for the Analysis of Surveys," *American Journal of Sociology*, **77**, 1035–1086.

Goodman, L. A. (1973a), "Causal Analysis of Data from Panel Studies and Other Kinds of Surveys," *American Journal of Sociology*, **78**, 1135–1191.

Goodman, L. A. (1973b), "The Analysis of Multidimensional Contingency Tables When Some Variables Are Posterior to Others: A Modified Path Analysis Approach," *Biometrika*, **61**, 215–231.

Greenlees, J. S., Reece, W. S., & Zieschang K. O. (1982), "Imputation of Missing Values When the Probability of Response Depends on the Variable Being Imputed," *Journal of the American Statistical Association*, **77**, 251–261.

Kalton, G. (1986), "Handling Wave Nonresponse in Panel Surveys," *Journal of Official Statistics*, **2**, 303–314.

Leaver, S. G. (1984), *Computing Global Maximum Likelihood Parameter Estimates for Product Models for Frequency Tables Involving Indirect Observation*, (Unpublished Ph.D. thesis), George Washington University, Washington, DC.

Little, R. J. A. (1982), "Models for Nonresponse in Sample Surveys," *Journal of the American Statistical Association*, **77**, 237–250.

Little, R. J. A. (1986), "Missing Data in Census Bureau Surveys," *Proceedings of the Second Annual Research Conference*, U.S. Bureau of the Census, 442–454.

Little, R. J. A., & Rubin, D. B. (1987), *Statistical Analysis with Missing Data*, Wiley, New York.

Oh, H. L., Scheuren, F., & Nisselson, H. (1980), "Differential Bias Impacts of Alternative Census Bureau Hot-Deck Procedures for Imputing Missing CPS Income Data," *Proceedings of the Section on Survey Research Methods*, American Statistical Association, 416–420.

Rosenbaum, P. R., & Rubin, D. B. (1983), "The Central Role of the Propensity Score in Observational Studies for Causal Effects," *Biometrika*, **70**, 41–55.

Rubin, D. B. (1976), "Inference and Missing Data," *Biometrika*, **63**, 581–592.

Rubin, D. B. (1978), "Multiple Imputations in Sample Surveys—A Phenomenological Bayesian Approach to Nonresponse," *Proceedings of the Section on Survey Research Methods*, American Statistical Association, 20–28.

Rubin, D. B. (1982), "Imputing Income in the CPS: Comments on 'Measures of the Aggregate Labor Cost in the United States,'" in *The Measurement of Labor Cost*, (J. Triplett, ed.), The University of Chicago Press, pp. 333–343.

Rubin, D. B. (1987), *Multiple Imputation for Nonresponse in Surveys*, Wiley, New York.

Scheuren, F. (1986), "Discussion," *Proceedings of the Second Annual Research Conference*, U.S. Bureau of the Census, 455–462.

Stasny, E. A. (1985), "Modeling Nonignorable Nonresponse in Panel Data," *Proceedings of the Section on Survey Research Methods*, American Statistical Association, 349–354.

Welniak, E. G., & Coder, J. F. (1980), "A Measure of the Bias in the March CPS Earnings Imputation Scheme," *Proceedings of the Section on Survey Research Methods*, American Statistical Association, 421–425.

NOTE

*The author would like to thank Charles H. Alexander and three anonymous reviewers for helpful comments.

Item Nonresponse in Panel Surveys*

Roderick J. A. Little and Hong-Lin Su

1. THE PROBLEM OF MISSING DATA IN PANEL SURVEYS

We consider item nonresponse in panel surveys, where an interview is conducted but responses to some questions are missing, because of refusal to answer, omission of the question by the interviewer, or deletion of an inappropriate response during data editing. Missing data destroys the rectangular structure of the data matrix, yielding a pattern such as that shown in Figure 1a. Panel data from three waves of interviews are displayed, column Y_{jk} representing data for variable k in wave j. The nonshaded areas represent missing values.

In discussing incomplete data methods it is useful to distinguish data with a *monotone* pattern of missingness from data with a haphazard pattern like that in Figure 1a. Given bivariate data on variables X and Y, we say X is *more observed* than Y if every case with Y observed has X observed as well. A monotone pattern occurs when the variables Y_1, ..., Y_K can be arranged so that Y_k is more observed than Y_{k+1} for $k = 1, ..., K-1$ (Rubin, 1974). Figure 1b illustrates monotone missingness for panel data; the blocks of variables in each wave have a monotone pattern, with earlier waves more observed than later waves; this pattern arises when wave nonresponse takes the form of attrition from the sample. Variables within each wave also have a monotone pattern, and furthermore the least observed variable in wave j is more observed than the most observed variable in wave $j + 1$. Intermediate patterns can be envisaged where the monotonicity property holds within but not between waves, or between but not within waves.

If data have a monotone or approximately monotone pattern, then special missing data methods can be developed that exploit the monotonicity (Anderson, 1957; Little & Rubin, 1987, Chapter 6; Rubin, 1974). However in panel surveys full monotonicity as in Figure 1b seems unlikely—the closest approximation to it might arise when wave nonresponse is monotone and item nonresponse is negligible. For a panel survey with substantial item nonresponse, methods for monotone patterns appear to have limited applicability.

Example 1. Missing Income Data in the Survey of Income and Program Participation: As an illustrative example here and elsewhere in the paper, we consider one form of item nonresponse in a large panel

400

FIGURE 1. DATA PATTERNS FOR A PANEL SURVEY WITH NONRESPONSE

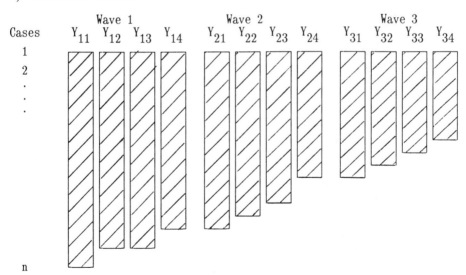

a) Haphazard Pattern

b) Monotone Pattern

survey conducted by the United States Bureau of the Census, the Survey of Income and Program Participation (SIPP). For an overview of the survey, see David (1985). One panel in the survey consists of about 30,000 individuals interviewed eight times, at four-month intervals, over a period of two and a half years. Detailed information on income and wealth is collected, yielding a large data base with over a thousand variables. The first two columns of Table 1 describe the missing data pattern for the amount of wages and salary income received on a monthly basis from the first job reported for the first twelve months of the survey (three survey interviews). The table is restricted to 12,609 individuals who reported working for all 12 months, and in particular excludes nonrespondents for an entire interview. Note that the pattern of missing data is not particularly monotonic. Also, for interviews 2 and 3, wages and salary are nearly always either missing or present for all four months reported in the interview.

There are three broad strategies for handling missing data in surveys: (a) to *impute* the missing values, that is fill them in with suitable estimates, and then analyze the filled-in data by complete data methods; (b) to analyze the incomplete data without adjustment; and (c) to discard nonrespondents and calculate *weights* for respondents to compensate for the loss of cases.

Option (c) is commonly applied to handle *unit* nonresponse, a special form of monotone nonresponse where entire questionnaires are missing. Although weighting can be extended to handle monotone patterns like Figure 1(b) (Kalton, 1986; Little & David, 1983), it does not appear very useful for haphazard patterns of missing data, which are usually the case with item nonresponse. Thus in this paper we confine attention to choices (a) and (b).

The organization of the paper is as follows. Section 2 discusses the mechanisms leading to missing data, and the biases that can result from analyzing incomplete data subject to such mechanisms. Section 3 discusses imputation for panel surveys, describing features of a good imputation method, practical methods of implementation, and a specific method for item nonresponse for repeated measures data like the wages and salary amounts in Example 1. Section 4 discusses methods that analyze the data directly without filling in the missing values. The value of simple complete-case and available-case methods is considered, and the literature on more sophisticated likelihood-based methods is outlined.

2. MISSING DATA MECHANISMS

When data are missing, a key issue is whether the nonrespondents and the respondents differ with respect to survey variables. If they do not, then analysis restricted to the respondents yields valid (though often in-

TABLE 1. MONTHLY WAGES AND SALARY AMOUNTS FOR
JOB 1 FROM THREE INTERVIEWS OF THE SIPP
(Pattern of missing data and mean and standard deviation of
adjusted monthly wages and salary averaged over available months,
classified by missing data pattern)

Missing Data Pattern					
Interview 1	Interview 2	Interview 3	Sample Size	Mean $(\overline{WS})$	SD $(\overline{WS})$
1111	1111	1111	10534	1344	984
1110	1111	1111	30	924	936
1011	1111	1111	429	1355	883
0111	1111	1111	22	1245	943
0110	1111	1111	413	1292	924
0101	1111	1111	408	1277	843
0000	1111	1111	321	1339	889
1111	1111	0000	124	1895	1435
1111	0000	1111	81	1827	1360
1111	0000	0000	60	2734	1895
0000	1111	0000	43	1426	738
0000	0000	1111	36	1541	932
Other Patterns					
xxxx	1111	1111	55	929	952
xxxx	1111	xxxx	25	1219	792
xxxx	xxxx	1111	13	1222	1081
xxxx	xxxx	xxxx	5	2047	1454
1111	1111	xxxx	5	432	336
1111	xxxx	1111	4	425	365
1111	xxxx	xxxx	1	1035	—

Key:
1 = Amount present for month.
0 = Amount missing for month.
x = Amount missing or present for month.

efficient) inferences. A major concern is that nonrespondents and respondents differ; for example, attritors differ from individuals who remain in the sample, or nonrespondents to income questions differ from respondents. Then analysis restricted to respondents is potentially biased.

Rubin (1976) and in a survey context Little (1982) formalize these ideas by introducing a response indicator matrix R, with the same dimensions as the data matrix Y, with entries 1 or 0 according to whether corresponding entries in Y are observed or missing. The missing data mechanism is then modeled in terms of the conditional distribution of R given Y. If this distribution does not depend on Y, then respondents and nonrespondents do not differ systematically, and analysis based on complete cases is valid. Rubin then describes the data as *missing completely at random* (MCAR).

If the data are not MCAR, an important issue is whether differences between respondents and nonrespondents can be explained in terms of variables recorded for both groups in the survey. For example, if age is fully recorded and income is sometimes missing, are differences in the income distributions of income nonrespondents and respondents attributable to differences in their age distributions? In other words, is nonresponse independent of income within age groups?

Rubin's (1976) definition of *missing at random* (MAR) addresses this issue in a more general setting. Data are MAR if the distribution of the response indicator matrix R given the data matrix Y depends on Y only through values Y_{obs} recorded in the survey. This would be the case in the above example if missingness to income depended only on the fully recorded variable age. If the data are MAR, Rubin shows that likelihood-based inferences using all the observed data Y_{obs} (not just the complete cases) yield valid inferences, without modeling the missing data mechanism. Note that full analysis based on all the data requires MAR, whereas analysis of complete cases generally requires the stronger MCAR assumption. A key objective of analyses that use all the data is to reduce nonresponse bias when the data are not MCAR.

If the data are not MAR, valid inferences require appropriate modeling of the response mechanism, which in our simple example would require a model for response to income as a function of age and income. We say then that the missing data mechanism is *nonignorable*, and call a model that includes the missing data mechanism a nonignorable model. Nonignorable models are generally both hard to specify correctly and sensitive to model misspecification (Little 1982, 1985; Little & Rubin, 1987, Chapter 11). Examples for normal data include Heckman (1976), Greenlees et al. (1982), Hausman and Wise (1979), and Winer (1983); nonignorable models for categorical data are discussed in Fay (1986) and references therein. We restrict attention in this paper to ignorable models.

Example 2. Assessing the MCAR Assumption for the SIPP Data: Tests of the MAR assumption have been proposed in the econometric literature, but themselves require strong assumptions. Tests of the MCAR assumption can be based on differences in the distributions of observed variables of respondents and nonrespondents, or more generally by differences in the distributions by pattern of missing data. For example, the last two columns of Table 1 present for each missing data pattern the mean and standard deviation of the mean average wages and salary over recorded months, adjusted for trend using the method discussed in (b) of Example 3 below. Differences in the means are not attributable to random fluctuations, implying that the data are not MCAR.

Table 2 illustrates another analysis of the MCAR assumption, using

TABLE 2. ANALYSIS OF MCAR ASSUMPTIONS: *T*-STATISTICS FOR DIFFERENCE IN MONTHLY MEAN WS OF COLUMN VARIABLE BY ABSENCE OR PRESENCE OF ROW VARIABLE

	WS1	WS2	WS3	WS4	WS5	WS6	WS7	WS8	WS9	WS10	WS11	WS12
WS1	—	-3.1	-1.9	-3.8	-2.1	-1.2	0.1	-2.1	-2.0	-0.6	-3.1	-1.4
WS2	-1.2	—	—	-1.2	1.5	-0.6	-1.5	1.3	-0.1	-1.3	-2.0	-0.4
WS3	-9.8	-4.0	—	-4.0	-2.6	-0.4	-1.5	-2.7	-1.1	-1.8	-2.9	-0.3
WS4	-5.3	-2.7	-2.5	—	-0.8	-2.4	-0.6	-0.6	-3.2	-0.6	-2.0	-3.3
WS5	6.2	5.7	5.7	5.1	—	*	*	*	2.5	2.6	2.8	2.5
WS6	6.3	5.8	5.8	5.2	*	—	*	*	2.5	2.5	2.7	2.4
WS7	6.4	5.8	5.9	5.3	*	*	—	*	2.7	2.7	2.9	2.6
WS8	6.4	5.8	6.0	5.5	*	*	*	—	2.8	2.8	3.0	2.6
WS9	6.1	5.9	6.2	5.9	3.7	3.6	3.4	3.6	—	*	*	*
WS10	6.1	5.8	6.2	5.9	3.7	3.6	3.3	3.5	*	—	*	*
WS11	6.0	5.7	6.1	5.8	3.6	3.5	3.3	3.4	*	*	—	*
WS12	6.2	5.8	6.2	5.9	3.7	3.6	3.4	3.5	*	*	*	—

* = less than 100 cases in absence group.

the BMDP8D program in Dixon (1983). Let WSj denote wages and salary in month j, for $j = 1, \ldots, 12$. For the row labeled WSj, cases are classified into two groups according to presence or absence of WSj. A two-sample t-test is conducted for each column variable WSk for $k \neq j$ ($k = j$ is impossible, since by definition there are no cases with WSj recorded in the absence group). Table 2 shows the resulting t-statistics, with signs that are positive if the mean income is larger for nonrespondents than for respondents. Statistics based on less than 100 cases in each group are not reported. If a particular variable is normal, independently distributed across cases (that is, design effects are not present) and the data are MCAR, the t-statistics are a (correlated) sample of standard normal deviates. In our example the WS amounts are far from normally distributed and design effects from cluster sampling are present, but the sample sizes are large and the t-statistics clearly large enough to invalidate the MCAR assumption. The pattern of signs of values is associated with lower observed income amounts. Missingness of income in later waves, however, is associated with higher observed income amounts, as can also be seen in Table 1.

Faced with data that are not MCAR, a primary objective of methods for handling missing data in surveys is to exploit information from incompletely recorded cases to limit the degree of nonresponse bias. One strategy for achieving this is to impute the missing values; this approach is the focus of the next section.

3. IMPUTATION

3.1. General Remarks

Imputation is a common method for handling item nonresponse in large surveys, and is sometimes used as a simple stop-gap measure in other settings. The method is particularly popular when a data producer is preparing a public use tape for a varied audience. The method might be viewed as a form of preliminary data editing, yielding a rectangular file convenient for data analysis. The producer may have access to information for creating imputations not available to the consumer, as when variables are omitted for reasons of confidentiality. Also the producer may be willing to devote resources to missing data adjustment not available to consumers, who regard missing values as a problem they would rather have someone else face. The creation of a set of "official" imputations tends to increase comparability of analyses of the same data, since differences in the way missing data are handled are eliminated.

The view of imputation as a preliminary to analysis is a deceptive organizational convenience; it is clear that imputation is part of the process of estimation, and should be viewed in terms of its impact on analysis. As analysis, good imputation methods (as discussed in Section 3.2) can improve naïve estimates based on complete or available cases — but bad imputation methods can make matters worse. Standard errors based on singly imputed data sets are typically wrong, as are associated tests and confidence intervals. Thus for inference, imputation leaves much to be desired. Multiple imputation (Section 3.6) can correct these deficiencies, at the expense of some increase in complexity.

3.2. A Basic Taxonomy of Imputation Methods

Imputation methods have a considerable literature; see, for example, Kalton and Kasprzyk (1982), Sande (1982), Madow et al. (1983), David et al. (1986), and Little (1988a). We find it useful to classify methods according to the three dimensions described below.

3.2.1. *Multivariate (M) or Univariate ($\bar{M}$).* Let $Y_{obs,i}$ denote the set of variables recorded in unit i and $Y_{mis,i}$ the set of variables missing. When $Y_{mis,i}$ contains more than one variable, a univariate imputation method treats each variable in $Y_{mis,i}$ separately, whereas a multivariate method imputes the variables in $Y_{mis,i}$ jointly, that is, taking into account associations between them. Univariate imputation can be simpler, but distorts relationships between variables in $Y_{mis,i}$, which is a concern if these relationships are the focus of subsequent analysis.

3.2.2. *Conditional (C) or Marginal ($\bar{C}$).* Marginal imputations for unit i are based on the marginal distribution of $Y_{mis,i}$; conditional imputations are based on the conditional distribution of $Y_{mis,i}$ given $Y_{obs,i}$, estimated from the data. An example of conditional imputation is regression imputation, where one variable (say Y_K) is sometimes missing and the other variables (say $Y_1, \ldots, Y_{K-1}$) are always present. The conditional distribution of Y_K given $Y_1, \ldots, Y_{K-1}$ is estimated by regressing Y_K on $Y_1, \ldots, Y_{K-1}$, using the cases with $Y_1, \ldots, Y_K$ observed. Then if Y_{iK} is missing it is imputed from $Y_{i1}, \ldots Y_{i, K-1}$ using the estimated regression equation. Other methods, like the adjustment cell methods discussed in the next section, are not based on an explicit model for the conditional distribution of $Y_{mis,i}$ given $Y_{obs,i}$, but nevertheless implicitly estimate this distribution from the data. Such methods are viewed as conditional under our taxonomy.

Conditional imputation is more work than marginal imputation, but by supplying more accurate predictions for the missing values, it can reduce nonresponse bias and increase the precision of estimates from the filled-in data. Marginal imputation is often worse than doing nothing. For example suppose the marginal mean of a missing variable is imputed. The sample mean of the filled-in data provides the same estimate of the mean of that variable as is obtained from available cases. Furthermore, the estimated standard error from the filled-in data is worse since it is based on an overstated sample size.

3.2.3. *Stochastic (S) or Mean (S̄).* Mean imputation imputes $Y_{mis,i}$ by an estimate of its mean, marginal or conditional depending on 3.2.2, above. Stochastic imputation imputes a value from the (estimated) marginal or conditional distribution of $Y_{mis,i}$. Mean imputation is generally not appropriate for categorical variables, since for such variables the mean is not a feasible outcome—for example, a fraction for a binary variable taking values 0 or 1. If quantitative variables are filled in by mean imputation, the resulting data can provide satisfactory estimates of means, but yields distorted estimates of dispersion or extreme percentiles of the distribution. For example, if missing incomes are imputed by mean imputation, estimates of percentage in poverty will be biased downward. Furthermore, mean imputation yields distortions when nonlinear transformations of the imputed variables are analyzed. For example, relationships between income and covariates are often estimated by linear regression on log income (see David et al., 1986; Greenlees et al., 1982). Mean imputations from such a model are conditional means of log income. If these imputations are exponentiated to provide an analysis of income, the resulting imputed incomes are biased downward. Mean imputation, like marginal imputation, distorts relationships between imputed variables; in fact, conditional mean imputation and marginal mean imputation yield the same imputations.

None of these problems apply to stochastic imputation, properly implemented. Thus stochastic imputation is generally preferable to mean imputation in multipurpose surveys. Methods of stochastic imputation that minimize the loss of precision relative to mean imputation are discussed in Ernst (1980) and Kalton and Kish (1984).

3.3. Practical Implementations

To summarize we prefer *multivariate conditional stochastic* (MCS) imputation methods to univariate marginal mean (M̄C̄S̄) imputation methods. In this section we provide a brief summary of common practical approaches.

3.3.1. *Adjustment Cell Methods.* A common conditional imputation method is to classify respondents and nonrespondents into adjustment cells with similar values of the covariates. Then missing values $Y_{mis,i}$ within each cell are imputed by respondent cell means (for mean imputation) or by values $Y_{mis,j}$ from respondents j in the same cell (for stochastic imputation). The CPS Hot Deck for imputing income fields in the Current Population Survey provides a complex example of the stochastic method (Oh & Scheuren, 1980; Welniak & Coder, 1980). In such multivariate nonresponse settings, the definition of adjustment cells needs to vary to take into account the fact that the set of observed variables varies from case to case. Also, if the number of observed covariates is large, the number of adjustment cells based on their joint distribution can become very large, and the cells need collapsing to insure that each nonrespondent has respondents in the same cell to provide an imputed value. In forming collapsing rules, the objective should be to include covariates that are highly predictive of the missing values. For a theoretical discussion of adjustment cell formation, see Little (1986b).

3.3.2. Matching Methods. In the stochastic version of the adjustment cell approach, nonrespondents are in effect matched to respondents by placing them in the same adjustment cell. A more general approach is to define a distance between nonrespondents and respondents, and then match each nonrespondent to the nearest respondent in that metric: the adjustment cell approach can be viewed as a special case where the distance takes value zero for cases in the same adjustment cell, and a value greater than zero otherwise. Examples of other distances are the Mahalanobis metric (for example, Vacek & Ashikaga, 1980) and the predictive mean metric (Little 1988a; Rubin, 1986). An interesting practical application is Colledge et al. (1978). Note that the matching approach does not require continuous variables to be grouped into categories, as is the case when adjustment cells are formed.

3.3.3. Regression Imputation. Since conditional imputation should be based on the conditional distribution of the missing variables $Y_{mis,i}$ given the observed variables $Y_{obs,i}$, it is natural to base imputations on the regression of $Y_{mis,i}$ on $Y_{obs,i}$. It has been observed (for example, Lillard, et al., 1986) that adjustment cell methods are based on an implicit regression model that includes all the interactions between the covariates used to form the adjustment cells, at the expense of all effects of any covariates excluded by a collapsing rule. Other regression models that drop high order interactions, include more main effects and allow continuous covariates, may make more èfficient use of the covariate information. Since the predictions from regression are estimated conditional means, if they are used directly to provide imputations the result is a conditional mean imputation method. Stochastic versions of regression imputation that add noise to these predictive means have also been considered (Scheuren, 1976). David et al. (1986) provide a comparison of stochastic regression methods with the CPS Hot Deck for imputing wages and salary amounts.

Note that if $Y_{mis,i}$ includes more than one variable, multivariate regression is required for imputation. Also, the set of outcome variables varies according to the set of variables missing in each case. One approach to this problem is to form the covariance matrix of all the variables using the complete cases, and then use the Sweep operator (for example, Clarke, 1982) to compute the regression $Y_{mis,i}$ on $Y_{obs,i}$ for each incomplete case i (Buck, 1960; Little & Rubin, 1987, Chapter 3).

Linear regression is not particularly appropriate when the missing variables are categorical or highly non-normal. Single binary variables might be handled by more appropriate methods like logistic or probit regression, but multivariate nonresponse with missing continuous and categorical data entails more complex modeling. One way of reducing the impact of the distributional assumptions of linear regression is to use the regressions to define a metric for matching, as in predictive mean matching (Little, 1988a).

3.4. Imputation in Panel Surveys

For imputation in panel surveys, another factor might be added to the taxonomy of Section 3.2. *Cross-sectional* imputation restricts information for predicting missing values in wave j to variables in wave j; *longitudinal* imputation allows information from other waves to be included. Longitudinal imputation clearly requires a longitudinal data file, with data for units from different waves matched. Advantages over cross-sectional imputation lie in the ability to use variables from other waves highly correlated with the missing variable to reduce nonresponse bias and improve precision. For example, if wage and salary information at wave j of the SIPP is missing, cross-sectional conditional imputation uses characteristics recorded at wave j that are mildly correlated with wages and salary, such as age, education, occupation, weeks worked. Longitudinal conditional imputation based on wage and salary data in previous or subsequent waves is likely to be much more effective, if these variables are observed (Heeringa & Lepkowski, 1986; Kalton & Miller, 1986).

A characteristic of longitudinal conditional imputation in many panel surveys is the large number of covariates from other waves potentially available for predicting the missing values. The adjustment cell approach seems unattractive in this setting, since the number of adjustment cells from the joint classification by covariates becomes extremely large, leading to a need for collapsing rules to find matches.

The work required in the creation of longitudinal imputes based on multivariate regression MCS method seems prohibitive, given the number of observed and missing variables involved. Practical limitations may lead to imputations that cannot satisfy all the diverse requirements of subsequent analyses. If so, the tailoring of missing data adjustments to specific analyses seems a more realistic approach.

Repeated measures on a single variable, such as the monthly wages and salary amounts in Example 1, are relatively common in panel data. For SIPP data of this form, Heeringa and Lepkowski (1986) consider simple longitudinal imputes based on direct substitution of values from adjacent observed waves. We propose a refinement that maintains most of the virtues of a multivariate stochastic conditional method, without the complexity of methods based on multivariate regression.

3.5. Imputation Based on Row and Column Fits: A Simple Stochastic Longitudinal Imputation Method for Repeated Measures Data

For repeated measures on a single variable, we believe that relatively efficient and simple imputations can often be based on row and column fits to the variable classified by row (unit) and column (period). If the missing values are well fitted by a model with additive row and column effects, then imputations may be based on an additive row + column fit:

$$\text{imputation} = (\text{row effect}) + (\text{column effect}) + (\text{residual}). \qquad (1)$$

If a multiplicative model (or equivalently an additive model for the logarithm of the variable) seems more appropriate, then imputations may be based on a multiplicative row × column fit:

$$\text{imputation} = (\text{row effect}) \times (\text{column effect}) \times (\text{residual}). \qquad (2)$$

The row and column effects in these equations are proportional to row and column means. The residual is found from a complete case, ideally as similar as possible to the incomplete case with respect to the variable being imputed. We now describe a specific implementation for the SIPP data in Example 1.

Example 3. Imputation of SIPP Wages and Salary Amounts by a Multiplicative Row × Column Fit: For positive variables our implementation of model (2) will, unlike model (1), insure that the imputed values are always positive as well. Because the monthly SIPP wages and salary amounts are positive, and income amounts are generally modeled on the log scale, a multiplicative model seems preferable, and hence we impute these variables by a multiplicative row × column fit. The method is implemented as follows:

(a) Column (period) effects of the form $c_j = 12\, \tilde{W}\tilde{S}_j / \sum_k \tilde{W}\tilde{S}_k$ are computed for each month $j = 1, \ldots, 12$, where $\tilde{W}\tilde{S}_j$ is the sample mean wages and salary amount for month j based on complete cases.

(b) Adjusted row (person) means $\tilde{W}\tilde{S}^{(i)} = m_i^{-1} \Sigma WS_{ij}/c_j$ are computed for both complete and incomplete cases. Here the summation is over recorded months for case i; m_i is the number of recorded months; WS_{ij} is the wages and salary for case i, month j; and c_j is the simple month correction from (a).

(c) Cases are ordered by $\tilde{W}\tilde{S}^{(i)}$, and incomplete case i is matched to the closest complete case, say l.

(d) Missing value WS_{ij} is imputed by

$$\hat{WS}_{ij} = [\tilde{W}\tilde{S}^{(i)}][c_j][WS_{lj}/(\tilde{W}\tilde{S}^{(l)}c_j)]$$

$$= WS_{lj}\tilde{W}\tilde{S}^{(i)}/\tilde{W}\tilde{S}^{(l)}, \tag{3}$$

where the three terms in square parentheses in equation (3) represent the row, column, and residual effects in equation (2), the first two terms estimate the predicted mean, and the last term is the stochastic component of the imputation from the matched case.

We imputed all the missing WS amounts for the 2081 complete cases in Example 1 by this method. To give a feel for the results, Table 3 shows the results for a subsample of 21 incomplete cases, chosen by ranking the incomplete cases by mean wages and salary for the months in which they were reported (column 2 of the table) and selecting every 100th case. For each case, the single entries are the observed WS amounts. The double entries in parentheses show imputations from two methods, the first from the row × column method, and the second from the cross-sectional hot deck imputation supplied by the Census Bureau. One would expect the row × column method to impute more consistent amounts across waves, as is evident in cases 5 and 21. The row × column method sometimes imputes the same amount within waves (cases 3, 13, 14, 17, 19, 21) and sometimes does not (cases 1, 5, 7, 9, 11, 12, 16, 18, 20), depending on the consistency of the amounts in the donor record. Note that the method can impute zero amounts when the donor record contains zeros (case 1)—zero amounts occur when the respondent has a job but does not receive wages for that month. The method provides a reasonable frequency of imputed zeros without the need to treat zero amounts separately, as may be necessary with other methods (for example, a maximum likelihood analysis of log amounts.) Aggregate estimates from the filled-in data are considered in Example 5 below.

Many other methods for imputing the WS amounts could be constructed, including the normal maximum likelihood method considered below in Example 5. The row × column method could be extended to allow conditioning on additional covariate information, either by applying it within adjustment cells defined by those covariates, or by including such information in the metric for defining the match to a responding case. However, the basic method has the following useful features: (a) the imputed values incorporate information about trend (from the column effects) and individual level (from the row effects) in a natural way; (b) the method does not require separate modeling for different patterns of missing data, dealing with all patterns simultaneously; and (c) the method is comparatively easy to implement, an important consideration with large complex data sets.

TABLE 3. OBSERVED AND [IMPUTED] VALUES FOR 21 OBSERVATIONS ON MONTHLY WAGES AND SALARY AMOUNTS FOR JOB 1 FROM (A) ROW × COLUMN METHOD AND (B) CROSS-SECTIONAL HOT-DECK METHOD[a]

OBS	MWS[b]	WS1	WS2	WS3	WS4	WS5	WS6	WS7	WS8	WS9	WS10	WS11	WS12
1	98	[199] [208]	[167] [167]	[0] [208]	167	80	80	80	100	85	85	85	50
2	270	400	[318] [500]	400	400	35	0	0	150	375	375	468	375
3	390	[180] [720]	[180] [900]	[180] [720]	[180] [720]	0	0	0	0	900	720	720	900
4	511	480	[425] [600]	480	480	600	480	480	600	480	480	600	480
5	628	615	615	615	615	615	615	615	615	[618] [1275]	[773] [1594]	[618] [1275]	[618] [1275]
6	720	700	700	700	700	1000	800	800	1000	520	520	650	520
7	817	[760] [875]	[648] [875]	700	[900] [875]	800	800	995	847	800	800	1000	800
8	902	1037	688	1037	1037	778	778	778	778	928	928	928	928
9	984	[1496] [860]	[800] [1296]	688	[748] [860]	688	688	860	688	1344	1680	1344	1344
10	1069	900	[1099] [1125]	900	900	1250	1000	1000	1250	1200	1024	1024	1280
11	1165	[1126] [900]	[1676] [720]	[1126] [900]	[1126] [720]	1400	1750	1400	1400	970	776	776	970
12	1271	[1406] [1625]	1300	1300	[1125] [1625]	1220	1220	1244	1244	1244	1555	1244	1244

continues

TABLE 3, continued

OBS	MWSb	WS1	WS2	WS3	WS4	WS5	WS6	WS7	WS8	WS9	WS10	WS11	WS12
13	1358	[1250][2073]	[1250][2592]	[1250][2073]	[1250][2073]	2200	1760	1760	2200	0	936	1170	936
14	1467	0	2000	2250	2250	1400	1120	1120	1400	[1499][1505]	[1499][1505]	[1499][1882]	[1499][1505]
15	1605	800	[1410][1000]	800	800	2875	2300	2300	2875	2000	2000	1000	0
16	1723	[1597][2341]	1873	1873	[1733][2341]	1600	1600	2000	1600	1600	2000	1600	1600
17	1897	[2110][2000]	1600	[2110][2000]	1600	1600	1600	2500	2000	2300	1840	1840	2300
18	2116	[4083][2440]	1952	[1856][2440]	1952	2400	3000	2400	2400	2000	1600	1600	2000
19	2377	[2280][3250]	2600	2600	2280 3250	2400	2400	3000	2400	2000	2500	2000	2000
20	2847	[2000][3000]	2400	[2501][3000]	2400	2680	3350	2680	2680	3500	2800	2800	3500
21	3825	3680	3680	3680	3680	[3804][3082]	[3804][2465]	[3804][2465]	[3804][3082]	[3814][1332]	[3814][1332]	[3814][1666]	[3814][1332]

a[   ] = imputed.

bMWS = mean wages and salary income for the months reported.

3.6. Inferences from Imputed Data

An important limitation of any imputation method was mentioned in Section 3.1. Estimates from the filled-in data may be satisfactory if an appropriate MCS imputation method is applied, but associated standard errors, tests, and confidence intervals will be invalid, since they do not take into account the fact that imputed values are estimates rather than data. Adjustments are needed to provide valid inferences, and finding appropriate adjustments can be a major task.

An interesting extension of MCS imputation that provides appropriate inferences from imputed data is multiple imputation (Rubin 1978, 1987; Rubin & Schenker, 1986). More than one (e.g., I) draws from the predictive distribution of Y_{mis} are created, and the I filled-in data sets are analyzed, with each of the I imputed values substituted in turn for each missing value. Estimates from the I analyses are then combined in a rather simple way to create an inference that allows for uncertainty introduced from imputation. Correct implementation of the method still requires modeling (implicit or explicit) of the data. Thus the method can be viewed as a means of communicating the results of the modeling in the form of multiply imputed data sets that can be analyzed by relatively simple extensions of complete data methods.

When the imputation is achieved by matching nonrespondents to respondents, as in the row × column method given above, a simple way of creating the multiple imputes is to match nonrespondents to more than one case (say the I closest neighbors). Although some modifications of this approach are needed to implement the method correctly (see Little, 1988a, for details), even if these modifications are omitted the method represents a useful step towards obtaining estimates of precision that allow for the lost information in the missing values. For an application see Oh and Scheuren (1980).

An alternative to multiple imputation that can yield valid inferences (at least in large samples) is to avoid filling in the data, and instead analyze the incomplete data directly using maximum likelihood methods (Little, 1982). We now consider direct methods of analysis.

4. DIRECT ANALYSIS OF THE INCOMPLETE DATA

4.1. Alternative Approaches

Suppose data are not imputed, but are left incomplete with a missing value code to identify the missing values. The analysis of such data often takes one of three forms, which using the terminology of Little and Rubin (1987) we shall call *complete-case* analysis, *available-case* analysis and *likelihood-based* analysis. Complete-case analysis of a set of variables $Y_1, \ldots, Y_K$ restricts attention to cases with all these variables observed. Available-case analysis uses the set of cases with Y_j observed to estimate parameters of the marginal distribution of Y_j, and the set of cases with both Y_j and Y_K observed to estimate parameters of two-way association, such as covariances or correlations. Likelihood-based analysis formulates a model for the data and missing data mechanism and uses the likelihood to estimate parameters, via maximum likelihood (ML) or Bayesian methods (Little, 1982; Little & Rubin, 1987).

Advantages of the latter approach are that (a) ML estimates are principled in the sense that they have known statistical properties under the assumed model, which can be clearly specified and subjected to model criticism; (b) ML under ignorable models provides appropriate inferences under the MAR rather than the stronger MCAR assumption (Rubin, 1976); (c) estimates are asymptotically efficient under the assumed model; and (d) a method for computing large sample standard errors that takes into account the fact that data are incomplete is available, based on the observed or expected information matrix. The main disadvantages are that (a) ML estimation for incomplete data often involves iterative algorithms that are expensive for large data sets and are not readily accessible in commercial statistical software; and (b) small sample inferences are not readily available. The latter, however, is a problem with most procedures that make use of the incomplete data, and is less of a concern with survey data sets, which are typically quite large. Also, the former becomes less of a concern as computing resources expand, and programs become more widely available. Thus ML estimation may supplant imputation, complete case or available case analysis as the method of choice when missing data are a serious problem.

Space precludes an adequate discussion of ML methods, so we shall limit ourselves to some references. (For a detailed review, see Little & Rubin, 1987, Part 2.) For continuous data, ML estimation under multivariate normality provides estimates of the mean and covariance matrix from incomplete data (see the BMDPAM program in Dixon, 1983). Robust estimation based on multivariate t and contaminated normal models is discussed in Little (1988b). For categorical data, ML estimates can be computed for loglinear models with partially classified contingency tables (Fuchs, 1982). Little and Schluchter (1985) discuss ML estimation for mixed categorical and continuous data with missing values, using a model that combines a loglinear model for the categorical variables with a multivariate analysis of variance model for the continuous variables given the categorical variables. Other extensions of the multivariate normal model, particularly important for panel survey data, are to repeated measures models (see Jennrich & Schluchter, 1986, and references therein). The model in the latter paper allows the mean to depend linearly on between-subject and within-subject design variables, and includes a flexible choice of covariance structures. For many of these models the EM algorithm (Dempster et al., 1977; Orchard & Woodbury, 1972) provides a useful computational approach.

4.2. Analysis of Complete and Available Cases: Loss of Information Assuming MCAR

Complete case analysis and available case analysis have the virtue of simplicity. How much is lost by using these methods rather than more complex methods such as maximum likelihood that use all the data efficiently? If the data are MCAR, then these methods do not have large sample biases, and we can assess the asymptotic loss of information by comparing the large sample variance of estimates based on complete or available cases with the large sample variance of (the fully efficient) ML estimates. Some general observations follow:

(1) *The loss of information depends on the fraction of incomplete cases.* Clearly if the fraction of incomplete cases is small, say on the order of 5 percent or less, then the missing data problem may be too inconsequential to devote resources to fully efficient methods.

(2) *The loss of information from using complete cases (that is, the information in the incomplete cases) depends crucially on the parameter of interest.* As an illustration, consider a bivariate normal sample on x and y with means (μ_x, μ_y), equal variances σ^2, correlation ρ, and missing values confined to y. Let one unit represent the information in a complete case. For inference about μ_x, the incomplete cases contain information 1, that is, as much information as the complete cases (since x is fully recorded). For inference about μ_y and $\mu_y + \mu_x$, the information in each incomplete case is ρ^2 and $(1 + \rho)/2$ respectively, and hence is large when x and y are highly positively correlated. For inference about $\mu_y - \mu_x$, the information in each incomplete case is $(1 - \rho)/2$, which is *small* when x and y are highly positively correlated. (For details on these results, see Little, 1976).

Suppose, for example, x and y are successive wages and salary amounts in the SIPP (with fairly high positive correlation). Incomplete cases contribute a lot to μ_x and $\mu_y + \mu_x$, but little to the change in income between months (intuitively, because the change is poorly correlated with the income in the observed month).

(3) *Imputation is generally less efficient than maximum likelihood.* How do imputation methods fare in these comparisons? Imputation of the estimated conditional mean of y given x yields ML estimates of μ_y, μ_x, and linear combinations of these parameters, but biased estimates of other parameters. Methods

that add noise to the conditional means are less efficient than ML, although some variants of these methods (like predictive mean matching) may provide protection against model misspecification.

(4) *Available case analysis is not always more efficient than complete case analysis, despite the fact that it appears to make more use of the available data.* Figure 2 sketches the large sample variances of estimates of $\mu_y - \mu_x$ from complete cases, available cases, and maximum likelihood, in the bivariate normal situation just considered. Available case analysis (that is, the sample mean of Y from complete cases minus the sample mean of X from all cases) is more efficient than complete case analysis with $\rho < 0.5$, but is less efficient when $\rho > 0.5$. Note that ML is fully efficient and hence dominates the other methods, but use of available cases for small ρ and complete cases for large ρ is close to fully efficient in this simple setting.

(5) *Multivariate analysis based on available cases can be problematic, particularly for highly correlated data.* For example, estimated covariance matrices based on available cases can yield estimated covariance matrices that are not positive semi-definite, and hence slopes that are undetermined (Haitovsky, 1968; Little & Rubin, 1987, Chapter 3).

4.3. Comparison of Methods When the Data Are Not MCAR

Since nonresponse often yields data that are not MCAR, it is important to evaluate methods under non-MCAR conditions. If the data are not MCAR but are MAR, then ML estimation ignoring the missing data mechanism is appropriate, whereas analysis of complete or available cases is subject to bias. If the data are not MAR, as when missingness of a variable depends on the value of that variable, then all these methods yield inconsistent estimates, but ML based on an ignorable model can still reduce the size of the bias by making some use of the incomplete cases.

We illustrate these points with two examples, one artificial and one involving the SIPP data.

Example 4. Estimates of Means from Incomplete Bivariate Normal Data: Table 4 shows 20 randomly generated bivariate normal observations on (X,Y) with sample means (.21, .22), standard deviations (.88, .97), and correlation .86. Three incomplete data sets are created by deleting values of Y according to each of three extreme non-MCAR mechanisms: Y is deleted in the ten cases with the smallest values of (a) X, (b) Y, and (c) $Y - X$. Note that mechanism (a) is MAR, since X is fully observed, and the mechanisms (b) and (c) are nonignorable. For each incomplete data set, Table 5 displays estimates of the means of X,

FIGURE 2. LARGE SAMPLE VARIANCE OF ESTIMATES OF $\mu_y - \mu_x$ FROM THREE METHODS, FOR BIVARIATE NORMAL DATA WITH MISSING VALUES ON Y AND MCAR MISSING DATA MECHANISM

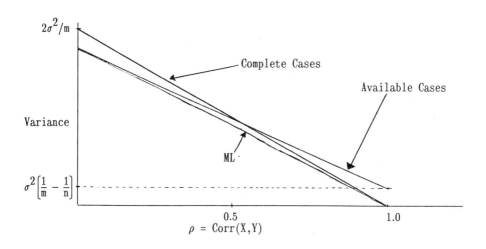

Note: m = number of complete cases; n = number of cases.

Y, and $Y - X$ from three methods—complete case analysis, available cases analysis, and ML assuming an ignorable mechanism. Target quantities are the sample estimates from the data before deletion, displayed in the last row of the table.

Complete case estimates of the means are biased, with a bias that depends on the size of the correlation between the variable whose mean is estimated and the variable underlying the missing data mechanism. This correlation is large for estimates in Table 5 indicated with an asterisk; these estimates are indeed poor. The ML method reduces the bias of estimates of μ_x and μ_y, particularly when missingness depends on X, when the method is truly ML and is in fact unbiased. The ML method is less successful in reducing bias for $\mu_y - \mu_x$ when missingness depends on $Y - X$, since in this case the information in the incomplete cases is low, since X and $Y - X$ are not highly correlated. Even so, ML will be no worse than complete cases on average, although it is slightly worse in this particular data set.

Note that for these non-MCAR mechanisms the available case estimates of change are markedly worse than those based on complete cases. This result applies more generally—see in particular the next

TABLE 4. BIVARIATE NORMAL SAMPLE FOR EXAMPLE 4

		Cases Deleted under		
X	Y	(a)	(b)	(c)
−.536	−.199	x	x	
.392	−.698		x	x
.038	−.429	x	x	x
.907	.590			x
−.696	−1.374	x	x	x
−.884	−1.009	x	x	x
−1.073	−.302	x	x	
1.288	1.183			x
.168	1.065	x		
−1.056	−.858	x	x	
1.696	1.831			
.717	−.071		x	x
−1.069	−1.170	x	x	x
.999	1.465			
.636	.809			
.224	.217	x		x
.494	.915			
.626	.641			x
−.204	.173	x	x	
1.584	1.679			

Notes:
   (a) is a non-MCAR mechanism in which Y is deleted in the ten cases with the smallest values of x.
   (b) is a non-MCAR mechanism in which y is deleted in the ten cases with the smallest values of y.
   (c) is a non-MCAR mechanism in which $\bar{Y}$ is deleted in the ten cases with the smallest values of $\bar{Y} - \bar{X}$.

example—and reflects the intuition that it is better to assess change from a fixed (equally unrepresentative) set of cases than to estimate change from a set of available cases that vary in degree of unrepresentativeness over time.

Example 5. Comparison of Estimates of Means and Correlations of SIPP Wages and Salary Amounts from Five Methods: Table 6 shows estimates of the 12 monthly mean wages and salary amounts for the data described in Example 1, calculated by five methods: complete case analysis, available case analysis, ML assuming multivariate normality with ignorable nonresponse, and sample means from the data imputed by the row × column method and the cross-sectional Census Bureau method in Example 3. The first three methods were computed using the BMDPAM program in Dixon (1983). To reduce the number of cases and sharpen distinctions between the methods, all incomplete cases and a 1-in-5 sample of the complete cases were included in the analysis, resulting in 4,183 cases, of which 2,102 were complete.

TABLE 5. ESTIMATES OF MEANS FOR DATA IN TABLE 4

	Parameter		
	μ_x	μ_y	$\mu_y - \mu_x$
Complete cases			
Missingness based on X	.93*	.83*	−0.10
Missingness based on Y	.86*	1.04*	−0.18
Missingness based on $Y - X$	.27	.66	.39*
Available cases			
Missingness based on X	.21	.83	.62
Missingness based on Y	.21	1.04	.83
Missingness based on $Y - X$	.21	.66	.45
ML assuming ignorable nonresponse			
Missingness based on X	.21	−.15	−.37
Missingness based on Y	.21	.55	.34
Missinginess based on $Y - X$	.21	.61	.39
No missing data	.21	.22	.01

*Denotes missing data mechanism highly correlated with parameter (see text).

TABLE 6. ESTIMATES OF MEAN AMOUNT OF MONTHLY WAGES AND SALARY INCOME (JOB 1) FROM FIVE METHODS

	Method				
Month	Complete Cases	Available Cases	Normal ML	Row × Column Fit	Cross-Sectional
1	1257	1312	1283	1284	1379
2	1298	1323	1315	1315	1362
3	1302	1346	1320	1329	1361
4	1319	1359	1341	1342	1385
5	1342	1349	1374	1374	1372
6	1346	1348	1374	1374	1370
7	1342	1341	1367	1369	1363
8	1338	1338	1365	1364	1362
9	1408	1382	1415	1409	1403
10	1412	1391	1424	1425	1413
11	1390	1373	1405	1401	1396
12	1385	1362	1394	1390	1385
Mean	1344	1352	1365	1365	1379

TABLE 7. AVERAGE ESTIMATED CORRELATIONS BETWEEN
WAGES AND SALARY INCOME AMOUNTS (JOB 1) IN
WAVES j and k, $1 \leq j, k \leq 3$, FROM FIVE METHODS

	Method				
Waves (j,k)	Complete Cases	Available[a] Cases	Normal ML	Row × Column Fit	Cross-Sectional
1,1	.91	.92	.92	.90	.92
2,2	.92	.92	.93	.92	.92
3,3	.88	.90	.90	.89	.90
1,2	.82	.81	.77	.81	.73
1,3	.81	.81	.81	.78	.73
2,3	.81	.80	.81	.80	.74

[a]Pairwise correlation method.

While an unequivocal verdict on which method is best is not possible without the true values of the missing data, we prefer the ML and row × column methods on theoretical terms, since they use wages and salary amounts for an individual from recorded months; the dependence of income on the pattern of missing data noted in Example 2 suggests that the use of observed income information will tend to reduce nonresponse bias. The ML and row × column methods both yield very similar estimated means. The complete-case method yields a slightly lower average level, but a very similar trend across months. Available-cases yields a reasonable average level, but a trend considerably smaller than the previous three methods, an indication of a bias similar to that found in Example 4. The cross-sectional estimates yield the highest level and the smallest trend. We believe the latter two methods may be worse than leaving the data incomplete.

Table 7 compares estimates of correlations from the five methods. Rather than display the full correlation matrix of all twelve variables, we have grouped the variable into the three waves and averaged the correlations between and within each wave. All the methods except the cross-sectional imputations yield fairly similar results, with larger correlation within than between waves. The cross-sectional imputes yield noticeably smaller correlations between waves, reflecting a downward bias when longitudinal information is not used to create the imputations.

5. CONCLUSION

The challenge in handling missing data in panel surveys lies in developing methods that incorporate longitudinal information on nonrespondents in a way that (a) reduces bias and (b) yields appropriate inferences. Widely used methods such as complete-case analysis, available-case analysis, and simple forms of imputation are generally deficient in this regard. Two key developments in the recent literature, namely, maximum likelihood methods and multiple imputation, lead to better methods. The model-based approach that underlies these methods is attractive, but fully specified models are time-consuming to develop and fit to extensive panel data with complex patterns of missing data. Thus there is scope for more research on simpler methods (like those of Section 3.5 based on row and column fits) that retain most of the features of fully efficient maximum likelihood procedures but are easier to implement.

REFERENCES

Anderson, T. W. (1957), "Maximum Likelihood Estimates for a Multivariate Normal Distribution When Some Observations Are Missing," *Journal of the American Statistical Association.* **52**, 200–203.

Buck, S. F. (1960), "A Method of Estimation of Missing Values in Multivariate Data Suitable for Use with an Electronic Computer," *Journal of the Royal Statistical Society*, Series B, **22**, 302–306.

Clarke, M. R. B. (1982), "The Gauss-Jordan Sweep Operator with Detection of Collinearity," *Applied Statistics*, **31**, 166–168.

Colledge, M. J., Johnson, J. H., Pare, R., & Sande, I. G. (1978), "Large-Scale Imputation of Survey Data," *Proceedings of the Section on Survey Research Methods*, American Statistical Association, 431–436.

David, M. (1985), "Introduction: The Design and Development of SIPP," *Journal of Economic and Social Measurement*, **13**, 215–224.

David, M., Little, R. J. A., Samuhel, M. E., & Triest, R. K. (1986), "Alternative Methods of CPS Income Imputation," *Journal of the American Statistical Association*, **81**, 29–41.

Dempster, A. P., Laird, N. M., & Rubin, D. B. (1977), "Maximum Likelihood from Incomplete Data via the EM Algorithm," *Journal of the Royal Statistical Society*, Series B, **39**, 1–38.

Dixon, W. J., ed. (1983), *BMDP Statistical Software*, University of California Press, Los Angeles.

Ernst, L. R. (1980), "Variance of the Estimated Mean for Several Imputation Procedures," *Proceedings of the Section on Survey Research Methods*, American Statistical Association, 716–720.

Fay, R. (1986), "Causal Models for Patterns of Nonresponse," *Journal of the American Statistical Association*, **81**, 354–365.

Fuchs, C. (1982), "Maximum Likelihood Estimation and Model Selection in Contingency Tables with Missing Data," *Journal of the American Statistical Association*, **77**, 270–278.

Greenlees, J. S., Reece, W. S., & Zieschang, K. O. (1982), "Imputation of Missing Values When the Probability of Response Depends on the Variable Being Imputed," *Journal of the American Statistical Association*, **77**, 251–261.

Haitovsky, Y. (1968), "Missing Data in Regression Analysis," *Journal of the Royal Statistical Society*, Series B, **30**, 67–81.

Hausman, J. A., & Wise, D. A. (1979), "Attrition Bias in Experimental and Panel Data: The Gary Income Maintenance Experiment," *Econometrica*, **47**, 455–73.

Heckman, J. (1976), "The Common Structure of Statistical Models of Truncation, Sample Selection, and Limited Dependent Variables and a Simple Estimator for Such Models," *Annals of Economic and Social Measurement*, **5**, 475–492.

Heeringa, S. G., & Lepkowski, J. M. (1986), "Longitudinal Imputation for the SIPP," *Proceedings of the Section on Survey Research Methods*, American Statistical Association, 206–210.

Jennrich, R., & Schluchter, M. (1986), "Unbalanced Repeated Measures Models with Structured Covariance Matrices," *Biometrics*, **42**, 805–820.

Kalton, G. (1986), "Handling Wave Nonresponse in Panel Surveys," *Journal of Official Statistics*, **2**, 303–314.

Kalton, G., & Kasprzyk, D. (1982), "Imputing for Missing Survey Responses," *Proceedings of the Section on Survey Research Methods*, American Statistical Association, 22–31.

Kalton, G., & Kish, L. (1984), "Some Efficient Random Imputation Methods," *Communications in Statistics—Theory and Methods*, **13**(16), 1919–1939.

Kalton, G., & Miller, M. (1986), "Effects of Wave Nonresponse on Panel Survey Estimates," *Proceedings of the Section on Survey Research Methods*, American Statistical Association, 194–199.

Lillard, L., Smith, J. P., & Welch, F. (1986), "What Do We Really Know about Wages: The Importance of Nonreporting and Census Imputation," *Journal of Political Economy*, **94**, 489–506.

Little, R. J. A. (1976), "Inference about Means from Incomplete Multivariate Data," *Biometrika*, **63**, 593–604.

Little, R. J. A. (1982), "Models for Nonresponse in Sample Surveys," *Journal of the American Statistical Association*, **77**, 237–250.

Little, R. J. A. (1985), "A Note about Models for Selectivity Bias," *Econometrica*, **53**, 1469–1474.

Little, R. J. A. (1986), "Survey Nonresponse Adjustments for Estimates of Means," *International Statistical Review*, **54**, 139–157.

Little, R. J. A. (1988a), "Missing Data in Large Surveys," *Journal of Business and Economic Statistics*, forthcoming (with discussion).

Little, R. J. A. (1988b), "Robust Estimation of the Mean and Covariate Matrix from Data with Missing Values," *Applied Statistics*, **37**, 23–38.

Little, R. J. A., & David, M. (1983), *Weighting Adjustments for Nonresponse in Panel Surveys* (Working paper), U.S. Bureau of the Census, Washington, DC.

Little, R. J. A., & Rubin, D. B. (1987), *The Statistical Analysis of Data with Missing Values*, John Wiley and Sons, New York.

Little, R. J. A., & Schluchter, M. D. (1985), "Maximum Likelihood Estimation for Mixed Continuous and Categorical Data with Missing Values," *Biometrika*, **72**, 497–512.

Madow, W. G., Olkin, I., & Rubin, D. B., eds. (1983), *Incomplete Data in Sample Surveys, Vol. 2: Theory and Bibliographies*, Academic Press, New York.

Oh, H. L., & Scheuren, F. (1980), "Estimating the Variance Impact of Missing CPS Income Data," *Proceedings of the Section on Survey Research Methods*, American Statistical Association, 408–415.

Orchard, T., & Woodbury, M. A. (1972), "A Missing Information Principle: Theory and Applications," *Proceedings of the Sixth Berkeley Symposium on Mathematical Statistics and Probability*, **1**, 697–715.

Rubin, D. B. (1974), "Characterizing the Estimation of Parameters in Incomplete Data Problems," *Journal of the American Statistical Association*, **69**, 467–474.

Rubin, D. B. (1976), "Inference and Missing Data," *Biometrika*, **63**, 581–592.

Rubin, D. B. (1978), "Multiple Imputations in Sample Surveys—A Phenomenological Bayesian Approach to Nonresponse," *Proceedings of the Section on Survey Research Methods*, American Statistical Association, 20–34.

Rubin, D. B. (1986), "Statistical Matching and File Concatenation with Adjusted Weights and Multiple Imputations," *Journal of Business and Economic Statistics*, **4**, 87–94.

Rubin, D. B. (1987), *Multiple Imputation in Sample Surveys and Censuses*, John Wiley and Sons, New York.

Rubin, D. B., & Schenker, N. (1986), "Multiple Imputation for Interval Estimation from Simple Random Samples with Ignorable Nonresponse," *Journal of the American Statistical Association*, **81**, 366–374.

Sande, I. G. (1982), "Imputation in Surveys: Coping with Reality," *The American Statistician*, **36**, 145–152.

Scheuren, F. (1976), *Preliminary Notes on the Partially Missing Data Problem—Some (Very) Elementary Considerations* (Working paper), Social Security Administration

Methodology Group.

Vacek, P. M., & Ashikaga, T. (1980), "An Examination of the Nearest Neighbor Rule for Imputing Missing Values," *Proceedings of the Statistical Computing Section,* American Statistical Association, 421–425.

Welniak, E. G., & Coder, J. F. (1980), "A Measure of the Bias in the March CPS Earnings Imputation Scheme," *Proceedings of the Section on Survey Research Methods,* American Statistical Association, 421–425.

Winer, R. S. (1983), "Attrition Bias in Econometric Models Estimated with Panel Data," *Journal of Marketing Research,* **20,** 177–186.

NOTE

*This research is supported by grant SES 84 11804 from the National Science Foundation. We would like to thank Martin David of the ISDP/SIPP Data Center, University of Wisconsin, and Graham Kalton of The University of Michigan for supplying data from the Survey of Income and Program Participation for the empirical work in the paper. We also want to thank the referees for useful comments on an earlier draft.

Nonresponse Adjustments: Discussion

Fritz Scheuren

The authors of the previous three papers have been particularly lucid in their approaches to the nonresponse problem in panel surveys. What can be added? Perhaps not a great deal beyond extending an appreciation for their efforts in bringing new insights to this topic.

One possible minor contribution, though, might be to spend some time looking at the paths each author has taken to the subject. Although these differ, in many cases their advice to practitioners would be very similar.

1. LEPKOWSKI PAPER

Basically, the Lepkowski paper takes an "ignorable" path to non-response adjustments in panel surveys; in particular, the author assumes various forms of missing at random (MAR) to motivate approaches. Along the way he talks about weighting and imputation strategies and contrasts them nicely. One of the things that is especially useful in the paper is the typology developed of how nonresponse affects different outcome measures from the survey (i.e., point-in-time or cross-section variables, change variables, or variables derived by cumulating over time).

The paper also provides a good bibliography of recent Survey of Income and Program Participation (SIPP) empirical work on the different approaches possible. In his presentation, Lepkowski rightly emphasizes that a compensation method that provides reasonably accurate results for one purpose may provide poor prediction for another (a point also made again in the other papers). One gets a realistic sense from Lepkowski's work of the extremely messy issues that have to be addressed. (For example, confining ourselves to monotone nonresponse patterns could leave out a lot of what happens.)

I find myself in strong agreement with him on the importance of weighting, especially multiple weighting. One suggestion here, though, is that he might have cited the raking literature more; multiple raking ratio estimation weights (and weighting in a nested way to deal with monotone patterns) comes quite naturally and can be done cheaply (e.g., Scheuren, 1981). The author's attempt to summarize the relative strengths of wave nonresponse compensation strategies is quite helpful.

In my opinion, however, while the objective is a worthy one, the attempt may be premature. For example, it would be worthwhile to add some question marks to Lepkowski's Table 4. Also, the author might want to use his excellent earlier discussion of analytic objectives to do the figure separately by type of analysis intended (e.g., are the results different if we are measuring change, cumulating from wave to wave, etc.?).

2. LITTLE–SU PAPER

The paper by Little and Su takes a more theoretical approach to many of the same issues raised by Lepkowski. In addition, paradoxically, the authors add some new SIPP empirical results to the discussion as a way of illustrating their concerns. On both counts, they need to be thanked for their contributions.

Little and Su strive for a more *principled* approach to the various trade-offs that beset the wave nonresponse problem. This is done, however, without really sacrificing the pragmatic, "do what works" flavor of the Lepkowski paper. Like Lepkowski, they confine themselves to models that assume some form of missing at random.

It was particularly good to see maximum likelihood estimation (MLE) ideas brought extensively into the discussion and used as a standard in evaluating ad hoc survey procedures. Unquestionably, the authors are making a key contribution in this paper to the reunification of survey statistics with other fields in our profession. The emphasis on MLE may also give the data user a better vehicle for participating with the data producer in the production of a tailored adjustment. There is little doubt that a team approach to the problem of missing data is key to better estimates both at the adjustment and, more important, at the design stage.

3. FAY PAPER

I cannot in the brief space available do anything more than summary justice to the fine paper by Fay. To begin with, I found his earlier work (Fay, 1986) on causal modeling quite interesting and see his extensions here as of considerable importance, too. This paper, like the previous one, is a nice blend of theory followed by meaty realistic data examples (as before, taken largely from SIPP). The contrast between Fay's work and that of Lepkowski and Little–Su is particularly important in that different models are invoked to deal with the missingness.

One of the best points in the paper arises in the author's asking (and answering) the question about the role of causal models in handling nonresponse. Let me quote from Fay (the emphasis is mine):

The correspondence between the ignorable model and a specific causal model for nonresponse for nested patterns raises interesting questions about the merits of each approach. *Should we regard causal models as worthwhile because they represent a possible route to generalize the assumption of ignorable nonresponse? Should we instead regard causal models as a natural representation of nonresponse processes, and conclude that the frequent reasonableness of the ignorable model derives from its ability to either be or to approximate the appropriate causal model in many situations?* An affirmative answer to the second question suggests missing data situations should be confronted as a problem of determining the appropriate mechanisms governing nonresponse, and that the model of ignorable nonresponse be regarded simply as one option among several.

I strongly support the need for more development of causal models and for their greater use. If I were to have one criticism of the Fay paper, it is that he seems to imply that others (notably Little) are satisfied with a status quo in which we continue to be confined to adjustment strategies that make some form of ignorable nonresponse assumption. I look to Fay to help us escape this possibility and suspect others will begin joining him in his path-breaking efforts.

4. SOME COMMON ELEMENTS AND CONCERNS

When taken together, all the papers share a commonality of elements, concerns, and settings. Little–Su has focused on item nonresponse and Lepkowski on wave nonresponse, while Fay's focus has been on general nonresponse. All have not only written on SIPP situations where lots is already known, but also where there is much that has to be added. Among the basic themes that could be drawn from the papers are: a comparison of weighting and imputation; comparisons of alternative imputation (weighting) strategies; and a discussion of the importance of models.

The authors collectively cover different types of variables (discrete, continuous, and, in Lepkowski's case, limited variables); different types of analyses (cross-section, change, and cumulation analyses); and, among other things, concerns about missingness patterns (monotone or attrition patterns versus the much-tougher-to-handle nonmonotone patterns).

Contrasting with this richness is the somewhat narrow setting of a single household survey that the authors have chosen. What about establishment surveys? How do these methods fare in highly skewed populations? The adjustment strategies are somewhat incomplete in that there is too little tie-in with the survey design implications of the sensitivity that might be seen in alternative techniques. Bailar, in her paper in this volume, makes this point quite nicely in a general context.

One other minor overall quibble is that more might have been done to link editing and response problems with nonresponse adjustments (e.g.,

Fellegi & Holt, 1976). There really is something to the point of view
that a continuum exists (e.g., Platek & Gray, 1978, 1983).

Incidentally, lest these comments be misunderstood, let me stress
that I am not trying to imply that the authors might have changed
their current papers, only to indicate that some important elements of
the discussion have been (necessarily) omitted that may have relevance
in other settings.

5. RECOMMENDATIONS

As part of this discussion it may be appropriate to offer some general
recommendations to practitioners on nonresponse adjustments in surveys. These follow from the three papers but are, however, also a
reflection of my own practice and thus undoubtedly suffer from gaps
and possibly misplaced emphases. In any case, my comments, such as
they are, have been divided into three groupings to reflect a possible
time ordering.

For the *near term*, more work needs to be done to develop better ways
to feed back nonresponse adjustment difficulties into survey redesign
(e.g., adding variables, altering procedures). Additionally, in adjusting
for nonresponse there is a need for procedures to focus more on process
variables (e.g., reasons for nonresponse) where survey practitioners
have a comparative advantage (e.g., Palmer, 1967) and to leave work
on nonresponse models that focus on outcome variables to users, or
work in partnership with users. Clearly, as the authors imply, there is
a need for *routine* use of multiple weighting and multiple imputation approaches (to do sensitivity analyses and to be responsive to different
groups of users).

For the *middle term*, survey processing systems should be altered to
bring forward in a usable way to the "adjuster" (whether producer or
user) more of the knowledge already available about the measurement
process (e.g., as noted above, the reasons cited by interviewers for nonresponse). After all, some forms of nonresponse could be MAR for certain analyses but *not* others. Tied in with a better use of existing
knowledge about the measurement process is the need to increase the
knowledge available, e.g., conduct surveys of interviewers and others in
the system, to learn more about reasons for nonresponse (Bailar et al.,
1977). A natural follow-up is to use this new knowledge not only to
make better nonresponse adjustments but, more important, to find
ways to reduce the nonresponse, i.e., improve the "system." As
Deming would say, 80% or more of the quality problems are in the system (Deming, 1986).

For the *longer term*, as I have advocated before, we need to build better computerized survey documentation, to help bridge user–producer language problems to a greater extent (David, 1985; Scheuren, 1978). It is also evident that the really successful panel surveys are the ones that have found a way to keep good people working on the survey for long periods, and in a way that allows such individuals to grow progressively more knowledgeable, even wise. There is no substitute for this human capital at any time; however, one technology that can support such people is to employ computer tools, like expert systems, that allow researchers to organize the data that are collected in the course of the panel in such a way that they can progressively improve their ability to adjust for various forms of missingness. Such methods might eventually make it affordable to routinely handle important nonresponse problems on a case-by-case, variable-by-variable, analysis-by-analysis basis (e.g., Greenberg & Surdi, 1984).

6. CONCLUSION

Unquestionably, as the authors make amply clear, there is a need for a principled but pragmatic approach to nonresponse in panel surveys. The days of ad hoc methods, unsupported by theory, are long over. Coupled with the appeal for principled, pragmatic methods is the need to find a broader set of principles, almost certainly including the causal modeling advocated by Fay.

REFERENCES

Bailar, B. A., Bailey, L., & Stevens, J. (1977), "Measures of Interviewer Bias and Variance," *Journal of Marketing Research*, 14, 337–343.

David, M. (1985), "Designing a Data Center for SIPP: An Observatory for the Social Sciences," *Proceedings of the Social Statistics Section*, American Statistical Association, 357–362.

Deming, W. E. (1986), *Out of the Crisis*, Massachusetts Institute of Technology, Center for Advanced Engineering Study.

Fay, R. (1986), "Causal Models for Patterns of Nonresponse," *Journal of the American Statistical Association*, 81, 354–365.

Fellegi, I. P., & Holt, D. (1976), "A Systematic Approach to Automated Edit and Imputation," *Journal of the American Statistical Association*, 71, 17–35.

Greenberg, B., & Surdi, R. (1984), "A Flexible and Interactive Edit and Imputation System for Ratio Edits," *Proceedings of the Section on Survey Research Methods*, American Statistical Association, 421–426.

Palmer, S. (1967), "On the Character and Influence of Nonresponse in the Current Population Survey," *Proceedings of the Social Statistics Section*, American Statistical Association, 73–81.

Platek, R., & Gray, G. B. (1978), "Nonresponse and Imputation," *Survey Methodology*, 4(2), 144–177.

Platek, R., & Gray, G. B. (1983), "Imputation Methodology: Total Survey Error," in *Incomplete Data in Sample Surveys* (W. G. Madow, I. Olkin, & D. B. Rubin, eds.), Academic Press. New York, Vol. 2, Part V, pp. 249–333.

Scheuren, F. (1978), "Discussion of 'Income Data Collection and Processing for the March Income Supplement to the Current Population Survey' by J. Coder," in *The Survey of Income and Program Participation: Proceedings of the Workshop on Data Processing* (D. Kasprzyk, ed.), U.S. Department of Health, Education and Welfare, pp. II-23-II-32.

Scheuren, F. (1981), "Methods of Estimation for the 1973 Exact Match Study," in *Methods of Estimation for the 1973 Exact Match Study* (F. Scheuren, H. Lock Oh, L. Vogel, R. Yuskavage, B. Kilss, & L DelBene) (Report No. 10 in the series Studies from Interagency Data Linkages), Social Security Administration, pp. 1–122.

Comparisons of Alternative Approaches to the Estimation of Simple Causal Models from Panel Data

Willard L. Rodgers

1. INTRODUCTION

The observed relationship between any pair of variables, call them X and Y, can be interpreted in terms of any of several causal mechanisms. The most straightforward possibilities include: (1) X causes Y; (2) Y causes X; or (3) the observed relationship is spurious, induced by a third variable, call it Z, which causes variation in both X and Y. The third of these possible explanations can be tested if a measure of the Z variable is available. In practice, however, we can never be sure that we have measures of all such Z variables—that is, that we can control for all sources of spurious covariation.

Panel data—i.e., repeated measurements on the same respondents— have often been regarded as the social scientist's answer to the experimental techniques available to natural and physical scientists. Unfortunately, panel data are not a panacea, and it has come to be recognized that the causal interpretation of relationships observed with panel data depends just as surely as in the case of cross-sectional data on the imposition of a causal model. For example, in the 1950s and 1960s social scientists commonly thought that panel data would permit empirical methods to distinguish between the first two possibilities listed in the first paragraph as possible explanations of the relationship between two variables: does X cause Y, or does Y cause X? Beginning with the analysis of turnover tables by Lazarsfeld (1948) and then extended to the analysis of continuous variables through the cross-lagged panel correlation technique developed by Campbell (1963) and by Pelz and Andrews (1964), the notion was that the relative strengths of two relationships—that between X at time 1 and Y at time 2 vs. that between Y at time 1 and X at time 2—are sufficient to establish the causal path. The fallacy of this notion was pointed out by Duncan (1969; see also Heise, 1970). In order for such an inference to be drawn, strong assumptions must be made, in particular about the relative stabilities over time of X and Y and about the lags in the effects of these variables on one another.

The cross-lagged panel correlation approach has more recently been popular as a means of testing, or even controlling, for spuriousness (Duncan, 1972; Kenny, 1973). Kessler and Greenberg (1981) review this approach, and conclude (p. 72) that it is only "a very limited test of a narrowly specified null hypothesis without any serious implications for parameter estimation in structural models" and that it is inappropriate as a method to control for spuriousness.

These past misuses of panel data for causal analysis should not be taken to mean that panel data are of no value for causal analysis. To the contrary, although panel data do not obviate the need for specification of a causal model and for the imposition of untestable assumptions implied by such a model, under certain circumstances they do offer important advantages over cross-sectional data in the estimation of the parameters of such a causal model. The objective of this paper is to present some of the conditions that affect the relative effectiveness of estimates from panel data relative to those from cross-sectional data.

Section 2 provides a brief overview of alternative approaches to the estimation of causal models from panel data. In subsequent sections, differences in the accuracy of parameter estimates provided by two of those alternatives are quantified, using both algebraic derivations and simulations. Particular attention is paid to the magnitude of sampling errors in estimates relative to two potential sources of bias in those estimates: misspecification of the causal model and errors in indicators of the conceptual variables. Consideration is limited to simple static models, in general involving just two or three variables at a time, and only linear and additive effects. Despite these limitations, some conclusions can be drawn with respect to the relative strengths and weaknesses of the alternative approaches to the analysis of panel data.

2. ALTERNATIVE APPROACHES TO THE ANALYSIS OF PANEL DATA

One distinction that can be made with respect to alternative causal models is whether or not a noticeable lag is assumed to intervene between cause and effect. More concretely, with respect to panel data collected with a constant interval of T units between successive waves of data collection, models can be distinguished according to whether the causal lag for the effect, say, of variable X on variable Y is assumed to be closer to 0 or to some multiple of T time units. Panel data are all but essential for the estimation of a model that specifies a sizable causal lag (unless one is willing to use retrospective data with respect to the lagged causal variables, a hazardous undertaking because of the likely presence of serious measurement error problems).

Even if it is assumed that the causal lags in a particular model are all (approximately) zero, panel data offer an analysis strategy that is not available with cross-sectional data: the analysis of change scores, rather than analysis of level scores. While frequently used in economic analyses, change scores have a rather unsavory reputation in other social sciences, primarily because of concern about sampling error. Unexplained variation in indicators of conceptual variables often dominates the variation that is explained by a causal model. One consequence is that the sampling variability in estimates of causal effects tends to be large, and this problem is generally compounded when those estimates are based on *changes* in the indicator variables. On the other side of the balance, variation in the endogenous variable attributable to unmeasured, but stable, characteristics of the respondents is eliminated when change scores are analyzed, thereby reducing the possibility that estimates of causal parameters are biased due to model misspecification; it is this virtue that has made change scores attractive to econometricians.

Another analysis strategy that has been so frequently advocated and used with panel data in the social sciences as to have become the standard (e.g., Harris, 1963; Cronbach & Furby, 1969; Kessler & Greenberg, 1981) is to include values of both the X and the Y variables from previous times as predictors to current levels of Y:

$$Y_{it} = \alpha + \beta_1 X_{i(t-T)} + \beta_2 X_{it} + \gamma Y_{i(t-T)} + \epsilon_{it} .$$

It should be emphasized that this is not merely a different method for estimating the coefficients in the model described later by expression (1), but a different model in that it introduces a prior measurement of the Y variable as a predictor of its current level. Various justifications for this re-specification of the causal model have been offered, including both concerns about measurement error and arguments based on the alleged necessity to take account of the correlation between the initial value of a variable and change on that variable. These justifications have been critically reviewed by Liker et al. (1985), who found them less than compelling for most types of data.

The focus of this paper is on the accuracy of estimates of parameters of static causal models of the following general form:

$$\mathbf{Y}_t = \alpha + \mathbf{X}_t \beta + \mathbf{Z} \gamma + \epsilon_t \tag{1}$$

where $\mathbf{Y}_t$ is an $N \times 1$ vector of values on the dependent variable at time t for a population of N cases; $\mathbf{X}_t$ is an $N \times k$ matrix of values on a set of k predictor variables; $\mathbf{Z}$ is an $N \times p$ matrix of variables; α, β, and γ are the parameters to be estimated; and ϵ_t is an $N \times 1$ vector of residuals (i.e., the part of $\mathbf{Y}_t$ that is independent of $\mathbf{X}_t$). It is assumed throughout this paper that the $\mathbf{Z}$ values for each individual are constant over some period of time; that the α, β, and γ parameters are time-invariant; that the variances and covariances of the exogenous variables ($\sigma^2_{x_k}$, $\sigma_{x_k x_{k'}}$, for all k and k') and the variance of the stochastic term (σ^2_ϵ) are also time-invariant; and that the stochastic term at time t is independent of the X terms at the same and every other time and of the stochastic terms at every other time (i.e., $\sigma_{x_{kt}\epsilon_t} = \sigma_{x_{kt}\epsilon_{t'}} = \sigma_{\epsilon_t\epsilon_{t'}} = 0$ for $t \neq t'$ and for all k).

To estimate these parameters, it is assumed that indicators of the Y and X variables have been measured for a simple random sample of n cases, and that the observed values, Y_t^* and X_t^*, are related to Y_t and X_t as follows:

$$Y_t^* = Y_t + u_t \, , \tag{2}$$

and

$$X_t^* = X_t + v_t \tag{3}$$

The Z variables are assumed to be unmeasured. The models that are actually estimated, then, are of the following general form:

$$Y_t^* = \alpha^* + X_t^*\beta^* + \epsilon_t^* \tag{4}$$

Typically, in the estimation of a model of the form given by expression (4), strong assumptions are made: for example, that measurement errors (u and v) are independent of true values (X and Y) and are also mutually independent, and that there are no Z variables. As seen in subsequent sections, violations of these assumptions have differential effects on the accuracy of parameter estimates from change scores and from cross-sectional (level) scores.

3. MODEL 1: ONE X VARIABLE, NO MEASUREMENT ERROR

Consider first a bivariate model that is properly specified (i.e., no unmeasured Z variables) and with no measurement error for either Y or X. This simplification of the general model (1) can be written for a particular time t and for individual i as:

$$y_{it} = \alpha + \beta x_{it} + \epsilon_{it} \, , \tag{5}$$

The objective, it is supposed, is to estimate the coefficient β, which represents the causal effect of X on Y. The data from which this coefficient is to be estimated are obtained from two waves of panel data, but initially the panel dimension of the data is ignored and the parameter is estimated from cross-sectional data. The ordinary least squares estimate of β is given by the following expression:

$$\hat{b} = s_{yx}/s_x^2 \, , \tag{6}$$

where s_{yx} is the sample covariance of the observed y and x values and s_x^2 is the sample variance of the x values. If the model is properly specified by expression (5), $\hat{b}$ is an unbiased estimator of β, and its variance is:

$$\sigma^2_{\hat{b}} = (\sigma^2_y - \beta^2\sigma^2_x)/n\sigma^2_x \tag{7}$$

Could the estimate of β be improved by taking into account the panel design? Specifically, consider change scores obtained from panel data by subtracting observations at one time (say $t = 1$) from those at another time ($t = 2$). The model can then be written as follows, directly from expression (5):

$$\Delta y_i = \beta\Delta x_i + \Delta\epsilon_i \tag{8}$$

where $\Delta y_i = y_{i2} - y_{i1}$ and Δx_i and $\Delta\epsilon_i$ are defined analogously. The least squares estimator of β from change data is:

$$\tilde{b} = s_{\Delta y \Delta x}/s^2_{\Delta x} , \tag{9}$$

where $s_{\Delta y \Delta x}$ is the sample covariance of changes on the Y and X variables and $s^2_{\Delta x}$ is the sample variance of changes on the X variable. Like the least squares estimator from cross-sectional data, the least squares estimator from these change data is unbiased if the model is properly specified by expression (5), and its variance is:

$$\sigma^2_{\tilde{b}} = (\sigma^2_{\Delta y} - \beta^2\sigma^2_{\Delta x})/n\sigma^2_{\Delta x} = \sigma^2_{\hat{b}}/(1 - \rho_{x_1 x_2}) . \tag{10}$$

where $\rho_{x_1 x_2}$ is the population value of the correlation between x_{i1} and x_{i2}.

Since $\rho_{x_1 x_2}$ is generally positive, expression (10) tells us that the causal influence of X on Y is more accurately (i.e., with less variance) estimated from cross-sectional data than from change scores if the model is correctly specified by expression (5). Indeed, the advantage of estimates from cross-sectional data is even greater than might first appear from inspection of expression (10). If two waves of panel data are available, the regression coefficient β can be estimated from either change scores or by treating the panel data as two cross sections. In the latter case, the total sample size is on the order of twice as large as that for change scores.[1] The relative advantage of cross-sectional data diminishes as the number of waves of data collection increases; for example, with three waves of data, the effective sample size for static score analysis would be on the order of $3n$ (i.e., three times the size of the sample at each data collection), while the effective sample for change score analysis across one time unit would be approximately $2n$, or 2/3 that for static score analysis.

It is important to remember that change scores do not provide evidence about the direction of causality. As pointed out in the Introduction, estimates of the parameters of static models based on change scores, as in expression (9), rest just as completely as do estimates based on cross-sectional scores on a strong assumption about causal direction, in this case that X causes Y.

4. MODEL 2: ONE X VARIABLE AND ONE Z VARIABLE, NO MEASUREMENT ERROR

Model 2 differs from Model 1 only by inclusion of a second predictor:

$$y_{it} = \alpha + \beta x_{it} + \gamma z_i + \epsilon_{it} . \tag{11}$$

The same assumptions concerning time invariance of the coefficients and variances are made here as for Model 1; and it is also assumed that the Z variable does not change over time—that is, it refers to a stable characteristic of the units of analysis. To complete the specification of Model 2, let the correlation of the X and Z variables be denoted by ρ_{xz}, which is assumed to be independent of t.

Now suppose that again our objective is to estimate the causal effect of X with respect to Y, and that we have panel data which include observations on X and Y for a set of n units, but that Z is unmeasured. The variance of the estimate of β from cross-sectional data has the following form:

$$\sigma^2_{\hat{b}} = (\sigma^2_y - \beta^2 \sigma^2_x - 2\beta\gamma\sigma_{xz})/n\sigma^2_x . \tag{12}$$

This differs from the variance of $\hat{b}$ if Model 1 applies, as given by expression (7), because of the model misspecification. Moreover, unlike the case for Model 1, this estimator of β is generally *not* unbiased. The expected value of this estimator is given by:

$$E(\hat{b}) = \beta + \gamma\rho_{xz}\sigma_z/\sigma_x ; \tag{13}$$

that is, the bias is zero only if Z has no causal influence on Y ($\gamma = 0$) *or* if X and Z are uncorrelated ($\rho_{xz} = 0$).

If change scores are used instead of static scores, the Z variable drops out of the equation for ΔY and the estimator, $\tilde{b}$, given by expression (9) is unbiased whether the true model is 1 or 2. Its variance is:

$$\sigma^2_{\tilde{b}} = (\sigma^2_y - \beta^2 \sigma^2_x - \gamma^2 \sigma^2_z - 2\beta\gamma\sigma_{xz})/[n\sigma^2_x(1 - \rho_{x_1 x_2})] . \tag{14}$$

We are confronted with a choice between an estimator ($\hat{b}$) that may be biased and another ($\tilde{b}$) that is unbiased but which generally has greater variance. The mean square errors (defined as the sum of the variance of the estimator and the square of its bias) of these two estimators permit their overall comparison and are given by the following expressions:

$$MSE(\hat{b}) = (\sigma^2_y - \beta^2 \sigma^2_x - 2\beta\gamma\sigma_{xz})/2n\sigma^2_x + (\gamma\rho_{xz}\sigma_z/\sigma_x)^2 \tag{15}$$

and

$$MSE(\tilde{b}) = \sigma^2_{\tilde{b}} = (\sigma^2_y - \beta^2 \sigma^2_x - \gamma^2 \sigma^2_z - 2\beta\gamma\sigma_{xz})/[n\sigma^2_x(1 - \rho_{x_1 x_2})], \tag{16}$$

if the effective size of the sample available for making the estimate of $\hat{b}$

is $2n$, twice as large as the sample available for estimating $\tilde{b}$. Given these conditions, $\hat{b}$ is a better estimator of β than is $\tilde{b}$ if the following condition holds:

$$n < \sigma_x^2[(\sigma_y^2 - \beta^2\sigma_x^2 - 2\beta\gamma\sigma_{xz})(1 + \rho_{x_1x_2})$$
$$- 2\gamma^2\sigma_z^2]/[2\gamma^2\sigma_{xz}^2(1 - \rho_{x_1x_2})]. \tag{17}$$

If, for greater intuitive clarity, the X, Y, and Z variables are all standardized, this inequality can be rewritten as:

$$n < [(1 - \beta^2 - 2\beta\gamma\rho_{xz})(1 + \rho_{x_1x_2}) - 2\gamma^2]/[2\gamma^2\rho_{xz}^2(1 - \rho_{x_1x_2})]. \tag{18}$$

From expression (18) it is seen that level score estimates are more accurate than change score estimates if any of the following conditions is true:

$$\gamma = 0, \quad \rho_{xz} = 0, \quad \text{or } \rho_{x_1x_2} = 1$$

—that is, if the omitted Z variable has no effect on Y, or if Z is uncorrelated with X, or if the X variable is completely stable across data collections. If none of these conditions is true, then as the sample size increases the change score estimates eventually become more accurate.[2] The minimum sample size at which the change score estimate is more accurate than the level score estimate depends on each of the parameters. It is generally smaller (and therefore change score estimates are more likely to be *better* than level score estimates) for *larger* absolute values of β and γ (the standardized regression coefficients for X and Z, respectively, on Y) and of ρ_{xz} (the correlation of X and Z); and for *smaller* algebraic values of $\rho_{x_1x_2}$ (the stability of X over time).

The use of change scores does not free us from untestable assumptions about unmeasured variables, but it does make these assumptions somewhat weaker: that each such variable (1) is unrelated to the dependent variable, or (2) is unrelated to the independent variable(s), or (3) does not change over the time between measurements. Even with apparently trivial levels of model misspecification (by exclusion of stable predictors), more accurate estimates are provided by change scores from moderately large samples than are provided by cross-sectional scores.

5. MEASUREMENT ERROR

In Models 1 and 2 it was assumed that the X and Y variables were measured without error. This is not a reasonable assumption for most social science data, so the consequences of measurement error on parameter estimates from cross-sectional and change data are considered in the remaining models.

The possible implications of measurement errors on estimates of all types of parameters from panel data are immense, matched only by our general ignorance about the nature and pervasiveness of such errors. Presser and Traugott (1983) report a study of response errors with respect to voting in each of three elections by respondents in a 1972–1974–1976 election panel study, but apart from that study and some others now in progress very little has been done to assess measurement errors in panel survey data.

To facilitate discussion of the implications of measurement errors in panel data, it is useful to distinguish between measurement errors that are random (i.e., errors that are distributed independently of any other variable of interest in the context of a particular analysis) from those that are systematic (i.e., those that are related to other pertinent variables, including the autocorrelation over time in measures of a single variable). It is well-known that while random errors do not introduce systematic biases into many statistics, including measures of central tendency and the covariances among different items, random errors in measures of the independent variables in causal models *do* introduce bias into correlations and estimates of causal effects. Random errors may be even more troublesome with respect to the analysis of panel data than is true for cross-sectional data. This follows from the fact that the focus of analysis with panel data is often on changes and trends at the level of individuals, and with many types of survey questions it may be found that true change during the interval between successive data collections is small compared to random error variance in the measure of change.

The potential consequences of systematic errors with respect to estimates from panel data are even more disturbing than those of random errors. It is also true, however, that in certain cases, if systematic errors are stable over time, they may actually improve the consistency of some estimators.

5.1. The Effects of Measurement Error on Estimates of Net Change

Before taking up the effects of measurement error on estimates of the parameters of causal models, it is useful to consider two more basic statistics: the net (aggregate) change in a variable from one time to another within a population, and the stability of a concept at the individual level across time.

Net change on a variable can be estimated either from two independent cross sections of the target population or from a panel study that collects data from a single sample at both times. In both cases, assume that the response of each individual i at time t with respect to the X variable includes measurement error v_{it}, so that the observed score is given by:

$$x_{it}^* = x_{it} + v_{it}. \tag{19}$$

Change from one time (say, $t = 1$) to another (say, $t = 2$) is estimated by the difference in the mean observed scores at times 1 and 2: $\Delta \bar{x}^* = \bar{x}_2^* - \bar{x}_1^*$, the expected value of which is:

$$E(\Delta \bar{x}^*) = E(\bar{x}_2 - \bar{x}_1) + E(\bar{v}_2 - \bar{v}_1). \tag{20}$$

(In this paper, which focuses on measurement error in the data from respondents, complications due to nonresponse, including panel attrition, are ignored.) If we assume that variances and covariances among these variables are equal at times 1 and 2, the variance of the estimated change is given by:[3]

$$\sigma_{\Delta \bar{x}^*}^2 = (2\sigma_{x_t}^2 - 2\sigma_{x1x2} + 2\sigma_{v_t}^2 - 2\sigma_{v1v2}$$
$$+ 2\sigma_{x_t v_t} - \sigma_{x2v1} - \sigma_{x1v2})/n. \tag{21}$$

With data from independent cross-sectional samples, the cross-time covariances in expression (21) have expected values of zero, but in general the covariance of the true scores, σ_{x1x2}, will be positive if the data are from a panel design, so that changes are estimated with greater precision (lower variances) if a panel design is used rather than successive cross sections. Moreover, positively correlated measurement errors ($\sigma_{v1v2} > 0$) also reduce the variance of the estimate of change. Presser and Traugott (1983), in the panel study of measurement error mentioned earlier, found that errors in reporting voting behavior in three elections were quite stable across a four-year period, so what little evidence is available suggests that measurement error may indeed enhance the advantage of panel designs over cross-sectional designs in the estimation of change.

5.2. The Effects of Measurement Error on Estimates of Stability

Data from successive cross sections can be used to estimate net change in a variable from one time to another, although, as shown in the previous section, generally not as accurately as is possible with panel data. Cross-sectional data, however, can tell us nothing about gross change—change at the individual level. To examine mobility at the individual level (or the converse, stability), data with temporal dimension (i.e., retrospective or panel data) are required. Measurement error, however, may bias estimates of mobility.

To show this, suppose again that the observed score for individual i is given by expression (19). Then it is straightforward to show that the stability of the observed scores from $t = 1$ to $t = 2$ is given by:

$$\rho_{x_1^* x_2^*} = (\rho_{x1x2}\sigma_x^2 + \rho_{v1v2}\sigma_v^2 + \rho_{x1v2}\sigma_x\sigma_v + \rho_{x2v1}\sigma_x\sigma_v)/\sigma_{x^*}^2 \tag{22}$$

(assuming again that the variances are the same at both times). If the correlations of measurement errors with the true scores are zero, the last two terms in the numerator of expression (22) drop out, yielding an expression that is equivalent to one given by Ashenfelter et al. [1986, p. 51, expression (20)]. As those authors note, this indicates that the stability in observed scores from time 1 to time 2 is a weighted average of the stability of the true scores and the stability of the measurement errors, with the weights equal to the ratios of valid and error variances, respectively, to the total variance of the measure. Under these assumptions, then, the estimated stability of X may be biased either upward or downward depending on the stability of the true scores relative to the stability of measurement error. An assumption that seems reasonable, but that lacks empirical support, is that the stability of the true scores is greater than the stability of the measurement errors; this assumption is made, for example, by Griliches and Hausman (1986) in their analysis of the effects of measurement errors on estimates from panel data. The possibility of correlations between measurement errors and true scores, which should not be overlooked, makes it even harder to predict the direction or size of the bias in estimated stability.

6. MODEL 3: ONE OR MORE X VARIABLES, MEASUREMENT ERROR IN THE DEPENDENT VARIABLE

6.1. Effects on Estimates from Cross-Sectional Scores

Measurement errors in the Y variable have consequences that are analogous to misspecification of Model 2 by omission of a relevant Z variable. Suppose that the true causal model is given by:

$$y_{it} = \alpha + \beta_1 x_{it} + \beta_2 w_{it} + \epsilon_{it} \tag{23}$$

but that the observed values of Y, y_{it}^*, contain errors as specified in expression (2). (The model includes just two exogenous variables, X and W, for ease of presentation, but the results are readily generalizable to multiple exogenous variables.) The expected value of the OLS estimator of β_1 is:

$$E(\hat{b}_1) = \beta_1 + (\rho_{xu} - \rho_{xw}\rho_{wu})\sigma_v/[(1 - \rho_{xw}^2)\sigma_x] \tag{24}$$

with an analogous expression for $E(\hat{b}_2)$, and where the ρ's are the population values of the correlations between the indicated variables and the σ's are standard deviations. If the measurement errors are independent of both predictor variables (i.e., $\rho_{xu} = \rho_{wu} = 0$), they do not introduce biases into estimates of regression parameters. If, however, the measurement errors are correlated with either of the predictor variables, biases may be introduced into estimates of both of the regression coefficients. Whether the estimates tend to over- or understate the true parameters depends on the particular configuration of covariances and variances in the above expression, and so cannot in general be taken into account without explicit information about the existence, direction, and magnitude of the covariances between the errors in the dependent variable and the values of all of the predictor variables. Rarely is such information available, however, and so it is generally simply assumed that these covariances are zero. Although such assumptions are nearly universal, Duncan and Hill (1985) provide evidence that it is *not* warranted in the case of a simple model of earnings determination. Specifically, a fairly sizable negative correlation between the measurement error of ln(earnings) and the true level of job tenure imparted a 30% bias to the regression coefficient for job tenure.

The mean square error in the estimator of β_1 from cross-sectional scores is:

$$MSE(\hat{b}_1) = \frac{\sigma_\epsilon^2 + \sigma_u^2}{2n\sigma_w^2(1 - \rho_{xw}^2)} + \frac{\sigma_u^2}{\sigma_x^2}\left(\frac{\rho_{xu} - \rho_{xw}\rho_{wu}}{1 - \rho_{xw}^2}\right)^2 . \tag{25}$$

6.2. Effects on Estimates from Change Scores

The estimate of β_1 in the above example as obtained from change scores has the following expected value:

$$E(\tilde{b}_1) = \beta_1 + \frac{\sigma_u}{\sigma_x}\left(\frac{(1 - \rho_{w_1w_2})\Omega_{xu} - \Omega_{xw}\Omega_{wu}}{(1 - \rho_{x_1x_2})(1 - \rho_{w_1w_2}) - \Omega_{xw}^2}\right) , \tag{26}$$

where $\Omega_{xu} = (\rho_{x_1u_1} + \rho_{x_2u_2} - \rho_{x_1u_2} - \rho_{x_2u_1})$, and Ω_{xw} and Ω_{wu} are defined in parallel fashion. From expression (26) and the analogous expression for $E(\tilde{b}_2)$ it can be seen that there is no bias in the estimate of either coefficient if (1) there is no measurement error (i.e., $\sigma_u^2 = 0$); or (2) the measurement errors are independent of *both* X and W (i.e., $\rho_{x_tu_{t'}} = \rho_{w_tu_{t'}} = 0$ for all t, t'); or (3) the measurement errors are perfectly autocorrelated across time (i.e., $\rho_{u_1u_2} = 1$, and therefore Ω_{xu} and Ω_{wu} both $= 0$). If none of these conditions holds, and if measurement error in Y is more strongly correlated with the true values of the X variable at the same time than at different times (i.e., $\Omega_{xu} > 0$), this introduces a positive bias into $\tilde{b}_1$ and a negative bias into $\tilde{b}_2$ (other things being equal).

The mean square error in the estimator of β_1 from change scores is:

$$MSE(\tilde{b}_1) = \left(\cfrac{\sigma_\epsilon^2 + \sigma_u^2(1 - \rho_{u_1 u_2})}{n\sigma_u^2(1 - \rho_{w_1 w_2}) - \cfrac{\Omega_{xw}^2}{4(1 - \rho_{x_1 x_2})}} \right)$$

$$+ \frac{\sigma_u^2}{\sigma_x^2} \left(\frac{(1 - \rho_{w_1 w_2})\Omega_{xu} - \Omega_{xw}\Omega_{wu}}{(1 - \rho_{x_1 x_2})(1 - \rho_{w_1 w_2}) - \Omega_{xw}^2} \right)^2 . \qquad (27)$$

7. MODEL 4: ONE INDEPENDENT VARIABLE, WITH MEASUREMENT ERROR

7.1. Effects on Estimates from Cross-Sectional Scores

Suppose that the true causal relationship between X and Y is given by expression (5), but that there are errors in the measurement, not of the Y variable but of the X variable, so that the observed values of X are related to the true values as in expression (3). Expression (5) can be rewritten in terms of the observed values, x_i^*, as follows:

$$y_i = \alpha + \beta x_i^* - \beta v_i + \epsilon_i . \qquad (28)$$

If the errors in the observed values of the X variable are independent of the true values of the X variable and of the stochastic term ϵ, and therefore of the Y variable as well, the limiting value of the OLS estimate of the regression coefficient, β, as the sample size increases is given by:

$$E(\hat{b}) = \beta \sigma_x^2 / (\sigma_x^2 + \sigma_v^2) . \qquad (29)$$

That is, the magnitude of the regression coefficient is underestimated in the ratio of the variance of the true scores to the variance of the observed scores on the X variable.

Expression (29) indicates the potentially devastating consequences of measurement error in the independent variable, even for the bivariate model considered here. Awareness of these consequences is reflected in the increasing emphasis placed on the specification of measurement models along with causal models and in the popularity of techniques, such as the LISREL computer program (Jöreskog & Sörbom, 1984), that allow simultaneous estimation of measurement and substantive models. To illustrate, consider the advantages of having two parallel ("tau equivalent," to use the terminology of psychometrics; see Novick & Lewis, 1967) indicators of the independent variable in a bivariate causal model. We rewrite expression (19) with an additional subscript, $j = 1, 2$, to differentiate between these indicators:

$$x_{ijt}^* = x_{it} + v_{ijt} \, . \tag{30}$$

We further assume, for the time being, that the measurement errors in these two indicators are independent of one another. Then an unbiased estimator of β is as follows:

$$\hat{B}_{jt} = s_{yx_{jt}^*}/s_{x_{1t}^* x_{2t}^*} \, . \tag{31}$$

7.2. Effects on Estimates from Change Scores

Now consider the effects of errors in the measurement of the X variable on estimates based on change scores from panel data. It is again assumed that the observed values of X are related to the true values as shown in expression (30). If the causal effect of X on Y is estimated by regressing change scores in Y on change scores in X^*, this estimate is biased:

$$E(\tilde{b}) = \beta \sigma_{\Delta x}^2/(\sigma_{\Delta x}^2 + \sigma_{\Delta v}^2) \, . \tag{32}$$

This can be rewritten, after some algebraic manipulations, as

$$E(\tilde{b}) = \beta \sigma_x^2/(\sigma_x^2 + \sigma_v^2 \psi) \, , \tag{33}$$

with $\psi = (1 - \rho_{v_1 v_2})/(1 - \rho_{x_1 x_2})$ and with $\rho_{x_1 x_2}$ being the correlation of the true scores on X at times 1 and 2, and $\rho_{v_1 v_2}$ the correlation of the measurement errors at these two times. ψ is never negative, so the expected value of $\tilde{b}$, like that of $\hat{b}$, is never greater in absolute value than β, given the stated assumptions. Note also that the bias in $\tilde{b}$ is less than that in $\hat{b}$ [i.e., $E(\hat{b}) < E(\tilde{b})$] only if $\psi < 1$, that is, only if true scores on the predictor variable are less stable than errors in the measurement of that variable.

As with cross-sectional scores, an unbiased estimate of β can be obtained if parallel measures of X are available at both waves of data collection:

$$\tilde{B}_j = s_{\Delta y \Delta x_j^*}/s_{\Delta x_1^* \Delta x_2^*} \, . \tag{34}$$

7.3. Comparison of Estimates from Level Scores
and from Change Scores

In the absence of tractable algebraic expressions for the mean square errors in these two types of corrected estimates, a series of simulations were performed to compare their accuracy. The following procedures were used in generating the simulated data. First, it was assumed that two waves of data have been collected, and that two fallible indicators of the X variable are available at each wave. These indicators, labeled x_{ijt}^* for case i, indicator j, and wave t, are related to the true values as shown by expression (30). Under these conditions, there are four estimators, of the form given by expression (31), which take into account measurement error in X (i.e., $j = 1, 2, t = 1, 2$), so in the simulations the average of these four estimates is taken. Similarly, there are two estimators based on change scores of the form given by expression (34) (i.e., $j = 1, 2$), and the average of the two estimates from each set of simulated data is taken. By taking the averages of the various types of estimates, the data are maximally exploited and at the same time the lack of independence of the different estimates is taken into account.

In the simulations, the following parameters were varied:

(1) The reliability of the indicators of X, where reliability is defined as the correlation between two parallel indicators of the same concept, was varied across the values 0.5, 0.7, and 0.9—that is, from a rather low to a rather high level of reliability. In the present context, the reliability of the indicators at time t is given by:

$$\rho_{x_{1t}^* x_{2t}^*} = \sigma_{x_t}^2 / (\sigma_{x_t}^2 + \sigma_{v_{jt}}^2) \,. \tag{35}$$

A similar expression could be written for the reliability of the measure of change in X.

(2) The stabilities of the true X values and of the measurement errors in the indicators of X were given values of 0, 0.5, or 0.8.

(3) The number of cases was 250, 500, 1,000, or 2,000.

One hundred simulations were done for each of 60 conditions. These and all other simulations reported in this paper were done on a microcomputer using the Gauss programming language (Edlefsen & Jones, 1986).

The results of the simulations are summarized in Table 1, which shows the root mean square error of the (corrected) estimates across the 100 simulations for each condition. (There is no systematic bias under any of the conditions simulated.) The change score estimates were less accurate, on the average, than the estimates based on cross-sectional data, under *all* of the conditions simulated. The accuracy of the cross-sectional estimates improves somewhat as the reliability of the indicators of the X variable improves, and also, of course, improves as the sample size increases, but does not depend on the stability of the true values of X or of the errors in measurement for that variable. The accuracy of the change-score estimates depends on the values of all of these parameters used in generating the data. The accuracy increases as the reliability of the indicators of X, as assessed by the correlation of two indicators within the same wave, increases. The accuracy is worse for more stable true values of the X variable, and is better for more stable measurement errors in the observed values of the indicators of the X variable. And again, the accuracy improves with sample size.

The primary observation to be made from these simulations, however, is that the cross-sectional estimates are more accurate than

TABLE 1. ROOT MEAN SQUARE ERROR IN ESTIMATES OF β

Stability of:		Cross-Sectional Scores			Change Scores		
		Reliability of Indicator			Reliability of Indicator		
True Scores	Measurement Errors	0.5	0.7	0.9	0.5	0.7	0.9
Sample size: 250							
.5	.5	.060	.051	.050	.138	.123	.093
.8	.5	.058	.046	.043	.294	.226	.154
.8	.0	.053	.046	.046	.436	.212	.153
.5	.8	.054	.055	.043	.137	.098	.087
.0	.8	.054	.049	.045	.084	.066	.058
Sample size: 500							
.5	.5	.039	.030	.032	.099	.063	.062
.8	.5	.040	.032	.030	.214	.142	.123
.8	.0	.039	.035	.032	.327	.146	.114
.5	.8	.039	.037	.032	.077	.084	.067
.0	.8	.040	.034	.028	.051	.051	.045
Sample size: 1,000							
.5	.5	.029	.025	.021	.069	.049	.042
.8	.5	.029	.025	.020	.150	.101	.071
.8	.0	.028	.027	.021	.159	.111	.071
.5	.8	.031	.023	.019	.066	.054	.042
.0	.8	.023	.023	.023	.033	.032	.031
Sample size: 2,000							
.5	.5	.019	.017	.015	.051	.036	.032
.8	.5	.017	.016	.016	.075	.074	.058
.8	.0	.018	.017	.017	.108	.068	.061
.5	.8	.019	:018	.015	.043	.043	.033
.0	.8	.021	.017	:018	.032	.024	.024

the change-score estimates across a wide range of conditions. It is only when the possibility of model misspecification is taken into account that change scores display their strength.

8. MODEL 5: ONE *X* VARIABLE AND ONE *Z* VARIABLE, WITH MEASUREMENT ERROR ON THE *X* VARIABLE

This model combines the features of Models 2 and 4. The structural model relating Y to the predictor variables X and Z is again given by expression (11), and it is again assumed that Z is unmeasured but stable. The measurement model for indicators of X is again given by expression (30). The expected value of the ordinary estimate of the β parameter, based on cross-sectional scores, is given by the following expression:

$$E(\hat{b}) = (\beta\sigma_x^2 + \gamma\rho_{xz}\sigma_x\sigma_z)/(\sigma_x^2 + \sigma_v^2) , \tag{36}$$

while that for change scores is again given by expression (32).

8.1. Comparison of Estimates from Level Scores and from Change Scores

Simulated data were again used to evaluate the relative accuracy of these two approaches to estimation of β. The data were generated in the following way. First, the values of the Z variable were generated from a pseudo-random normal distribution, $N(0, 1)$. Then the values of X_1 and X_2 (corresponding to waves 1 and 2 of a panel study) were generated:

$$x_{ti} = \rho z_i + \tau r_{ti} + v_{ti} , \tag{37}$$

where the values of r_{ti} and v_{ti} were generated from $N(0, 1)$ and $N(0, 1 - \rho^2 - \tau^2)$ distributions, respectively, independent of one another and of the other variables. Two indicators of the X variable at each wave were then constructed as follows:

$$x_{jti}^* = x_{ti} + v_{jti} , \tag{38}$$

with the v_{jti} generated from $N(0, \lambda^2)$ distributions, independent of one another and of every other variable. Finally, the values of Y were generated:

$$y_{ti} = \beta x_{ti} + \gamma z_i + \epsilon_{ti} , \tag{39}$$

with ϵ_{ti} generated from $N(0, \sigma_\epsilon^2)$ distributions, independent of one another and of every other variable. Note that the x values are constructed to have a population variance of 1, and so that the stability of the x values across the two waves is $\rho^2 + \tau^2$. The stability of the measurement errors in the x_{jti}^* values is zero, and the reliability of x_{jt}^* as an indicator of x_t is given by:

$$\rho_{x_{1t}^* x_{2t}^*} = 1/(1 + \lambda^2) . \tag{40}$$

As with the preceding model, for each simulation the estimates from cross-sectional data were averaged across indicators and waves ($j = 1$, 2; $t = 1$, 2), and the estimates from change scores were averaged across indicators ($j = 1$, 2).

To illustrate the findings from the simulations, consider Figure 1, which displays the root mean square error for estimates from cross-sectional and change scores (corrected, in each case, for unreliability of the indicators of X; that is, the errors in $\hat{B}$ and $\tilde{B}$ are shown, rather

FIGURE 1. ROOT MEAN SQUARE ERROR FOR ESTIMATES
BASED ON CROSS-SECTIONAL AND CHANGE SCORES

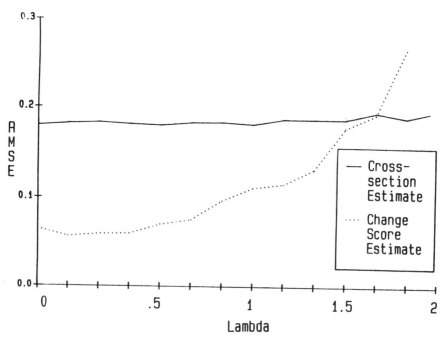

than the errors in $\hat{b}$ and $\tilde{b}$). For all of the simulations represented in this figure, $\rho = 0.6$ and $\tau = 0.6$, so that the stability of X across the two waves is 0.72; $\beta = 0.3$, and $\gamma = 0.3$. The number of cases generated per simulation was 1,000, and 100 simulations were done for each of ten values of λ. The errors in the estimates from cross-sectional data are dominated by the bias due to the unmeasured Z variable ($\rho\gamma = 0.6 \times 0.3 = 0.18$); even with highly unreliable indicators of X, the root mean square error does not climb markedly higher than the bias term. (Note that high values of λ correspond to low reliabilities; when $\lambda = 2$, the reliability of the indicators is only 0.2.) The errors in the estimates from the change scores, on the other hand, are due entirely to sampling variability, since there is no bias in these estimates, and the accuracy of these estimates deteriorates as the value of λ increases (i.e., as the reliability of the indicators falls). The crossover point, where the two estimates are about equally accurate, occurs for $\lambda \simeq 1.6$ (i.e., for reliability $\simeq .28$). That is, given the fixed parameters, change scores provide a more accurate estimate of the effect of X on Y than do cross-sectional scores unless the indicators of X are highly unreliable. This is true despite the fact that the estimates use all available information: the cross-sectional estimate from each simulation is actually the average of four estimates, while the change-score estimate is the average of just two estimates. See Table 2 for a more detailed display of the data summarized in Figure 1.[4]

The crossover point at which cross-sectional estimates tend to provide more accurate estimates of β than change-score estimates depends on each of the parameters that were held fixed in Figure 1: the values of ρ, τ, and γ, and the number of cases per simulation. Figure 2 provides a summary of a large number of simulations in which each of the values of two of these parameters—ρ and sample size—were varied. As in Figure 1, β and γ are fixed at 0.3, and τ is fixed at 0.6, for all of these simulations. The abscissa in this figure shows the values of ρ, and the ordinate shows values of λ. The darkened regions represent values of ρ and λ for which change scores provide more accurate estimates of β than do cross-sectional scores, with the darkening shades of gray representing decreasing numbers of cases per simulation.

The simulations summarized in Figure 2 again show that, in the absence of bias due to model misspecification, cross-sectional scores provide more accurate estimates than do change scores. On the other hand, even rather small biases due to model misspecification quickly tip the balance toward change scores. This is especially true with large samples, since with smaller samples sampling variance can more readily drown out the inaccuracy due to misspecification. Also important is the amount of measurement error in the indicators of the independent variable. The smaller the proportion of measurement error—that is, the higher the validity of the indicators—the more accurate are the estimates from both cross-sectional and change scores, but also the more likely it is that the change scores provide relatively more accurate estimates than do cross-sectional scores. Finally, the U-shaped relationship between the value of the rho parameter and the relative accuracy of the two types of estimates arises because of the importance of the rho parameter with respect to the stability of the independent variable (i.e., stability = $\rho^2 + \tau^2$). If the X variable has very little stability, there is necessarily little relationship between X and the omitted Z variable, and thus little bias due to misspecification; while if the X variable is highly stable, most of the change in X reflects only measurement error rather than true change.

8.2. Consequences of Covariances Involving Measurement Errors

All the models we have estimated to this point have made strong assumptions about the independence of the measurement errors in the indicators of the X variable relative to one another and to the X and Y variables. There is increasing empirical evidence (Duncan & Hill, 1985; Duncan & Mathiowetz, 1985; Rodgers & Herzog, 1987) that assumptions of this type are often violated in actual survey data. There are many possible types of nonindependence involving the measurement er-

CROSS-SECTIONAL AND CHANGE PARAMETERS

TABLE 2. AVERAGE ESTIMATE OF REGRESSION COEFFICIENT, SAMPLING VARIABILITY, AND ROOT MEAN SQUARE ERROR FOR ESTIMATES BASED ON CROSS-SECTIONAL AND CHANGE SCORES

	Cross-Sectional Estimates		Change Score Estimates	
	Uncorrected	Corrected	Uncorrected	Corrected
Lambda: 0.00				
Average estimate	.478	.478	.301	.301
Sampling error	.026	.026	.055	.055
Root mean square error	.180	.180	.055	.055
Lambda: 0.20				
Average estimate	.463	.481	.267	.305
Sampling error	.022	.022	.050	.057
Root mean square error	.164	.183	.060	.057
Lambda: 0.30				
Average estimate	.442	.482	.229	.304
Sampling error	.024	.026	.045	.060
Root mean square error	.144	.184	.084	.060
Lambda: 0.40				
Average estimate	.414	.481	.189	.297
Sampling error	.018	.021	.038	.061
Root mean square error	.115	.182	.117	.061
Lambda: 0.50				
Average estimate	.382	.478	.157	.299
Sampling error	.024	.029	.036	.071
Root mean square error	.086	.181	.147	.071
Lambda: 0.60				
Average estimate	.354	.482	.127	.290
Sampling error	.018	.025	.033	.076
Root mean square error	.057	.184	.176	.076
Lambda: 0.80				
Average estimate	.293	.481	.089	.293
Sampling error	.020	.034	.028	.097
Root mean square error	.021	.184	.213	.097
Lambda: 1.00				
Average estimate	.240	.480	.067	.318
Sampling error	.014	.027	.021	.110
Root mean square error	.062	.182	.234	.111
Lambda: 1.20				
Average estimate	.197	.484	.049	.311
Sampling error	.015	.038	.017	.115
Root mean square error	.104	.187	.252	.115
Lambda: 1.40				
Average estimate	.163	.484	.038	.311
Sampling error	.012	.037	.013	.131
Root mean square error	.137	.187	.263	.132
Lambda: 1.50				
Average estimate	.147	.482	.033	.327
Sampling error	.010	.042	.013	.176
Root mean square error	.153	.187	.267	.178

TABLE 2, continued

	Cross-Sectional Estimates		Change Score Estimates	
	Uncorrected	Corrected	Uncorrected	Corrected
Lambda: 1.60				
Average estimate	.136	.490	.030	.350
Sampling error	.010	.044	.012	.186
Root mean square error	.164	.195	.270	.192
Lambda: 1.70				
Average estimate	.124	.482	.027	.348
Sampling error	.012	.052	.012	.264
Root mean square error	.177	.189	.273	.268
Lambda: 2.00				
Average estimate	.096	.486	.019	.279
Sampling error	.008	.063	.011	1.086
Root mean square error	.205	.197	.281	1.086

Notes: The values shown are based on 100 to 300 simulations for each value of lambda, with 1,000 cases per simulation. The regression coefficient is fixed at $\beta = 0.3$, so the average estimate should be compared to this value. The parameter values used in generating the simulated data (cf. expressions (37) to (39) in the text) were: $\rho = 0.6$, $\gamma = 0.3$, and $\tau = 0.6$.

rors, and this paper only begins to explore some of the consequences of violations of the standard assumptions about their independence with respect to the accuracy of estimates from cross-sectional and change score data.

The results of simulations of data into which such covariances have been incorporated are summarized in Table 3. The top set of rows shows the root mean square error for the situation just examined, with *no* nonzero covariances involving the measurement errors, shown for comparison with the remaining sets of rows. The second set of rows shows the root mean square error from simulations in which there is a positive covariance across waves in measurement errors for each of the indicators of X. This type of covariance probably occurs frequently in actual panel studies because of individual differences among respondents in their interpretation of specific questions; for example, individuals may differ in whether or not they include informal sources of income (such as exchanges of services with friends and relatives, or illegal sources) when reporting total household income, or in their tendencies to under- or over-report income from savings and investments. To build such a covariance into simulated data, the v_{ijt} term in expression (30) was generated as follows:

$$v_{jti} = \mu S_{ji} + \delta_{jti} , \tag{41}$$

where the S_{ji} and δ_{jti} values were generated from $N(0, 1)$ and $N(0, 1 -$

FIGURE 2. PARAMETER VALUES FOR WHICH CHANGE SCORES
PROVIDE MORE ACCURATE ESTIMATES OF REGRESSION
COEFFICIENTS THAN DO CROSS-SECTIONAL SCORES
(Sample sizes, top to bottom: 10,000; 2,000; 1,000; 500; 250)

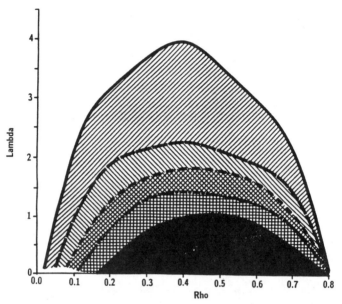

μ^2) distributions, respectively, independently of everything else. For
the simulations shown in Table 3, the value of μ was set to either the
square root of 0.5 or to 0.5, implying a covariance in the measurement
errors of 0.5 or 0.25, respectively. Comparing the entries in the second
and third sets of rows to the corresponding entries in the first set of
rows shows that this type and level of correlated measurement errors
has little or no effect on the accuracy of the cross-sectional estimates,
but results in somewhat *more* accurate estimates from change scores.

The next simulations introduced covariances into time-specific
measurement errors. This type of covariance might arise if, for example, interviewers differ in their ability to obtain complete reports of,
say, sick days, and if respondents generally are interviewed by different
interviewers at different waves of data collection. For these simulations, this type of covariance was introduced as follows:

$$v_{jti} = \nu T_{ti} + \delta_{jti}, \tag{42}$$

with the T_{ti} and δ_{jti} values generated from independent $N(0, 1)$ and $N(0,
1 - \nu^2)$ distributions. For the simulations shown in the fourth and fifth
set of rows of Table 3, the value of ν was fixed at either the square root
of 0.5 or at 0.5, implying correlations of 0.5 and 0.25. This type of
covariance introduces a bias into the corrected estimates of β from
change scores, but also reduces the sampling variance of those es-

TABLE 3. EFFECTS OF CORRELATED MEASUREMENT ERROR
ON ROOT MEAN SQUARE ERROR IN ESTIMATES OF β

Lambda	Cross-Sectional Estimates		Change Score Estimates	
	Uncorrected	Corrected	Uncorrected	Corrected
Independent Measurement Errors				
.30	.142	.182	.090	.065
.50	.083	.177	.150	.074
.80	.026	.185	.215	.107
Item-Specific Covariance of Measurement Errors ($\mu^2 = 0.25$)				
.30	.141	.180	.076	.064
.50	.087	.182	.129	.072
.80	.027	.185	.189	.093
Item-Specific Covariance of Measurement Errors ($\mu^2 = 0.50$)				
.30	.142	.181	.059	.055
.50	.083	.179	.107	.064
.80	.017	.181	.164	.068
Time-Specific Covariance of Measurement Errors ($v^2 = 0.25$)				
.30	.146	.175	.093	.070
.50	.086	.153	.155	.095
.80	.025	.119	.206	.118
Time-Specific Covariance of Measurement Errors ($v^2 = 0.50$)				
.30	.143	.162	.080	.061
.50	.087	.129	.144	.102
.80	.019	.064	.215	.173

Notes: The values shown are each based on 100 simulations, with 1,000 cases per simulation. The parameter values used in generating the simulated data (cf. expressions (37) to (39) in the text) were: $\rho = 0.6$, $\gamma = 0.3$, and $\tau = 0.6$. The μ and τ parameters are set either to the indicated values or to 0; these parameters are defined in expressions (41) and (42).

timates, and the net effect on the root mean square error is a slight deterioration in the accuracy of these estimates. The estimates from cross-sectional scores, on the other hand, have less bias (i.e., there are now two partially offsetting biases with the particular parameter values used in these simulations) as well as less sampling variability, so the overall accuracy is considerably improved. In practical situations, however, there is no assurance that the biases would not be in the same direction, rather than in the opposite direction as was the case in these simulations.

9. MODEL 6: MULTIPLE INDEPENDENT VARIABLES, WITH MEASUREMENT ERROR

If there are measurement errors with respect to any of the independent variables in an explanatory model, the estimates of the regression coefficients for *all* of the independent variables will generally be biased. This is true to a greater extent if the errors of measurement are correlated with the true values of any of the explanatory variables, but also holds true even if the measurement errors are randomly distributed. The algebra gets lengthy even if there are just two predictor variables, and little would be gained from inspection of the expressions for the limiting values of the estimates in the presence of measurement error aside from renewing the conclusion that measurement errors can have important consequences on the accuracy of the estimates of explanatory models and that for this reason it is important to increase our understanding of measurement error.

Errors in the measurement of any of the predictors in a causal model may also cause biases in the estimated coefficients of all of the predictors based on change scores. Again, the algebra gets lengthy even with just two variables, so explicit expressions are not given for the biases. As in the bivariate case, the effects of measurement errors are reduced, however, if these errors are stable from one data collection to the next.

10. CONCLUSIONS

As social scientists, we would like to have models that are correctly specified, indicators of the concepts that are part of those models that include no measurement error, and samples that are large enough to make sampling variances trivial. In practice, we often are confronted with theories and data that afford none of these conditions, and our goal is to make optimal use of less than perfect data. Panel data bring a time-dimension that is available only in imperfect form, at best, in cross-sectional studies. One way to exploit this added dimension is to estimate the same type of static models that can be estimated from cross-sectional data, using change scores to reduce or eliminate spurious effects due to person-characteristics that remain more or less constant over the course of the study. At the same time, however, estimates based on change scores often have greater sampling variance than do those based on cross-sectional scores, and this will often render the change score estimates less accurate than the cross-sectional estimates. This paper has explored these trade-offs for several types of situations, with special emphasis on the consequences of measurement errors with respect to estimates of the effect of one variable on another.

My overall conclusion is that unless there is good reason for confidence that one's explanatory model is properly specified, at least with respect to enduring characteristics of the respondents, change-scores are preferable to cross-sectional data for the estimation of static causal models. A penalty is paid for this conservatism, however, in the form of increased sampling variance of the parameter estimates. Moreover, we dare not overlook the fact that the use of change scores by no means eliminates the possibilities for model misspecification. For example, spurious correlations may exist across change scores as well as across cross-sectional scores; change scores only eliminate spurious effects to the extent that the source of the spurious relationship is fairly constant across time.

Another conclusion is that we must continue efforts to understand and reduce errors in measurement, particularly with respect to the explanatory variables in causal models. At present we know very little about the covariates of measurement error even with respect to cross-sectional data, and even less with respect to panel data. In the absence of such information, we must make arbitrary assumptions, and thus leave open the possibility that estimates of population characteristics may be biased.

REFERENCES

Ashenfelter, O., Deaton, A., & Solon, G. (1986), "Collecting Panel Data in Developing Countries: Does it Make Sense?" (Living Standards Measurement Study, Working Paper No. 23), The World Bank, Washington, DC.

Campbell, D. T. (1963), "From Description to Experimentation: Interpreting Trends as Quasi-Experiments," in *Problems in Measuring Change* (C. W. Harris, ed.), University of Wisconsin Press, Madison, pp. 212–242.

Cronbach, L. J., & Furby, L. (1970), "How We Should Measure 'Change'—Or Should We?," *Psychological Bulletin*, 74, 68–80.

Duncan, G. J., & Hill, D. H. (1985), "An Investigation of the Extent and Consequences of Measurement Error in Labor-Economic Survey Data," *Journal of Labor Economics*, 3, 508–522.

Duncan, G. J., & Mathiowetz, N. A. (1985), *A Validation Study of Economic Survey Data*, Institute for Social Research, Ann Arbor, MI.

Duncan, O. D. (1969), "Some Linear Models for Two-Wave, Two-Variable Panel Analysis," *Psychological Bulletin*, 72, 177–182.

Duncan, O. D. (1972), "Unmeasured Variables in Linear Models for Panel Analysis," in *Sociological Methodology 1972* (H. L. Costner, ed.), Jossey-Bass, San Francisco, pp. 36–82.

Edlefsen, L. E., & Jones, S. D. (1986), *Gauss: Programming Language Manual*, Aptech Systems, Inc., Kent, WA.

Griliches, Z., & Hausman, J. A. (1986), "Errors in Variables in Panel Data," *Journal of Econometrics*, 31, 93–118.

Harris, C. W. (ed.) (1963), *Problems in Measuring Change*, University of Wisconsin Press, Madison.

Heise, D. R. (1970), "Causal Inference from Panel Data," in *Sociological Methodology 1970* (E. F. Borgatta & G. W. Bohrnstedt, eds.), Jossey-Bass, San Francisco, pp. 3–27.

Jöreskog, K. G., & Sörbom, D. (1984), *LISREL VI: Analysis of Linear Structural Relationships by the Method of Maximum Likelihood*, Scientific Software, Inc., Mooresville, IN.

Kenny, D. A. (1973), "Cross-Lagged and Synchronous Common Factors in Panel Data," in *Structural Equation Models in the Social Sciences* (A. S. Goldberger &

O. D. Duncan, eds.), Seminar Press, New York, pp. 153–165.

Kessler, R. C., & Greenberg, D. F. (1981), *Linear Panel Analysis: Models of Quantitative Change*, Academic Press, New York.

Lazarsfeld, P. F. (1948), "The Use of Panels in Social Research," *Proceedings of the American Philosophical Society*, **92**, 405–410.

Liker, J., Augustyniak, S., & Duncan, G. J. (1985), "Panel Data and Models of Change: A Comparison of First Difference and Conventional Two-Wave Models," *Social Science Research*, **14**, 80–101.

Novick, M. R., & Lewis, C. (1967), "Coefficient Alpha and the Reliability of Composite Measurements," *Psychometrika*, **32**, 1–13.

Pelz, D. C., & Andrews, F. M. (1964), "Detecting Causal Priorities in Panel Study Data," *American Sociological Review*, **29**, 836–848.

Presser, S., & Traugott, M. W. (1983), "Correlated Response Errors in a Panel Study of Voting," paper presented at the Annual Meeting of the American Association for Public Opinion Research, Buck Hill Falls, PA.

Rodgers, W. L., & Herzog, A. R. (1987), "Covariances of Measurement Errors in Survey Responses," *Journal of Official Statistics*, **3**, 403–418.

NOTES

[1] The effective sample size for the cross-sectional analysis is not, in general, fully twice as large as that for change score analysis; since the observations in the two waves are not independent, the effective sample size is reduced. On the other hand, change scores can only be calculated for cases with nonmissing observations at both times, so if there is substantial sample attrition at the second data collection, or if there is a substantial proportion of item missing data, the sample size for change scores will be considerably smaller than the sample size for either wave of static scores.

[2] This is true because while the variances of both estimators of β are inversely proportional to sample size, the bias in the estimator from static scores is independent of sample size.

[3] A similar derivation is given in Ashenfelter et al. (1986), except that those authors assume that the covariance terms involving X and V are all zero—that is, that measurement errors are unrelated to the true values of the variable.

[4] Examination of the average estimates of the β parameter reveals that the bias due to misspecification (omission of the Z variable) may be counteracted by the bias due to measurement error in the simple (uncorrected) estimation from cross-sectional data. Specifically, for the particular values of ρ and γ used in these simulations, these biases just about balance each other out when $\lambda = 0.8$. Since the correction for attenuation (i.e., the B estimates) eliminates the bias due to measurement error but does nothing to eliminate the bias due to misspecification, the corrected estimates actually are more biased than the uncorrected (i.e., the b) estimates. This is not a particularly helpful observation, however, since we are assuming that we know nothing about the bias due to the omitted Z variable.

Estimation of Level and Change Using Current Preliminary Data*

J. N. K. Rao, K. P. Srinath, and B. Quenneville

1. INTRODUCTION

The usual practice in repeated surveys conducted by government agencies is to report a preliminary estimate Y^p_{T+1} of the total (or mean) of a characteristic of interest for the current period $(T + 1)$ and a final estimate Y_T for the previous period T. The estimated change between the current and preliminary periods is given by $Y^p_{T+1} - Y_T$, ignoring the difference between the preliminary and final estimates. There is thus the possibility that $Y^p_{T+1} - Y_T$ can be biased and very different from the final estimate of change $Y_{T+1} - Y_T$, apart from the fact that it is not the best estimate of change.

Preliminary estimates may be subject to nonresponse bias or coverage errors or both. Bias could also arise due to revisions made to the preliminary data on the basis of more accurate information available only at the time of final estimates. It is useful, therefore, to examine the problem of constructing improved preliminary estimates for the current period that will, in turn, lead to improved estimates of change.

In this article, two approaches to constructing improved preliminary estimates of level and change are considered. The first method, called the sampling approach and described in Section 2, uses a weighted average of the estimate Y^p_{T+1} and the estimate Y_T adjusted for the change $Y^p_{T+1} - Y^p_T$. The "optimal" weights are determined by minimizing the mean square error of the estimates of level and change. The second approach views the problem of improving the preliminary estimates as a prediction problem. It uses the methods of time series analysis by modeling the final estimates Y_t and the measurement errors $Y^*_t = Y^p_t - Y_t$, $t = 1, \ldots, T$. The models are cast in a state space form and "optimal" predictors are then derived using the Kalman filter, as described by Harvey (1984). In Section 3, "optimal" predictors of level and change are obtained, ignoring sampling errors in the final estimates. In other words, the intention here is to get the best possible predictors of Y_{T+1} and $Y_{T+1} - Y_T$ given $(Y_1, \ldots, Y_T)$ and the current preliminary estimate Y^p_{T+1} and the structures of final estimates Y_t and measurement errors Y^*_t, $t = 1, \ldots, T$. The results are extended in Section 4 to the case of sampling errors in the final estimates, by

457

modeling the sampling errors and the unknown population values X_t rather than the final estimates Y_t, $t = 1, \ldots, T$. The parameters in the model are estimated by using the methods given in Miazaki (1985), to arrive at an "optimal" predictor of X_{T+1}. A more general approach is formulated in Section 5. Finally, the results given in Sections 2–4 are applied in Section 6 to data from the quarterly survey of industrial corporations conducted by Statistics Canada.

2. SAMPLING APPROACH

2.1. Estimates of Level and Change

To simplify the notation, let $T + 1 = 2$ so that the current preliminary estimate is Y_2^p and the final and preliminary estimates for the previous period are Y_1 and Y_1^p, respectively. If the distinction between preliminary and final estimates is ignored in estimating change, then we estimate the change as

$$\Delta_1 = Y_2^p - Y_1. \tag{1}$$

Another estimate of change that is perhaps less biased than Δ_1 is given by

$$\Delta_0 = Y_2^p - Y_1^p. \tag{2}$$

This estimate would, however, introduce the inconsistency (and possibly user inconvenience) in that Δ_0 and Y_1 do not add up to the current preliminary estimate Y_2^p, unlike Δ_1 and Y_1. A third option that assures consistency is to report the estimate of change as

$$\Delta_\alpha = Y_{2,\alpha} - Y_1 \tag{3}$$

and the estimate of current level as

$$Y_{2,\alpha} = \alpha Y_2^p + (1 - \alpha)[Y_1 + (Y_2^p - Y_1^p)] \tag{4}$$

where α is a weight, $0 \le \alpha \le 1$ (Fellegi, 1985). Note that $Y_{2,\alpha}$ is of the form of a composite estimate since it is a weighted average of Y_2^p and Y_1 adjusted for the change $Y_2^p - Y_1^p$. The estimate Δ_α reduces to equation (1) if $\alpha = 1$ and to equation (2) if $\alpha = 0$.

2.2. Choice of α

To determine the "optimal" α, we first derive the bias and the mean square error (MSE) of Δ_α under the following reasonable assumptions (Fellegi, 1985):

Assumption 1: $E(Y_2^p - Y_1^p) = E(Y_2 - Y_1)$ \hfill (5)

where Y_2 is the final estimate for the current period.

Assumption 2: $|B(Y_i^p)| > |B(Y_i)|$, $i = 1, 2$ (6)

where $B(Y_i^p)$ and $B(Y_i)$ denote the biases of Y_i^p and Y_i respectively. Further, we assume, for simplicity, that $B(Y_1) = B(Y_2) = 0$ so that $B(Y_1^p) = B(Y_2^p) = \delta$, say, under assumption 1.

Under assumptions 1 and 2, it immediately follows that

$$|B(\Delta_\alpha)| = \alpha|\delta| .$$ (7)

The absolute bias in Δ_α, therefore, is minimized at $\alpha = 0$ so that the best estimate of change from the point of view of minimizing the bias is given by equation (2), i.e., $Y_2^p - Y_1^p$.

Similarly, noting that $Y_{2,\alpha} = \Delta_\alpha + Y_1$ and $B(Y_1) = 0$, we see that $|B(Y_{2,\alpha})|$ is also equal to $\alpha|\delta|$, which is again minimized at $\alpha = 0$. Since $\alpha = 0$ minimizes both $|B(\Delta_\alpha)|$ and $|B(Y_{2,\alpha})|$, the consistency of figures is also assured. For this choice of α, the estimates $\Delta_0 = Y_2^p - Y_1^p$ and $Y_{2,0} = Y_2^p + (Y_1 - Y_1^p)$ are unbiased for current change and level respectively, under assumptions 1 and 2.

Turning to the variance of Δ_α, we get

$$V(\Delta_\alpha) = V(Y_2^p) + \alpha^2 V(Y_1) + (1 - \alpha)^2 V(Y_1^p) - 2\alpha \text{Cov}(Y_2^p, Y_1)$$

$$- 2(1 - \alpha)\text{Cov}(Y_2^p, Y_1^p) + 2\alpha(1 - \alpha)\text{Cov}(Y_1^p, Y_1).$$ (8)

It now follows from equations (7) and (8) that the value of α minimizing $\text{MSE}(\Delta_\alpha) = V(\Delta_\alpha) + [B(\Delta_\alpha)]^2$ is given by

$$\alpha_c = [V(Y_1^p) - \text{Cov}(Y_1^p, Y_1) + \text{Cov}(Y_2^p, Y_1) - \text{Cov}(Y_2^p, Y_1^p)]$$

$$\times [V(Y_1 - Y_1^p) + \delta^2]^{-1} .$$ (9)

Similarly, the "optimal" value of α minimizing $\text{MSE}(Y_{2\alpha})$ is given by

$$\alpha_l = [V(Y_1 - Y_1^p) + \text{Cov}(Y_2^p, Y_1) - \text{Cov}(Y_2^p, Y_1^p)]$$

$$\times [V(Y_1 - Y_1^p) + \delta^2]^{-1} .$$ (10)

It follows from equations (9) and (10) that α_l can be expressed in terms of α_c as

$$\alpha_l = \alpha_c + [V(Y_1) - \text{Cov}(Y_1^p, Y_1)] [V(Y_1 - Y_1^p) + \delta^2]^{-1} .$$ (11)

It is clear from equation (11) that the "optimal" α's for minimizing the MSE's of change and level are in general different and that both depend on the variances and covariances of the preliminary and final estimates. These variances and covariances depend in turn on the procedures followed to obtain the preliminary estimates. For example, if undercoverage is present at the time of reporting the preliminary estimate, it may be reasonable to assume that

Assumption 3: $Y_1 = Y_1^p + Z_1$; $\text{Cov}(Y_1^p, Z_1) = \text{Cov}(Y_2^p, Z_1) = 0$. (12)

It follows that

$$\text{Cov}(Y_1^p, Y_1) = V(Y_1^p), \ \text{Cov}(Y_2^p, Y_1) = \text{Cov}(Y_2^p, Y_1^p) \qquad (13)$$

and $V(Y_1) > V(Y_1^p)$ under assumption 3, and the "optimal" α minimizing $\text{MSE}(\Delta_\alpha)$ reduces to $\alpha_c = 0$, the value also minimizing $|B(\Delta_\alpha)|$ as shown before. For estimating the level, however, the "optimal" choice of α reduces to

$$\alpha_l = [1 + \delta^2/A]^{-1}, \ A = V(Y_1) - V(Y_1^p). \qquad (14)$$

If the squared bias, δ^2, is much bigger than the difference in the variances, A, then $\alpha_l \doteq 0$, the value also minimizing the absolute bias of $Y_{2,\alpha}$. Thus $\alpha = 0$ is the best choice for both level and change under assumption 3, provided $\delta^2/A >> 1$.

If the bias in the preliminary estimates is due to nonresponse and a simple weighting adjustment is made, then it can be shown that no simplification of the expressions (9) and (10) for α_c and α_l can be achieved.

If α_l is not approximately equal to α_c and change is deemed more important than level, then α_c should be used for both change and level to ensure consistency of figures. The loss in efficiency of the resulting estimate of level relative to the "optimal" estimate of level is given by

$$L = [V(Y_1) - \text{Cov}(Y_1^p, Y_1)]^2$$
$$\times \{[V(Y_1 - Y_1^p) + \delta^2]V(Y_1^p) + \delta^2 V(Y_1 - Y_1^p)\}^{-1}, \qquad (15)$$

which reduces to

$$L = A[V(Y_1^p)(1 + \delta^2/A) + \delta^2]^{-1} \qquad (16)$$

in the case of undercoverage.

3. TIME SERIES APPROACH I

The sampling approach of Section 2 uses only the previous period estimates $\{Y_T^p, Y_T\}$, in conjunction with the current preliminary estimate Y_{T+1}^p, to arrive at improved preliminary estimates of current level and change. Better preliminary estimates, however, can be obtained by using all the available data $\{Y_t, Y_t^p\}$ or $\{Y_t, Y_t^*\}$, $t = 1, \ldots, T$ and Y_{T+1}^p, and by taking account of the structure of the revisions and combining this with a model of the final estimates $\{Y_t\}$. Using a state space formulation of this model and the Kalman filter, "optimal" predictors of level Y_{T+1} and change $Y_{T+1} - Y_T$ are obtained in this section, following Harvey (1984). First, a general formulation of the Kalman filter is given below.

3.1. Kalman Filter

The state space model consists of a *measurement equation*

$$w_t = z_t'\beta_t + \xi_t, \ t = 1, \ldots, T \tag{17}$$

and a *transition equation*

$$\beta_t = G_t\beta_{t-1} + \eta_t, \ t = 1, \ldots, T, \tag{18}$$

where β_t is an $(m \times 1)$ state vector, z_t is an $(m \times 1)$ fixed vector, G_t is a fixed $(m \times m)$ matrix, and the errors ξ_t and η_t are independent. It is further assumed that $\xi_t \sim \text{NID}(0,h_t)$ and $\eta_t \sim \text{NID}(0,Q_t)$ where h_t is a fixed scalar and Q_t is a fixed $(m \times m)$ matrix. The matrices G_t, Q_t, the vector z_t, and the scalar h_t may depend on unknown parameters.

Let b_t be the minimum mean square estimator (MMSE) of β_t based on all the information up to and including time t, and let P_t be the MSE matrix of b_t, i.e., the covariance matrix of $b_t - \beta_t$. The MMSE of β_{t+1}, given b_t and P_t, is then given by

$$b_{t+1|t} = G_{t+1}b_t \tag{19}$$

with MSE matrix

$$P_{t+1|t} = G_{t+1}P_tG_{t+1}' + Q_{t+1}. \tag{20}$$

Once w_{t+1} becomes available, this estimator of β_{t+1} can be updated as follows:

$$b_{t+1} = b_{t+1|t} + P_{t+1|t}z_{t+1} (w_{t+1} - z_{t+1}'b_{t+1|t})/f_{t+1} \tag{21}$$

$$P_{t+1} = P_{t+1|t} - P_{t+1|t}z_{t+1}z_{t+1}' P_{t+1|t}/f_{t+1} \tag{22}$$

with

$$f_{t+1} = z_{t+1}'P_{t+1|t}z_{t+1} + h_{t+1}. \tag{23}$$

Starting values b_0 and P_0 are needed to implement the Kalman filter given by equations (19)–(23).

3.2. Minimum Mean Square Estimator of Y_{T+1}

Following Harvey (1984), we assume that the measurement errors in the preliminary estimates, $Y_t^* = Y_t^p - Y_t$, follow a stationary AR(1) process, i.e.,

$$Y_t^* = \psi Y_{t-1}^* + \zeta_t, \ t = 1, \ldots, T \tag{24}$$

and that Y_t itself follows a (possibly nonstationary) AR(1) process:

$$Y_t = \phi Y_{t-1} + \epsilon_t, \ t = 1, \ldots, T. \tag{25}$$

To simplify the presentation, we have assumed an AR(1) model for $\{Y_t\}$ but the results readily extend to an AR(p) process for any $p \geq 1$. The error vectors $(\zeta_t,\epsilon_t)'$ are assumed to be NID(0, diag$[\sigma_0^2,\sigma^2]$). In the state space form, we can write the transition equation (18) with terms

$$\beta_t = \begin{bmatrix} Y_t \\ Y_t^* \end{bmatrix}, \quad G_t = G = \begin{bmatrix} \phi & 0 \\ 0 & \psi \end{bmatrix}, \quad \eta_t = \begin{bmatrix} \epsilon_t \\ \zeta_t \end{bmatrix}, \quad Q_t = Q = \begin{bmatrix} \sigma^2 & 0 \\ 0 & \sigma_0^2 \end{bmatrix}.$$

(26)

The measurement equation is needed only for time $(T + 1)$ since the Kalman filter in our problem can be initiated at time T with $b_T = \beta_T$ and $P_T = 0$. It is given by equation (17) with terms

$$w_{T+1} = Y_{T+1}^p, \quad z_{T+1}' = (1, 1), \quad \xi_{T+1} = 0.$$

(27)

It follows from equations (19) and (20) that

$$b_{T+1|T} = \begin{bmatrix} \phi Y_T \\ \psi Y_T^* \end{bmatrix}, \quad P_{T+1|T} = Q.$$

(28)

Further, from equations (21) and (22), the first element of MMSE b_{T+1} is given by

$$\hat{Y}_{T+1} = \lambda(\phi Y_T) + (1 - \lambda)(Y_{T+1}^p - \psi Y_T^*)$$

(29)

with $\mathrm{MSE}(\hat{Y}_{T+1}) = \sigma_0^2 \sigma^2/(\sigma^2 + \sigma_0^2)$, where $\lambda = \sigma_0^2/(\sigma^2 + \sigma_0^2)$. In the special case when $\psi = 0$, the "optimal" predictor given by equation (29) of Y_{T+1} reduces to Harvey's (1984) formula:

$$\hat{Y}_{T+1} = \lambda(\phi Y_T) + (1 - \lambda)Y_{T+1}^p,$$

(30)

which is a simple weighted average of the preliminary estimate Y_{T+1}^p and the "optimal" predictor from the time series model in equation (25) fitted to the final estimates $\{Y_t\}$, $t = 1, \ldots, T$.

For the general AR(p) model, $Y_t = \phi_1 Y_{t-1} + \ldots + \phi_p Y_{t-p} + \epsilon_t$, the "optimal" predictor of Y_{T+1} is given by

$$\hat{Y}_{T+1} = \lambda(\phi_1 Y_T + \ldots + \phi_p Y_{T+1-p}) + (1 - \lambda)(Y_{T+1}^p - \psi Y_T^*).$$ (31)

The "optimal" predictor of change is given by $\hat{Y}_{T+1} - Y_T$.

The parameters (ϕ, σ^2) and (ψ, σ_0^2) can be estimated by applying standard time series methods to the two series $\{Y_t\}$ and $\{Y_t^*\}$, $t = 1, \ldots, T$ separately (e.g., see Fuller, 1976).

4. TIME SERIES APPROACH II

In the previous section, sampling errors in the final estimates were ignored by considering the prediction of Y_{T+1}, rather than the unknown population value X_{T+1} for the current period. We now extend these results to the case of sampling errors by modeling the X_t, rather than the final estimates Y_t, and the sampling errors u_t, where $Y_t = X_t + u_t$

and $E(u_t) = 0$, $t = 1, \ldots, T$.

4.1. "Optimal" Predictor of X_{T+1}

To simplify the presentation we again assume an AR(1) model but on the X_t's, i.e., X_t follows a (possibly nonstationary) AR(1) process:

$$X_t = \phi X_{t-1} + \epsilon_t, \quad t = 1, \ldots, T. \tag{32}$$

The model on the measurement errors in the preliminary estimates, Y_t^*, is as before, i.e., a stationary AR(1) process

$$Y_t^* = \psi Y_{t-1}^* + \zeta_t, \quad t = 1, \ldots, T. \tag{33}$$

The error vectors $(\zeta_t, \epsilon_t, u_t)$ are assumed to be NID$(0, \text{diag}[\sigma_0^2, \sigma^2, \sigma_u^2])$. Extension of this model to a moving average process for $\{u_t\}$ can be made, following Miazaki (1985). The latter model is more realistic for rotating samples where the sampling unit stays in the sample for a fixed number of occasions.

In the state space form, we can write the transition equation (18) with terms

$$\beta_t = \begin{bmatrix} X_t \\ Y_t \\ Y_t^* \end{bmatrix}, \quad G_t = G = \begin{bmatrix} \phi & 0 & 0 \\ \phi & 0 & 0 \\ 0 & 0 & \psi \end{bmatrix}, \quad \eta_t = \begin{bmatrix} \epsilon_t \\ \epsilon_t + u_t \\ \zeta_t \end{bmatrix},$$

$$\tag{34}$$

$$Q_t = Q = \begin{bmatrix} \sigma^2 & \sigma^2 & 0 \\ \sigma^2 & \sigma^2 + \sigma_u^2 & 0 \\ 0 & 0 & \sigma_0^2 \end{bmatrix}$$

The Kalman filter cannot be initiated at time T, as in Section 3, since X_T is unknown. The measurement equation is given by equation (17) with terms

$$w_{t+1} = Y_{t+1}^p, \quad z_{t+1}' = (0\ 1\ 1), \quad \xi_{t+1} = 0. \tag{35}$$

The MMSE of β_t, based on all the information up to and including time t, is $b_t = (\hat{X}_t, Y_t, Y_t^*)'$ with MSE matrix

$$
P_t = \begin{bmatrix} V(\hat{X}_t - X_t) & 0 & 0 \\ 0 & 0 & 0 \\ 0 & 0 & 0 \end{bmatrix}.
$$

(36)

The MMSE of β_{t+1}, given b_t and P_t, is then given by

$$
b_{t+1|t} = \begin{bmatrix} \hat{X}_{t+1|t} \\ \hat{Y}_{t+1|t} \\ \hat{Y}^*_{t+1|t} \end{bmatrix} = Gb_t = \begin{bmatrix} \phi\hat{X}_t \\ \phi\hat{X}_t \\ \psi Y^*_t \end{bmatrix}
$$

(37)

with MSE matrix

$$
P_{t+1|t} = GP_tG' + Q = \begin{bmatrix} \tilde{\sigma}_t^2 & \tilde{\sigma}_t^2 & 0 \\ \tilde{\sigma}_t^2 & \tilde{\sigma}_t^2 + \sigma_u^2 & 0 \\ 0 & 0 & \sigma_0^2 \end{bmatrix},
$$

(38)

where

$$
\tilde{\sigma}_t^2 = \phi^2 V(\hat{X}_t - X_t) + \sigma^2 .
$$

(39)

Once the preliminary estimate Y^p_{t+1} becomes available, the estimate $b_{t+1|t}$ can be updated using equations (21) and (22). The updated estimate $b^p_{t+1} = (\hat{X}^p_{t+1}, \hat{Y}_{t+1}, \hat{Y}^*_{t+1})$ is given by

$$
\hat{X}^p_{t+1} = \phi\hat{X}_t + \frac{\tilde{\sigma}_t^2}{\tilde{\sigma}_t^2 + \sigma_u^2 + \sigma_0^2}[Y^p_{t+1} - (\phi\hat{X}_t + \psi Y^*_t)]
$$

(40)

$$
\hat{Y}_{t+1} = \phi\hat{X}_t + \frac{\tilde{\sigma}_t^2 + \sigma_u^2}{\tilde{\sigma}_t^2 + \sigma_u^2 + \sigma_0^2}[Y^p_{t+1} - (\phi\hat{X}_t + \psi Y^*_t)]
$$

(41)

and

$$
Y^*_{t+1} = \psi Y^*_t + \frac{\sigma_0^2}{\tilde{\sigma}_t^2 + \sigma_u^2 + \sigma_0^2}[Y^p_{t+1} - (\phi\hat{X}_t + \psi Y^*_t)]
$$

(42)

with MSE matrix

$$P_{t+1}^p = (\tilde{\sigma}_t^2 + \sigma_u^2 + \sigma_0^2)^{-1} \begin{bmatrix} \sigma_t^2(\sigma_u^2 + \sigma_0^2) & \tilde{\sigma}_t^2\sigma_0^2 & -\tilde{\sigma}_t^2\sigma_0^2 \\ \tilde{\sigma}_t^2\sigma_0^2 & \sigma_0^2(\tilde{\sigma}_t^2 + \sigma_u^2) & -\sigma_0^2(\tilde{\sigma}_t^2 + \sigma_u^2) \\ -\tilde{\sigma}_t^2\sigma_0^2 & -\sigma_0^2(\tilde{\sigma}_t^2 + \sigma_u^2) & \sigma_0^2(\tilde{\sigma}_t^2 + \sigma_u^2) \end{bmatrix}.$$

(43)

We denote $\hat{X}_{T+1}^p$ as the preliminary MMSE of X_{t+1}. Once Y_{t+1} becomes available, this estimate can be updated to get the final MMSE of X_{t+1}, namely $\hat{X}_{t+1}$, noting that the corresponding measurement equation is given by equation (17) with terms

$$w_{t+1} = Y_{t+1}, \quad z'_{t+1} = (0, 1, 0), \quad \xi_{T+1} = 0. \tag{44}$$

Using b_{t+1}^p and P_{t+1}^p in place of $b_{t+1|t}$ and $P_{t+1|t}$ in equation (21) we get the final estimate b_{t+1} with first element given by

$$\hat{X}_{t+1} = \hat{X}_{t+1}^p + \frac{\tilde{\sigma}_t^2}{\tilde{\sigma}_t^2 + \sigma_u^2}(Y_{t+1} - \hat{Y}_{t+1}).$$

(45)

The MSE of $\hat{X}_{T+1}$ is obtained from equation (22):

$$\text{MSE}(\hat{X}_{t+1}) = V(\hat{X}_{t+1} - X_{t+1}) = (\tilde{\sigma}_t^2\sigma_u^2)/(\tilde{\sigma}_t^2 + \sigma_u^2). \tag{46}$$

It can be shown that the same estimate, $\hat{X}_{t+1}$, is obtained by applying a generalized Kalman filter simultaneously to two measurement equations involving Y_{t+1}^p and Y_{t+1}, i.e., equations (35) and (44).

Using starting values b_0 and P_0, the above procedure is recursively applied until the current time $T + 1$ to get the "optimal" predictor $\hat{X}_{T+1}^p$, using the current preliminary estimate Y_{T+1}^p and the estimates for previous periods: $\{Y_t^p, Y_t\}$, $t = 1, \ldots, T$. In the empirical study of Section 6, we took $b_0' = (Y_0, Y_0, 0)$ and $P_0 = 0_3$, where Y_0 is the final estimate for period 0 and 0_3 is the null matrix of order 3.

The "optimal" predictor of true change, $X_{t+1} - X_T$, is given by $\hat{X}_{T+1}^p - \hat{X}_T$.

4.2. Estimation of Parameters

The parameters (ψ, σ_0^2) can be estimated by applying standard time series methods to the series $\{Y_t^*\}$, $t = 1, \ldots, T$, as before. Miazaki (1985) has given two methods for estimating the parameters ϕ, σ^2 and σ_u^2 from the series $\{Y_t\}$, $t = 1, \ldots, T$, assuming that the $\{X_t\}$ process is stationary. Since the series $\{X_t\}$ is possibly nonstationary in our application, we first eliminate the trend $k_t'\alpha$ (e.g., $k_t'\alpha = \alpha_0 + \alpha_1 t$) from Y_t and Y_t^p and then apply the methods to the residuals $\tilde{e}_t = Y_t - k_t'\alpha$ and $\tilde{e}_t^p = Y_t^p - k_t'\tilde{\alpha}$ where $\tilde{\alpha}$ is the ordinary least squares estimator of

α based on Y_t's. That is, $X_t - k_t'\alpha$ follows a stationary AR(1) process $X_t - k_t'\alpha = \phi(X_{t-1} - k_{t-1}'\alpha) + \epsilon_t$ with $|\phi| < 1$, and $Y_t - k_t'\alpha = X_t - k_t'\alpha + u_t$.

The first method (Walker, 1960), based on the well-known method of moments, is simple and provides the estimates $\hat{\phi}$, $\hat{\sigma}^2$, $\hat{\sigma}_u^2$ directly from the $\{Y_t\}$ series. Letting $e_t = Y_t - k_t'\alpha$, it is easily seen that

$$\gamma_e(0) = V(e_t) = \sigma_u^2 + \sigma^2/(1 - \phi^2),$$

$$\gamma_e(1) = \text{Cov}(e_t, e_{t+1}) = \phi\sigma^2/(1 - \phi^2) \tag{47}$$

and

$$\gamma_e(2) = \text{Cov}(e_t, e_{t+2}) = \phi\gamma_e(1). \tag{48}$$

Replacing $\gamma_e(h)$, $h = 0,1,2$ by their estimates

$$\hat{\gamma}_e(0) = \frac{1}{T}\sum_1^T \tilde{e}_t^2, \quad \hat{\gamma}_e(1) = \frac{1}{T-1}\sum_1^{T-1} \tilde{e}_t\tilde{e}_{t+1},$$

$$\hat{\gamma}_e(2) = \frac{1}{T-2}\sum_1^{T-2} \tilde{e}_t\tilde{e}_{t+2} \tag{49}$$

we get the method of moments estimators of ϕ, σ_u^2, and σ^2:

$$\hat{\phi} = \hat{\gamma}_e(2)/\hat{\gamma}_e(1), \quad \hat{\sigma}^2 = \frac{[\hat{\gamma}_e(1)]^2}{\hat{\gamma}_e(2)} - \hat{\gamma}_e(2) \tag{50}$$

and

$$\hat{\sigma}_u^2 = \hat{\gamma}_e(0) - \frac{[\hat{\gamma}_e(1)]^2}{\hat{\gamma}_e(2)}. \tag{51}$$

The second method is more efficient than the previous method, but it requires the knowledge of the sampling variance σ_u^2 estimated by standard sample survey methods from cross-sectional data. It uses the fact that the series $\{e_t\}$ follows a stationary ARMA(1,1) process

$$e_t - \phi e_{t-1} = z_t - \theta z_{t-1} \tag{52}$$

where $z_t \sim \text{NID}(0, \sigma_z^2)$ and the parameters satisfy the following restrictions:

$$R_1 = \sigma^2 + (1 - \phi^2)\sigma_u^2 - (1 + \theta^2)\sigma_z^2 = 0 \tag{53}$$

$$R_2 = -\phi\sigma_u^2 + \theta\sigma_z^2 = 0. \tag{54}$$

The model given by equation (52) with parameters satisfying equations (53) and (54) is called the restricted model. Now using an asymptotically efficient estimate of $(\phi, \theta, \sigma_z^2)$ for the unrestricted model, say $(\tilde{\phi}, \tilde{\theta},$

$\tilde{\sigma}_z^2$), and its estimated covariance matrix, say $\tilde{G}$, obtained from $\{\tilde{e}_t\}$, a generalized least squares estimate $(\tilde{\phi}, \tilde{\theta}, \tilde{\sigma}_z^2, \tilde{\sigma}^2)$ of the parameters in the following model is obtained:

$$\begin{bmatrix} \tilde{\phi} \\ \tilde{\theta} \\ \tilde{\sigma}_z^2 \end{bmatrix} = \begin{bmatrix} \phi \\ \theta \\ \sigma_z^2 \end{bmatrix} + \begin{bmatrix} a_1 \\ a_2 \\ a_3 \end{bmatrix} \tag{55}$$

together with equations (53) and (54), where $\tilde{G}$ is the estimated covariance matrix of $a = (a_1\ a_2\ a_3)'$. Using the restriction given in equation (54), we get $\phi = \theta\sigma_z^2/\sigma_u^2$. The resulting model in equation (55) is next reduced to the standard form with NID$(0, I)$ errors by premultiplying by the upper triangular matrix T given by the Cholesky decomposition of $\tilde{G}^{-1}$, i.e., $T'T = \tilde{G}^{-1}$. The least squares estimates $(\bar{\theta}, \bar{\sigma}_z^2)$ are obtained iteratively by the well-known Gauss–Newton method for nonlinear regression. Substituting these estimates into the restrictions (53) and (54), the estimates of ϕ and σ^2, denoted by $\bar{\phi}$ and $\bar{\sigma}^2$, are finally obtained. The estimate $(\bar{\phi}, \bar{\theta}, \bar{\sigma}_z^2, \bar{\sigma}^2)$ is called the preliminary estimate by Miazaki (1985).

Following Miazaki (1985), an improved estimate of $(\phi, \theta, \sigma_z^2, \sigma^2)$ is obtained by applying nonlinear least squares to the following system:

$$\tilde{e}_t = \phi\tilde{e}_{t-1} - \theta z_{t-1} + z_t, \quad t = 1, \ldots, T \tag{56}$$

$$[2(T-2)\bar{\sigma}_z^2]^{1/2} = [2(T-2)\sigma_z^2]^{1/2} + b_1 \tag{57}$$

together with the restrictions $0 = cR_1 + a_4$ and $0 = cR_2 + a_5$, where c is an arbitrarily large number and $a_4, a_5, b_1 \sim$ NID$(0,1)$. The preliminary estimate $(\bar{\phi}, \bar{\theta}, \bar{\sigma}_z^2, \bar{\sigma}^2)$ provides the starting value for the iterative procedure to get the improved estimate $(\hat{\phi}, \hat{\theta}, \hat{\sigma}_z^2, \hat{\sigma}^2)$. Starting with $\tilde{e}_0 = Y_0 - k_0'\tilde{\alpha}$ and $z_0 = 0$, the values of z_{t-1} in equation (56) are obtained successively:

$$z_1 = \tilde{e}_1 - \phi\tilde{e}_0, \quad z_2 = \tilde{e}_2 - (\phi\tilde{e}_1 - \theta z_1), \quad z_3 = \tilde{e}_3 - (\phi\tilde{e}_2 - \theta z_2), \ldots.$$

5. TIME SERIES APPROACH III

In the time series approach II we assumed that the sampling errors u_t are NID$(0, \sigma_u^2)$ to obtain the "optimal" predictor $\hat{X}_{T+1}^p$. We now extend this result to the case of an arbitrary covariance matrix of $(u_1, u_2, \ldots u_{T+1})'$, say Σ_u. We assume that Σ_u is known or can be estimated from cross-sectional data for the $(T + 1)$ periods by standard sample survey methods. The state space formulation in this general case is somewhat cumbersome and in any case offers no computational advantage compared to the straightforward stochastic least squares method. We use the latter method to derive the "optimal" predictor of X_{T+1}, using the data $\{Y_t^p, Y_t\}$, $t = 1, \ldots, T$, and Y_{T+1}^p.

5.1. Stochastic Least Squares

We assume that $X_t = \phi X_{t-1} + \epsilon_t$ and $Y_t^* = \psi Y_{t-1}^* + \xi_t$ as before, where $Y_t = X_t + u_t$. We can express the model as a linear model with stochastic parameters:

$$Y = Z\beta + Gv \tag{58}$$

where

$$Y = (Y_1, \ldots, Y_T, Y_1^p, \ldots, Y_T^p, Y_{T+1}^p)',$$

$$v = (u_1, \ldots, u_T, u_{T+1}, Y_1^*, \ldots, Y_{T+1}^*)',$$

$$\beta = (X_1, \ldots, X_{T+1})',$$

and Z is a $(2T + 1) \times (T + 1)$ matrix given by

$$Z = \begin{bmatrix} I_T & 0 \\ I_T & 0 \\ 0' & 1 \end{bmatrix} \tag{59}$$

and G is a $(2T + 1) \times (2T + 2)$ matrix given by

$$G = \begin{bmatrix} I_T & 0 & 0_T & 0 \\ I_T & 0 & I_T & 0 \\ 0' & 1 & 0 & 1 \end{bmatrix} \tag{60}$$

where I_T is the $T \times T$ identity matrix, 0 is a T-vector of zeros, and 0_T is $T \times T$ matrix of zeros. The covariance matrix of v is given by

$$\Sigma_v = \begin{bmatrix} \Sigma_u & 0_{T+1} \\ 0_{T+1} & \Sigma^* \end{bmatrix} \tag{61}$$

where Σ^* is the covariance matrix of $(Y_1^*, \ldots, Y_{T+1}^*)'$:

$$\Sigma^* = \frac{\sigma_0^2}{1 - \psi^2} \begin{bmatrix} 1 & \psi & \ldots & \psi^T \\ \psi & 1 & \ldots & \psi^{T-1} \\ \cdots & \cdots & \cdots & \cdots \\ \psi^T & \ldots & \ldots & 1 \end{bmatrix}.$$

(62)

Since the series $\{X_t\}$ is possibly nonstationary, we assume that $X_t = k_t' \alpha + a_t$ where $\{a_t\}$ is a stationary AR(1) process: $a_t = \phi a_{t-1} + \epsilon_t$ so that the covariance matrix of $a = (a_1, \ldots, a_{T+1})'$ is also of the form of equation (63) with σ_0^2 and ψ replaced by σ^2 and ϕ respectively. Denote this covariance matrix by Σ_a. The model may now be written as

$$Y = ZK\alpha + [ZG]\begin{bmatrix} a \\ u \end{bmatrix} = Z^*\alpha + b, \text{ say,}$$

(63)

where $K' = (k_1, \ldots, k_{T+1})$. The error vector b has mean zero and covariance matrix $\Sigma_b = Z\Sigma_a Z' + G\Sigma_u G'$.

It now follows from Harville (1976) that the "optimal" predictor of a is given by

$$\hat{a} = \Sigma_a Z' \Sigma_b^{-1}(Y - Z^*\hat{\alpha})$$

(64)

where

$$\hat{\alpha} = (Z^{*'}\Sigma_b^{-1}Z^*)^{-1}Z^{*'}\Sigma_b^{-1}Y$$

(65)

Hence, the "optimal" estimate of level X_{T+1} is given by $\hat{X}_{T+1}^p = k_{T+1}'\hat{\alpha}_{T+1} + \hat{a}_{T+1}$. The previous month estimate is revised to $\hat{X}_T = k_T'\hat{\alpha} + \hat{a}_T$, and the "optimal" estimate of change is given by $\hat{X}_{T+1}^p - \hat{X}_T$.

The parameters σ_0^2 and ψ can be estimated from the series $\{Y_t^*\}$ as before, but the estimation of σ^2 and ϕ presents computational difficulties. Binder and Dick (1986) proposed the EM algorithm for estimating the parameters of models with arbitrary survey error structure. This algorithm could be applied to the series $\{\tilde{e}_t\}$, where $\tilde{e}_t = Y_t - k_t'\tilde{\alpha}$ and $\tilde{\alpha}$ is the least squares estimator of α, but a computational procedure has not been worked out.

5.2. An Alternative, Computationally Simpler Method

An alternative, computationally simpler approach assumes models on the population values X_t^* and X_t, where $X_t^* = X_t^p - X_t$ with $X_t^p = E(Y_t^p)$ and $X_t = E(Y_t)$, and E denotes the expectation with respect to the sampling design. Assuming that $X_t^* = \psi X_{t-1}^* + \zeta_t^*$, $|\psi| < 1$, and $a_t = \phi a_{t-1} + \epsilon_t$, $|\phi| < 1$, where $a_t = X_t - k_t'\alpha$, the model can be written as a linear model with stochastic parameters:

$$Y = X + u \tag{66}$$

and

$$X = K_1\alpha + a \tag{67}$$

where

$$Y = (Y_1, \ldots, Y_T, Y_1^*, \ldots, Y_T^*, Y_{T+1}^p)',$$

$$X = (X_1, \ldots, X_T, X_1^*, \ldots, X_T^*, X_{T+1} + X_{T+1}^*)'.$$

Further,

$$K_1\alpha = \begin{bmatrix} K\alpha \\ 0 \\ k'_{T+1}\alpha \end{bmatrix} \tag{68}$$

$K' = (k_1, \ldots, k_T)$, $u = (u_1, \ldots, u_T, u_1^*, \ldots, u_t^*, u_{T+1} + u_{T+1}^*)'$ is the vector of sampling errors of Y with covariance matrix Σ_u, say, and

$$a = (a_1, \ldots, a_T, X_1^*, \ldots, X_T^*, a_{T+1} + X_{T+1}^*)'$$

with covariance matrix Σ_a, say. We assume that a_i's are uncorrelated with X_j^*'s. It is also reasonable to assume that the u_i's are uncorrelated with u_j^*'s, and that the u_j^*'s are uncorrelated with mean zero and variance σ_u^{*2}. The covariance matrix of the sampling errors $(u_1, \ldots, u_{T+1})'$ is assumed to be known (or estimated from cross-sectional data for the $T + 1$ periods) so that Σ_u is also known except for σ_u^{*2}.

The "optimal" predictor of $X_{T+1} + X_{T+1}^* = X_{T+1}^p$, denoted by $\hat{X}_{T+1}^p$, is given by the last element of

$$\hat{X} = Y - \Sigma_u \Sigma_Y^{-1}(Y - K_1\hat{\alpha}) \tag{69}$$

where Σ_Y is the unconditional covariance matrix of $Y - K_1\alpha = u + a$ and

$$\hat{\alpha} = (K_1' \Sigma_Y^{-1} K_1)^{-1} K_1^{-1} \Sigma_Y^{-1} Y \tag{70}$$

(see, e.g., Jones, 1980). It follows from equations (69) and (70) that we only need the estimates of Σ_Y and σ_u^{*2} to compute $\hat{X}$, apart from the estimated covariance matrix of sampling errors $(u_1 \ldots u_{T+1})'$. The "optimal" estimate of level X_{T+1} is then obtained as $\tilde{X}_{T+1} = \hat{X}_{T+1}^p - \hat{\psi}\hat{X}_T^*$, where $\hat{\psi}$ is the estimate of ψ.

The previous month's estimate is revised to $\hat{X}_T$, the t-th element of $\hat{X}$ given by equation (69), and the "optimal" estimate of change is given by $\tilde{X}_{T+1} - \hat{X}_T$.

Miazaki's (1985) method of moments can be applied to the data $(Y_1^*, \ldots, Y_T^*)$ in conjunction with the model $X_T^* = \psi X_{t-1}^* + \zeta_t^*$ and $Y_t^* = X_t^* + u_t^*$ to get the estimates of ψ, σ_u^{*2} and σ^{*2}, where $V(\zeta_t^*) = \sigma^{*2}$. Assuming now that $\{u_t\}$ is stationary, it follows that $\{u_t + a_t\} = \{Y_t - k_t'\alpha\}$ is also stationary and the elements of its covariance matrix can be estimated using the autocovariance estimators

$$c_k(Y) = (T - k)^{-1} \sum_{t=1}^{T-k} \tilde{e}_t \tilde{e}_{t+k}, \quad k = 0,1,2, \ldots \tag{71}$$

where $\tilde{e}_t = Y_t - k_t'\tilde{\alpha}$ and $\tilde{\alpha}$ is the least squares estimator of α. In practice, $c_k(Y)$ is taken as zero for small values of $T - k$. The same method can be used to estimate the elements of the covariance matrix of $\{u_t^* + X_t^*\} = \{Y_t^*\}$, using

$$c_k(Y^*) = (T - k)^{-1} \sum_{t=1}^{T-k} Y_t^* Y_{t+k}^*, \quad k = 0,1,2, \ldots \tag{72}$$

Alternatively, noting that $V(Y_t^*) = \sigma_u^{*2} + \sigma^{*2}/(1 - \psi^2)$ and $\text{Cov}(Y_t^*, Y_{t+k}^*) = \psi^k \sigma^{*2}/(1 - \psi^2)$, we can substitute the moment estimates of σ_u^{*2}, σ^{*2} and ψ to get the estimated covariance matrix of $\{Y_t^*\}$. The variance of $Y_{T+1}^p - k_{T+1}'\alpha$ is estimated as $T^{-1}\Sigma(\tilde{e}_t^p)^2$, where $\tilde{e}_t^p = Y_t^p - k_t'\tilde{\alpha}$. Finally, the covariance of $Y_l - k_l'\alpha$ and $Y_{T+1}^p - k_{T+1}'\alpha$ is estimated as $(l - 1)^{-1}\Sigma\tilde{e}_t\tilde{e}_{t+T+1-l}^p$, $l = T, T - 1, \ldots$ and taking it as zero for large values of $T+1-l$, and $\text{Cov}(Y_l^*, Y_{T+1}^p - k_{T+1}'\alpha) = \text{Cov}(X_l^*, X_{T+1}^*)$ is estimated as either $(l - 1)^{-1}\Sigma Y_t^* \tilde{e}_{t+T+1-l}^p$ or by substituting the moment estimates of ψ and σ^{*2} in the formula $\psi^{T+1-l}\sigma^{*2}/(1 - \psi^2)$. Note also that $\text{Cov}(Y_i, Y_j^*) = 0$. Thus all the elements of the covariance matrix Σ_Y can be estimated.

6. EXAMPLE

The results given in Section 2–4 are now applied to data from the quarterly surveys of industrial corporations conducted by Statistics Canada. Table 1 gives the preliminary estimates of profits before taxes of industrial corporations (excluding real estate and construction), Y_t^p, and the first revisions, Y_t, for the periods 1974-quarter 4 to 1985-quarter 4. Only the final estimates were reported before 1974-quarter 3. Annual benchmark adjustments were also made to the revisions, but we do not consider the benchmarked series in this paper. Missing values Y_t^p for four quarters were replaced by the corresponding revisions, Y_t, and the benchmarked value was used to replace missing Y_t^p and Y_t for one quarter.

The resulting graphs of Y_t^p and Y_t do not exhibit any significant dif-

TABLE 1. ESTIMATES OF PROFITS BASED ON THE QUARTERLY SURVEY
OF INDUSTRIAL CORPORATIONS (IN MILLIONS OF DOLLARS)

Obs.	Quarter		Y_t^p	Y_t
1	1974	4	3931	3915
2	1975	1	3304	3313
3		2	3774	3779
4		3	3509	3491
5		4	4028	3944
6	1976	1	3374	3378
7		2	3838	3856
8		3	3393	3402
9		4	3546[a]	3546
10	1977	1	3681	3676
11		2	3913	3913
12		3	3581	3637
13		4	4365	4366
14	1978	1	3908	3904
15		2	5385[a]	5385[b]
16		3	5428[a]	5428
17		4	6582	6626
18	1979	1	6425	6420
19		2	7411	7437
20		3	7650	7743
21		4	8543	8621
22	1980	1	8178[a]	8178
23		2	7853	8021
24		3	7802	7759
25		4	8617	8549
26	1981	1	8430	8415
27		2	8626[a]	8626
28		3	6660	6615
29		4	6311	6155
30	1982	1	4913	4888
31		2	4646	4430
32		3	4296	4257
33		4	4677	4950
34	1983	1	5735	5677
35		2	6798	6951
36		3	7433	7376
37		4	8492	8520
38	1984	1	8142	8255
39		2	9038	8925
40		3	8717	8738
41		4	9479	9694
42	1985	1	9440	9485
43		2	9628	9710
44		3	8760	8812
45		4	10270	10342

[a]Preliminary estimates (Y_t^p) are not available for these quarters. Therefore final
estimates (Y_t) are imputed.

[b]Final estimate (Y_t) is not available for this quarter. Annual benchmarked estimate is imputed.

ferences, and, in fact, maximum relative error of Y_t^p, i.e., $\max|Y_t^p - Y_t|/Y_t$ for $1 \leq t \leq 45$, is equal to 5.5%. The data, however, are useful for illustrating the different methods of estimating level and change.

Lefrançois (1986) analyzed these data using the Box-Jenkins transfer function models, and concluded that the preliminary estimate Y_{T+1}^p provides the best predictor of the final estimate Y_{T+1}. He used the data for the periods 1975-quarter 2 to 1984-quarter 4, while we have also included the data for 1985 and the data for 1975-quarter 1 and 1974-quarter 4.

6.1. Estimates of Level

The estimate of level given by equation (4) from the sampling approach with $\alpha = 0$ is used in this study since the estimates of variances and covariance for calculating the optimal α are not available. Moreover, the results in Section 2 indicate that $\alpha = 0$ is a reasonable choice. Denoting this estimator by $\hat{Y}_{T+1}(0)$ [$= Y_T + (Y_{T+1}^p - Y_T^p)$], the relative error of $\hat{Y}_{T+1}(0)$ with respect to Y_{T+1} for $T + 1 = 36$ to 45 (i.e., $|\hat{Y}_{T+1}(0) - Y_{T+1}|/Y_{T+1}$) is given in column 3 of Table 2. The relative error of Y_{T+1}^p with respect to Y_{T+1} is given in column 2 of Table 2.

Turning to time series approach I, the simple AR(1) model $Y_t = \phi Y_{t-1} + \epsilon_t$, fitted by standard least squares (see Fuller, 1976, Section 8.5) to all the 45 data points Y_t, gives $\hat{\phi} = 1.0179$ with standard error of 0.0167, error variance

$$\hat{\sigma}^2 = \sum_{1}^{44}(Y_{t+1} - \hat{\phi}Y_t)^2/43 = 534624$$

and $R^2 = 0.988$. The fit is good, but Lefrançois (1986) obtained a better fit by fitting ARIMA models to his series Y_t. He obtained the model

$$(1 - 0.60B^2 + 0.23B^3)(1 - B)Y_t = 156.6 - 1908D_t + \hat{\epsilon}_t \quad (73)$$

with error variance $\hat{\sigma}^2 = 209123$, where B is the backward shift operator, D_t is a dummy variable taking the value $D = 0$ except for $t = 1981$ − quarter 3 for which $D = 1$. The dummy variable was used to correct for the effect of a large residual on the estimates of parameters.

The fitting of the AR(1) model $Y_t^* = \psi Y_{t-1}^* + \zeta_t$ to the measurement errors $Y_t^*(t = 1, \ldots, 45)$ gives $\hat{\psi} = -0.0015$ with standard error of 0.1520. The hypothesis $\psi = 0$ is not rejected, and henceforth we assume $\psi = 0$ in which case $\hat{\sigma}_0^2 = 7268$. Substituting the estimates of ϕ, σ^2 and σ_0^2 in equation (30), we get the "optimal" predictor of Y_{T+1} for Harvey's model:

TABLE 2. PERCENTAGE RELATIVE ERROR WITH RESPECT TO Y_{T+1}

$T+1$	Y^p_{T+1}	$\hat{Y}_{T+1}(0)$	$\hat{Y}_{T+1}(1)$	$\hat{Y}_{T+1}(2)$	$\hat{Y}_{T+1}(3)$	$\hat{Y}_{T+1}(4)$	$\hat{Y}_{T+1}(5)$
				Relative Error of			
36	0.77	2.85	0.71	0.75	1.17	0.70	0.70
37	0.32	1.00	0.48	0.47	0.23	0.48	0.49
38	1.37	1.03	1.28	1.17	1.13	1.31	1.31
39	1.27	2.53	1.17	1.25	1.80	1.17	1.17
40	0.24	1.53	0.18	0.25	0.06	0.20	0.20
41	2.22	2.00	2.30	2.26	1.71	2.31	2.31
42	0.47	1.79	0.41	0.42	0.14	0.45	0.45
43	0.84	0.38	0.84	0.63	0.44	0.85	0.85
44	0.59	0.34	0.42	0.30	0.43	0.43	0.43
45	0.70	0.19	0.86	1.06	0.04	0.86	0.86
Average (ARE)	0.88	1.36	0.87	0.86	0.72	0.88	0.88

$$\hat{Y}_{T+1}(1) = 0.0133(1.0179Y_T) + 0.9867Y^p_{T+1}, \qquad (74)$$

where $\hat{\lambda} = \hat{\sigma}_0^2/(\hat{\sigma}_0^2 + \hat{\sigma}^2) = 0.0133$. Similarly, for the Lefrançois model given by equation (73), the "optimal" predictor of Y_{T+1} is

$$\hat{Y}_{T+1}(2) = 0.0334 (Y_T + 0.60Y_{T-1} - 0.83Y_{T-2} + 0.23Y_{t-3}$$
$$+ 156.6 - 1908D_{T+1}) + 0.9666Y^p_{T+1} \qquad (75)$$

using the result in equation (31) for the general AR(p) model, where $\hat{\lambda} = 0.0334$. The relative errors of $\hat{Y}_{T+1}(1)$ and $\hat{Y}_{T+1}(2)$ with respect to Y_{T+1} for $T + 1 = 36$ to 45 are given in columns 4 and 5 respectively of Table 2.

It is clear from equations (74) and (75) that the "optimal" predictor of Y_{T+1} gives essentially a zero weight to the predictors $\tilde{Y}_{T+1}$ from time series models fitted to the final estimates Y_t, and hence puts all the weight on the current preliminary estimate Y^p_{T+1}. This result is in agreement with the conclusion of Lefrançois (1986), but the Kalman filter approach clearly exhibits the relative importance of $\tilde{Y}_{T+1}$ and Y^p_{T+1} in predicting the final estimate Y_{T+1}.

A similar conclusion is also reached by fitting the model $Y_t = \beta_0 + \beta_1 Y_{t-1} + \beta_2 Y^p_t + \epsilon_t$ by ordinary least squares to the data $\{Y^p_t, Y_t\}$:

$$\hat{Y}_{T+1}(3) = -34.53 - 0.017\ Y_T + 1.0248\ Y^p_{T+1}$$
$$\quad\ (38.41) \quad\ \ (0.018) \qquad (0.018) \qquad\qquad (76)$$

The least squares predictor, $\hat{Y}_{T+1}(3)$, however, cannot be interpreted as a weighted average. The relative error of $\hat{Y}_{T+1}(3)$ with respect to Y_{T+1} for $T + 1 = 36$ to 45 is given in column 6 of Table 2.

Turning now to time series approach II for predicting the unknown population value X_{T+1} for the current period by taking account of sampling errors, we obtained the following parameter estimates by the method of moments, taking $k'_t\alpha = \alpha_0 + \alpha_1 t$: $\tilde{\alpha}_0 = 3214.9$, $\tilde{\alpha}_1 = 133.8$,

$$\hat{\phi} = 0.887,\ \hat{\sigma}^2 = 377697,\ \hat{\sigma}_u^2 = 74679$$

$$\hat{\sigma}_0^2 = 7268 \text{ (as before)}. \qquad (77)$$

We used $b'_0 = (Y_0 - k'_0\tilde{\alpha},\ Y_0 - k'_0\tilde{\alpha},\ 0) = (Y_0 - \tilde{\alpha}_0,\ Y_0 - \tilde{\alpha}_0,\ 0)$ and $P_0 = 0_3$ as the starting values for the Kalman filter, where $Y_0 = 3811$ is the final estimate for period 0 (1974-quarter 3). The "optimal" predictor $\hat{X}^p_{T+1}(1) = \hat{e}^p_{T+1} + k'_{T+1}\tilde{\alpha}$, based on the data $\{Y^p_t, Y_t\}$, $t = 1$, ..., T and Y^p_{T+1}, was computed for $T + 1 = 36$ to 45, where $\hat{e}^p_{T+1}$ was obtained by replacing Y^p_t and Y_t by $\tilde{e}^p_t$ and $\tilde{e}_t$, respectively, in equation (40). Similarly, the predictor $\hat{Y}_{T+1}(4)$ of Y_{T+1} for $T + 1 = 36$ to 45 was computed from equation (41). The final estimate $\hat{X}_{T+1}(1)$ was computed from equation (45), using Y_{T+1} to update $\hat{X}^p_{T+1}(1)$. Note that $\hat{X}_{T+1}(1)$ cannot be computed in practice since only Y^p_{T+1} for the current period $T + 1$ will be available, but these values can be used to compare the accuracies of different predictors of X_{T+1}, i.e., $\hat{X}_{T+1}(1)$ can be used as a proxy for the true X_{T+1}. The relative error of $\hat{Y}_{T+1}(4)$ with respect to Y_{T+1} for $T + 1 = 36$ to 45 is given in column 7 of Table 2, while the relative errors of Y^p_{T+1}, $\hat{Y}_{T+1}(0)$, ..., $\hat{Y}_{T+1}(3)$ and $\hat{X}^p_{T+1}(1)$ with respect to $\hat{X}_{T+1}(1)$ for $T + 1 = 36$ to 45 are given in Table 3.

TABLE 3. PERCENTAGE RELATIVE ERROR WITH RESPECT TO $\hat{X}_{T+1}(1)$

$T + 1$	Y^p_{T+1}	$\hat{Y}_{T+1}(0)$	$\hat{Y}_{T+1}(1)$	$\hat{Y}_{T+1}(2)$	$\hat{Y}_{T+1}(3)$	$\hat{X}^p_{T+1}(1)$
			Relative Error of			
36	1.40	3.50	1.34	1.38	1.80	0.60
37	1.40	0.71	1.24	1.25	1.97	0.42
38	1.78	1.43	1.69	1.57	1.54	1.11
39	2.11	3.40	2.02	2.10	2.64	1.00
40	0.57	1.87	0.52	0.60	0.27	0.17
41	0.98	0.76	1.06	1.02	0.47	2.00
42	0.67	1.60	0.61	0.62	0.33	0.38
43	0.61	0.15	0.61	0.40	0.21	0.73
44	2.12	1.20	1.96	1.83	1.96	0.36
45	1.11	1.62	0.93	0.74	1.77	0.74
Average (ARE)	1.28	1.63	1.20	1.15	1.30	0.75

We have also implemented Miazaki's second method, but using the estimate $\hat{\sigma}^2_u = 74769$ from method 1 as a proxy for σ^2_u since the survey estimate of σ^2_u is not available. We obtained the following parameter estimates, again taking $k'_t\alpha = \alpha_0 + \alpha_1 t$:

$$\hat{\phi} = 0.903, \quad \hat{\sigma}^2 = 366043, \quad \hat{\sigma}^2_0 = 7268 \text{ (as before).} \qquad (78)$$

Proceeding along the lines of method 1, we computed the predictors $\hat{X}^p_{T+1}(2)$ and $\hat{Y}_{T+1}(5)$ of X_{T+1} and Y_{T+1} respectively for $T + 1 = 36$ to 45. The final estimate $\hat{X}_{T+1}(2)$ was computed from equation (45) using Y_{T+1} to update $\hat{X}^p_{T+1}(2)$. The relative error of $\hat{Y}_{T+1}(5)$ with respect to Y_{T+1} for $T + 1 = 36$ to 45 is given in the last column of Table 2, while the relative errors of Y^p_{T+1}, $\hat{Y}_{T+1}(0)$, ..., $\hat{Y}_{T+1}(3)$ and $\hat{X}^p_{T+1}(2)$ with respect to $\hat{X}_{T+1}(2)$ for $T + 1 = 36$ to 45 are given in Table 4.

The approaches developed in Section 5 were not implemented since the required covariance matrix of sampling errors is not available.

6.2. Comparison of Estimates of Level

The accuracy of the estimates of level may be judged by their average relative error (ARE) over $T + 1 = 36$ to 45. The ARE values are given in the last rows of Tables 2-4. For estimating Y_{T+1}, the ARE

TABLE 4. PERCENTAGE RELATIVE ERROR WITH RESPECT TO $\hat{X}_{T+1}(2)$

$T+1$	$\hat{Y}^p_{T+1}$	$\hat{Y}_{T+1}(0)$	$\hat{Y}_{T+1}(1)$	$\hat{Y}_{T+1}(2)$	$\hat{Y}_{T+1}(3)$	$\hat{X}^p_{T+1}(2)$
			Relative Error of			
36	1.47	3.55	1.40	1.45	1.87	0.59
37	1.46	0.78	1.31	1.32	2.04	0.42
38	1.78	1.44	1.70	1.58	1.55	1.11
39	2.13	3.41	2.04	2.12	2.66	1.00
40	0.59	1.88	0.54	0.61	0.29	0.17
41	0.96	0.74	1.04	1.00	0.45	1.98
42	0.68	1.57	0.62	0.64	0.36	0.37
43	0.63	0.16	0.62	0.42	0.22	0.72
44	2.17	1.25	2.01	1.89	2.02	0.35
45	1.14	1.65	0.97	0.77	1.80	0.74
Average (ARE)	1.30	1.65	1.23	1.18	1.33	0.75

values in Table 2 are all very small, ranging from 0.72% to 1.36%. The ARE's of the time series predictors $\hat{Y}_{T+1}(1)$, $\hat{Y}_{T+1}(2)$, $\hat{Y}_{T+1}(4)$, $\hat{Y}_{T+1}(5)$, and the preliminary estimate Y^p_{T+1} are essentially equal (about 0.9%), while the ARE of the least squares predictor $\hat{Y}_{T+1}(3)$ is slightly smaller (0.72%). The sampling approach gives the largest ARE, 1.36%.

Turning to the estimation of $\hat{X}_{T+1}(1)$, a proxy for X_{T+1}, the ARE values in Table 3 are also small, but the predictor $\hat{X}^p_{T+1}(1)$ from time series approach II performs better than the predictors that ignore sampling errors: it has an ARE value of 0.75% compared to 1.15%–1.30%. The preliminary estimate Y^p_{T+1} leads to a slightly larger ARE, 1.28%, while the sampling approach gives the largest ARE, 1.63%.

Similar results are obtained if $\hat{X}_{T+1}(2)$ is used as a proxy for X_{T+1}, as can be seen from the last row of Table 4.

6.3. Estimates of Change

The results in Section 6.2 have shown that $\hat{X}^p_{T+1}(1)$ or $\hat{X}^p_{T+1}(2)$ is a good predictor of the population level X_{T+1}. The corresponding estimate of the change $X_{T+1} - X_T$ is given by $\hat{X}^p_{T+1}(1) - \hat{X}_T(1)$ or $\hat{X}^p_{T+1}(2) - \hat{X}_T(2)$, assuming that $\hat{X}^p_T(i)$ is updated to $\hat{X}_T(i)$ at time $T+1$ ($i = 1, 2$). In the sampling approach, Y^p_T is revised to Y_T at time $T+1$ and the estimate of change is taken as $\hat{Y}_{T+1}(0) - Y_T = Y^p_{T+1} - Y^p_T$. We have obtained the following ARE values in estimating $X_{T+1} - X_T$, taking $\hat{X}_{T+1}(1) - \hat{X}_T(1)$ or $\hat{X}_{T+1} - \hat{X}_{T+1}(2)$ as a proxy for $X_{T+1} - X_T$:

$$\text{ARE}[\hat{Y}_{T+1}(0) - Y_T^p, \hat{X}_{T+1}(1) - \hat{X}_T(1)] = 77.3\%$$

$$\text{ARE}[\hat{X}_{T+1}^p(1) - \hat{X}_T(1), \hat{X}_{T+1}(1) - \hat{X}_T(1)] = 27.8\%$$

$$\text{ARE}[\hat{Y}_{T+1}(0) - Y_T^p, \hat{X}_{T+1}(2) - \hat{X}_T(2)] = 81.4\%$$

$$\text{ARE}[\hat{X}_{T+1}^p(2) - \hat{X}_T(2), \hat{X}_{T+1}(2) - \hat{X}_T(2)] = 28.8\%.$$

Here

$$\text{ARE}[\hat{Y}_{T+1}(0) - Y_T^p, \hat{X}_{T+1}(1) - \hat{X}_T(1)] =$$

$$\frac{1}{10} \sum_{T+1=36}^{45} |(\hat{Y}_{T+1}(0) - Y_T^p) - (\hat{X}_{T+1}(1) - \hat{X}_T(1))|$$

$$\times |\hat{X}_{T+1}(1) - \hat{X}_T(1)|^{-1}$$

and the other ARE's are similarly defined.

The ARE values for all the methods are large compared to the corresponding ARE values for level since the denominator $|\hat{X}_{T+1}(1) - \hat{X}_T(1)|$ or $|\hat{X}_{T+1}(2) - \hat{X}_T(2)|$ in the ARE formula for change is very small compared to the denominator $\hat{X}_{T+1}(1)$ or $\hat{X}_{T+1}(2)$ in the ARE formula for level. The ARE from the sampling approach, however, is about three times larger than the ARE from time series approach II.

6.4. Summary of Empirical Results

The main results of the application to profits data may be summarized as follows:

(1) The "optimal" time series predictors in equations (74) and (75) of the final estimate Y_{T+1} for the current period give essentially a zero weight to the predictor obtained from fitting time series models to the final estimates for the previous periods, and hence they put all the weight on the current preliminary estimate Y_{T+1}^p. In other words, the current preliminary estimate essentially provides the best predictor of Y_{T+1} for the profits data.

(2) For estimating the unknown population value X_{T+1} for the current period taking sampling errors into account, the "optimal" predictor $\hat{X}_{T+1}^p$ from time series approach II, given by equation (40), performs better than the predictor that ignores sampling error. It is also much better than the estimate obtained from the sampling approach using only the previous period estimates.

(3) For estimating the true change $X_{T+1} - X_T$, the average relative error (ARE) values for all the methods are large compared to the corresponding ARE values for level. The ARE from the sampling approach, however, is about three times larger than the ARE from the time series approach II.

REFERENCES

Binder, D. A., & Dick, P. (1986), *Modeling and Estimation for Repeated Surveys* (Technical report), Statistics Canada.

Fellegi, I. P. (1985), "Estimating Change" (Internal memorandum), Statistics Canada.

Fuller, W. A. (1976), *Introduction to Statistical Time Series*, John Wiley and Sons, New York.

Harvey, A. C. (1984), "A Unified View of Statistical Forecasting Procedures," *Journal of Forecasting*, 3, 245–275.

Harville, D. A. (1976), "Extension of the Gauss–Markov Theorem to Include the Estimation of Random Effects," *The Annals of Statistics*, 4, 384–395.

Jones, R. G. (1980), "Best Linear Unbiased Estimators for Repeated Surveys," *Journal of the Royal Statistical Society*, B, 42, 221–226.

Lefrançois, B. (1986), *An Example of the Application of Time Series Analysis to the Study of Revisions for Estimating Changes* (Technical report), Statistics Canada.

Miazaki, E. S. (1985), *Estimation for Time Series Subject to the Error of Rotation Sampling*, (Unpublished Ph.D. thesis), Iowa State University.

Walker, A. M. (1960), "Some Consequences of Superimposed Error in Time Series Analysis," *Biometrika*, 47, 33–43.

NOTE

*Our thanks are due to David R. Brillinger and a referee for constructive comments and suggestions.

Estimation of Cross-Sectional and Change Parameters: Discussion

Wayne A. Fuller

The two topics—correlation over time (time series) and measurement error—covered in the two preceding papers have long been of interest. Time series is a large field and there are extensive studies of response error in sample surveys (Dalenius, 1977). However, the literature that combines time series and measurement error models with survey sampling is limited. The melding of these areas makes the papers particularly stimulating.

Rao, Srinath, and Quenneville (RSQ) consider the problem of improving preliminary estimates for statistics reported from a continuing survey, and they differentiate among several approaches. Perhaps the most distinctive difference is the difference between the "sampling approach" and the "time series approach." In the sampling approach, the unknown true values at times 1 and 2 are considered to be fixed. In the time series approach, the unknown true values are considered to be random variables. More specifically, the time series approach considers the two values to be part of a realization of some stochastic process.

Therefore, the variance for the sampling approach is the variance arising from sampling, while the variance for the time series approach is the variance measured with respect to all possible realizations. This variance can also be the limit for a particular realization.

To emphasize some of these ideas, let us consider one of the RSQ time series models. In that model, $Y_T^* = Y_T^P - Y_T$ is the preliminary measurement error and it is assumed that

$$Y_T^* = \psi Y_{T-1}^* + \zeta_t ,$$

$$Y_T = \phi Y_{T-1} + \epsilon_t ,$$

$$(\zeta_t , \epsilon_t) \sim NI [0, \text{diag}(\sigma_0^2 , \sigma^2)].$$

If we substitute from the two equations, we obtain

$$Y_{T+1} = Y_{T+1}^P - Y_{T+1}^* = Y_{T+1}^P - \psi Y_T^* - \zeta_t .$$

Because we know Y_{T+1}^P and Y_T^* at the time of the preliminary estimate for $T + 1$, it is natural to use Y_T^* to predict the preliminary measurement error in order to estimate Y_{T+1}. This produces the estimator

$$\tilde{Y}_{T+1} = Y_{T+1}^P - \psi(Y_T^P - Y_T).$$

In this framework, $\tilde{Y}_{T+1}$ is the best estimator of Y_{T+1} treating the unknown Y_{T+1} as fixed. In constructing the predictor, only the preliminary measurement error is treated as random. If we permit $\{Y_T\}$ to be random, then there is a second predictor of Y_{T+1} given by ϕY_T. Furthermore, the error in the predictor based upon Y_T is ϵ_t, which is independent of the error, ζ_t, in the predictor of the preliminary measurement error. Therefore, the best predictor, under the random model, is the weighted average of the two predictors

$$\hat{Y}_{T+1} = \lambda \phi Y_T + (1 - \lambda)\tilde{Y}_{T+1}$$

$$= \lambda \phi Y_T + (1 - \lambda)[Y_{T+1}^P - \psi(Y_T^P - Y_T)]$$

$$= \alpha_1 Y_T + \alpha_2 Y_{T+1}^P + \alpha_3 Y_T^P,$$

where $\lambda = (\sigma^2 + \sigma_0^2)^{-1}\sigma_0^2$, $\alpha_1 = \lambda \phi + (1 - \lambda)\psi$, $\alpha_2 = 1 - \lambda$, and $\alpha_3 = (\lambda - 1)\psi$. The predictor $\hat{Y}_{T-1}$ is given in equation (29) of RSQ. Because this predictor is a linear function of (Y_T, Y_T^P, Y_{T+1}^P), a simple way to estimate the model for prediction purposes is to compute the ordinary least-squares regression of Y_{T+1} on (Y_T, Y_T^P, Y_{T+1}^P). The regression method may not be fully efficient under the model, but it does not require the measurement model to be true in order to provide a good predictor.

Rodgers describes the effects of measurement error and of omitted variables on the estimation of socioeconomic models from data containing two repeated measures on the same individuals. He demonstrates that measurement error can have sizable effects upon regression coefficients.

One of Rodgers' main points is that taking differences is a way of removing the effect of an omitted variable on the estimated regression coefficient. Critical to this result is the assumption that the omitted variable is constant over time. That is, taking differences has the desired effect in a study of people if the omitted variable represents a person effect. Rodgers is led by his simulations to a preference for the use of change scores in the estimation of some static models. His description of the effects of measurement error and of omitted variables is very useful.

I will emphasize below the importance of using measurement error methods in the analysis of differences because of the potential variance inflation associated with differences. I will also illustrate methods of comparing within-cluster and between-cluster regressions.

Let us consider a model of the type described by Rodgers. Three variables are observed at two points in time. The vectors are (Y_1, X_1^*, W_1) and (Y_2, X_2^*, W_2). In Rodgers' paper, W is a second determination on x, but I have changed the name as a notational aid in separating the two determinations from the two time periods. The model is

$$Y_{ti} = x_{ti}\beta + e_{ti}, \qquad t = 1, 2$$

$$X_{ti}^* = x_{ti} + u_{ti}, \qquad t = 1, 2$$

$$(e_{1i}, e_{2i})' \sim NI(0, \Sigma_{ee}),$$

$$(u_{1i}, u_{2i}) \sim NI(0, \Sigma_{uu}),$$

where $Cov\{(e_{ti}, u_{ti}), W_{rj}\} = (0, 0)$ for all t, r, i, j, and $Cov\{x_{ti}, W_{ti}\} \neq 0$. This is an instrumental variable model and is a slight generalization of Rodgers' model. The e_{ti} is the error in the equation and the u_{ti} are the measurement errors in the explanatory variable. The intercept is omitted to simplify the model. The W_{ti} can be, but are not required to be, a second determination on x_{ti}. The critical assumptions are that the covariance between (e_{ti}, u_{ti}) and W_{rj} is $(0, 0)$ and that there is a nonzero covariance between W_{ti} and x_{ti}.

Under this model there is a test with power to identify the presence of an omitted variable correlated with x_{ti}. To construct the test, we transform the model and write

$$Y_{2i} - Y_{1i} = (x_{2i} - x_{1i})\beta_d + e_{2i} - e_{1i},$$

$$Y_{2i} + Y_{1i} = (x_{2i} + x_{1i})\beta_s + e_{2i} + e_{1i},$$

$$X_{2i}^* \pm X_{1i}^* = (x_{2i} \pm x_{1i}) + (u_{2i} \pm u_{1i}).$$

The regression equation for the differences can be termed the within-person regression if the observational units are people. In the same way, the regression equation for the sums might be called the among-persons regression. The model can be estimated by three-stage least-squares with and without the restriction that $\beta_d = \beta_s$. A test of the hypothesis that $\beta_d = \beta_s$ furnishes a test of the hypothesis that a variable correlated with x_{ti} is omitted from the model.

To illustrate these ideas we generated two samples of size 1,000. Each sample was generated according to the following equations:

$$(X_{1i}, W_{1i}) = (x_{1i}, x_{1i}) + (u_{11i}, u_{21i}),$$

$$(X_{2i}, W_{2i}) = (x_{2i}, x_{2i}) + (u_{12i}, u_{22i}),$$

$$x_{ti} = \rho z_i + \pi_i + v_{ti}, \qquad t = 1, 2$$

$$Y_{ti} = \beta x_{ti} + \gamma z_i + e_{ti}, \qquad t = 1, 2.$$

The random components were generated as normal independent zero-mean random variables. The variances of r_i and z_i were one and

$$v_{ti} \sim NI(0, 1 - \rho^2 - \tau^2), \qquad t = 1, 2$$

$$u_{jti} \sim NI(0, \lambda^2), \qquad j = 1, 2; t = 1, 2$$

$$e_{ti} \sim \text{NI}(0,\ 1 - \beta^2 - \gamma^2),\qquad t = 1,\ 2.$$

The parameter vector for the first sample is $\{\rho = 0,\ \tau = 0.8485,\ \lambda = 0.5,\ \beta = 0.3,\ \gamma = 0.3\}$ and the parameter vector for the second sample is $\{\rho = 0.6,\ \tau = 0.6,\ \lambda = 0.5,\ \beta = 0.3,\ \gamma = 0.3\}$. Thus, in the first sample the "true" x_{ti} is uncorrelated with the "model misspecification term" z_i, while in the second sample t_{ti} and z_i are correlated. The parameters are those defined by Rodgers.

The ordinary least-squares regressions of $Y_{2i} - Y_{1i}$ on $X_{2i} - X_{1i}$ and of $Y_{1i} + Y_{2i}$ on $X_{1i} + X_{2i}$ gave the following results.

$$\textit{First Sample:}\quad D\hat{Y} = 0.172\ DX,$$
$$(0.040)$$

$$S\hat{Y} = 0.286\ SX,$$
$$(0.024)$$

$$\textit{Second Sample:}\quad D\hat{Y} = 0.114\ DX,$$
$$(0.039)$$

$$S\hat{Y} = 0.446\ SX,$$
$$(0.021)$$

where $DY = Y_2 - Y_1$ and $SY = Y_2 + Y_1$, and DX and SX are defined similarly. We will ignore the intercept terms throughout.

The coefficients in the first two equations are biased estimators of 0.3 because of the presence of measurement error. The bias due to measurement error is more severe in the first equation because the variation in the differences of x_{ti} is smaller than the variation of the sums of the x_{ti}. The reliability ratio for X is 0.8, the reliability ratio for DX is 0.53, and the reliability ratio for SX is 0.87. The reliability ratio is the ratio of the variance of the true values to the variance of the observed values (Fuller, 1987).

In the second sample the bias in the first coefficient is due only to measurement error and is the same as the bias of the first coefficient of the first sample. The bias in the second coefficient of the second sample is due to measurement error and to the correlation of x_t with the omitted variable z_t. The effect of the omitted variable dominates the effect of measurement error and the net bias is positive. For this model, the true relative bias is a positive 48%, where the measurement error bias is a negative 13% and the omitted variable bias is a positive 63%.

If the two equations of the first model are estimated by instrumental variables using the three-stage least-squares option of a package such as SAS, we obtain

$$D\hat{Y} = 0.345 \, DX,$$
$$(0.075)$$

$$S\hat{Y} = 0.336 \, SX,$$
$$(0.027)$$

where we used W_1 and W_2 as instrumental variables for both equations. The coefficients of both of these models are consistent estimators under the correct specification for this sample. Note, however, that the coefficient estimated with the differences has a variance about eight times that of the coefficient estimated with the sums. It is possible to test the hypothesis that the true coefficients are equal and the F-value is $F = 0.01$. If the two equations are estimated subject to the restriction that the coefficients are equal, the estimated coefficient is 0.337 with a standard error of 0.025.

The three-stage least-squares estimates for the second sample are

$$D\hat{Y} = 0.214 \, DX,$$
$$(0.070)$$

$$S\hat{Y} = 0.497 \, SX.$$
$$(0.024)$$

The estimator constructed with the differences is a consistent estimator, but with a large standard error. The estimator constructed with the sums is seriously biased. The estimate differs from the true value by more than eight standard errors. In this case, the test of the hypothesis that the two coefficients are equal gives an F-value of $F = 14.5$, and one easily rejects the equal coefficient model. Therefore, with a sample such as this one, the investigator must use the high variance estimator or attempt to identify an indicator for the variable z_t.

Rodgers mentions that measurement error may be related to the true values of the variables. The model in which the mean of the measurement error and/or the variance of the measurement error are related to the true values has been called the model with error functionally related to the true values. Estimation procedures for such models are available in the program EV CARP (Fuller, 1984; Schnell & Fuller, 1987).

While measurement error cannot be ignored in any analysis of survey data, the paper by Rodgers emphasizes the particular importance of measurement error in the analysis of change. It follows that a portion of the resources for any survey, and particularly of a panel survey, should be used for the estimation of the parameters of the measurement error process.

The effort devoted to the measurement process will depend upon the size of the original project. For small studies, effort may be restricted to searching the literature for estimates of the parameters of the measurement process. A good source of raw data on response error for socioeconomic variables is the Census Bureau comparison of Census and Current Population Survey responses (U.S. Department of Commerce, 1975). A review of existing studies summarizing the available estimates of measurement error variances would be a very useful project.

Either large or small surveys can include multiple indicators for some important items. It is not necessary that all indicators be unbiased for the item of interest in order that they be used as instrumental variables.

Large interviewer surveys should be planned to include reinterviews. Reinterviews serve as quality control on interviewers and provide replicate measures that can be used to estimate the error variances. Under reasonable assumptions for the levels of error variances, one-third of the total resources can be profitably used to generate replicate observations for the estimation of error variances.

Information external to the survey, such as administrative records, can be used to provide information on the parameters of measurement error distributions. Possible sources include records on residence characteristics such as age and type of construction, utility records on energy use, and company records on hours worked.

It is not enough to collect data on the nature and magnitude of response error. The information should be used in estimation procedures of the kinds under discussion. Software for a number of procedures is available. For example, the program EV CARP provides the user with estimation techniques for measurement error models for complex survey samples.

REFERENCES

Dalenius, T. (1977), "Bibliography on Nonsampling Errors in Surveys—I," *International Statistical Review*, **45**, 71–89, 181–197, 303–317.

Fuller, W. A. (1984), "Measurement Error Models with Heterogeneous Error Variances," in *Topics in Applied Statistics* (Y. P. Chaubey & T. D. Dwivedi, eds.), Concordia University, Montreal.

Fuller, W. A. (1987), *Measurement Error Models*, Wiley, New York.

Schnell, D., & Fuller, W. A. (1987), *EV CARP*, Statistical Laboratory, Iowa State University, Ames.

U.S. Department of Commerce (1975), *1970 Census of Population and Housing: Accuracy of Data for Selected Population Characteristics as Measured by the 1970 CPS-Census Match* (PHE(E) – 11), U.S. Government Printing Office, Washington, DC.

The Value of Panel Data in Economic Research*

Gary Solon

The last two decades have seen the development of numerous major panel surveys that collect data useful for economic research. These surveys have been described in detail elsewhere (Borus, 1982), but the most frequently used by economists have been the Panel Study of Income Dynamics, the National Longitudinal Surveys of Labor Market Experience, longitudinally matched records from the Current Population Survey, and the series of federally funded negative income tax experiments. Section 1 of this paper outlines how economists have used these data and focuses particularly on types of research that would have been impossible without longitudinal information. No attempt is made to survey the almost countless substantive issues that have been addressed with panel data, but Section 1 does provide several examples to illustrate the *types* of analysis for which longitudinal data are essential. Section 2 discusses special estimation problems that have arisen in longitudinal economic research and considers some methodological implications. Section 3 summarizes the findings.

1. ANALYTICAL BENEFITS OF PANEL DATA

Panel surveys collect longitudinal data by repeatedly interviewing the same individuals or families. Longitudinal information can be assembled instead through one-time retrospective interviews, but retrospective surveys are notoriously susceptible to recall error. Panel surveys lessen (though, as discussed in the next section, do not nearly eliminate) measurement error by reducing the time lapse between the interview and the time period to which the questions refer.

Relative to cross-sectional data, panel data offer two main benefits for economic research: (1) they make possible the measurement and analysis of micro-level dynamics and (2) they facilitate controlling for the effects of unobserved variables. This section discusses these benefits in turn and provides examples from the economics literature.

1.1. Dynamic Analysis

Many economic issues involve relationships between an individual or

486

family's value of a variable at one time and their value of the same or another variable at a different time. Measurement and analysis of such dynamic relationships necessarily require longitudinal data. An obvious example is the measurement of income mobility. Cross-sectional data can tell, for instance, what proportion of a population is in poverty at a point in time, and repeated cross sections can show how that proportion changes over time. But only longitudinal data can tell what proportion of those who are poor in one period remain poor in another.

In fact, just such information on income mobility in general and persistence of poverty in particular has been one of the main contributions of the Panel Study of Income Dynamics (PSID). Duncan and Morgan (1984), for example, use the PSID data to describe movements among quintiles in the family income distribution. They report that 56% of the individuals whose family incomes placed in the bottom quintile in 1971 remained in the bottom quintile in 1978. Of those in the top quintile in 1971, 48% remained in the top quintile in 1978. Overall, 40% of the 1971 sample stayed in the same quintile in 1978, 37% moved one quintile, and 23% moved at least two quintiles. Although Duncan and Morgan note a "substantial and perhaps surprising degree of income mobility" in these statistics, there is also quite a substantial degree of serial correlation.

More detailed analyses typically have focused on the low end of the income distribution. Hill (1981), for example, examines the persistence of poverty status as defined in official poverty statistics. Roughly speaking, the official poverty threshold for a family of given size and composition starts with an estimated minimal annual food budget, multiplies it by three, and makes some additional adjustments for family size and farm/nonfarm status. Applying this poverty standard to PSID data for 1969–78, Hill finds that the cross-sectional poverty rate averaged 7.7% over this period and that about 60% of those poor each year remained poor the next year. Seven-tenths of 1% (0.7%) of the sampled individuals were poor in all ten years, and 2.6% were poor in at least eight years. Again, the data reveal considerable turnover in the poverty population but also a significant degree of chronic poverty. Many longitudinal analyses of poverty have also investigated the characteristics of the persistently and temporarily poor. Some of the most striking findings pertain to racial composition. Hill, for instance, reports that 42% of the poor in the 1978 cross section were black (as compared to 12% black in the U.S. population). Furthermore, 61% of those poor in all of the 1969–78 years and 62% of those poor in at least eight years were black. Thus, the concentration of poverty among blacks is even greater when the focus is on chronic poverty.

Much of the economic research on income dynamics has examined individual earnings rather than family income. An outstanding example is Lillard and Willis' (1978) study of PSID data on working male heads of households. One part of their analysis involves a variance components model that views the natural logarithm of an individual's annual earnings as the sum of a permanent component and a transitory (but possibly serially correlated) component. Their estimates of this model, based on 1967–73 data, measure the degree of persistence in individual earnings status. Their full-sample results indicate that the serial correlation in log earnings is .84 from one year to the next, declines to .78 at a two-year interval and .75 at a three-year interval, and falls asymptotically to .73. When they stratify their sample by race, they find less earnings mobility among blacks than whites. For whites, they estimate a one-year serial correlation of .83 that declines asymptotically to .71. For blacks, they estimate a one-year serial correlation of .89 that declines asymptotically to .81.

Like many of the studies of family income, Lillard and Willis' analysis of earnings devotes special attention to the low end of the distribution. They define low earnings in a particular year as earnings below half the median observed for males in that year's Current Population Survey. By this standard, 3% of the whites and 13% of the blacks in the 1967 PSID sample had low earnings. Lillard and Willis estimate that, of the low-earning men in a given year, about 45% of the whites and 65% of the blacks will have low earnings the next year, and about 25% of the whites and 50% of the blacks will have low earnings in both of the next two years.

The results cited above give summary descriptions of the extent of income mobility. An important additional question is what factors are associated with significant changes in economic status. Research on this question has found that, although fluctuations in individual earnings are the main single source of family income changes, changes in family structure—especially divorce and marriage—also are very important. Bane and Ellwood's (1986) study of poverty spells, for example, concludes:

> . . . a fall in a head's earnings explained spell beginning in only a minority of cases. In nearly half the cases family structure and life cycle events were associated with the start of a poverty stay. Our research suggests, therefore, that models which concentrate only on the earnings dynamics of household heads will miss a great deal of the dynamics of poverty.

All these results pertain to income mobility over time for the same individuals or families. Another important aspect of income mobility is the intergenerational transmission of status. Numerous studies by sociologists have investigated this issue, but most of these studies have measured parents' status on the basis of their offspring's retrospective reports, which appear to be extremely unreliable. (See Massagli & Hauser, 1983, for evidence and further references.) Fortunately, the PSID has now been under way for so long that some of the children in the original sample families are now well into adulthood. It therefore has become possible to relate their current status to that of their parents as contemporaneously reported at the outset of the survey by the parents themselves. Work in progress by Mary Corcoran, Roger Gordon, Deborah Laren, and myself is exploiting this intergenerational span of the PSID to produce new evidence on the extent of intergenerational mobility.

1.2. Controlling for Unobserved Variables

The second main contribution of panel data has been their use in controlling for the effects of unobserved variables. This use can be illustrated with a simple generic model. Suppose one wishes to estimate the slope coefficient β in the regression model

$$y_{it} = \alpha + \beta x_{it} + u_i + \epsilon_{it} \tag{1}$$

where $i = 1, \ldots, N$ indexes individuals and t indexes time periods. The error component u_i reflects the effects of time-invariant unobserved individual characteristics, and ϵ_{it} is a transitory error term. Assume that $\text{Cov}(x_{it}, \epsilon_{is}) = \text{Cov}(u_i, \epsilon_{it}) = 0$ for any t and s, but $\text{Cov}(x_{it}, u_i) \neq 0$; that is, at least some of the unobserved characteristics are correlated with x_{it}. Then, if one estimates an ordinary least squares (OLS) regression of y_{it} on x_{it} with cross-sectional data for any particular period, the probability limit (as $N \to \infty$) of the estimator of β is

$$\operatorname*{plim}_{N \to \infty} \hat{\beta}_{OLS} = \beta + [\text{Cov}(x_{it}, u_i)/\text{Var}(x_{it})]. \tag{2}$$

Thus the cross-sectional estimator is inconsistent as long as x_{it} and u_i are correlated.

But now suppose one has panel data from two different time periods t and s. Then one may "difference" equation (1) to obtain

$$y_{it} - y_{is} = \beta(x_{it} - x_{is}) + \epsilon_{it} - \epsilon_{is}. \tag{3}$$

This procedure "differences out" u_i, and applying OLS to equation (3) consequently yields consistent estimation of β. In effect, this approach controls for unobserved characteristics by considering how the same individual's y changes as his x changes.[1]

This type of use of panel data has been very widespread in economic research. One example is the longitudinal estimation of union–nonunion wage differences. In that case, y_{it} is some wage measure (usually the natural logarithm of the hourly wage rate), and x_{it} is a dummy variable equal to 1 if the ith worker is a union member in period t and 0 otherwise. The coefficient β then represents the proportional union–nonunion wage gap. If unobserved factors such as ability or motivation influence wages and are correlated with union status, cross-sectional estimation of equation (1) is inconsistent. Longitudinal studies, such as Freeman

(1984) and Mellow (1981), have therefore used panel data to estimate instead the differenced equation (3).

A similar example is the estimation of wage premiums paid in dangerous or otherwise unpleasant jobs. For instance, x_{it} might then be a measure of the risk of injury associated with the ith worker's job in period t. In this case, too, several researchers, such as Brown (1980) and Duncan and Holmlund (1983), have used panel data in efforts to eliminate inconsistency due to unobserved worker characteristics.

2. SPECIAL PROBLEMS

Longitudinal surveys have facilitated economic research on a wide variety of topics. Indeed, some issues, such as the extent of income mobility, simply cannot be addressed without longitudinal information. Nevertheless, the use of panel data in economic research has generated a set of special estimation problems.

One of the most obvious problems, amply discussed elsewhere, is the problem of sample attrition. The response rate in the PSID's first year (1968) was 76%. By 1983, cumulative attrition had reduced this rate to 45% (Survey Research Center, 1985). Not only does such attrition reduce sample size, but, if it is nonrandom, it may undermine the representativeness of the sample remaining in the survey. Becketti et al. (1983) give a detailed analysis of the attrition problem in the PSID. Bias due to nonrandom attrition can be viewed as a special case of the sample selection bias discussed by Heckman (1979), and Hausman and Wise (1979) have analyzed it as such with regard to one of the negative income tax experiments.

The remainder of this section discusses two other sources of bias: measurement error and dynamic self-selection. Neither is peculiar to panel data, but the types of analyses performed with panel data may be especially sensitive to these problems.

2.1. Measurement Error

Although panel surveys presumably generate more reliable data than would be obtained from single retrospective interviews, panel data are still subject to considerable measurement error. This error is probably no greater than in cross-sectional surveys (and may be less),[2] but the types of longitudinal analysis overviewed in the previous section may be especially susceptible to bias from measurement error. Several examples follow.

First, suppose that, in a panel study of income mobility, one wishes to estimate the first-order serial correlation of y_{it}, which is the ith family's income in year t.[3] Let $\mathrm{Var}(y_{it}) = \sigma_y^2$ and $\mathrm{Cov}(y_{it}, y_{i,t-1}) = \rho_y \sigma_y^2$. Then ρ_y is the true value of the parameter to be estimated, and a large value of ρ_y signifies a *low* degree of year-to-year mobility. Now suppose that the panel survey obtains not the true y_{it}, but instead the error-ridden measure

$$y_{it}^* = y_{it} + v_{it} \tag{4}$$

where v_{it} is a random measurement error with $\text{Var}(v_{it}) = \sigma_v^2$, $\text{Cov}(v_{it}, v_{i,t-1}) = \rho_v \sigma_v^2$, and $\text{Cov}(y_{it}, v_{is}) = 0$ for any t and s.

Then, if one estimates the true serial correlation parameter ρ_y with the sample correlation r_{y^*} of the error-ridden y_{it}^*, the probability limit of this estimator is

$$\underset{N \to \infty}{\text{plim}}\ r_{y^*} = \text{Cov}(y_{it}^*, y_{i,t-1}^*)/\text{Var}(y_{it}^*)$$

$$= (\rho_y \sigma_y^2 + \rho_v \sigma_v^2)/(\sigma_y^2 + \sigma_v^2) \tag{5}$$

$$= \rho_y - [\sigma_v^2(\rho_y - \rho_v)/(\sigma_y^2 + \sigma_v^2)].$$

If measurement error is present ($\sigma_v^2 > 0$), the estimator is consistent only in the improbable case that y and v have the same serial correlation. It is inconsistent downward or upward as ρ_y is greater or less than ρ_v.

Given the substantial persistence of measured income status indicated in Section 1, unless ρ_v is remarkably high, measurement error probably biases downward the estimation of ρ_y and hence exaggerates the extent of income mobility. Since there is virtually no evidence on serial correlation in measurement error, however, this is only a conjecture.

A second example concerns the estimation of wage premiums for undesirable job characteristics. Suppose that, in equation (1), y_{it} is a wage measure for the ith worker's job in period t and x_{it} is the risk of injury in that job. In addition to the assumptions previously made for eq'tion (1), assume that x_{it} is imperfectly measured by

$$x_{it}^* = x_{it} + v_{it} \tag{6}$$

where $\text{Var}(x_{it}) = \sigma_x^2$, $\text{Cov}(x_{it}, x_{i,t-1}) = \rho_x \sigma_x^2$, v_{it} is a random measurement error with $\text{Var}(v_{it}) = \sigma_v^2$, $\text{Cov}(v_{it}, v_{i,t-1}) = \rho_v \sigma_v^2$, and $\text{Cov}(x_{it}, v_{is}) = \text{Cov}(u_i, v_{it}) = \text{Cov}(\epsilon_{it}, v_{is}) = 0$ for any t and s.

Then, if one applies OLS to a cross-sectional regression of y_{it} on x_{it}^* in any particular period, the probability limit of the slope coefficient estimator is

$$\underset{N \to \infty}{\text{plim}}\ \hat{\beta}_{OLS} = \beta + [\text{Cov}(x_{it}, u_i)/(\sigma_x^2 + \sigma_v^2)] - [\beta\sigma_v^2/(\sigma_x^2 + \sigma_v^2)]. \tag{7}$$

The inconsistency of this estimator involves two terms. The first, as discussed in Section 1, arises from the correlation between x_{it} and the unobserved individual-specific effect u_i. The second comes from measurement error.

Now suppose that one has panel data from two adjacent periods, t and $t - 1$. One may then apply OLS to the differenced equation (3) where $s = t - 1$. The probability limit of this panel estimator of β is

$$\operatorname*{plim}_{N \to \infty} \hat{\beta}_{DIFF} = \beta - \frac{\beta \sigma_v^2}{[(1 - \rho_x)/(1 - \rho_v)] \sigma_x^2 + \sigma_v^2}. \tag{8}$$

Exploiting the panel data to difference out u_i thus eliminates the first source of inconsistency, but may either increase or decrease the second. Specifically, if $\rho_x > \rho_v$, the inconsistency due to measurement error is larger for the panel estimator $\hat{\beta}_{DIFF}$ than for the cross-sectional estimator $\hat{\beta}_{OLS}$. This occurs because, if the serial correlation of the true x is greater than that of the measurement error, differencing increases the noise-to-signal ratio for the measured explanatory variable. On the other hand, if $\rho_x < \rho_v$, the conclusion is reversed.

According to Duncan and Holmlund (1983), the latter condition applies to their study of wage premiums because, "with respondent self-reports of working conditions, there will likely be persistent tendencies of some respondents to over- or understate 'true' working conditions." But again, in the absence of direct evidence on serial correlation in measurement error, this is merely a conjecture.

A third example is the literature on labor supply behavior over the life cycle.[4] In this literature, exemplified by MaCurdy (1981) and Altonji (1986), y_{it} typically is the logarithm of annual hours worked, x_{it} is the logarithm of the hourly wage rate, and the slope coefficient β is estimated with the differenced data as in equation (3). As discussed in detail by Ashenfelter (1984), measurement error appears to be a particularly serious problem in this research area. Because the wage measure is usually computed by dividing annual earnings by annual hours, measurement error in either contributes to error in the wage variable. On the basis of comparisons with an alternative wage measure, Altonji has reached the shocking conclusion that the year-to-year change in the measured log wage, the relevant explanatory variable in equation (3), may have as high as a .9 ratio of error variance to total variance. Related evidence appears in Duncan and Hill's (1985) report on a recent study at The University of Michigan's Institute for Social Research. The Institute conducted a PSID-like survey of the employees of a firm that already had accurate records on the employees' earnings and employment. A comparison of the survey responses and the firms' records revealed that, although survey responses for some variables were reliable, an hourly wage measure based on the survey responses for annual earnings and hours displayed over a .7 ratio of error variance to total variance.

Growing recognition of the importance of measurement error in panel analyses has led several researchers to account for such error in their estimation procedures. Altonji, for example, estimated his labor supply model with an instrumental-variable approach in which he used one wage measure as an instrument for the other. The consistency of the resulting estimator depends on the untested assumption of a zero correlation between the measurement errors in the two wage variables. In a different context, Griliches and Hausman (1986) have suggested using lagged values of error-ridden regressors as instruments for the current values. Consistency then depends on an assumption of zero serial correlation in the measurement error. Similarly, Abowd and Zellner (1985) and Poterba and Summers (1986) have estimated transition rates among labor force categories under the untested assumption that the process generating classification errors is serially independent.

Clearly, the reliability of the techniques that attempt to account for measurement error depends heavily on the error's correlation structure. Furthermore, serial correlation in measurement error is crucial in determining the direction and magnitude of inconsistency for panel estimators that do not account for measurement error. In particular, as illustrated in the examples on income mobility and wage premiums, panel estimation is especially sensitive to measurement error when the "noise" is less serially correlated than the "signal."

It therefore is remarkable that virtually nothing is known about correlation, especially serial correlation, among measurement errors. Methodological survey research on this topic is urgently needed. One worthwhile idea, suggested by Abowd and Zellner, is to introduce a panel aspect to the Census Bureau's reinterview program for the Current Population Survey. Another is the Institute for Social Research's project that will reinterview the above-described PSID validation study sample. Both would enable much-needed investigation of measurement error's serial correlation properties.

2.2. Dynamic Self-Selection

As detailed in Section 1, panel data have been widely used to control for unobserved variables in regression models. For example, longitudinal studies of the wage effects of union status and job hazards have controlled for the error component u_i in equation (1) by differencing it out as in equation (3). Such procedures have been motivated by the belief that cross-sectional OLS estimation of (1) is biased by correlation between u_i and x_{it}. One reason often given for expecting such correlation is that the individual's status with respect to x is endogenously determined by factors including some of those underlying u. For instance, Brown's (1980) study of wage premiums argues that workers with high u may "'spend' some of their greater earning capacity" on reducing their x, that is, on choosing less dangerous jobs.

Most longitudinal analyses of this genre have failed to note that panel estimation also is subject to a similar self-selection bias. Identification of β in panel estimation of equation (3) requires sample variation in the regressor $x_{it} - x_{is}$. This, in turn, typically requires that some of the sampled workers have changed jobs. Since any plausible model of job-changing decisions contains an element of self-selection, $x_{it} - x_{is}$ is likely to be correlated with the error term in equation (3).

This point can be vividly illustrated with an extreme example. Suppose a labor market contains two types of jobs, safe ones and dangerous ones. Assume (implausibly, but for simplicity) that both job types offer identical stochastic wage distributions, so that there is no systematic wage premium for dangerous work. If, however, workers prefer safe jobs to dangerous jobs, a worker with a safe job will switch to a dangerous job only if it happens to offer a sufficiently higher wage. On the other hand, a worker with a dangerous job will switch to a safe job if it offers *any* wage increase or even a sufficiently small decrease. Consequently, one would tend to estimate a positive β in equation (3) because workers choosing to move from safe to dangerous jobs will tend to exhibit larger wage gains than workers moving in the opposite direction. This tendency to estimate a positive β arises entirely from the self-selection of job changers and not from any underlying wage premium.

This example is quite contrived, of course, but the point it highlights is more generally applicable. A formal analysis in Solon (1986) shows that, even where the wage gap between the two job types is nonzero and regardless of whether workers prefer one job type to the other, self-selection causes conventional panel estimators of the wage gap to be inconsistent.[5]

3. SUMMARY

Panel data have facilitated economic research on a wide variety of topics. Some of these topics, such as the extent and correlates of income mobility, simply could not have been addressed without longitudinal information. Nevertheless, the sorts of analyses conducted with panel data have raised new estimation issues. Among the most important of these are potential biases due to sample attrition, measurement error, and dynamic self-selection. Estimation procedures commonly used with panel data are especially vulnerable to bias from measurement error if the measurement error is less serially correlated than is the true value of the measured variable. Despite the practical importance of serial correlation in measurement error, almost nothing is known about its actual properties. Methodological survey research on this topic is urgently needed.

REFERENCES

Abowd, J. M., & Zellner, A. (1985), "Estimating Gross Labor-Force Flows," *Journal of Business and Economic Statistics*, 3, 254–83.

Altonji, J. G. (1986), "Intertemporal Substitution in Labor Supply: Evidence from Micro Data," *Journal of Political Economy*, 94, S176–215.

Altonji, J. G., & Siow, A. (1987), "Testing the Response of Consumption to Income Changes with (Noisy) Panel Data," *Quarterly Journal of Economics*, 102, 293–328.

Ashenfelter, O. (1984), "Macroeconomic Analyses and Microeconomic Analyses of Labor Supply," *Carnegie-Rochester Conference Series on Public Policy*, 21, 117–56.

Ashenfelter, O., Deaton, A., & Solon, G. (1986), *Collecting Panel Data in Developing Countries: Does It Make Sense?* (Living Standards Measurement Study Working Paper No. 23), World Bank.

Bane, M. J., & Ellwood, D. T. (1986), "Slipping into and out of Poverty: The Dynamics of Spells," *Journal of Human Resources*, 21, 1–23.

Becketti, S., Gould, W., Lillard, L., & Welch, F. (1983), *Attrition from the PSID*, Unicon Research Corporation.

Borus, M. E. (1982), "An Inventory of Longitudinal Data Sets of Interest to Economists," *Review of Public Data Use*, 10, 113–26.

Brown, C. (1980), "Equalizing Differences in the Labor Market," *Quarterly Journal of Economics*, 94, 113–34.

Chamberlain, G. (1985), "Heterogeneity, Omitted Variable Bias, and Duration Dependence," in *Longitudinal Analysis of Labor Market Data* (J. J. Heckman & B. Singer, eds.), Cambridge University Press.

Duncan, G. J., & Hill, D. H. (1985), "An Investigation of the Extent and Consequences of Measurement Error in Labor-Economic Survey Data," *Journal of Labor Economics*, 3, 508–32.

Duncan, G. J., & Holmlund, B. (1983), "Was Adam Smith Right after All? Another Test of the Theory of Compensating Wage Differentials," *Journal of Labor Economics*, 1, 366–79.

Duncan, G. J., Juster, F. T., & Morgan, J. N. (1984), "The Role of Panel Studies in a World of Scarce Research Resources," in *The Collection and Analysis of Economic and Consumer Behavior Data* (S. Sudman & M. A. Spaeth, eds.), Bureau of Economic and Business Research, Champaign, IL.

Duncan, G. J., & Morgan, J. N. (1984), "An Overview of Family Economic Mobility," in *Years of Poverty, Years of Plenty* (G. J. Duncan, ed.), Institute for Social Research, The University of Michigan, Ann Arbor.

Freeman, R. B. (1984), "Longitudinal Analyses of the Effects of Trade Unions," *Journal of Labor Economics*, 2, 1–26.

Griliches, Z., & Hausman, J. A. (1986), "Errors in Variables in Panel Data," *Journal of Econometrics*, 31, 93–118.

Hall, R. E., & Mishkin, F. S. (1982), "The Sensitivity of Consumption to Transitory Income: Estimates from Panel Data on Households," *Econometrica*, 50, 461–81.

Hausman, J. A., & Wise, D. A. (1979), "Attrition Bias in Experimental and Panel Data: The Gary Income Maintenance Experiment," *Econometrica*, 47, 455–73.

Heckman, J. J. (1979), "Sample Selection Bias as a Specification Error," *Econometrica*, 47, 153–61.

Heckman, J. J., & Robb, R. (1985), "Alternative Methods for Evaluating the Impact of Interventions," in *Longitudinal Analysis of Labor Market Data* (J. J. Heckman & B. Singer, eds.), Cambridge University Press.

Heckman, J. J., & Singer, B. (1985), "Social Science Duration Analysis," in *Longitudinal Analysis of Labor Market Data* (J. J. Heckman & B. Singer, eds.), Cambridge University Press.

Hill, M. S. (1981), "Some Dynamic Aspects of Poverty," in *Five Thousand American Families—Patterns of Economic Progress* (M. S. Hill, D. H. Hill, & J. N. Morgan, eds.), Institute for Social Research, The University of Michigan, Ann Arbor, Vol. 9.

Hill, M. S., & Duncan, G. J. (1987), "Parental Family Income and the Socioeconomic Attainment of Children," *Social Science Research*, 16, 39–73.

Lillard, L. A., & Willis, R. J. (1978), "Dynamic Aspects of Earning Mobility," *Econometrica*, **46**, 985–1012.

MaCurdy, T. E. (1981), "An Empirical Model of Labor Supply in a Life-Cycle Setting," *Journal of Political Economy*, **89**, 1059–85.

Massagli, M. P., & Hauser, R. M. (1983), "Response Variability in Self- and Proxy Reports of Paternal and Filial Socioeconomic Characteristics," *American Journal of Sociology*, **89**, 420–31.

Mellow, W. (1981), "Unionism and Wages: A Longitudinal Analysis," *Review of Economics and Statistics*, **63**, 43–52.

Poterba, J. M., & Summers, L. H. (1986), "Reporting Errors and Labor Market Dynamics," *Econometrica*, **54**, 1319–38.

Solon, G. (1986), *Bias in Longitudinal Estimation of Wage Gaps* (Technical Working Paper No. 58), National Bureau of Economic Research.

Survey Research Center (1985), *A Panel Study of Income Dynamics: Procedures and Tape Codes, 1983 Interviewing Year—Wave XVI, a Supplement*, The University of Michigan, Ann Arbor.

NOTES

[*]The author thanks Greg J. Duncan, Charles Brown, and an anonymous referee for their helpful comments.

[1]Panel data can facilitate the treatment of unobserved variables in nonlinear models also, but the models and estimation procedures are considerably more complex. See Chamberlain (1985) and Heckman and Singer (1985) for examples.

[2]Duncan, Juster, and Morgan (1984) argue that, as panel respondents become more accustomed to the interviewing process and more aware of its purposes, they provide more accurate information.

[3]The ensuing analysis is similar to that in the appendix to Ashenfelter et al. (1986).

[4]Another relevant, closely related literature is that on goods consumption over the life cycle. See Hall and Mishkin (1982) and Altonji and Siow (1987).

[5]Heckman and Robb (1985) make a similar point with respect to estimating the earnings impact of federally funded training programs.

Sample Dynamics: Some Behavioral Issues*

Lee A. Lillard

1. INTRODUCTION

Panel surveys are most useful for the analysis of models of dynamic behavior, change over time, latent variable models, and measurement error—the kinds of models discussed in the paper by Solon. Heckman discusses circumstances when independent cross sections may do as well or better. Panel data are usually treated as observations of a given economic unit for a fixed number of periods. The purpose of this paper is to point out some of the ways that panel samples themselves are dynamic in ways that may infiuence or be influenced by the behaviors under study.

Almost every empirical study, whatever the behavioral issue, is based on a sample containing relevant information from some target population. In panel surveys, the dynamics of sample membership can play a potentially important role in the content of any particular sample and thus in the consistency and precision of empirical results. While this is true of all longitudinal surveys, it is particularly true of family- or household-oriented surveys like the Panel Study of Income Dynamics (PSID) and the Survey of Income and Program Participation (SIPP). The central focus of this discussion is the PSID because of its unique position as a long-running (20 uninterrupted years), family-oriented survey that attempts to follows all members of the original sample households, and which has collected valuable supplemental information at various points along the way. The purpose of this paper is to point out some aspects of sample dynamics that may have implications for the estimation of behavioral models and that may not be widely recognized. This discussion relies heavily on my own experience with the PSID[1] and other panel data, and in particular on a study of the representativeness of the PSID after 14 years. See Becketti et al. (1983, 1987). The PSID data are described more fully in the appendix to this paper.

Section 2 introduces the issues of sample dynamics that arise in family-oriented panel surveys. These are primarily related to patterns of entry into and exit from the panel. While the implications of sample dynamics should be considered systematically for each type of behavior under study, some general empirical results may be of value. First,

497

Section 3 discusses the potential influence of entry and exit of sample members on behavioral outcomes. The approach is applied to earnings and labor supply models. Second, Section 4 considers the influence of recent behavior on the likelihood of leaving the panel, both for original sample members and new entrants. Section 5 summarizes our conclusions.

2. IN WHAT SENSE IS A LONGITUDINAL SAMPLE DYNAMIC?

First, consider the initial sampling frame, which interacts with the other dimensions of sample chance. Are households or individuals selected at random or on the basis of a probability sample, or are certain groups oversampled? The PSID, for example, initially oversampled low-income households. In 1968, the PSID interviewed 4,802 households/families, of whom 2,930 were selected from the Survey Research Center's master sampling frame. These families are called the SRC sample. Another 1,872 families were drawn from the Bureau of the Census's Survey of Economic Opportunity and are called the SEO sample. Thus one portion (about three-fifths) of the resulting sample (SRC) was a self-weighting nationally representative sample and the other portion (SEO) was a low-income sample based on family income prior to the initial 1968 interview in relation to twice the poverty line. Weights are provided that made the initial full sample nationally representative. Similarly, the SIPP provides weights that made that sample representative. There is currently a controversy about whether these weights should be used in the estimation of behavioral models, as is evidenced by discussion at this conference. Without judging whether weighting is appropriate or not, I will make some comments about the implications of sample dynamics for weighting if it is used.

What then happens to sample composition in a household- or family-oriented survey that runs for many years? The original families or household units do not remain as they were. Couples split up, individuals going their own ways to form new family units; children grow up, leave home, and form their own families; individuals die and the survivors form new household units; some individuals or families decide to leave the survey; and some new individuals join the households of members of the original survey, and perhaps later leave those household units. The implication is that there is entry into and exit from sample households, that these households split to form many households, and that some households exit altogether.

2.1. Attrition

A well-known major source of sample dynamics is attrition from the sample. Attrition of individuals is drastically reduced when all individuals of the original households are followed. This is extremely important for studies of living arrangements, the emergence of children into adulthood, migration, and poverty among female-headed households, for example. However, not all individuals can be found or will agree to participate in the survey repeatedly as time goes on.

The attrition rate experienced by the PSID has been quite low. After an initial loss of about 14% between the first and second waves, the annual full sample attrition rate has been about 2% among the original sample members. The rates are slightly higher for male heads and slightly lower for wives and female heads (including childless women). See Table 1 for the distribution by years in the sample. The effect of attrition accumulates, leaving only about 60% of the original sample members present after 14 years for the 1981 survey. These attrition rates are similar to and compare favorably with the experience of other annual panel surveys such as the National Longitudinal Surveys (NLS). The NLS surveys are individual- rather than family-oriented, so that the comparison is with male heads or wives and female heads. See Rhoton (1986).

There is some evidence appearing at this conference and elsewhere that response rates are "per survey requested of respondents." That is, for example, the attrition rates are not dramatically lower after two- or three-year intervals in the NLS than after the single-year intervals, and the attrition rates are not much better between the much more frequent surveys in the SIPP. More will be presented later concerning attrition in the PSID.

These rates of attrition, resulting in a loss of 40% of the original sample over 14 years, introduce substantial potential for the remaining sample to become less and less representative over time. Since the issue of the effects of attrition must be addressed separately for each behavioral study, it is very important that the data for those individuals who leave the sample over time be retained and made available for analysis.[2] These observations may be valid for some analyses, thus almost doubling sample size, and may be used to assess the impact of attrition for others.

2.2. Entry of Nonsample Individuals

A second major source of sample dynamics is much less widely recognized. This is the introduction of "nonsample" individuals who joined the households of original sample members, but who were not a part of the survey in 1968. While they are not a part of the official "sample," the full set of survey information is collected for these individuals. The information gain from using the additional data can be substantial. However, current procedures do not include following these individuals once they leave the household of a sample member. Therefore, the attrition rate for them is higher.

TABLE 1. DURATION IN THE PSID PANEL FOR THE ORIGINAL SAMPLE

Duration (Years)	Full Sample		Male Heads	Wives and Female Heads
	SRC	SEO		
1	14.5	10.1	14.3	11.8
2	2.7	3.4	3.9	2.9
3	2.2	1.9	2.6	2.1
4	1.8	2.6	3.1	1.7
5	2.2	2.5	2.7	2.1
6	2.1	2.1	3.0	1.9
7	1.8	2.3	2.2	1.7
8	2.1	3.0	2.6	2.4
9	1.6	2.5	2.1	2.1
10	1.9	1.6	2.0	1.9
11	1.6	2.4	2.0	1.6
12	1.6	2.7	1.7	1.8
13	2.3	2.7	2.2	1.9
14	61.6	60.4	55.7	65.1

Several issues arise immediately. First, how do these individuals affect the weighting scheme developed to adjust for the initial sampling frame? This an open subject for study. The current procedure in the PSID is to give these individuals a weight of zero so that they are implicitly excluded from any behavioral analysis using the individual weights.[3] On the other hand, an unweighted analysis includes these individuals with full weight.

How many nonsample individuals are we talking about? Table 2 gives this information for male heads and for wives and female heads in the PSID. Compared to the 4,888 sample member male heads ever present in the PSID, the sample size can be enhanced by 32% to 6,452 by the inclusion of nonsample male heads. Similarly, the sample of wives and female heads can be increased from 6,292 by 26% to 7,903. Note, however, that this is virtually all wives. Since nonsample individuals are not followed when a split occurs, there are no nonsample female heads. Therefore, the sample composition is clearly and directly affected by the behavior of the individuals involved, namely marriage and separation. Wives are clearly overrepresented in an unweighted analysis of adult women, and the nonsample wives of sample members are completely excluded in a weighted analysis.

This introduces the second issue related to nonsample individuals. The inclusion of these individuals is prima facie not unrelated to certain behavior. They have moved into the household of a sample individual and thus have experienced events related to migration, living arrangements, and possibly marriage. The exit of these individuals is clearly

TABLE 2. SAMPLE SIZES, SAMPLE MEMBERSHIP, AND ATTRITION

Subsample	Total Sample	Male Heads	Wives and Female[a] Heads
Total ever present	30,903	6,452	7,903
Percentage sample members	76.5	75.8	79.6
Percentage remaining in 1981	66.5	66.5	71.7
Percentage sample members and remaining in 1981		49.4	57.2

[a]Includes single persons.

more directly related in an inverse way to these same events.

2.3. Implications for Cross-Sectional Samples

Cross-sectional sample sizes, as they would appear on each year's distribution tape, are presented in Table 3. The sample sizes are remarkably stable across the years for both male heads and women who are wives or female heads, as well as for the separate wife and female head subgroups. If the cross sections were independent, they approximately replicate a CPS-type survey. However, observations are not independent over time and the stability of sample size masks a great deal of compositional change in them over time. In fact, there is a tendency for the cross-sectional sample sizes to grow over time. Increasing sample sizes will not occur in individual-oriented samples like the NLS.

The proportion who are sample members declines substantially over the 14-year period. This is due to the marriage of children who split off from the parental household as they grow up, and to the dissolution and reformation of family and household relationships that involve new nonsample persons. Analyses based on weighted data would implicitly include only the sample individuals, and thus the increasing sample size is turned into a decline.

Those present in 1981 (and thus present in the file) are a declining proportion of the earlier years' potential full samples. The resulting sample sizes for earlier years increasingly underrepresent the corresponding years' samples. These represent the samples for earlier years that would be obtained from the 1981 distribution tape if the nonresponse records (as of 1981) were not incorporated.

The fraction representing the SRC sample is relatively stable, with some decline due to the split off of male children from the SEO sample (who had larger families) and due to greater marital instability among low-income households (resulting in more nonsample additions).

TABLE 3. CROSS-SECTIONAL SAMPLE SIZES AND COMPOSITIONS

Year	Male Head			Wives and Female Heads		
	Sample Size	Sample Members	Pres. 1981	Sample Size	Sample Members	Pres. 1981
1968	3253	100.0	57.3	4286	100.0	65.3
1969	2955	97.2	66.0	4007	98.0	73.1
1970	3054	93.8	68.6	4136	95.7	75.3
1971	3175	90.7	70.8	4294	93.6	76.7
1972	3326	87.7	73.2	4480	90.9	77.9
1973	3479	85.0	75.6	4655	88.8	79.6
1974	3620	83.2	78.3	4814	87.3	81.6
1975	3754	82.0	81.1	4950	86.3	83.5
1976	3831	80.2	84.9	5067	84.9	85.9
1977	3950	78.6	87.4	5141	83.8	88.4
1978	4069	77.1	90.2	5249	82.6	90.7
1979	4214	75.5	93.4	5450	80.2	96.2
1980	4351	74.1	97.2	5585	80.2	96.2
1981	4129	75.4	100.0	5372	81.1	100.0
Person Years	51,160	83.4	81.7	67,486	87.5	84.3

Year	Wives			Female Heads		
	Sample Size	Sample Members	Pres. 1981	Sample Size	Sample Members	Pres. 1981
1968	3012	100.0	67.3	1346	100.0	60.9
1969	2811	97.2	75.2	1269	99.8	68.2
1970	2870	94.1	76.9	1309	99.5	71.6
1971	2932	90.9	78.3	1362	99.5	73.3
1972	3038	87.1	79.4	1416	99.3	74.7
1973	3134	83.8	80.8	1468	99.4	76.8
1974	3206	81.2	82.9	1521	99.7	78.4
1975	3248	79.5	84.8	1568	99.6	80.8
1976	3287	77.1	87.1	1596	99.6	84.0
1977	3305	75.1	89.0	1627	99.8	87.5
1978	3371	73.2	90.7	1630	99.9	91.0
1979	3477	70.5	92.9	1693	100.0	93.0
1980	3552	69.1	95.7	1701	100.0	97.6
1981	3353	69.9	100.0	1743	100.0	100.0
Person Years	44,596	81.4	84.9	21,249	87.5	82.3

2.4. Implications for Longitudinal Samples

A particular advantage of panel data is the ability to study behavior in a longitudinal context. This includes components of variance models for panel data, dynamic models for individual time series, event dating and duration models, and many others. Frequently, the particular period of time for which a behavioral unit is observed is only indirectly relevant (e.g., as an explanatory factor). All relevant intervals of observation may be used whether or not the beginning and ending dates of the time interval are the same. Similarly, unequal numbers of replication, or different interval lengths, may be a computational nuisance, but may present no conceptual difficulty.

There are some basic choices to be made in choosing a longitudinal sample, each of which may affect the appropriate choice of behavioral model and estimation procedure (see Table 4). First, are current (1981) nonrespondents to be included? Doing so enhances sample sizes substantially, by 50% for men and by 40% for women (assuming nonsample individuals are included). One implication of including the nonrespondents for behavioral modeling is that individuals are not observed for the same length of time (duration), and some are observed for nonoverlapping time periods. This can introduce substantial complexity into the development of statistical software for dynamic models. Even this use of the data assumed that these individuals are missing at random.

The second choice is whether to include nonsample individuals for whom longitudinal data are available. This is related to the choice of whether to do weighted or unweighted analysis. There are potentially 6,452 male heads and 7,903 female heads and wives. If weighted analysis is used, then the longitudinal data for 1,564 nonsample men and 1,611 nonsample women receives zero weight. This is a substantial loss of information for these important groups. Unweighted analysis using the SRC (self-weighting) sample results in an even larger loss of sample size.

2.5. Individual Status Changes over Time

A third sense in which a panel sample is dynamic is that the individuals or households in a sample may change their "status" over time. For example, children become adults, heads of households, wives, and parents over the course of time. Single individuals marry, and couples divorce so that wives become female heads of households, and so forth. Therefore, the composition of a sample of individuals of a particular type can vary substantially over time and samples can be expanded to include individuals "ever as" the type. This is particularly important in the analysis of transitions and other dynamic behavior. To illustrate the substantial amount of change in status, Table 5 reports marital status and transitions between wife and female head (including childless) for women over the period that they are either.

3. ARE BEHAVIOR AND SAMPLE DYNAMICS RELATED?

This question must be answered in the context of each particular be-

TABLE 4. LONGITUDINAL SAMPLE SIZES

| | Male Heads[a] | | Wives and Female Heads[a] | |
Subsample	Total	SRC	Total	SRC
Total ever present	6452	3928	7903	4554
Sample members	4888	3107	6292	3618
Nonsample	1564	821	1611	936
Present in 1981[b]	4291	2638	5670	3250
Sample members	3189	2029	4522	2546
Nonsample heads	1102	609	1148	704
Nonrespondents in 1981[c]	2161	1290	2233	1304
Sample members	1699	1078	1770	1182
Nonsample heads	462	212	463	232

[a]Includes single persons.

[b]Present on the 1981 Wave 14 public use tape.

[c]Missing from the 1981 Wave 14 public use tape.

havioral model and application of panel data. In this section we take two simplistic approaches, in the context of the PSID panel. The first is to determine if the parameters of a behavioral model are different for those individuals entering or exiting. We consider earnings and labor supply equations here. The second is to determine whether attrition from the sample is related to demographic characteristics and recent behavior. These results are only meant to be suggestive of potential areas for further research. More rigorous approaches should be undertaken in the context of particular applications.[4]

If an analysis is to be based on individual time series for the behavioral variables or on event dates over the interval, then it is important for the duration of the interval of observation to be unrelated to the behavioral process being studied. In the case of event histories, censoring of event durations should be random. In the case of individual time series, or pooled replications, the outcomes for each period should be unrelated to attrition.

3.1. Is Behavior Related to Entry and Exit from a Panel?

One approach to considering the relationship between behavior and sample dynamics is to compare the behavior of individuals who have different patterns of entry and exit.[5]

Separate comparisons were made for male heads, for female heads,

TABLE 5. MARITAL STATUS AND TRANSITIONS FOR ADULT WOMEN

Duration as Wife or Head (Years)	No Change		One Change	Two Changes	Three or More
	Always a Wife	Always a Head			
1	67.6	32.4			
2	46.4	44.7	8.7		
3	41.3	40.0	17.0	1.6	
4	40.3	38.5	18.8	2.2	0.0
5	40.5	33.1	19.8	4.3	2.1
6	43.5	23.0	22.6	7.2	3.4
7	44.2	28.1	21.2	4.1	2.3
8	40.0	25.0	24.2	7.9	2.8
9	37.1	27.4	24.4	7.5	3.4
10	39.6	20.1	29.9	7.7	2.6
11	36.8	19.4	24.8	13.9	5.0
12	34.3	22.5	26.4	12.2	4.4
13	36.0	17.1	29.5	10.6	5.8
14	51.4	20.7	19.0	6.4	2.3

and for wives in the PSID. The behavior is represented by a standard log annual labor earnings equation and a simple log annual hours of work equation for those who worked in each group. Standard regression procedures are used and no attempt is made to correct for self-selection of workers. Both weighted and unweighted comparisons were made. The regression equations include race (indicators for black and for other nonwhite), years of schooling (linear spline below and above high school), potential work experience and its square, and census regions. The entry and exit patterns to be compared here are represented by a comparison of (1) those who subsequently leave the panel versus those who do not, and (2) sample versus nonsample individuals.

To compare those who leave the panel with those who do not, we compare the estimated equations for 1967 earnings and hours (from the original 1968 survey) for three distinct subsets of the original 1968 sample members: (1) those who remain throughout the panel to 1981, (2) those who leave after 1975 but before 1981, and (3) those who leave at or before 1975. Various degrees of difference are allowed and tested (using a standard F-test), including a simple intercept difference and a fully different set of coefficients. For the most part, these results suggest that labor income and hours of work are related to these variables in the same way for those who stay and for those who leave. This was true for male heads, female heads, and wives.

To consider entry into a panel, we compare the estimated equations for 1980 earnings and hours (from the 1981 survey) for sample and for nonsample individuals. Again, various degrees of difference are allowed and tested (using a standard F-test), including a simple intercept difference and a fully different set of coefficients. However, only unweighted comparisons were made, since nonsample individuals are assigned zero weight. And again, no significant differences were found.

3.2. Is the Attrition Rate Related to Recent Behavior?

Another approach to considering the relationship between behavior and sample dynamics is to compare the rate of attrition among individuals with different behavior. This introduces a wider range of behavior than the earnings and hours of work equations considered above.

In this section we consider a simple descriptive model of attrition from the sample and thus of duration in the panel. We use a simple (Weibull) proportional hazards model. In each period t, an individual has a probability of leaving the survey before the next period $t + 1$ given by a Weibull hazard function. The Weibull specification allows for an increasing, decreasing, or constant rate of exit as the duration in the sample increases. The proportional hazard formulation allows for the introduction of time-varying covariates; a log-linear regression specification is used here.

The estimates of the Weibull parameters and the regression coefficients are presented in Table 6 for both sample and nonsample individuals and for both male heads and for female heads and wives. These estimates are based on the longest period of continuous participation in the survey as a male head or as a wife or female head. Children, other household members, and nonsample persons enter into these categories after 1968.

First, the estimates indicate the significantly greater exit rate after the first interview. This is a widely recognized phenomenon in most panel surveys. The Weibull parameters (gamma 2 in particular) indicate further that the hazard rate is constant thereafter, indicating that an exponential model is sufficient. The level of the hazard rate (gamma 1) across groups indicates that women have a lower hazard rate than men—the rate is exp(gamma 1)—and that sample individuals have a lower hazard rate than nonsample individuals.

Some general patterns emerge for both men and women. Two clear results are that nonwhites are more likely to leave the panel, and that leaving the sample is related to mobility, having moved, or planning to move in the near future. Studies focusing on migration and related behavior may be affected by attrition.

The rate of exit increases with age and is especially great for men who have recently retired or become disabled. Therefore, behaviors related to aging and retirement may be affected.

The exit rate is greater for low-income individuals as measured by

TABLE 6. PROPORTIONAL HAZARD MODEL FOR SAMPLE ATTRITION

Variable	Male Heads		Female Heads and Wives	
	Sample	Nonsample	Sample	Nonsample
First interview	1.5006***	−0.2127	1.5092***	−1.1929***
	(0.1479)	(0.2147)	(0.1351)	(0.2481)
SEO subsample	−0.0177	0.0301	−0.1211*	0.2611**
	(0.0674)	(0.1093)	(0.0660)	(0.1269)
Age spline < 25	−0.0141	−0.0402	0.0207	0.0346
	(0.0243)	(0.0329)	(0.0199)	(0.0262)
Age spline 25 < A < 45	−0.0092*	−0.0022	−0.0116***	0.0322***
	(0.0047)	(0.0093)	(0.0043)	(0.0107)
Age spline > 45	0.0265***	−0.0072	0.0367***	0.0147
	(0.0036)	(0.0117)	(0.0029)	(0.0154)
Nonwhite	0.3277***	0.5704***	0.3021***	0.4726***
	(0.0650)	(0.1085)	(0.0645)	(0.1282)
Taxable income ($1,000)	−0.0182***	0.0026	−0.0427***	−0.0120***
	(0.0034)	(0.0045)	(0.0055)	(0.0033)
Might move	0.2645***	0.1886**	0.2532***	0.4970***
	(0.0598)	(0.0925)	(0.0560)	(0.1036)
Just moved	0.3185***	0.1199	0.4037***	1.1744***
	(0.0717)	(0.0971)	(0.0672)	(0.1062)
Welfare received	0.0316	0.4150**	−0.0508	−0.5949**
	(0.1121)	(0.1666)	(0.0791)	(0.2548)
Retired or disabled	0.3539***	0.8493***		
	(0.0905)	(0.2151)		
Unemployed	0.5899***	0.5517***		
	(0.1130)	(0.1763)		
Hazard Parameters				
Gamma 1	−3.2989***	−2.1600**	−4.5286***	−4.1105***
	(0.5655)	(0.8956)	(0.5056)	(0.6875)
Gamma 2	0.0164	−0.1483	−0.0557	0.0202
	(0.0491)	(0.0970)	(0.0466)	(0.0958)
Log l	−6347.3	−1878.1	−7287.6	−1542.8
Sample size	4,888	1,564	6,292	1,611
Mean number periods	8.1	4.8	8.7	4.6

Notes: Standard errors in parentheses; *, **, and *** denote significance levels of .10, .05, and .01, respectively, from zero.

the level of taxable family income. However, receipt of welfare is not associated with attrition except for nonsample men who are more likely to leave and nonsample women who are less likely to leave. This may be related to eligibility requirements for welfare. While the SEO subsample was a lower-income group in 1967, there is no difference in the exit rate of SRC and SEO men. SRC and SEO women are different, but in opposite directions for sample and nonsample women. The signs are in a reasonable direction.

Men who were unemployed in the last year are significantly more likely to leave the sample before the next interview than those who were not. This raises the possibility that models of unemployment spells and job change will be affected by attrition. If there is heterogeneity in the probability of unemployment, then studies of unemployment based on detailed data collected in special surveys in the eighties may under represent the level of unemployment. These issues deserve further study.

4. CONCLUDING REMARKS

This paper has been concerned with possible interrelationships between the entry and exit of individuals to and from a panel survey and the behavior of the individuals in the panel. First, we described the most important dimensions of entry and exit, namely the entry of nonsample individuals who become members of the households of sample households and the exit of individuals through attrition. It was shown that the inclusion of nonsample individuals or of the earlier observations for those who leave the panel can dramatically affect the size and character of the sample used for the estimation of a particular behavioral model.

These issues are particularly important for a long-running family-oriented panel like the PSID, which we consider in detail, but are also relevant for shorter household-oriented panels like the SIPP and for long-running individual-oriented panels like the NLS.

Even the simple descriptive information about the dynamics of sample size in a panel survey that we presented here suggest some implications concerning weighting and imputation. Remember that the original weights at the beginning of panel are designed to adjust for stratified sampling procedures. Since panel data are most useful for models requiring longitudinal data, let us consider that case. We would also argue that all individuals with relevant panel intervals be included, whether or not an individual remains in the panel in the most recent survey (assuming for now that attrition is random). If one believes that weighting is appropriate for behavioral models, then the original 1968 sample weights would seem most appropriate. There is no need to update the weights over time. This begs the question of how to weight nonsample individuals should they be included. One possibility is to weight them the same as the other (sample) members of the 1968 households from which they are generated. If unweighted analysis is used, then the issue does not arise.

Some analysts have suggested imputation procedures to "fill-in" the missing values generated by wave nonresponse. While imputation procedures may be useful for generating missing values for random item nonresponse, the idea seems futile in the case of wave nonresponse generated by attrition, or the entry of nonsample individuals. Consider the following implication of full imputation. There are 4,888 male heads who are sample members in the PSID. While there are potentially 68,432 person years of data, only 39,555 are actually observed, or 57.8%. Imputation would almost double the number of person years without adding any real information. While the suggestion is clearly inappropriate in the context of a very long panel like the PSID, it is less obvious in the context of shorter panels like the rotating groups in the SIPP. These issues can only become more important as time progresses.

In addition, we examined the relationship between entry and exit of individuals and their behavior. For simple earnings and hours-of-work regression equations we found no evidence of differences for those who subsequently leave the sample and those who do not and we found no difference for those nonsample individuals who entered the sample from the original sample members. However, we did find that the exit rate from the PSID panel after a given interview was related to certain individual "behavior" reported in that interview. That is, leaving the panel is related to planned and actual mobility, to age, retirement and disability, to income (but only partially to receipt of welfare), and to having experienced unemployment. This suggests that further, more careful, study should be given to these issues.

APPENDIX: FEATURES OF THE PSID DATA

The basic character of the PSID will be important to our discussion. In this section we outline the essential features. While the PSID data are distributed in a number of forms, we will limit our discussion to the individual longitudinal tape.

In 1968, the PSID interviewed 4,802 households/families. Of this group, 2,930 families were selected from SRC's master sampling frame. These families (and/or the members of these families) are called the SRC sample. The other 1,872 families were drawn from the Bureau of the Census's Survey of Economic Opportunity (SEO) and are called the SEO sample.

A major goal of the PSID was to facilitate study of the determinants of poverty. Since a random sample of 5,000 families would include too few poverty and minority families, the SEO sample was added to the PSID. The SEO sample is a subset of the approximately 30,000 families interviewed in 1966 and 1967 for the Survey of Economic Opportunity. One important criterion for inclusion in the PSID was that the household had family income in 1966 less than or equal to twice the

1966 poverty line for the corresponding family size.

The 1968 SRC sample may be treated as approximately a random sample of U.S. families. The User Guide to the PSID (SRC, 1984) suggests unweighted analysis if only the SRC sample is included and weighted analysis if the SRC and SEO are combined. In 1968, the PSID calculated weights for each family that represent the ex-ante probability that a family appears in the PSID. Each individual in the sample in 1968 is assigned the weight corresponding to the family. Since 1978 the individual weights have been recalculated annually for each individual in the sample to account for differential nonresponse rates in the succeeding waves of the PSID.

A substantial number of individuals in each wave after 1968 are assigned a weight of zero. These are the so-called nonsample individuals, that is, persons who entered the sample after 1968 through marriage or a living arrangement that placed them in the same household with a sample person. The PSID assigns a weight of zero to these individuals to indicate that they are not a part of the original panel design. However, the same survey information is collected for them. A child born into the sample household is given an individual weight that is the average of the parents' weights. Therefore, any weighted analysis implicitly omits them.

Another important feature of nonsample persons is that no attempt is made to continue interviewing them if they stop residing with a sample person. Sample persons, on the other hand, are pursued even if they leave their original family. Such sample persons are called split-offs.

Until recently, behavioral researchers have not had access to the records of those individuals or families who leave the PSID sample— nonrespondents. Any family or individual that does not respond to the most current wave survey is removed from the distribution tape.

The most frequently used analysis samples are (1) the SRC probability sample and (2) all sample members (with positive weights) from the combined SRC and SEO samples. Analysis has almost always been limited to those individuals present on the current distribution tape (1981 in this case). Therefore we consider each of these subgroups. The results reported in this paper are based on a nonrespondent file constructed for an evaluation of the representativeness of the PSID as of 1981 (Wave 14). The data thus refer to the period 1968–1981 rather than 1983 as in the Non-Respondent File recently released by the PSID.

REFERENCES

Becketti, S., Gould, W., Lillard, L., & Welch, F., (1983), *Attrition from the PSID*, Unicon Report.
Becketti, S., Gould, W., Lillard, L., & Welch, F., (1987, rev.), "The PSID after Four-

teen Years: An Evaluation," *Journal of Labor Economics*, forthcoming.

Lillard, L., & Waite, L. (1987), *Children and Marital Stability* (RAND working paper), RAND Corporation, Santa Monica, CA.

Lillard, L. (1983), "A Model of Wage Expectations in Labor Supply," in *Panel Data on Incomes* (A. B. Atkinson & F. A. Cowell, eds.), ICERD Volume.

Lillard, L., & Willis, R., (1978), "Dynamic Aspects of Earning Mobility," *Econometrica*, **46**.

Rhoton, P. (1986), "Attrition and the National Longitudinal Surveys of Labor Market Experience: Avoidance, Control and Correction," *IASSIST Quarterly*, 23–35.

Survey Research Center (1984), *User Guide to the PSID*, Inter-university Consortium for Political and Social Research, Institute for Social Research, The University of Michigan, Ann Arbor.

NOTES

*Support for this paper was provided by the RAND Corporation, but it builds on earlier work by Becketti, Gould, Lillard, and Welch (1983, 1987) funded by the National Science Foundation.

[1] See Lillard & Willis (1978) and Lillard (1983) for examples of models of earnings, poverty, and labor supply dynamics, and see Lillard & Waite (1987) for an application to marital stability and childbearing. Ongoing research includes analysis of the relationship of attrition to marital stability and the relationship between retrospective reports of marital status and contemporaneous reports over the period of a panel.

[2] The PSID has, for example, recently constructed a file of data for nonrespondents who had left the survey prior to 1983.

[3] An unsuspecting analyst might even overstate the sample size he has used since nonsample individuals are not otherwise indicated without a careful look at the documentation.

[4] One example, under study by the author, is the relationship between attrition and divorce. Individuals experiencing a divorce may be less willing to respond to a detailed survey about their behavior. If so, then estimated divorce rates will be understated unless the estimation procedure addresses the issue.

[5] These results are taken largely from the analysis of Becketti et al. (1983, 1987). That study also compared these behavioral relationships in the PSID with those in data from the CPS.

The Value of Longitudinal Data for Solving the Problem of Selection Bias in Evaluating the Impact of Treatments on Outcomes*

James J. Heckman and Richard Robb

This paper considers the following problem. Persons are given (or else select) "treatments" but the assignment of persons to treatments is nonrandom. Differences in measured outcomes among persons are due to the treatment and to factors that would make people different on outcome measures even if there were no causal effect of the treatment. "Treatment" as we use the term may be a drug trial, a training program, attending school, joining a union, or migrating to a region. We assume that it is not possible to simultaneously observe the same person in the treated and untreated states. If it is possible to observe the same person in both states, the problem of selection bias disappears. By observing the same person in the treated and untreated states, it is possible to isolate the treatment in question without having to invoke further assumptions.

We consider minimal identifying assumptions required to isolate the parameters of interest. We present assumptions that exploit three types of widely available data to solve the problem of estimating the impact of treatments on outcomes free of selection bias: (1) a single cross section of posttreatment outcome measures, (2) a temporal sequence of cross sections of unrelated people (repeated cross-sectional data), and (3) longitudinal data in which the same individuals are followed over time. These three types of data are listed in order of their availability and in inverse order of their cost of acquisition (total and not necessarily per observation). Assuming random sampling techniques are applied to collect all three types of data, the three sources form a hierarchy: longitudinal data can be used to generate a single cross section or a set of temporal cross sections in which the identities of individuals are ignored, and repeated cross sections can be used as single cross sections.

Although longitudinal data are widely regarded in the social science and statistical communities as a panacea for selection, causality, and simultaneity problems, there is no need to use longitudinal data to identify the impact of treatments on outcomes if conventional specifications of outcome equations are adopted. Estimators based on repeated cross-sectional data for unrelated persons identify the same parameter. This is true for all conventional longitudinal estimators. A major conclusion of this paper is that the relative benefits of longitudinal data have been overstated, because the potential benefits of cross-sectional and repeated cross-sectional data have been understated. This conclusion has the positive implication that conventional longitudinal estimators can be implemented on more widely available repeated cross-sectional data.

When minimal identifying assumptions are explored for models fit on the three types of data, we find that *different* and not necessarily more plausible assumptions can be invoked in longitudinal analyses than in cross-sectional and repeated cross-sectional analyses. The fact that more types of minimal identifying assumptions can be invoked with longitudinal data (since the longitudinal data can be used as a cross section or a repeated cross section) does not make more plausible those assumptions that uniquely exploit longitudinal data.

We demonstrate that unless explicit distributional assumptions are invoked, all cross-sectional estimators require the presence of at least one regressor variable in the decision rule determining the assignment of persons to treatment. This requirement may seem innocuous, but it rules out a completely nonparametric cross-sectional approach. Without a regressor in the assignment equation, it is necessary to invoke distributional assumptions. Longitudinal and repeated cross-sectional estimators do not require a regressor in the assignment equation. This is a major advantage of such data.

The focus of this paper is on model identification and not on estimation or on the efficiency of alternative estimators.[1] Identification is a rather sharp concept that may not be all that helpful a guide to what will "work" in practice. Different identified estimators may perform quite differently in practice, depending on the degree of overidentification or on the nature of the identifying restrictions. However, if a parameter is not identified, an estimator of that parameter cannot have any desirable statistical properties. Securing identification is a necessary first step toward construction of a desirable estimator but certainly is not the last step. This paper concentrates on the necessary first step. Our focus on identification and on the trade-offs in assumptions that secure identification should make clear that we are not offering a nostrum for selection bias that "works" in all cases.

We focus on identification because the current literature in social statistics is unclear on this topic. A major goal of this paper is to demonstrate that previous work on selection bias has often imposed unnecessarily strong assumptions (e.g., normality). Part of the great variability in estimates obtained in some analyses using selection-bias procedures may be due to the imposition of different types of "extra conditions" not required to identify the parameters of interest. Separating out essential from inessential assumptions is a main goal of this paper and is a necessary step in evaluating the merits of longitudinal data.

We use large-sample theory in our analysis. Given the size of many social science data sets with hundreds and thousands of independent observations and given the available Monte Carlo evidence, large-sample methods are reliable. For this reason, we view our large-sample analysis as the natural point of departure for research on selection models. We discuss efficiency or variance questions in the conclusion to this paper.

This paper draws heavily on our previous work (Heckman & Robb, 1985, 1986). To avoid repetition and to focus on essential points, we refer the reader to our longer companion papers for technical details of many arguments and a more comprehensive treatment of the issues discussed in this paper.

The organization of this paper is as follows. Section 1 describes the notation and a behavioral model of the assignment of persons to treatment. Section 2 discusses the definition of the appropriate causal or structural parameter of interest. Sections 3–7 present a discussion of alternative estimation methods for different types of data. The paper concludes with a brief summary of the main points and a discussion of the efficiency of alternative estimators.

1. NOTATION AND A MODEL OF PROGRAM PARTICIPATION

1.1. Outcome Functions

To focus on essential aspects of the problem, we assume that individuals have only one opportunity to take the treatment. This opportunity occurs in period k. Treatment takes one period for participants to complete.

Denote the latent outcome of individual i in period t by Y_{it}^*. This is the outcome measure of an individual in the absence of any treatment programs. Latent outcomes depend on a vector of observed characteristics, X_{it}. Let U_{it} represent the error term in the latent outcome equation and assume that

$$E(U_{it} \mid X_{it}) = 0 .$$

Adopting a linear specification, we write the latent outcome as

$$Y_{it}^* = X_{it}\beta + U_{it}$$

where β is a vector of parameters. Linearity is adopted only as a convenient starting point and is not an essential aspect of any of the methods presented in this paper. Throughout this paper we assume that the mean of U_{it} given X_{it} is the same for all X_{it}. Sometimes we will assume independence between X_{it} and current, future, and lagged values of U_{it}. When X_{it} contains lagged values of Y_{it}^*, we assume that the equation for Y_{it}^* can be solved for a reduced form expression involving only exogenous regressor variables and we use this expression in our paper. Under standard conditions it is possible to estimate the original structure from the reduced form so defined.

Under these assumptions β is the coefficient of X in the conditional expectation of Y^* given X. Observed outcome Y_{it} is related to the latent outcome Y_{it}^* in the following way:

$$Y_{it} = Y_{it}^* + d_i\alpha, \quad t > k$$

$$Y_{it} = Y_{it}^*, \quad t \leq k$$

where $d_i = 1$ if the person takes treatment and $d_i = 0$ otherwise, and where α is one definition of the causal or structural effect of treatment on outcomes. Observed outcomes are the sum of latent outcomes and the structural shift term $d_i\alpha$ that is a consequence of treatment. Y_{it} is thus the sum of two random variables when $t > k$.

The problem of selection bias arises because d_i may be correlated with U_{it}. This is a consequence of selection decisions by agents. Thus selection bias is present if

$$E[U_{it}d_i] \neq 0 .$$

All selection bias models must impose a new assumption onto the original model to undo the impact of this nonzero covariance.

Observed outcomes may be written as

$$Y_{it} = X_{it}\beta + d_i\alpha + U_{it}, \quad t > k \tag{1}$$

$$Y_{it} = X_{it}\beta + U_{it}, \quad t \leq k$$

where β and α are parameters. Because of the covariance between d_i and U_{it}

$$E(Y_{it} \mid X_{it}, d_i) \neq X_{it}\beta + d_i\alpha.$$

Throughout much of this paper we ignore effects of treatment that grow or decay over time. (See our companion papers for a discussion of this topic.) We next develop the stochastic relationship between d_i and U_{it} in equation (1).

1.2. Assignment Rules

The decision to be treated can be described in terms of an index function framework. Let IN_i be an index of net benefits to the appropriate decision maker from taking the treatment. It is a function of observed (Z_i) and unobserved (V_i) variables. Thus

$$IN_i = Z_i\gamma + V_i . \tag{2}$$

In terms of this function,

$$d_i = 1 \text{ iff } IN_i > 0$$

$$d_i = 0 \text{ otherwise.}$$

The distribution function of V_i is denoted as $F(v_i) = Pr(V_i < v_i)$. V_i is assumed to be independently and identically distributed across persons. Let $p = E[d_i] = Pr[d_i = 1]$ and assume $1 > p > 0$. Assuming that V_i is distributed independently of Z_i (a requirement not needed for most of the estimators considered in this paper), we may write $Pr(d_i = 1 \mid Z_i) = 1 - F(-Z_i\gamma)$, which is sometimes called the "propensity score" in statistics. (See, e.g., Rosenbaum & Rubin, 1983.)

The condition for the existence of selection bias

$$E[U_{it}d_i] \neq 0$$

may occur because of stochastic dependence between U_{it} and the unobservable V_i in equation (2) (selection on the unobservables) or because of stochastic dependence between U_{it} and Z_i in equation (2) (selection on the observables).

2. RANDOM COEFFICIENTS AND THE STRUCTURAL PARAMETER OF INTEREST

In seeking to determine the impact of treatment on outcomes in the presence of nonrandom assignment of persons to treatment, it is useful to distinguish two questions that are frequently confused in the literature.

1. What would be the mean impact of treatment on outcomes if people were randomly assigned to treatment?
2. How do the postprogram mean outcomes of the treated compare to what they would have been in the absence of treatment?

The second question makes a hypothetical contrast between the postprogram outcomes of the treated in the presence and in the absence of treatment. This hypothetical contrast eliminates factors that would make the outcomes of the treated different from those of the untreated even in the absence of any treatment. The two questions have the same answer if equation (1) generates outcomes so that treatment has the same impact on everyone. The two questions also have the same answer if there is random assignment to treatment and attention centers on estimating the *population* mean response to treatment.

In the presence of nonrandom assignment and variation in the impact of treatment among persons, the two questions have different answers. Question (2) is the appropriate one to ask if interest centers on forecasting the change in the mean of the posttreatment outcomes of participants compared to what *they* would have earned in the absence of treatment when the same selection rule governs past and future participants. It is important to note that the answer to this question is all that is required to estimate the future program impact if future selection criteria are like past criteria and it is all that is required to evaluate the gross return from treatment.

To clarify this distinction, we consider a random-coefficient version of equation (1) in which α varies in the population. In this model, the impact of treatment may differ across persons and may even be negative for some people. We write in place of equation (1)

$$Y_{it} = X_{it}\beta + d_i\alpha_i + U_{it}, \quad t > k.$$

Define $E[\alpha_i] = \bar{\alpha}$ and $\epsilon_i = \alpha_i - \bar{\alpha}$ so $E[\epsilon_i] = 0$. With this notation, we can rewrite the equation above as

$$Y_{it} = X_{it}\beta + d_i\bar{\alpha} + \{U_{it} + d_i\epsilon_i\} \qquad (3)$$

Note that the expected value of the term in braces is nonzero. X_{it} is assumed to be independent of (U_{it}, ϵ_i).

In models without regressors in the decision rule we can always work with the redefined model

$$Y_{it} = X_{it}\beta + d_i\alpha^* + \{U_{it} + d_i(\epsilon_i - E[\epsilon_i \mid d_i = 1])\} \qquad (4)$$

where

$$\alpha^* = \bar{\alpha} + E[\epsilon_i \mid d_i = 1]$$

and content ourselves with the estimation of α^*.[2] If everywhere we replace α with α^*, the analysis of fixed-coefficient equation (1) applies to equation (4), provided that account is taken of the new error component in the disturbance.

The parameter α^* answers question (2). It addresses the question of determining the effect of training on the people selected as trainees. This parameter is useful in making forecasts when the same selection rule operates in the future as has operated in the past. In the presence of nonrandom selection into training it does not answer question (1) (see Heckman & Robb, 1985, 1986). Without regressors in the decision rule given by equation (2), question (1) cannot be answered, so far as we can see, unless specific distributional assumptions are invoked.

Much of the statistical literature assumes that $\bar{\alpha}$ is the parameter of interest. (See Fisher, 1953; Lee, 1978; and Rosenbaum & Rubin, 1983.) In the context of estimating the impact of nonrandom treatments that are likely to be nonrandomly assigned in the future, $\bar{\alpha}$ is not an interesting policy or evaluation parameter since it does not recognize selection decisions by agents. Only if random assignment is to be followed in the future is there interest in this parameter. Of course, α^* is interesting for prediction purposes only to the extent that current selec-

tion rules will govern future participation. In this paper we do not address the more general problem of estimating future policy impacts
when selection rules are to be changed. To answer this question requires stronger assumptions on the joint distribution of ϵ_i, U_{it}, and V_i
than are required to estimate $\bar{\alpha}$ or α^*.[3]

3. CROSS-SECTIONAL PROCEDURES

3.1. Without Distributional Assumptions a Regressor Is Needed

Let $\bar{Y}_t^{(1)}$ denote the sample mean of participant outcomes and let $\bar{Y}_t^{(0)}$
denote the sample mean of nonparticipant outcomes:

$$\bar{Y}_t^{(1)} = \frac{\sum d_i Y_{it}}{\sum d_i}, \quad \bar{Y}_t^{(0)} = \frac{\sum (1 - d_i) Y_{it}}{\sum (1 - d_i)}, \quad 0 < \sum d_i < I_t$$

where I_t is the sample size in year t. We retain the assumption that
the data are generated by a random sampling scheme. If no regressors
appear in equation (1), then $X_{it}\beta = \beta_t$ and

$$\text{plim } \bar{Y}_t^{(1)} = \beta_t + \alpha + E[U_{it} \mid d_i = 1]$$

$$\text{plim } \bar{Y}_t^{(0)} = \beta_t + E[U_{it} \mid d_i = 0]$$

Thus

$$\text{plim } (\bar{Y}_t^{(1)} - \bar{Y}_t^{(0)}) = \alpha + \frac{E[U_{it} \mid d_i = 1]}{1 - p}, \quad 0 < p < 1,$$

since $pE[U_{it} \mid d_i = 1] + (1 - p)E[U_{it} \mid d_i = 0] = 0$. Even if p were
known, α cannot be separated from $E[U_{it} \mid d_i = 1]$ using cross-sectional
data on sample means. Sample variances do not aid in securing identification unless $E[U_{it}^2 \mid d_i = 0]$ or $E[U_{it}^2 \mid d_i = 1]$ are known a priori.
Similar remarks apply to the information available from higher mo-
ments unless they are restricted in some a priori fashion.

3.2. Overview of Cross-Sectional Procedures That Use Regressors

If, however, $E[U_{it} \mid d_i = 1, Z_i]$ is a nonconstant function of Z_i, it is possible (with additional assumptions) to solve this identification problem.
Securing identification in this fashion explicitly precludes a fully nonparametric strategy in which both the outcome function equation (1)
and decision rule equation (2) are estimated in each (X_{it}, Z_i) stratum.
For within each stratum, $E[U_{it} \mid d_i = 1, Z_i]$ is a constant function of Z_i
and α is not identified from cross-sectional data. Restrictions across
strata are required.

If $E[U_{it} \mid d_i = 1, Z_i]$ is a nonconstant function of Z_i, it is possible to exploit this information in a variety of ways depending on what else is assumed about the model. Here we simply sketch alternative strategies. In our earlier papers we present a systematic discussion of each approach.

(a) Suppose Z_i or a subset of Z_i is exogenous with respect to U_{it}. Under conditions specified more fully below, the exogenous subset may be used to construct an instrumental variable for d_i in equation (1), and α can be consistently estimated by instrumental-variables methods. No explicit distributional assumptions about U_{it} or V_i are required (Heckman, 1978). The assignment rule (equation 2) need not be fully specified.

(b) Suppose that Z_i is distributed independently of V_i and the functional form of the distribution of V_i is known. Under standard conditions, γ in equation (2) can be consistently estimated by conventional methods in discrete-choice analysis (Amemiya, 1981). If Z_i is distributed independently of U_{it}, $F(-Z_i\hat{\gamma})$ can be used as an instrument for d_i in equation (1) (Heckman, 1978).

(c) Under the same conditions as specified in (b),

$$E[Y_{it} \mid X_{it}, Z_i] = X_{it}\beta + \alpha(1 - F(-Z_i\gamma)).$$

Then β and α can be consistently estimated using $F(-Z_i\hat{\gamma})$ in place of $F(-Z_i\gamma)$ in the equation (Heckman, 1976, 1978) or else the equation can be estimated by nonlinear least squares, estimating β, α, and γ jointly (given the functional form of F) (Barnow et al., 1980).

(d) If the functional forms of $E[U_{it} \mid d_i = 1, Z_i]$ and $E[U_{it} \mid d_i = 0, Z_i]$ as functions of Z_i are known up to a finite set of parameters, it is sometimes possible to consistently estimate β, α, and the parameters of the conditional means from the (nonlinear) regression function

$$E[Y_{it} \mid d_i, X_{it}, Z_i] = X_{it}\beta + d_i\alpha + d_iE[U_{it} \mid d_i = 1, Z_i]$$
$$+ (1 - d_i)E[U_{it} \mid d_i = 0, Z_i]. \tag{5}$$

One way to acquire information about the functional form of $E[U_{it} \mid d_i = 1, Z_i]$ is to assume knowledge of the functional form of the joint distribution of (U_{it}, V_i) (e.g., that it is bivariate normal), but this is not strictly required. Note further that this procedure does not require that Z_i be distributed independently of V_i in equation (2) (Heckman, 1980).

(e) Instead of (d), it is possible to perform a two-stage estimation procedure if the joint density of (U_{it}, V_i) is assumed known up to a finite set of parameters and Z_i is distributed independently of V_i and U_{it}. In stage 1, $E[U_{it} \mid d_i = 1, Z_i]$ and $E[U_{it} \mid d_i = 0, Z_i]$ are determined up to some unknown parameters by conventional discrete-choice analysis. Then regression equation (5) is run using estimated E values in place of population E values on the right-hand side of the equation (Heckman, 1976).

(f) Under the assumptions of (e), use maximum likelihood to consistently estimate α (Heckman, 1978).

Note that a separate value of α may be estimated for each cross section so that, depending on the number of cross sections, it is possible to estimate growth and decay effects in treatment.

Conventional selection bias approaches (d)–(f) as well as (b)–(c) rely on strong distributional assumptions or else assumptions about nonlinearities in the model, but in fact these are not required. Distributional assumptions are usually not motivated by behavioral theory. Given that a regressor appears in decision rule equation (2), if it is uncorrelated with U_{it}, the regressor is an instrumental variable for d_i. It is not necessary to invoke strong distributional assumptions, but if they are invoked, Z_i need not be uncorrelated with U_{it}. In many papers, however, Z_i and U_{it} are usually assumed to be independent. The imposition of overidentifying "information," if false, may lead to considerable bias and instability in the estimates. However, the overidentifying assumptions are testable and so false restrictions need not be imposed. Conventional practice imposes these overidentifying restrictions without testing them. We next discuss the instrumental-variable procedure in greater detail.[4]

3.3. The Instrumental-Variable Estimator

This estimator is the least demanding in the a priori conditions that must be satisfied for its use. It requires the following assumptions:

(a) There is at least one variable in Z_i, Z_i^e, with a nonzero γ coefficient in equation (2), such that for some known transformation of Z_i^e, $g(Z_i^e)$,

$$E[U_{it}g(Z_i^e)] = 0. \tag{6a}$$

(b) Array X_{it} and d_i into a vector $J_{1it} = (X_{it}, d_i)$. Array X_{it} and $g(Z_i^e)$ into a vector $J_{2it} = (X_{it}, g(Z_i^e))$. In this notation, it is assumed that

$$E\left[\sum_{i=1}^{I_t} \frac{J'_{2it}J_{1it}}{I_t}\right] \tag{6b}$$

has full column rank uniformly in I_t for I_t sufficiently large, where I_t denotes the number of individuals in period t.

With these assumptions and conventional technical conditions (see Heckman & Robb, 1985, pp. 184–185), the instrumental variable (I.V.) estimator

$$\begin{bmatrix} \hat{\beta} \\ \hat{\alpha} \end{bmatrix}_{\text{I.V.}} = \left[\sum_{i=1}^{I_t} \frac{J'_{2it} J_{1it}}{I_t} \right]^{-1} \sum_{i=1}^{I_t} \frac{J'_{2it} Y_{it}}{I_t}$$

is consistent for (β, α) regardless of any covariance between U_{it} and d_i.

It is important to notice how weak these conditions are. The functional form of the distribution of V_i need not be known. Z_i need not be distributed independently of V_i. Moreover, $g(Z_i^e)$ may be a nonlinear function of variables appearing in X_{it} as long as assumptions (6a) and (6b) are satisfied.

The instrumental variable, $g(Z_i^e)$, may also be a lagged value of time-varying variables appearing in X_{it}, provided the analyst has access to longitudinal data. The rank condition (6b) will generally be satisfied in this case as long as X_{it} exhibits serial dependence. Thus longitudinal data (on exogenous characteristics) may provide a natural source of instrumental variables.

3.4. Identification through Assumptions about the Marginal Distribution of U_{it}

If no regressor appears in decision rule equation (2), the estimators summarized up to this point in this section cannot be used to consistently estimate α. Heckman (1978) demonstrates that if (U_{it}, V_i) are normally distributed, α is identified even if there is no regressor in decision rule equation (2). His conditions are overly strong.

If U_{it} has zero third and fifth central moments, α is identified even if no regressor appears in the enrollment rule. This assumption about U_{it} is implied by normality or symmetry of the density of U_{it} but it is weaker than either provided that the required moments are finite. The fact that α can be identified by invoking distributional assumptions about U_{it} illustrates the more general point that there is a trade-off between assumptions about regressors and assumptions about the distribution of U_{it} that must be invoked to identify α. It also demonstrates that conventional practice overparameterizes the problem (i.e., assuming that U_{it} is normal is unnecessary).

We have established that, under the following assumptions, α in equation (1) is identified:

$$E[U_{it}^3] = 0 \tag{7a}$$

$$E[U_{it}^5] = 0 \tag{7b}$$

$$\{U_{it}, V_i\} \text{ is } iid. \tag{7c}$$

A consistent method-of-moments estimator can be devised that exploits these assumptions. (See Heckman & Robb, 1985.) Find $\hat{\alpha}$ that sets a weighted average of the sample analogs of $E[U_{it}^3]$ and $E[U_{it}^5]$ as close to zero as possible.

To simplify the exposition, suppose that there are no regressors in the outcome function equation (1), so $X_{it}\beta = \beta_t$. The proposed estimator finds the value of $\hat{\alpha}$ that sets

$$\frac{1}{I_t} \sum_{i=1}^{I_t} [(Y_{it} - \bar{Y}_t) - \hat{\alpha}(d_i - \bar{d})]^3 \tag{8a}$$

and

$$\frac{1}{I_t} \sum_{i=1}^{I_t} [(Y_{it} - \bar{Y}_t) - \hat{\alpha}(d_i - \bar{d})]^5 \tag{8b}$$

as close to zero as possible in a suitably chosen metric where, as before, $\bar{Y}_t$ and $\bar{d}$ denote sample means. A pair of moments is required in order to pick the unique consistent root. In our companion paper (Heckman & Robb, 1985), we establish the existence of a unique root that sets equations (8a) and (8b) to zero in large samples. Obviously, other moment restrictions could be used to identify α.[5]

3.5. Selection on Observables

In the special case in which

$$E(U_{it} \mid d_i, Z_i) = E(U_{it} \mid Z_i),$$

selection is said to occur on the observables. Such a case can arise if U_{it} is distributed independently of V_i in equation (2) but U_{it} and Z_i are stochastically dependent (i.e., some of the observables in the assignment equation are correlated with the unobservables in the outcome equation). In this case, U_{it} and d_i can be shown to be conditionally independent given Z_i. If it is further assumed that U_{it} and V_i conditional on Z_i are independent, then U_{it} and d_i can be shown to be conditionally independent given Z_i. In the notation of Dawid (1979), as used by Rosenbaum and Rubin (1983),

$$U_{it} \parallel d_i \mid Z_i,$$

i.e., given Z_i, d_i is strongly ignorable. In a random-coefficient model the required condition is

$$(U_{it} + \epsilon_i d_i) \parallel d_i \mid Z_i.$$

The strategy for consistent estimation presented in Section 3.2 must be modified. In particular, methods (a)–(c) are inappropriate. However, method (d) still applies and simplifies because

$$E(U_{it} \mid d_i = 1, Z_i) = E(U_{it} \mid d_i = 0, Z_i) = E(U_{it} \mid Z_i)$$

so that we obtain in place of equation (5)

$$E(Y_{it} \mid d_i, X_{it}, Z_i) = X_{it}\beta + d_i\alpha + E(U_{it} \mid Z_i) . \tag{9}$$

Specifying the joint distribution of U_{it}, Z_i or just the conditional mean of U_{it} given Z_i, produces a formula for $E(U_{it} \mid Z_i)$ up to a set of parameters. The model can be estimated by nonlinear regression. Conditions for the existence of a consistent estimator of α are presented in our companion papers. (See also the discussion in Barnow et al., 1980.)

Method (f) produces a consistent estimator provided that an explicit probabilistic relationship between U_{it} and Z_i is postulated. (See notes 4 and 5.)

4. REPEATED CROSS-SECTIONAL METHODS FOR THE CASE WHEN PARTICIPANT STATUS IDENTITY OF INDIVIDUALS IS UNKNOWN

Assuming a time-homogenous environment and access to repeated cross-sectional data and random sampling, it is possible to identify α (a) without any regressor in the decision rule, (b) without need to specify the joint distribution of U_{it} and V_i, and (c) without any need to know which individuals in the sample participated in treatment (but the population proportion of participants must be known or consistently estimable).

To see why this claim is true, suppose that no regressors appear in the outcome function.[6] In the notation of equation (1), $X_{it}\beta = \beta_t$. Then, assuming a random-sampling scheme generates the data,

$$\text{plim } \bar{Y}_t = \text{plim } \frac{\sum Y_{it}}{I_t} = E[\beta_t + \alpha d_i + U_{it}] = \beta_t + \alpha p, \quad t > k$$

$$\text{plim } \bar{Y}_{t'} = \text{plim } \frac{\sum Y_{it'}}{I_{t'}} = E[\beta_{t'} + U_{it'}] = \beta_{t'}, \quad t' < k.$$

In a time-homogenous environment, $\beta_t = \beta_{t'}$ and

$$\text{plim } \frac{\bar{Y}_t - \bar{Y}_{t'}}{\hat{p}} = \alpha$$

where $\hat{p}$ is a consistent estimator of $p = E[d_i]$.

With more than two years of repeated cross-sectional data, one can apply the same principles to identify α while relaxing the time-homogeneity assumption. For instance, suppose that population-mean outcomes lie on a polynomial of order $L-2$:

$$\beta_t = \pi_0 + \pi_1 t + \ldots + \pi_{L-2} t^{L-2}.$$

From L temporally distinct cross sections, it is possible to estimate consistently the $(L - 1)$ π-parameters and α, provided that the number of observations in each cross section becomes large and there is at least one preprogram and one postprogram cross section.

If the effect of treatment differs across periods, it is still possible to identify α_t, provided that the environment changes in a "sufficiently regular" way. For example, suppose

$$\beta_t = \pi_0 + \pi_1 t$$
$$\alpha_t = \phi_0(\phi_1)^{t-k} \text{ for } t > k.$$

In this case, π_0, π_1, ϕ_0, and ϕ_1 are identified from the means of four cross sections, so long as at least two of these means come from a preprogram period and two come from postprogram periods.

Our companion papers state more rigorously the conditions required to consistently estimate α using repeated cross-sectional data that do not record the treatment identity of individuals. Section 7 examines the sensitivity of this class of estimators to violations of the random-sampling assumption.

5. LONGITUDINAL PROCEDURES

We now consider three examples of estimators that are based on longitudinal data.

5.1. The Fixed-Effects Method

This method was developed by Mundlak (1961, 1978) and refined by Chamberlain (1982). It is widely used in recent social science data analyses. It is based on the following assumption:

$$E[U_{it} - U_{it'} \mid d_i, X_{it} - X_{it'}] = 0 \text{ for all } t, t', \quad t > k > t'. \quad (10)$$

As a consequence of this assumption, we may write a difference regression as

$$E[Y_{it} - Y_{it'} \mid X_{it} - X_{it'}, d_i] = (X_{it} - X_{it'})\beta + d_i\alpha, \quad t > k > t'.$$

Suppose that equation (10) holds and the analyst has access to one year of preprogram and one year of postprogram outcomes. Regressing the difference between postprogram outcomes in any year and latent outcomes in any preprogram year on the change in regressors between those years and a dummy variable for treatment status produces a consistent estimator of α. Although this estimator is widely used, the behavioral motivation for it is weak. (See Heckman & Robb, 1985.)[7]

5.2. U_{it} Follows a First-Order Autoregressive Process

Suppose next that U_{it} follows a first-order autoregression:

$$U_{it} = \rho U_{i,\,t-1} + \nu_{it} \tag{11}$$

where $E[\nu_{it}] = 0$ and the ν_{it} are mutually independently (not necessarily identically) distributed random variables with $\rho \neq 1$. Substitution using equations (1) and (11) to solve for $U_{it'}$ yields

$$Y_{it} = [X_{it} - (X_{it'}\rho^{t-t'})]\beta + (1 - \rho^{t-t'})d_i\alpha \tag{12}$$

$$+ \rho^{t-t'}Y_{it'} + \{ \sum_{j=0}^{t-(t'+1)} \rho^j \nu_{i,\,t-j}\}, \quad t > t' > k.$$

Assume $E(\nu_{it} \mid d_i = 1) = 0$ for all $t > k$. Heckman and Wolpin (1976) invoke a similar assumption in their analysis of affirmative action programs. If X_{ij} is independent of $\nu_{ij'}$ for all j, j' (an overly strong condition), then (linear or nonlinear) least squares applied to equation (12) consistently estimates α as the number of observations becomes large. (The appropriate nonlinear regression increases efficiency by imposing the implied cross-coefficient restrictions.) As is the case with the fixed-effect estimator, increasing the length of the panel and keeping the same assumptions, the model becomes overidentified (and hence testable) for panels with more than two observations per person.[8]

5.3. U_{it} Is Covariance-Stationary

The next procedure invokes an assumption implicitly used in many papers in evaluation analysis (e.g., Ashenfelter, 1978; Bassi, 1983; and others) but exploits the assumption in a novel way. We assume

(a) U_{it} is covariance stationary, so $E[U_{it}U_{i,\,t-j}] = \sigma_j$, for $j \geq 0$. (13a)

(b) Access to at least two observations on preprogram earnings in t' and $t' - j$ as well as one period of postprogram earnings in t where $t - t' = j$. (13b)

(c) $pE[U_{it'} \mid d_i = 1] \neq 0$. (13c)

We lose no essential generality by suppressing the effect of regressors in equation (1). (See note 6.) Thus, let

$$Y_{it} = \beta_t + d_i\alpha + U_{it}, \quad t > k$$

$$Y_{it'} = \beta_{t'} + U_{it'}, \quad t' < k$$

where β_t and $\beta_{t'}$ are period-specific intercepts.

From a random sample of preprogram outcomes from periods t' and $t' - j$, σ_j can be consistently estimated from the sample covariances between $Y_{it'}$ and $Y_{i,\,t'-j}$:

$$m_1 = \frac{\sum (Y_{it'} - \bar{Y}_{t'})(Y_{i,\,t'-j} - \bar{Y}_{t'-j})}{I_{t'}}$$

$$\text{plim } m_1 = \sigma_j \,.$$

If $t > k$ and $t - t' = j$ so that the postprogram outcome data are as far removed in time from t' as t' is removed from $t' - j$, form the sample covariance between Y_{it} and $Y_{it'}$:

$$m_2 = \frac{\sum (Y_{it} - \bar{Y}_t)(Y_{it'} - \bar{Y}_{t'})}{I_t}$$

which has the probability limit

$$\text{plim } m_2 = \sigma_j + \alpha p \, E[U_{it'} \mid d_i = 1], \quad t > k > t'.$$

Form the sample covariance between d_i and $Y_{it'}$,

$$m_3 = \frac{\sum (Y_{it'} - \bar{Y}_{t'}) d_i}{I_t}$$

with probability limit

$$\text{plim } m_3 = p \, E[U_{it'} \mid d_i = 1], \quad t' < k.$$

Combining this information and assuming $p \, E[U_{it'} \mid d_i = 1] \neq 0$ for $t' < k$,

$$\text{plim } \hat{\alpha} = \text{plim } \frac{m_2 - m_1}{m_3} = \alpha.$$

For panels of sufficient length (e.g., more than two preprogram observations or more than two postprogram observations per person) the stationarity assumption can be tested. Thus, as before, increasing the length of the panel converts a just-identified model to an overidentified one.[9]

6. REPEATED CROSS-SECTIONAL ANALOGUES OF LONGITUDINAL PROCEDURES

The previous section presented longitudinal estimators of α. For all cases considered there, α can actually be identified with repeated cross-sectional data. Our earlier paper gives additional examples of longitudinal estimators that can be implemented on repeated cross-sectional data.

6.1. The Fixed-Effect Model

As in Section 5.1, assume that equation (10) holds so

$$E[U_{it} \mid d_i = 1] = E[U_{it'} \mid d_i = 1]$$
$$E[U_{it} \mid d_i = 0] = E[U_{it'} \mid d_i = 0], \quad t > k > t'$$

for all t, t'. Let $X_{it}\beta = \beta_t$ and define $\hat{\alpha}$ in terms of the notation of Section 3.1 as

$$\hat{\alpha} = [\bar{Y}_t^{(1)} - \bar{Y}_t^{(0)}] - [\bar{Y}_{t'}^{(1)} - \bar{Y}_{t'}^{(0)}].$$

Assuming random sampling, consistency of $\hat{\alpha}$ follows immediately from equation (10):

$$\text{plim } \hat{\alpha} = (\alpha + \beta_t - \beta_t + E[U_{it} \mid d_i = 1] - E[U_{it} \mid d_i = 0])$$
$$- (\beta_{t'} - \beta_{t'} + E[U_{it'} \mid d_i = 1] - E[U_{it'} \mid d_i = 0]) = \alpha.$$

As in the case of the longitudinal version of this estimator, with more than two cross sections, the hypothesis of equation (10) is subject to test (i.e., the model is overidentified).

6.2. U_{it} Follows a First-Order Autoregressive Process

In one respect the preceding example is contrived. It assumes that in preprogram cross sections we know the identity of future program participants. Such data might exist but this seems unlikely. One advantage of longitudinal data for estimating α in the fixed-effect model is that if the survey extends before period k, the identity of future participants is known.

The need for preprogram earnings to identify α is, however, only an artifact of the fixed-effect assumption, equation (10). Suppose instead that U_{it} follows a first-order autoregressive process given by equation (11) and that

$$E[\nu_{it} \mid d_i] = 0, \quad t > k \tag{14}$$

as in Section 5.2. With three successive postprogram cross sections in which the identity of participants is known, it is possible to identify α.

To establish this result, let the three postprogram periods be t, $t+1$, and $t+2$. Assuming, as before, that no X regressor appears in equation (1),

$$\text{plim } \bar{Y}_j^{(1)} = \beta_j + \alpha + E[U_{ij} \mid d_i = 1]$$
$$\text{plim } \bar{Y}_j^{(0)} = \beta_j + E[U_{ij} \mid d_i = 0].$$

From equation (11),

$$E[U_{i,\,t+1} \mid d_i = 1] = \rho\, E[U_{it} \mid d_i = 1]$$

$$E[U_{i,\,t+1} \mid d_i = 0] = \rho\, E[U_{it} \mid d_i = 0]$$

$$E[U_{i,\,t+2} \mid d_i = 1] = \rho^2\, E[U_{it} \mid d_i = 1]$$

$$E[U_{i,\,t+2} \mid d_i = 0] = \rho^2\, E[U_{it} \mid d_i = 0].$$

Using these formulae, it is straightforward to verify that $\hat{\rho}$ defined by

$$\hat{\rho} = \frac{(\bar{Y}^{(1)}_{t+2} - \bar{Y}^{(0)}_{t+2}) - (\bar{Y}^{(1)}_{t+1} - \bar{Y}^{(0)}_{t+1})}{(\bar{Y}^{(1)}_{t+1} - \bar{Y}^{(0)}_{t+1}) - (\bar{Y}^{(1)}_{t} - \bar{Y}^{(0)}_{t})}$$

is consistent for ρ and that $\hat{\alpha}$ defined by

$$\hat{\alpha} = \frac{(\bar{Y}^{(1)}_{t+2} - \bar{Y}^{(0)}_{t+2}) - \hat{\rho}\,(\bar{Y}^{(1)}_{t+1} - \bar{Y}^{(0)}_{t+1})}{1 - \hat{\rho}}$$

is consistent for α.[10]

For this model, the advantage of longitudinal data is clear. Only two time periods of longitudinal data are required to identify α, but three periods of repeated cross-sectional data are required to estimate the same parameter. However, if Y_{it} is subject to measurement error, the apparent advantages of longitudinal data become less clear. Repeated cross-sectional estimators are robust to mean-zero measurement error in the variables. The longitudinal regression estimator discussed in Section 5.2 does not identify α unless the analyst observes outcomes without error. Given three years of longitudinal data and assuming that measurement error is serially uncorrelated, one could instrument $Y_{it'}$ in equation (12), using the outcome measurement in the earliest year as an instrument. This requires one more year of data. Thus one advantage of the longitudinal estimator disappears in the presence of measurement error.[11] With four or more repeated cross sections, the model is obviously overidentified and hence subject to test.

6.3. Covariance Stationarity

For simplicity we assume that there are no regressors in the outcome equation and let $X_{it}\beta = \beta_t$. (See Heckman & Robb, 1985, for the case in which regressors are present.) Assume that conditions (13a) and (13c) are satisfied. Before presenting the repeated cross-sectional estimator, it is helpful to record the following facts:

$$\text{Var}(Y_{it}) = \alpha^2(1 - p)p + 2\alpha\, E[U_{it} \,|\, d_i = 1]p + \sigma_u^2, \quad t > k \quad (15a)$$

$$\text{Var}(Y_{it'}) = \sigma_u^2, \quad t' < k \quad (15b)$$

$$\text{Cov}(Y_{it}, d_i) = \alpha\, p(1 - p) + p\, E[U_{it} \,|\, d_i = 1]. \quad (15c)$$

Note that $E[U_{it}^2] = E[U_{it'}^2]$ by virtue of assumption (13a). Then

$$
\hat{\alpha} = (p(1 - p))^{-1} \left(\frac{\sum(Y_{it} - \bar{Y}_t)d_i}{I_t} - \left\{ \left(\frac{\sum(Y_{it} - \bar{Y}_t)d_i}{I_t} \right)^2 \right. \right.
$$

$$
\left. \left. - p(1 - p)\left(\frac{\sum(Y_{it} - \bar{Y}_t)^2}{I_t} - \frac{\sum(Y_{it'} - \bar{Y}_{t'})^2}{I_{t'}} \right) \right\}^{1/2} \right)
\quad (16)
$$

is consistent for α.

This expression arises by subtracting equation (15b) from equation (15a). Then use equation (15c) to get an expression for $E[U_{it} \,|\, d_i = 1]$ that can be substituted into the expression for the difference between equations (15a) and (15b). Replacing population moments by sample counterparts produces a quadratic equation in $\hat{\alpha}$, with the negative root given by equation (16). The positive root is inconsistent for α.[12]

Notice that the estimators of Sections 5.3 and 6.3 exploit different features of the covariance-stationarity assumptions. The longitudinal procedure only requires that $E[U_{it}U_{i,t-j}] = E[U_{it'}U_{i,t'-j}]$ for $j > 0$; variances need not be equal across periods. The repeated cross-sectional analogue above only requires that $E[U_{it}U_{i,t-j}] = E[U_{i,t'}U_{i,t'-j}]$ for $j = 0$; covariances may differ among equispaced pairs of the U_{it}. With more than two cross sections, the covariance-stationarity assumption is overidentifying and hence subject to test.

6.4. Anomalous Features of First-Difference Estimators

Nearly all of the estimators require a control group (i.e., a sample of nonparticipants). The only exception is the fixed-effect estimator in a time-homogenous environment. In this case, if condition equation (10) holds, if we let $X_{it}\beta = \beta_t$ to simplify the exposition, and if the environment is time homogenous so $\beta_t = \beta_{t'}$, then

$$\hat{\alpha} = \bar{Y}_t^{(1)} - \bar{Y}_{t'}^{(1)}, \quad t > k > t'$$

consistently estimates α. The claim that "if the environment is stationary, you don't need a control group" (see, e.g., Bassi, 1983) is false except for the special conditions that justify use of the fixed-effect estimator.

Most of the procedures considered here can be implemented using only postprogram data. The estimators exploiting covariance-stationarity, certain repeated cross-sectional estimators, and first-difference methods constitute an exception to this rule. In this sense, those estimators are anomalous.

7. NONRANDOM SAMPLING PLANS

The data available for analyzing the impact of treatment on outcomes are often nonrandom samples. Frequently they consist of pooled data from two sources: (a) a sample of treated subjects selected from program records and (b) a sample of nontreated subjects selected from some national sample. Typically, such samples overrepresent participants relative to their proportion in the population. This creates the problem of choice-based sampling analyzed by Manski and Lerman (1977) and Manski and McFadden (1981).

A second problem, contamination bias, arises when the participant status of certain individuals is recorded with error. Many control samples do not reveal whether or not persons have received treatment.

Both of these sampling situations combine the following types of data:

(A) Y_{it}, X_{it}, and Z_i for a sample of participants ($d_i = 1$)

(B) Y_{it}, X_{it}, and Z_i for a sample of nonparticipants ($d_i = 0$)

(C) Y_{it}, X_{it}, and Z_i for a national "control" sample of the population where the treatment status of persons is not known.

If type (A) and (B) data are combined and the sample proportion of participants does not converge to the population proportion of participants, the combined sample is a choice-based sample. If type (A) and (C) data are combined with or without type (B) data, there is contamination bias because the participant status of some persons is not known.

Most procedures developed in the context of random sampling can be modified to consistently estimate α using choice-based samples or contaminated-control groups (i.e., groups in which treatment status is not known for individuals). In some cases, a consistent estimator of the population proportion of participants is required. We illustrate these claims by showing how to modify the instrumental-variable estimator to address both sampling schemes. Our companion paper gives explicit case-by-case treatment of these issues for each estimator developed there.

7.1. The I.V. Estimator: Choice-Based Sampling

If condition (6a) is strengthened to read

$$E[X'_{it}U_{it} \mid d_i] = 0$$

$$E[g(Z^e_i)U_{it} \mid d_i] = 0 \qquad (17)$$

and condition (6b) is also met, the I.V. estimator is consistent for α in choice-based samples.

To see why this is so, write the normal equations for the I.V. estimator in the following form:

$$
\begin{bmatrix}
\dfrac{\sum X'_{it}X_{it}}{I_t} & \dfrac{\sum X'_{it}d_i}{I_t} \\[2.5ex]
\dfrac{\sum g(Z^e_i)X_{it}}{I_t} & \dfrac{\sum g(Z^e_i)d_i}{I_t}
\end{bmatrix}
\begin{bmatrix} \hat{\beta} \\ \hat{\alpha} \end{bmatrix}
=
\begin{bmatrix}
\dfrac{\sum X'_{it}Y_{it}}{I_t} \\[2.5ex]
\dfrac{\sum g(Z^e_i)Y_{it}}{I_t}
\end{bmatrix}
\qquad (18)
$$

$$
=
\begin{bmatrix}
\dfrac{\sum X'_{it}X_{it}}{I_t} & \dfrac{\sum X'_{it}d_i}{I_t} \\[2.5ex]
\dfrac{\sum g(Z^e_i)X_{it}}{I_t} & \dfrac{\sum g(Z^e_i)d_i}{I_t}
\end{bmatrix}
\begin{bmatrix} \beta \\ \alpha \end{bmatrix}
+
\begin{bmatrix}
\dfrac{\sum X'_{it}U_{it}}{I_t} \\[2.5ex]
\dfrac{\sum g(Z^e_i)U_{it}}{I_t}
\end{bmatrix}
$$

Since equation (17) guarantees that

$$\operatorname*{plim}_{I_t \to \infty} \frac{\sum X'_{it}U_{it}}{I_t} = 0 \quad \text{and} \quad \operatorname*{plim}_{I_t \to \infty} \frac{\sum g(Z^e_i)U_{it}}{I_t} = 0 \qquad (19)$$

and rank condition (6b) holds, the I.V. estimator is consistent.

In a choice-based sample, let the probability that an individual has participated in the program be p^*. Even if conditions (6a) and (6b) are satisfied, there is no guarantee that condition (19) will be met in populations generated from choice-based samples without invoking equation (17). This is so because

$$\text{plim}_{I_t \to \infty} \frac{\sum X'_{it} U_{it}}{I_t} = E[X'_{it} U_{it} \mid d_i = 1]p^* + E[X'_{it} U_{it} \mid d_i = 0] (1 - p^*)$$

$$\text{plim}_{I_t \to \infty} \frac{\sum g(Z_i^e) U_{it}}{I_t} = E[g(Z_i^e) U_{it} \mid d_i = 1]p^*$$

$$+ E[g(Z_i^e) U_{it} \mid d_i = 0] (1 - p^*).$$

These expressions are generally not zero if $p \neq p^*$, so the I.V. estimator is generally inconsistent.

In the case of random sampling, $p^* = Pr[d_i = 1] = p$ and the above expressions are identically zero. They are also zero if equation (17) is satisfied. However, it is not necessary to invoke equation (17). Provided p is known, it is possible to reweight the data to secure consistent estimators under the assumptions of Section 3. Multiplying equation (1) by the weight

$$\omega_i = d_i \frac{p}{p^*} + (1 - d_i) \left(\frac{1 - p}{1 - p^*} \right)$$

and applying I.V. to the transformed equation produces an estimator that satisfies equation (19). It is straightforward to check that weighting the sample at hand back to random-sampling proportions causes the I.V. method to consistently estimate α and β. (See Heckman & Robb, 1985.)

7.2. The I.V. Estimator: Contamination Bias

For data of type (C), d_i is not observed. Applying the I.V. estimator to pooled samples (A) and (C) assuming that observations in (C) have $d_i = 0$ produces an inconsistent estimator. However, it is possible to construct a consistent estimator for this case.

In terms of the I.V. equations (18), from sample (C) it is possible to generate the cross-products

$$\frac{\sum X'_{it} X_{it}}{I_c}, \quad \frac{\sum g(Z_i^e) X_{it}}{I_c}, \quad \frac{\sum X'_{it} Y_{it}}{I_c}, \quad \frac{\sum g(Z_i^e) Y_{it}}{I_c},$$

which converge to the desired population counterparts where I_c denotes the number of observations in sample (C). Missing is information on the cross-products

$$\frac{\sum X'_{it} d_i}{I_c} , \quad \frac{\sum g(Z^e_i) d_i}{I_c} .$$

Notice that if d_i were measured accurately in sample (C),

$$\operatorname*{plim}_{I_c \to \infty} \frac{\sum X'_{it} d_i}{I_c} = p \, E[X'_{it} \mid d_i = 1]$$

$$\operatorname*{plim}_{I_c \to \infty} \frac{\sum g(Z^e_i) d_i}{I_c} = p \, E[g(Z^e_i) \mid d_i = 1].$$

But the means of X_{it} and $g(Z^e_i)$ in sample (A) converge to $E[X_{it} \mid d_i = 1]$ and $E[g(Z^e_i) \mid d_i = 1]$, respectively. Hence, inserting the sample (A) means of X_{it} and $g(Z^e_i)$ multiplied by p in the second column of the matrix I.V. equations (18) produces a consistent I.V. estimator provided that in the limit the size of samples (A) and (C) both become large at the same rate.

7.3. Repeated Cross Section Methods with Unknown Participant Status and Choice-Based Sampling

The repeated cross-sectional estimators discussed in Section 4 are inconsistent when applied to choice-based samples unless additional conditions are assumed. For example, when the environment is time homogenous and equation (10) also holds, $(\bar{Y}_t - \bar{Y}_{t'})/p^*$ is a consistent estimator of α in choice-based samples as long as the same proportion of participants is sampled in periods t' and t. If a condition such as equation (10) is not met, it is necessary to know the identity of participants in order to weight the sample back to the proportion of participants that would be produced by a random sample in order to obtain consistent estimators. Hence the class of estimators that does not require knowledge of individual participant status is not robust to choice-based sampling.

7.4. Control-Function Estimators

Some cross-sectional and longitudinal procedures are robust to choice-based sampling. Those procedures construct a control function, K_{it}, with the following properties:

K_{it} depends on variables $\ldots, Y_{i, t+1}, Y_{it}, Y_{i, t-1}, \ldots,$
$X_{i, t+1}, X_{it}, X_{i, t-1}, \ldots, d_i$ and parameters ψ and $\qquad$ (20)

(a) $E[U_{it} - K_{it} \mid d_i, X_{it}, K_{it}, \psi] = 0$
(b) ψ is identified.

When inserted into the outcome function equation (1), K_{it} purges the equation of dependence between U_{it} and d_i. Rewriting equation (1) to incorporate K_{it},

$$Y_{it} = X_{it}\beta + d_i\alpha + K_{it} + \{U_{it} - K_{it}\}. \tag{21}$$

The purged disturbance $\{U_{it} - K_{it}\}$ is orthogonal to the right-hand side variables in the new equation. Thus regression applied to equation (21) consistently estimates the parameters (α, β, ψ). Moreover, equation (20) implies that $\{U_{it} - K_{it}\}$ is orthogonal to the right-hand side variables conditional on d_i, X_{it}, and K_{it}:

$$E[Y_{it} \mid X_{it}, d_i, K_{it}] = X_{it}\beta + d_i\alpha + K_{it}.$$

Thus if type (A) and (B) data are combined in any proportion, least squares estimators of equation (21) consistently estimate (α, β, ψ), provided the number of participants and nonparticipants in the sample both become large. The class of control-function estimators that satisfies equation (20) can be implemented without modification to produce consistent estimators in choice-based samples.

The sample-selection bias methods (d)–(e) described in Section 3.2 exploit the control-function principle. So do methods that control for selection solely on the observables. Our companion paper gives further examples of control-function estimators. Fixed-effect and autoregressive estimators are also robust to choice-based sampling. (See Heckman & Robb, 1985.)

7.5. Summary and Conclusions on Robustness Properties

Repeated cross-sectional estimators that do not exploit knowledge of the training identity of persons are generally not robust to choice-based sampling nor can they be weighted to produce consistent estimators of α. (Repeated cross-sectional estimators with training identity known can obviously be reweighted to produce consistent estimators.) However, these estimators are robust to contamination bias provided that the population proportion of participants is known or can be consistently estimated.

Using robustness to contamination bias or choice-based sampling as a criterion for selecting estimators does not suggest a clear ordering of cross-sectional, repeated cross-sectional, or longitudinal estimators that require knowledge of the participant status of individual observations.

8. SUMMARY AND CONCLUSIONS

This paper presents alternative methods for estimating the impact of treatments on outcomes when nonrandom selection characterizes the assignment of persons into treatment categories. In the absence of genuine experimental data, assumptions must be invoked to solve the problem of selection bias.

We have defined the parameters of behavioral interest for a prototypical problem of estimating the impact of treatments on outcomes. We have explored the benefits of having access to cross-sectional, repeated cross-sectional, and longitudinal data by considering the assumptions required to use a variety of estimators to identify the behavioral parameters of interest. Because many samples are choice-based samples and because the problem of measurement error is pervasive, we examine the robustness of estimators to choice-based sampling and measurement error.

We find that cross-sectional selection bias estimators do not require the elaborate distributional assumptions frequently invoked in practice. Such conventional overidentifying assumptions are in principle testable.

A key conclusion of our analysis is that the benefits of longitudinal data for identifying interventions have been overstated in the recent literature because a false comparison has been made. Repeated cross-sectional data can often be used to identify the same parameters as can be identified in longitudinal data. Both longitudinal and repeated cross-sectional estimators invoke different and not necessarily weaker assumptions to solve the selection bias problem than do cross-sectional estimators. No estimator has a monopoly on robustness to choice-based sampling or measurement error.

Throughout this analysis we have deliberately refrained from discussing the efficiency of alternative estimators. Much of the literature on longitudinal data focuses on efficiency questions in spite of the fact that the identification problem discussed in this paper should logically take priority.

A discussion of efficiency makes sense only within the context of a fully specified model. The focus in this paper has been on the trade-offs in assumptions that must be imposed in order to estimate a single coefficient when the analyst has access to different types of data. Since different assumptions about the underlying model are invoked in order to justify the validity of alternative estimators, an efficiency or sampling variance comparison is often meaningless. Under the assumptions about an underlying model that justify one estimator, properties of another estimator may not be defined. Only by postulating a common assumption set that is unnecessarily large for any single estimator is it possible to make efficiency comparisons. For the topic of this paper— model identification—the efficiency issue is a red herring.

Even if a common set of assumptions about the underlying model is invoked to justify efficiency comparisons for a class of estimators, conventional efficiency comparisons are often meaningless. First, the frequently stated claim that longitudinal estimators are more efficient than cross-sectional estimators is superficial. It ignores the relative sizes of the available cross-sectional and longitudinal samples. Because of the substantially greater cost of collecting longitudinal data *free of attrition bias*, the number of persons followed in longitudinal studies rarely exceeds 500 in most social science analyses. In contrast, the available cross-sectional and repeated cross-sectional samples have thousands of observations. Given the relative sizes of the available cross-sectional and longitudinal samples, "inefficient" cross-sectional and repeated cross-sectional estimators might have much smaller sampling variances than "efficient" longitudinal estimators fit on much smaller samples. In this sense, our proposed cross-sectional and repeated cross-sectional estimators might be feasibly efficient given the relative sizes of the samples for the two types of data sources.

Second, many of the cross-sectional and repeated cross-sectional estimators proposed in this paper require only sample means of variables. They are thus very simple to compute and are also robust to mean-zero measurement error in all of the variables. More sophisticated longitudinal and cross-sectional estimators are often computationally complex and in practice are implemented on only a fraction of the available data to save computing cost. Simple methods based on means use all of the data and thus, in practice, might be more efficient.

REFERENCES

Amemiya, T. (1981), "Qualitative Response Models: A Survey," *Journal of Economic Literature*, 19, 1483–1536.

Barnow, B., Cain, G., & Goldberger, A. (1980), "Issues in the Analysis of Selectivity Bias," in *Evaluation Studies* (E. Stromsdorder & G. Farkas, eds.), Sage Publications, Beverly Hills, CA, Vol. 5.

Barros, R. (1987), *Two Essays on Selection and Identification Problems in Economics*, (Unpublished Ph.D. thesis), University of Chicago, IL.

Bassi, L. (1983), *Estimating the Effect of Training Programs with Nonrandom Selection*, (Unpublished Ph.D. thesis), Princeton University, NJ.

Chamberlain, G. (1982), "Multivariate Regression Models for Panel Data," *Journal of Econometrics*, 18, 1–46.

Dawid, A. P. (1979), "Conditional Independence in Statistical Theory (with Discussion)," *Journal of Royal Statistical Society*, Series B, 41, 1–31.

Fisher, R. A. (1953), *The Design of Experiments*, Hafner, London.

Heckman, J. (1976), "Simultaneous Equations Models with Continuous and Discrete Endogenous Variables and Structural Shifts," in *Studies in Nonlinear Estimation* (S. Goldfeld & R. Quandt, eds.), Ballinger, Cambridge, MA.

Heckman, J. (1978), "Dummy Endogenous Variables in a Simultaneous Equations System," *Econometrica*, 46, 931–961.

Heckman, J. (1980), "Addendum to Sample Selection Bias as a Specification Error," in *Evaluation Studies* (E. Stromsdorfer & G. Farkas, eds.), Sage, San Francisco, Vol. 5.

Heckman, J., & Robb, R. (1985), "Alternative Methods for Evaluating the Impact of Interventions," in *Longitudinal Analysis of Labor Market Data* (J. Heckman & B. Singer, eds.), Cambridge University Press, New York.

Heckman, J., & Robb, R. (1986), "Alternative Methods for Solving the Problem of Selection Bias in Evaluating the Impact of Treatment on Outcomes," in *Drawing Inferences from Self-Selected Samples* (H. Wainer, ed.), Springer-Verlag, New York, Berlin.

Heckman, J., & Wolpin, K. (1976), "Does the Contract Compliance Program Work?: An Analysis of Chicago Data," *Industrial and Labor Relations Review*, 19, 415–433.

Lee, L. F. (1978), "Unionism and Wage Rates: A Simultaneous Equations Model with Qualitative and Limited Dependent Variables," *International Economic Review*, **19**, 415–433.

MaCurdy, T. (1982), "The Use of Time-Series Processes to Model the Error Structure of Earnings in a Longitudinal Data Analysis," *Journal of Econometrics*, **18**(1), 83–114.

Manski, C., & Lerman, S. (1977), "The Estimation of Choice Probabilities from Choice-Based Samples," *Econometrica*, **45**, 1977–1988.

Manski, C., & McFadden, D. (1981), "Alternative Estimators and Sample Designs for Discrete Choice Analysis," in *Structural Analysis of Discrete Data with Econometric Applications* (C. Manski & D. McFadden, eds.), MIT Press, Cambridge, MA.

Mundlak, Y. (1961), "Empirical Production Functions Free of Management Bias," *Journal of Farm Economics*, **43**, 45–56.

Mundlak, Y. (1978), "On the Pooling of Time-Series and Cross Section Data," *Econometrica*, **46**, 69–85.

Rosenbaum, P., & Rubin, D. (1983), "The Central Role of the Propensity Score in Observational Studies for Causal Effects," *Biometrika*, **70**, 41–55.

NOTES

[*]This research was supported by NSF SES-8107963, NIH-1-R01-HD16846–01 and NIH Grant #HD-19226 to the Quantitative Economics Group at NORC. We have benefited from helpful comments made by Stephen Stigler and members of the Statistics Workshop at the University of Chicago. We also thank John Tukey for his comments at an ETS conference and for very helpful correspondence. Rod Little, Don Rubin, and Gary Solon are also thanked for their comments. Ricardo Barros made especially helpful comments on several drafts of this paper.

[1]The way we have written this paper may cause confusion. We establish identifiability by establishing the existence of consistent estimators. Thus we combine two topics that might fruitfully be decoupled. By establishing the consistency of a variety of estimators we present a large sample guide to estimation under a variety of assumptions. However, the price of this approach to model identification is that we invoke assumptions not strictly required for identification alone. (See Barros, 1987, where this type of separation of assumptions is done.) Fewer assumptions are required for identification than are required for consistent estimation. Clarity would be served if the reader mentally substituted "c-identified" (for consistency identified) for "identified" everywhere the subject of identification is discussed in this paper.

[2]In a model with regressors in the decision rule, it is possible to define $\alpha^*(X, Z) = \bar{\alpha} + E(\epsilon_i \mid d_i = 1, X, Z)$. The modifications required in the text to accommodate this case are negligible.

[3]It is also important to note that any definition of the structural treatment coefficient is conditioned on the stability of the environment in which the program is operating. In the context of a training program, a tenfold expansion of training activity may affect the labor market for the trained and raise the cost of the training activity (and hence the content of programs). For either $\bar{\alpha}$ or α^* to be interesting parameters it must be assumed that such effects are not present in the transition from the sample period to the future. If they are present, it is necessary to estimate how the change in the environment will affect these parameters. In this paper we abstract from these issues, as well as other possible sources of interdependence among outcomes. The resolution of these additional problems would require stronger assumptions than we have invoked here.

[4]Notice that in a transition from a fixed-coefficient model to a random-coefficient model the analysis of this section focuses on estimation of α^* for the latter. Clearly it is possible to redefine U_{it} in equation (1) to include $d_i(\epsilon_i - E(\epsilon_i \mid d_i = 1))$. With this modification, all of our analysis in the text remains intact.

[5]The remark in note 4 applies with full force to this section. Different assumptions are being made in the case of estimating α^* than are invoked in the case of estimat-

ing the fixed-coefficient model (i.e., third and fifth moment assumptions are being invoked about $U_{it} + d_i(\epsilon_i - E(\epsilon_i \mid d_i = 1))$ in the former case). The main point—that if *some* moment assumptions are being invoked it is possible to estimate α or α^*—remains intact.

[6]If regressors appear in the outcome function, the following procedure can be used. Rewrite equation (1) as $Y_{it} = \beta_t + X_{it}\pi + d_i\alpha + U_{it}$. It is possible to estimate π from preprogram data. (This assumes there are no time-invariant variables in X_{it}. If there are such variables, they may be deleted from the regressor vector and π appropriately redefined without affecting the analysis.) Replace Y_{it} by $Y_{it} - X_{it}\hat{\pi}$ and the analysis in the text goes through provided $\hat{\pi}$ is a consistent estimator of π. Note that we are assuming that no X_{it} variables become nonconstant after period k.

[7]We repeat the point made in notes 4 and 5 that if α^* is the coefficient of interest, U_{it} is redefined to be $U_{it} + d_i(\epsilon_i - E(\epsilon_i \mid d_i = 1))$.

[8]In the context of estimating α^* in the random coefficient model, it is not natural to specify equation (11) in the text for the redefined U_{it}. In general, if U_{it} has an autoregressive representation, $U_{it} + d_i(\epsilon_i - E(\epsilon_i \mid d_i = 1))$ will not. A more natural specification models error component $d_i(\epsilon_i - E(\epsilon_i \mid d_i = 1))$ as a permanent postprogram component in the error term. In place of the error term in braces in equation (12), write

$$\left\{ \sum_{j=0}^{t-(t'+1)} \rho^j \nu_{i,\,t-j} + \phi_i(1 - \rho^{t-t'}) \right\} \text{ for } t,\, t' > k$$

where $\phi_i = d_i(\epsilon_i - E(\epsilon_i \mid d_i = 1))$. Orthogonality conditions will *not* be satisfied between ϕ_i and $Y_{it'}$ and an instrument for lagged $Y_{it'}$ will be required to consistently estimate α^* or else the time-series methods of MaCurdy (1982) will have to be invoked to obtain consistent estimators.

[9]As in notes 4 and 5, we emphasize that different assumptions are being made in the random coefficient version of the model than are made in the fixed coefficient version. Note, however, that in this subsection we do not require that variances be equal in preprogram and postprogram periods so that the estimator $\hat{\alpha}$ is still appropriate as an estimator for α^* if, e.g., U_{it} is uncorrelated with $d_i(\epsilon_i - E(\epsilon_i \mid d_i = 1))$ for all t.

[10]This estimator is obviously consistent for either the fixed-coefficient (α) or random-coefficient (α^*) model since $E(d_i(\epsilon_i - E(\epsilon_i \mid d_i = 1)) \mid d_i = 1) = 0$.

[11]Recall from our discussion in note 8 that in the random-coefficient model developed there an instrument for $Y_{it'}$ is required even in the absence of measurement error.

[12]This estimator requires that the variance of U_{it} ($t > k$) be the same as the variance of $U_{it'}$ ($t' < k$). Thus in the random-coefficient model, if U_{it} has a constant variance for all t, $U_{it} + d_i(\epsilon_i - E(\epsilon_i \mid d_i = 1))$ will not have the same variance as $U_{it'}$. It is possible, but artificial, to invoke equality of the variances for the two disturbance terms. Thus, in this sense, our proposed covariance-stationary estimator is *not* robust when applied to estimate α^* in repeated cross-sectional data.

The Issue of Weights in Panel Surveys of Individual Behavior*

Jan M. Hoem

1. INTRODUCTION

Many social scientists and most survey statisticians display great unease when anyone analyzes survey data without using conventional sampling weights. Such unease is certainly warranted when the analysis consists in estimating population-level statistics in the finite population from which the sample was drawn. Currently, however, the feeling also extends to cases where it is misplaced, for instance to situations where the analysis is evidently based on probabilistic models of human behavior, say, and where you must really force your imagination to find counterparts to model parameters and other characteristics in finite population statistics. This is particularly clear in sample-based event-history analyses and panel studies. It is hard to see what meaningful finite population statistics are estimated by occurrence/exposure rates, Nelson-Aalen plots, logistic regression coefficients, or the estimated coefficients of hazard regression analysis, all of which are used with samples as well as with complete sets of life history segments. Introducing sampling weights in such analyses can complicate matters, for the standard statistical theory of inference procedures of this nature then collapses and special theory must be applied. Such theory has been developed for models of contingency tables (see Rao & Scott, 1984; Smith, 1984, Section 4.2; and their references; for a computer program, see Fay, 1982) and indeed for generalized linear models (Binder, 1983; Chambless & Boyle, 1985). Some conscientious empirical investigators have computed values for estimators and test statistics both with and without weights, and have often been relieved to find that the outcomes have largely been the same (Schirm et al., 1982; Rindfuss et al., 1986, footnote 1; and others). The position taken in this paper is that such worries and such computing exercises can be superfluous in situations where the investigator is really involved in *modeling* human behavior rather than in calculating descriptive statistics for the finite population, and that they may divert attention away from more important concerns of modeling and analysis.

In the considerable disagreement on the issue of weighting in the current statistical literature, there is a standing dispute between those who would apparently really like to see standard weights (reciprocal selection probabilities) applied in "most" circumstances and others who feel that "if the analyst wishes to use the [sample] data to estimate a properly specified model, then the case for weighting is much weaker, since the model presumably 'controls for' the effects of the factors which lead to the need for weights in the first place" except apparently for particular dependent variables in the model. (This typical quotation from the *User Guide for the Panel Study of Income Dynamics* [Survey Research Center, 1983], p. A13, catches the spirit of many other formulations in the literature.) Some weighting protagonists seem to see the latter position as quite lenient in this spectrum, perhaps as a concession to model builders. To me (and to some others), even that standpoint is too much in favor of the use of sampling weights in the situations of this paper. Our approach is to follow general statistical notions, to regard the sampling mechanism as part of the total model of the "random experiment" that produces the sample data, and to incorporate it into the likelihood, with normal consequences for the statistical analysis.

Since I take issue with much that is found in contributions from sample theory practitioners and teachers (Kalton, 1981; O'Muircheartaigh & Wong, 1981; Hansen et al., 1983; Survey Research Center, 1983; Kish, 1981; and others), I should perhaps make clear from the outset that I realize that they do not all have the same position on all relevant issues, that opinions may develop over time, and that no one can be expected to present the full breadth of his reasoning on any single occasion, let alone the reasoning of others. There is no collective responsibility for arguments and recommendations presented. Let me also state unequivocally that I do not question the appropriateness of common weighting procedures in inference to finite population statistics. However, I want to line up with those who feel that the thinking on the matters raised here has been unduly dominated by the spirit of finite population descriptions. One cannot allow comments on the modeling of behavior to be confined to brief asides (see Section 5.1 of Hansen et al., 1983, for a typical example), nor is it sensible to relegate infinite population modeling concepts to the role of motivators of definitions of new finite population parameters to be studied by the design-based approach, as we are invited to do by Binder (1983) and Chambless and Boyle (1985) as well as by Folsom et al. in their paper in this volume.

The modeling approach induces us to focus on issues that have received insufficient attention or less than lucid treatment in the sampling literature. For instance, it is important to distinguish between the various elements of the comprehensive model of the real-life phenomenon investigated by means of the sample data. Likewise, one needs to keep these elements separate from the various statistical procedures available and from the functions that the procedures have in the analysis. We have in mind collections of records of segments of individual life histories, so one part of the total model is the submodel of individual behavior. Submodel misspecification is one issue and the use of sampling weights is another, and an operative connection between them remains to be demonstrated. Sampling weights have not been devised to correct for or protect against such misspecification in model-based analysis, and we know of no proof that they can serve this function in general, as seems implied by many formulations in the literature. Whether the sampling mechanism is informative (i.e., whether it depends on the random outcome of the real-life phenomena under analysis) is a separate question again. For instance, the examples offered by Hansen et al. (1983, Section 2) and Duncan (1982, Appendix 2) in support of the supremacy of the design-based approach have sampling mechanisms that are manifestly informative and therefore are not relevant for the issue of robustness against behavioral model misspecification nor for the separate issue of robustness against informatory status.

The informatory status of the sampling plan depends on the model, for the model defines which variables are seen as stochastic. Model misspecification at *this* level may result in an unrealistic declaration that the sampling plan is noninformative when it is not, which may lead into the well-known dangers of outcome-based sampling. In this connection, weighting does seem to have a function as a guard against model misspecification in certain cases. Holt et al. (1980) and Nathan and Holt (1980) have shown that weighting may produce a robust though inefficient estimator for a linear regression coefficient when the sampling plan is informative. (See also Jewell, 1985.) Unfortunately, one does not seem to really know why this is so nor to what extent current results can be generalized.

Outcome-based sampling does not appear as much of a problem in the selection of the initial target sample for prospective panel surveys of individuals, for the sample is typically drawn at the beginning of the observational period (at time 0) and therefore cannot be influenced by the later behavior of the (potential) respondents. On the other hand, since the extent of nonresponse may depend on such behavior, it may introduce an element of outcome-dependence in the effective sample, as is well recognized. (See, for instance, the paper by Fay in this volume.) This gives leeway to all the usual ingenuity of survey samplers in estimating nonresponse probabilities and applying *their* reciprocals to the various response groups, but in itself it gives no opening for the reciprocals of the sampling probabilities of the target sample.

This reserved attitude to the use of sampling weights is not weakened by the fact that it is sensible to use all information available about the members of the target population at the time of sample selection. If we condition on whatever has happened up to and including time 0 and concentrate on investigating whatever happens after that point, the sampling plan can be outcome-*independent* even if population characteristics at time 0 are used extensively when the sample is drawn. To the extent that such "starting data" have a bearing on the topic of the investigation, the information should be included in the model and thus be one of the guides of subsequent analysis. Poststratification into behaviorally distinct groups can be sensible, across prior stratum boundaries if this is suitable. Concomitant variables should be exploited as usual. (See Sugden & Smith, 1984, for a discussion of problems that arise if the analyst has less information than the sampler.)

It is important to be careful about features like poststratification according to behavior or other outcomes *after* time 0, however. Such procedures are prone to introduce selection biases of a form characteristic of outcome-dependent sampling. In an example below, we discuss the role of sampling weights in counteracting such biases in panel studies.

One must exercise similar care in the analysis of data for respondents who are included in the sample *after* time 0. In general, it will be unproblematic to utilize individual-level data for periods after entry of such individuals provided they are homogeneous with those who have been in the study from the outset. (Technically speaking, entry should be representable by a noninformative left-truncation mechanism; see Wellek, 1986, and Keiding & Gill, 1987, Section 5a.) The use of retrospective data for periods before entry may be another matter when entry into the sample is part of the life course outcome under analysis, as Lillard evidently suspects (see his paper in this volume). In the PSID, for instance, nonsample individuals may enter the sample because they become members of existing sample households. When such entry is through marriage to a sample household member, then the analysis of the process leading up to marriage needs to take into account that it actually ends in the upcoming marriage. Sample entry then represents a form of outcome-based sampling. (See Hoem, 1969, and Kelding & Gill, 1987, for some discussion of technical aspects.) If one is able to compute the real inclusion probability of such an entrant, there is a legitimate place for reciprocal probability weights in the analysis (Hoem, 1985, Section 2.2). This probability will be a much more complex entity than merely the initial inclusion probability of the household entered, however.

The panel medium is typically geared to the collection of data about the respondents' situation at fixed times during the observational period. This sets the stage for a particularly simple model presentation, and we use the time-discrete Markov chain as an uncomplicated prototype of panel models to convey the essentials of our notions with a minimum of disturbance by mathematical or circumstantial complexity. We extend this simplicity further by mostly assuming that data collection is by a two-wave panel only. In practical applications, more complex models and more extensive observational plans are bound to be needed. (A broader catalogue over various issues of design and usages of panel data has been given by Duncan & Kalton, 1985.) Indeed, much richer analyses can be made if data are obtained for continuous life histories, for then the whole tool-bag of event-history studies is available. This is particularly useful if the timing of events or the duration of spells are important for an understanding of the dynamics of behavior, as it is bound to be in most fields. (See Allison, 1982, for a discussion of the pros and cons of continuous-time and discrete-time methods.) The panel vehicle can be used to obtain retrospective information for periods before time 0 and between other times of data collection, and is used in this manner by some major data agencies. The issue of weighting is not changed in character by such an extension.

Much of the reasoning presented here has basically been given before, though usually with different emphases and not with the special issues of panel studies in mind. Some recent references are Fienberg (1980, pp. 335–338), Little (1982), Smith (1983, 1984), and Hoem (1985). Lest I be accused like so many others of being a meddling theorist with no ground contact and therefore nothing useful to tell sampling practitioners (cf. O'Muircheartaigh & Wong, 1981, p. 487; Kish, 1981), let me note as a credential that I became interested in the weighting issue in connection with my own empirical research in demography based on panel and event-history data, and as a discussion partner with a number of colleagues working in demography, sociology, economics, and epidemiology.

2. SAMPLING MARKOV CHAIN SAMPLE PATHS

2.1. Framework

Here is the very simple mathematical framework in which our arguments will be dressed. Assume that the N individuals of a population move independently between states in the finite state space J of a time-homogeneous Markov chain model. Suppose that the state x_t occupied by individual i at time t is observed at discrete times $t = 0, 1, 2, \ldots$ and let the unit time transition probability be

$$p_{jk} = P\{x_{t+1} = k \mid x_t = j\}.$$

Suppose to begin with that x_t is observed for each individual only at times 0 and 1, and regard x_0 as exogenous, i.e., as determined before the "experiment" whose outcome is observed by the investigator. Let $\chi_i(j) = 1$ if individual i is in state j at time 0, and let $\chi_i(j) = 0$ otherwise. Then the $\chi_i(j)$ are nonstochastic indicators, and the number

$$N_j = \sum_{i=1}^{N} \chi_i(j)$$

of population members in state j at time 0 is also nonstochastic and perhaps known to the investigator. Let N_{jk} of the latter be in state k at time 1, and suppose for now that the $\{p_{jk}\}$ have no particular structure, i.e., that all that is essentially known about these probabilities is that $\Sigma_{k \in J} p_{jk} = 1$ for all $j \in J$, with the possible exception that some transitions may be impossible and the corresponding p_{jk} may equal 0. If all population data $\{N_{jk}\}$ were available, then the maximum likelihood estimator of a (nonzero) p_{jk} would of course be the multinomial proportion

$$p_{jk}^* = N_{jk}/N_j, \tag{1}$$

and we would have

$$\mathrm{Var}(p_k^*) = p_{jk}(1 - p_{jk})/N_j . \tag{2}$$

Now suppose, however, that a sample S of individuals is drawn at time 0 according to some known sampling plan $p(s) = P\{S = s\}$. (If sampling is with replacement, then our S is the sample after the removal of any doubles.) The probability that individual i is a member of S is $\pi_i = \Sigma_{\{s:i \in s\}} p(s)$, which we assume to be positive for all i. Suppose that the estimation of the $\{p_{jk}\}$ is to be based on the sample data

$$D = \{S; [N_{ijk} : i \in S, j \in J, k \in J]\},$$

where $N_{ijk} = 1$ if individual i is in state j at time 0 as well as in state k at time 1, with $N_{ijk} = 0$ otherwise. Assume that the individual transitions after time 0 are independent of membership in the sample, i.e., that the sampling (and subsequent observation) does not affect individual behavior. Then strict adherence to conventional sampling theory would lead to the estimation of N_{ij} by

$$\tilde{N}_{jk} = \sum_{i \in S} N_{ijk}/\pi_i,$$

and correspondingly to $\tilde{p}_{jk}^* = \tilde{N}_{jk}/N_j$ as an estimator (predictor) of the population statistic p_{jk}^* (if N_j is known).

In a superpopulationist vein one may note that

$$E\{N_{ijk} \mid S = s\} = \chi_i(j) \, p_{jk} \text{ if } i \in S. \tag{3}$$

If we define

$$\tilde{N}_j = \sum_{i \in S} \chi_i(j)/\pi_i,$$

then,

$$E(\tilde{p}_{jk}^* \mid S = s) = p_{jk}\tilde{N}_j/N_j,$$

so $\tilde{p}_{jk}^*$ is not an an unbiased estimator for the parameter p_{jk} when the sample is given, in contrast to, say,

$$\tilde{p}_{jk} = \tilde{N}_{jk}/\tilde{N}_j, \tag{4}$$

which may be used even if N_j is unknown to the investigator. On the other hand,

$$E\tilde{N}_j = \sum_s p(s) \sum_{i \in s} \chi_i(j)\pi_i^{-1} = \sum_{i=1}^{N} \chi_i(j)\pi_i^{-1} \sum_{\{s:i \in s\}} p(s) = N_j,$$

so

$$E(\tilde{p}_{jk}^*) = E(\tilde{p}_{jk}) = p_{jk}$$

when E denotes the expectation operator in the model that accounts for the randomness of *both* S *and* the population data $\{N_{ijk}: i = 1, \ldots, N;$ $j \in J;\ k \in J\}$, unless a conditional expectation is indicated explicitly. In the total model, therefore, (i.e., when all currently random elements are included), both estimators $\tilde{p}_{jk}^*$ and $\tilde{p}_{jk}$ are unbiased.

The two-wave set-up does not really exploit the Markovian properties of the chain model. All we have used so far is its notation. In reality, we are only dealing with a set of related contingency tables, one for each starting state $j \in J$. As we noted at the beginning of Section 1, the mathematics for dealing with inference in such models, with or without weighting, is already available.

2.2. The Likelihood Approach

These estimators are based on survey sampling notions and their super-populationist extension. A classical statistical approach would be to establish the total likelihood corresponding to the observed data, and to maximize it. The likelihood is $\Lambda' = p(S)\,\Lambda$, where

$$\Lambda = \prod_{i \in S} \prod_{j \in J} \prod_{k \in J} [p_{jk}]^{N(i,j,k)}, \tag{5}$$

with $N(i,j,k) = N_{ijk}$ in the exponent. If $N_S(j,k) = \sum_{i \in S} N_{ijk}$ is the number of transitions from j to k observed in the *sample*, and if

$$n_S(j) = \sum_{k \in J} N_S(j,k) = \sum_{i \in S} \chi_i(j) \tag{6}$$

is the number of *sample* members who start out in state j at time 0, then the MLE of p_{jk} is

$$\hat{p}_{jk} = N_S(j,k)/n_S(j), \tag{7}$$

provided that $p(.)$ is functionally independent of the $\{p_{jk}\}$, as we will assume throughout. Note that formula (7) has precisely the same structure as formula (2), i.e., $\hat{p}_{jk}$ is the estimator that elementary statistical theory will lead to if the sample is treated as if it where the whole population. Beyond the fact that the sample is the vehicle that provides the data, the form of the estimator is not influenced by the sampling mechanism. In particular, no sampling weights are involved.

By formulas (3) and (6), we easily derive the unbiasedness results

$$E\{\hat{p}_{jk} \mid S = s\} = E\{\hat{p}_{jk}\} = p_{jk}.$$

In parallel with formula (2),

$$\text{Var}(\hat{p}_{jk} \mid S = s) = p_{jk}(1 - p_{jk})/n_S(j),$$

and consequently

$$\text{Var}(\hat{p}_{jk}) = p_{jk}(1 - p_{jk}) \, E\{1/n_S(j)\}, \tag{8}$$

with similar results for covariances. Formula (8) shows that the *properties* of $\hat{p}_{jk}$ are certainly influenced by the sampling mechanism, for $p(.)$ determines the final item in the formula. The likelihood approach allows us to construct estimators whose *form* is not influenced by the sampling, but we cannot ignore the fact that a sample has been drawn when we study (unconditional) estimator *properties*.

2.3. A Weighted "Likelihood"

The likelihood approach is sometimes interpreted in a manner different from the one that lead to formula (5), namely as follows. (See, for example, Chambless & Boyle, 1985.) If all population data were available, then the likelihood would be

$$\Lambda_{tot} = \prod_{j \in J} \prod_{k \in J} [p_{jk}]^{N(j,k)},$$

with $N(j,k) = N_{jk}$. When one is restricted to the sample data D, then each member i in the *sample* "represents" $1/\pi_i$ members in the *population*. It seems logical then to maximize an "estimate" of Λ_{tot} given by

$$\tilde{\Lambda}_{tot} = \prod_{i \in S} (\Lambda_i)^{1/\pi(i)},$$

where $\pi(i) = \pi_i$, and where

$$\Lambda_i = \prod_{j \in J} \prod_{k \in J} [p_{jk}]^{N(i,j,k)}$$

is the likelihood contribution of individual i. Such maximization leads to the estimator $\tilde{p}_{jk}$ of formula (4), which thus gets a kind of legitimization as a pseudo-maximum-likelihood estimator. However, given S, $\hat{p}_{jk}$ of formula (7) is known from general theory to have minimal variance among unbiased estimators, and this property carries over to the unconditional variance in formula (8). Using $\tilde{p}_{jk}$ instead of $\hat{p}_{jk}$ must entail some loss in efficiency.

2.4. Concomitant Information

We have kept to the very simple situation above to minimize the effort of presentation. The main outcome of our argument is retained in cases with more extensive observational plans or a more complex structure in the transition probabilities. Assume for instance that for individual i, the $j \rightarrow k$ transition probability is $p_{ijk} = \phi_{jk}(z_i, \theta)$, where z_i is this individual's value on a vector of concomitant variables; θ is some unknown multidimensional parameter; and the function ϕ_{jk} may perhaps specify a logistic regression, it may have a simple form as in Section 4 below, or it may be of some quite different complexity.

Assume that the z_i are exogenous and that the sampling mechanism is independent of θ, though it may depend on $z_1, \ldots, z_N$, perhaps through some system of stratification of the members of the population. Then the likelihood of D continues to have the form $\Lambda' = p(S)\Lambda$, where Λ is given in formula (5), except that p_{jk} is now replaced by $\phi_{jk}(z_i, \theta)$. Maximization again proceeds without regard to any sampling weights $(1/\pi_i)$ and the MLE for θ is constructed as if the sample members constitute the whole population of interest.

Some investigations will oversample certain minorities. This in itself is hardly a sufficient reason to use sampling weights in the estimation of parameters of behavioral models. Let us distinguish three situations:

(1) If the behavior of the minority is the same as that of other people, then applying reciprocal sampling probabilities just gives more weight to some observations than to others of the same kind, which cannot be efficient under any approach.

(2) If minority behavior differs from other behavior (and of course knowledge or a suspicion of this is the reason why the minority was oversampled in the first place), then it should be reflected in sufficiently

accurate modeling. The model may have one or more parameters whose values are different for the minority, in which case likelihood maximization (or something similar) will pick up these differentials. Alternatively, a different model may be needed for the minority, in which case separate analysis is more sensible. Who has a proof that weighting can overcome the inferential errors of an inadequate model that tries to account for behavioral differences?

(3) If minority behavior differs from other behavior and a model is fit that does *not* have features to pick this up, then the model is misspecified and weighting will not be of much use. Instead of parameter values (or separate models) that identify the behavioral differences between population groups, one gets fitted parameter values that represent some fictitious "mean behavior" that no group has. One has then lost sight of interesting behavioral differentials. Modeling them is more sensible.

Note that we discuss the role of the sampling weights as an issue separate from the question whether the model ϕ_{jk} is correctly specified. The latter question is certainly important for the empirical analysis, but it must be addressed directly. There is no a priori reason why the use of sampling weights can be expected to compensate for an incorrect specification of individual behavior. Anyone who feels that weights may give some protection against model misspecification of this kind should demonstrate it and explore why it works, if it does.

2.5. Three Waves

The character of these arguments does not change if the observational plan is more extensive than the one above. To take a single step in such a direction, let us revert to nonspecified transition probabilities p_{jk} of a Markov chain but let the state x_t of each individual be observed at time 2 as well as at time 0 and 1. For sample member i, let $N(i,j,k,l) = 1$ if this individual has the state sequence $x_0 = j$, $x_1 = k$, and $x_2 = l$, and let $N(i,j,k,l) = 0$ otherwise. With a time-homogeneous Markov chain model, the likelihood of the sample data now becomes

$$p(S) \prod_{i \in S} \prod_{j \in J} \prod_{k \in J} \prod_{l \in J} \{p_{jk}\, p_{kl}\}^{N(i,j,k,l)},$$

which is maximized by

$$\hat{p}_{jk} = \frac{N_S(j,k) + \Sigma_{i \in S}\Sigma_{l \in J}N(i,l,j,k)}{n_S(j) + \Sigma_{i \in S}\Sigma_{l \in J}\Sigma_{k \in J}N(i,l,j,k)}. \tag{9}$$

Properties of this kind of estimator were studied by Anderson and Goodman (1957). There is no role for sampling weights $(1/\pi_i)$ in this estimation procedure either.

In what follows, we revert to the case where individuals are observed in two waves only.

3. UNIT NONRESPONSE

Let us now address the issue of characteristic-dependent nonresponse. Consider the simple Markov chain model again, let sample individuals be observed at times 0 and 1, and let us make the assumptions

(1) that whether an individual responds at time 1 is independent of the outcome at time 0 as well as of the transition behavior between times 0 and 1;

(2) that both at time 0 and 1 each individual in state j has a response probability of β_j; and

(3) that individuals choose to respond or abstain independently of each other.

The response model above is sufficient to serve as an illustration for our purposes. In practice, a more complex response model will surely be needed. For instance, one must often expect the response outcome at time 1 to depend on what has happened before that time. For a more complete model in a three-state set-up, see Stasny (1986a,b) and her references. Marini et al. (1979) study a situation with normally distributed variables. Both papers use the maximum likelihood approach and no weighting.

To establish a likelihood, we introduce $A(i,t)$, which equals 1 if individual i responds when approached at time t, and which equals 0 if this individual is a nonrespondent at time t. (A is for "answer".) Suppose that the state at time 0 is known for all (sample) members; the state at time 1 is obtained only for respondents. Then $\{\chi_i(j) : i \in S, j \in J\}$ is exogenous and the sample data consist of S, $\{A(i,t) : i \in S; t = 0,1\}$, and $\{N(i,j,k): i \in S; j \in J; k \in J; A(i,1) = 1\}$. The likelihood of these data is

$$p(S)\prod_{i \in S} \prod_{j \in J} [[\beta_j]^{A(i,0)}[1 - \beta_j]^{1-A(i,0)}]$$

$$\times \left[\left(\prod_{k \in J} [p_{jk}\beta_k]^{A(i,1)N(i,j,k)} \right) [\gamma_j]^{1-A(i,1)} \right] \chi_i(j).$$

where $\gamma_j = 1 - \sum_{k \in J} p_{jk}\beta_k$ is the probability that an individual who is in state j at time 0 will be a nonrespondent at time 1. We introduce $\beta_{jk} = p_{jk}\beta_k$, note that $\gamma_j = 1 - \Sigma_k\beta_{jk}$, and reorganize the likelihood, which becomes

$$p(S) \ \left(\prod_{j \in J} [\beta_j]^{R(j,0)} \, [1 \, - \, \beta_j]^{n(j) - R(j,0)} \right)$$

$$\times \prod_{j \in J} \ \prod_{k \in J} \left([\beta_{jk}]^{T(j,k)} \, [\gamma_j]^{n(j) - R(j,1)} \right),$$

where

$$R(j,t) \, = \, \sum_{i \in S} \chi_i(j) \, A(i,t)$$

is the number of respondents at time t among sample members who were in state j at time 0; where

$$T(j,k) \, = \, \sum_{i \in S} A(i,1) N(i,j,k)$$

is the number of $j \rightarrow k$ transitions actually observed; and where we have written $n(j)$ for $n_S(j)$ to facilitate the typing of exponents. Since $\Sigma_k T(j,k) = \Sigma_{i \in S} A(i,1) \chi_i(j) = R(j,1)$, maximizing the likelihood is straightforward, and we get the MLEs

$$\hat{\beta}_j \, = \, R(j,0) / n_S(j) \qquad\qquad (10)$$

and

$$\hat{\beta}_{jk} \, = \, T(j,k) / n_S(j),$$

which makes $\hat{\beta}_{jk} / \hat{\beta}_k$ the current maximum likelihood estimator of p_{jk}. Unfortunately, these estimators do not add up to 1 when we take the sum $\Sigma_{k \in J}$, so the adjusted estimator

$$\hat{p}'_{jk} \, = \, \frac{T(j,k) / \hat{\beta}_k}{\displaystyle\sum_{l \in J} T(j,l) \hat{\beta}_l} \qquad\qquad (11)$$

is perhaps preferable. As the population size and sample size go to infinity together, the denominator of formula (11) will converge to 1 in probability under any reasonable asymptotic.

Adjustment by means of reciprocal response probabilities is of course an old trick in survey sampling analysis. It appears so easily above because we have made things simple for ourselves through our assumptions. More complex response models will lead to results of a similar nature, however, and the main message conveyed again is that sampling probabilities just do not enter into the formula for the estimator.

4. OUTCOME-DEPENDENT SAMPLING

4.1 Basic Notions

We have assumed that the sample was drawn at time 0, and of course that it could only be based on information available at that point. This is the natural situation in prospective panel surveys. In retrospective surveys, by contrast, one has the option of using whatever information is available when the sample is drawn at the *end* of the study period. (In a retrospective *panel* study, information would be obtained concerning the situation of the respondents at fixed times prior to the time of selection.) If any information concerning the respondents' behavior during the study period is used in the sample selection, then the sample S is outcome-dependent, and subsequent data analysis must be made with great care to avoid the many pitfalls inherent in such a set-up. Even if the original sample S is outcome-*in*dependent, subsequent poststratification according to the value of an outcome-variable may introduce similar effects.

Properties of the sampling plan will generally enter into a likelihood analysis if the sampling is outcome-dependent, and they may help provide a guard against selection biases otherwise produced. In some situations, the influence of the sampling plan then works via the (reciprocal) sampling fractions in outcome-based strata.

We discuss a simple example in Sections 4.2 to 4.4 below. The formal model there goes back to Colding-Jørgensen and Simonsen (1940), and it has been used for purposes similar to ours by Aalen et al. (1980) and by Hoem and Funck Jensen (1982, Section 5.3). It is a time-continuous Markov chain model used for statistical inference from panel data. A review of such issues has been given recently by Kalbfleisch and Lawless (1985), who also address computational aspects as well as the incorporation of covariates. Among the references that they do not give are Singer and Spilerman (1977), Singer (1981), and Allison (1982). Formulas given by Funck Jensen (1982) for transition probabilities in terms of transition intensities will be useful in such analyses.

Outcome-dependent observational plans and the biases they produce appear in many shapes in most fields of statistics and have correspondingly many names, such as length biased sampling, prevalence sampling, selection biases, restriction biases, selection by virtue of survival, purged sampling, anticipatory observation, and choice-based sampling. We have reviewed them in the Markov chain setting elsewhere (Hoem & Funck Jensen, 1982, Section 6; Hoem, 1985, Sections 2.2 and 2.3). Cohen and Cohen (1984) recently discussed them for clinical trials. For an account of their appearance in sociology, building mainly on previous work in economics by Goldberger and Heckman, see Berk and Ray (1982) and Berk (1983). Some further references are Hoem (1969), Cosslett (1981), Manski and McFadden (1981), Vardi (1985), Rao (1985), Rindfuss et al. (1985), and Hoem et al. (1986).

4.2. Example 1: Childbearing and Promotion

We now turn to the simple example in Figure 1, which (in one of its guises) reflects some central features of the interaction between childbearing and promotion of women in a bureaucratic hierarchy where the employer is not permitted to let promotion to a higher-grade job be influenced by the employee's private life.

FIGURE 1. MARKOV CHAIN REPRESENTATION OF FIRST CHILDBEARING AND PROMOTION TO A HIGHER-GRADE JOB IN A BUREAUCRATIC HIERARCHY [AS WELL AS OF MOTHER-AND-INFANT MORTALITY]

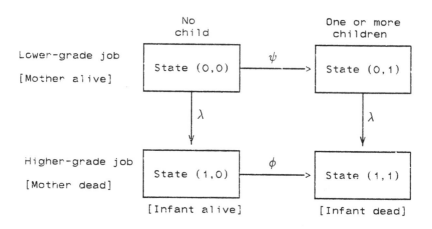

In this model, the states are denoted by a two-digit code (x,y), where x indicates whether she has a lower-grade job (code $x = 0$) or a higher-grade job (code $x = 1$). The second element y indicates whether a woman has had a child (code $y = 1$) or not (code $y = 0$). Thus a woman is in state $(0,0)$ if she works in a lower-grade job and has not yet had a child. At the birth of her first child, she moves to state $(0,1)$ if she still has a lower-grade job, and so on. A woman cannot have a child and move to a higher-grade job at the same time. Otherwise, she can have her first child or be promoted at any time. Let (x_t, y_t) be a woman's state at time t. The transition intensities in our time-continuous Markov chain model for a particular group of women are the constant parameters λ, ψ, and ϕ indicated in the figure. Thus, ψ and ϕ are birth intensities for women in lower- and higher-grade jobs, respectively, and λ is the rate at which women are promoted to higher-grade jobs. We assume that the bureaucratic rules ensure that the latter is not influenced by the presence or arrival of a child. (In the terminology introduced by Schweder, 1970, x_t is locally independent of y_t.) On the other hand, suppose that women in higher-grade jobs have specific motives for reducing their natality, so $\psi \geq \phi$. We take (x_0, y_0) to be exogenous and let

$$p_{ab}(x,y) = P\{x_1 = x, y_1 = y \mid x_0 = a, y_0 = b\} \text{ for } a,b,x,y \in J,$$

with $J = \{0,1\}$. If $\delta = \psi - \phi$, it follows readily that

$$p_{00}(0,0) = e^{-(\lambda+\psi)}, \ p_{00}(0,1) = e^{-\lambda}(1 - e^{-\psi}),$$

$$p_{00}(1,0) = e^{-\phi}(1 - e^{-(\lambda+\delta)})\,\lambda/(\lambda+\delta),$$

$$p_{01}(0,1) = e^{-\lambda}, \ p_{10}(1,0) = e^{-\phi},$$

and so on. In particular,

$$P\{x_1 = 1 \mid x_0 = 0, y_0 = b\} = 1 - e^{-\lambda} \text{ for } b = 0,1, \tag{12}$$

i.e., the probability of getting promoted to a higher-grade job by time 1 for a woman who is not there at time 0 equals $1 - e^{-\lambda}$, irrespective of her childbearing status at time 0.

Assume that a Scandinavian type population register is available, so that the target population can be stratified by childbearing status whenever needed, and suppose that information on job status at times 0 and 1 is collected from the members of a sample. Assume that the respondents are grouped according to childbearing status at *time 1*, either because the sample was selected this way in the first place or through poststratification when the data are prepared for analysis. As part of the investigation, one may then estimate transition probabilities for the promotion variable x_t, given the outcome on the childbearing variable, i.e., conditional probabilities of the form

$$\eta_{by} = P\{x_1 = 1 \mid x_0 = 0, y_0 = b, y_1 = y\}$$

$$= p_{0b}(1,y)/[p_{0b}(0,y) + p_{0b}(1,y)].$$

As we demonstrate below, it turns out that

$$\eta_{01} < \eta_{11} = 1 - e^{-\lambda} < \eta_{00} \text{ when } \psi > \phi. \tag{13}$$

This looks as if the arrival of a first child ($y_0 = 0, y_1 = 1$) reduces the chances of getting promoted to a higher-grade job over the unit time period, and as if *not* having a child ($y_0 = 0, y_1 = 0$) improves those chances, despite the fact that we have *postulated* no influence of childbearing status on the promotion variable in our model. The childbearing-status *in*dependence of the promotion variable is well reflected in the unconditional probabilities in formula (12), but it gets garbled in the conditional transition probabilities, as is apparent from inequalities (13). The analysis based on such poststratification easily induces the investigator to conclude that the arrival of a first child is a hindrance to further promotion even when it is not. Conditioning on the outcome of one variable in the investigation of another in life course analysis is a risky business.

To prove relation (13), it simplifies matters to study $\bar{\eta}_{by} = 1 - \eta_{by}$ and to demonstrate the equivalent relation

$$\bar{\eta}_{00} < \bar{\eta}_{11} = e^{-\lambda} = \bar{\eta}_{01} \quad \text{when } \delta > 0.$$

First note that

$$\bar{\eta}_{00} = e^{-\lambda} v/w, \quad \bar{\eta}_{11} = e^{-\lambda},$$

$$\text{and } \bar{\eta}_{01} = e^{-\lambda}(1 - v)/(1 - w),$$

with $v = e^{-\psi}$ and $w = p_{00}(0,0) + p_{00}(1,0) = e^{-\psi} f_\lambda(\delta)$,

where

$$f_\lambda(\delta) = e^{-\lambda} + (e^\delta - e^{-\lambda})\lambda/(\lambda+\delta).$$

Since both w and $1 - w$ are positive probabilities, $0 < w < 1$. Because $f_\lambda(0) = 1$ and $\partial f_\lambda(\delta)/\partial\delta > 0$, we get $f_\lambda(\delta) > 1$ and therefore all in all $0 < v < w < 1$ when $\delta > 0$. From this our inequalities follow directly.

The above formulas for the $\bar{\eta}_{by}$ also show that $\eta_{00} = \eta_{11} = \eta_{01} = 1 - e^{-\lambda}$ if $\psi = \phi$. This is the unconditional promotion probability in formula (12). When childbearing is *not* influenced by job grade, therefore, one is relieved of the dangers of systematic errors in conclusions (within this model) otherwise inherent in outcome-dependent analysis. For the particular value 0 of the parameter δ, outcome-dependence is still present but the selection biases it causes vanish.

Klevmarken (1986) has an example in which not only do the selection biases vanish when a particular parameter has the value 0, but the entire outcome-dependence disappears at the same time.

4.3. Example 1 Continued: Likelihood Analysis

The selection biases just described arise because the investigator supposedly has not used the full information contained in the data, as summarized by their likelihood. To see how the likelihood approach would work in this particular example, let us assume that all the N members of a target population start in state $(0,0)$ at time 0, and introduce $p_{00} = p_{00}(0,0)$, $p_{01} = p_{00}(0,1)$, $p_{10} = p_{00}(1,0)$, and $p_{11} = p_{00}(1,1)$. Let $q = p_{00} + p_{10}$ be the probability that a woman has had no child by time 1, and let us drop the initial subscript 0 in the ηs now involved and write $\eta_y = \eta_{oy}$ for $y = 0,1$. Assume that at time 1, the women are grouped into two strata, stratum 0 for those who have had no children and stratum 1 for the rest. Suppose that a simple random sample is then drawn from each of these strata at time 1, and that the job grade status at time 1 is obtained (without misclassification error or non-response) from all members of the sample. We now want to establish the likelihood of these data. We usually follow the convention of denoting random variables by capital Latin letters and their values by the corresponding lower-case letters, but in the present connection this would lead us to cover up some correspondences with survey sampling theory that we want to display. For the remainder of this section, therefore, let random variables be in boldface capitals or lower-case letters, and let their values be in corresponding nonboldface capitals or lower-case letters. At time 1, $\boldsymbol{M}_k$ members of the target population are in state k, for $k = (0,0)$, $(0,1)$, and so on. Of the members, $\boldsymbol{N}_0 = \boldsymbol{M}_{00} + \boldsymbol{M}_{10}$ have not had a child, and $\boldsymbol{N}_1 = N - \boldsymbol{N}_0$ of them have had a child. A random sample of $\boldsymbol{n}_0$ out of the $\boldsymbol{N}_0$ is drawn from stratum 0, and another random sample of $\boldsymbol{n}_1$ is drawn from the $\boldsymbol{N}_1$ members of stratum 1. (Since the sample sizes are restricted to not exceed the random number of members of the two strata, they are random variables themselves.) The numbers $\boldsymbol{m}_k$ of members of the sample who are in each state k at time 1 are registered. The likelihood of obtaining these data is then the observed value of the probability

$$P\{\boldsymbol{N}_0 = N_0, \boldsymbol{m}_{00} = m_{00}, \boldsymbol{m}_{01} = m_{01}, \boldsymbol{m}_{10} = m_{10}, \text{ and } \boldsymbol{m}_{11} = m_{11}\}$$

$$= \Sigma_0 \Sigma_1 \frac{N!}{M_{00}! \, M_{01}! \, M_{10}! \, M_{11}!} \left[\prod_{k=(0,0)}^{(1,1)} (p_k)^{M_k} \right] \tag{14}$$

$$\times \frac{\displaystyle\prod_{k=(0,0)}^{(1,1)} \binom{M_k}{m_k}}{\dbinom{M_{00} + M_{10}}{n_0} \dbinom{M_{01} + M_{11}}{n_1}},$$

where Σ_0 is the sum over all pairs (M_{00}, M_{01}) for which $M_{00} + M_{01} = N_0$ and Σ_1 is the sum over all pairs (M_{01}, M_{11}) for which $M_{01} + M_{11} = N_1$. We show below that the likelihood can be rewritten as

$$\binom{N}{N_0} q^{N_0}(1 - q)^{N-N_0} \binom{n_0}{m_{10}} (\eta_0)^{m_{10}}$$

$$\text{(15)}$$

$$\times (1 - \eta_0)^{n_0 - m_{10}} \binom{n_1}{m_{11}} (\eta_1)^{m_{11}}(1 - \eta_1)^{n_1 - m_{11}},$$

which is maximized by the natural estimators

$$\hat{q} = N_0/N, \ \hat{\eta}_0 = m_{10}/n_0, \text{ and } \hat{\eta}_1 = m_{11}/n_1.$$

We substitute these into the one-to-one relations that connect the parameters p_k with our current equivalent parameters q, η_0, and η_1, namely the relations $p_{00} = q(1 - \eta_0)$, $p_{10} = q\eta_0$, and so on, and get the maximum likelihood estimators

$$\hat{p}_{00} = m_{00}/(Nf_0), \ \hat{p}_{01} = m_{01}/(Nf_1),$$

$$\hat{p}_{10} = m_{10}/(Nf_0), \text{ and } \hat{p}_{11} = m_{11}/(Nf_1),$$

where $f_k = n_k/N_k$ is the sampling fraction in stratum k, for $k = 0,1$. (Let $f_k = 1$ in the unlikely case that $N_k = 0$.) Provided n_0 and n_1 are sensible functions of N and N_0, as we can safely assume, these estimators will be consistent for their respective estimands as $N \to \infty$.

To bring out the close connection between these estimators and what survey sampling theory would suggest without any appeal to the likelihood approach, note that $X_0 = m_{10}$ is the number of members of stratum 0 who have the property of being in state $(1,0)$ at time 1. The corresponding number of members from stratum 1 is $X_1 = 0$. A Horvitz-Thompson (HT) estimator of the proportion M_{10}/N of the target population that is in state $(1,0)$ at time 1 can then be written as

$$\hat{M}_{10}/N = N^{-1}\Sigma_k X_k/f_k,$$

which is $\hat{p}_{10}$. (The sum is taken from $k = 0$ to $k = 1$.) By symmetry, the other probability estimators can be represented in a similar manner. It turns out, therefore, that in this particular case, the maximum likelihood estimators and the HT-estimators coincide. The reciprocal sampling fractions serve to balance the sampling biases otherwise inherent in the outcome-dependent sampling plan.

Note, however, that the sampling fractions f_0 and f_1 are not a priori inclusion probabilities. Every member of the target population has the same probability of ending up in the sample, and that probability is

$$q E(f_0) + (1 - q) E(f_1).$$

We need to know what functions n_0 and n_1 are to compute this probability. The sampling fraction f_k is "only" the conditional selection probability, given that a population member has ended up in stratum k by time 1.

To demonstrate the transition between formulas (14) and (15), note that

$$P\{N_0 = N_0\} = \binom{N}{N_0}(q)^{N_0}(1 - q)^{N-N_0} \tag{16}$$

and

$$P\{\overset{(1,1)}{\underset{k=(0,0)}{\cap}} (M_k = M_k) \mid N_0 = N_0\} \tag{17}$$

$$= \binom{N_0}{M_{10}}(\eta_0)^{M_{10}}(1 - \eta_0)^{N_0-M_{10}}\binom{N_1}{M_{11}}(\eta_1)^{M_{11}}(1 - \eta_1)^{N_1-M_{11}}.$$

Therefore, the expression in formula (14) can be written as that of formula (16) multiplied by the product of two sums corresponding to Σ_0 and Σ_1 in formula (14). After some minor rearrangement, the first sum can be written as

$$\binom{n_0}{m_{10}}\sum_{M_{10}=0}^{N_0}\binom{N_0 - n_0}{M_{10} - m_{10}}(\eta_0)^{M_{10}}(1 - \eta_0)^{N_0-M_{10}}$$

$$= \binom{n_0}{m_{10}}\sum_{x=-m_{10}}^{N_0-m_{10}}\binom{N_0 - n_0}{x}(\eta_0)^{x+m_{10}}(1 - \eta_0)^{N_0-m_{10}-x}.$$

Items in the latter sum are 0 for $x < 0$ and for $x > N_0 - n_0$ (remember that $m_{10} \leq n_0$), so this expression reduces to

$$\binom{n_0}{m_{10}}(\eta_0)^{m_{10}}(1 - \eta_0)^{n_0-m_{10}}. \tag{18}$$

The sum based on Σ_1 is quite similar. Therefore, formula (15) results if you multiply together the expressions in (16), (18), and the Σ_1-based expression corresponding to (18), and subsequently substitute N_0, n_0, n_1, m_{10}, and m_{11} for N_0, n_0, n_1, m_{10}, and m_{11}.

4.4. Example 2: Maternal and Infant Mortality

The model in Figure 1 may be reinterpreted in a manner that makes it suitable for the analysis of the impact of the death of a mother on the survival of her baby in historical data. Consider a mother-and-infant pair, let x_t be 0 as long as the mother is alive, and let x_t jump to 1 if the mother dies. Similarly, let y_t be an indicator of whether the child is dead at time t. Let the interval between times 0 and 1 be such that both mother and child have (acceptably) constant mortality during it, let λ be the mother's force of mortality, and let ψ and ϕ be the forces of mortality of the infant before and after any death of the mother. Suppose that the two cannot die simultaneously. Assume that the child is sufficiently dependent on the mother's personal care that the infant's (force of) mortality jumps to a higher level if the mother dies, i.e., $\phi > \psi$, but that the mother's mortality is not similarly influenced by the death of the child. Then all inequalities in relation (13) are reversed. If the data for the mother-and-infant pairs are sorted according to whether the child is alive at time 1, therefore, the investigator is invited to conclude erroneously that the death of the child adversely affects the mother's chances of survival in this model. Furthermore, the mother's survival chances will be estimated as better than they really are from data on pairs with infants surviving to time 1. Prior stratification or poststratification according to the infants' survival status at time 1 should be avoided.

Sorting the pairs according to the *mother's* survival status at time 1 is less dangerous for conclusions about the infant's survival chances. Such grouping corresponds to working with conditional probabilities of the form

$$\zeta_{ax} = P\{y_1 = 1 \mid x_0 = a, y_0 = 0, x_1 = x\}$$

$$= p_{a0}(x,1)/[p_{a0}(x,0) + p_{a0}(x,1)]$$

instead of $\theta_a = P\{y_1 = 1 \mid x_0 = a, y_0 = 0\}$. A simple argument similar to the one that established relation (13) shows that

$$\zeta_{00} = 1 - e^{-\psi} < \theta_0 < \zeta_{01} < \zeta_{11} = \theta_1 = 1 - e^{-\phi} \text{ when } \psi < \phi. \quad (19)$$

Thus, it is less dangerous for the infant to have a surviving mother than to lose her during the unit interval, and the latter event is less adverse than being without the mother throughout the interval, all of which conclusions are correct.

5. CONCLUDING DISCUSSION

The previous sections embed some very general issues into some very straightforward settings. The question of whether to use weights in the investigation of models for panel data is itself a special case of a more general issue concerning the role of weights in any analysis of sample data that involve the parameters of statistical models. Let us reiterate some general results here for the case of panel studies.

In a population of N independent individuals, let Y_i be a description of the life course of member i over a finite set of time points $t_0 = 0 < t_1 < \ldots < t_m < \infty$. Since we have sample paths of time-discrete Markov chains particularly in mind, assume for simplicity that for each Y_i there are only a finite number of possible paths, let y be one of them, and introduce $\xi_i(y) = P\{Y_i = y\}$. Note that $y = \{y(t_0), \ldots, y(t_m)\}$, so the probability law ξ_i is a multidimensional distribution function. It reflects our notions about the possible behavior of individual i over the finite time set $\{t_k\}$. It may depend in any way whatsoever on individual-specific values of exogenous variables, including the value $Y_i(0)$ of Y_i at time 0, which we take as nonstochastic but not necessarily known to the investigator before the sample is drawn. (If real individuals come in independent clusters with internal interaction, like households, then redefine an "individual" to be such a cluster for the purposes of the general theory. If clusters can recombine over time, we need a more general framework than the present one. Conditioning on the $Y_i(0)$ is convenient and useful in panel studies, but it is not essential for our argument.)

From this population, a sample S is selected according to the sampling mechanism $p(s) = P\{S = s\}$. We allow $p(.)$ to depend on the same exogenous variables as the ξ_i do, including the given values $Y_i(0)$ if they are known, but $p(.)$ may not depend on the outcomes of the Y_i *after* time 0. Essentially, the sample is drawn *at* time 0, and then the investigator cannot utilize anything that happens later. We assume that observation is unobtrusive, in the sense that membership in the sample does not influence an individual's behavior. (Alternatively, the investigator can have a theory for the influence of observation on behavior. The problem of obtrusive observation is common to all statistical analysis. See Section 4 of Duncan & Kalton, 1985, for a brief review of current experience with it in panel studies.) On the other hand, we allow for nonresponse by letting this feature be an integral part of Y_i, i.e., one of the possible values for $Y_i(t_k)$ at any time t_k may be an indication that data are missing because of nonresponse.

Let $S = \{I_1, \ldots, I_{n(S)}\}$ and $s = \{i_1, \ldots, i_{n(s)}\}$, and let us write $I(j)$ for I_j and $i(j)$ for i_j in subscripts. Then the sample data $D = \{S; Y_i : i \in S\}$ have a probability distribution given by

$P\left[\{S = s\} \,\&\, \{Y_i = y_i \text{ for all } i \in S\}\right]$

$$= P\left[\{n(S) = n(s)\} \,\&\, \bigcap_{j=1}^{n(s)} \{I_j = i_j\} \,\&\, \bigcap_{j=1}^{n(s)} \{Y_{I(j)} = y_{I(j)}\}\right]$$

$$= P\left[\{n(S) = n(s)\} \,\&\, \bigcap_{j=1}^{n(s)} \{I_j = i_j\} \,\&\, \bigcap_{j=1}^{n(s)} \{Y_{i(j)} = y_{i(j)}\}\right]$$

$$= P\left[\{n(S) = n(s)\} \,\&\, \bigcap_{j=1}^{n(s)} \{I_j = i_j\} \,\Big|\, \bigcap_{j=1}^{n(s)} \{Y_{i(j)} = y_{i(j)}\}\right] \quad.$$

$$\times P\left[\bigcap_{j=1}^{n(s)} \{Y_{i(j)} = y_{i(j)}\}\right]$$

$$= P\{S = s\} \times \bigcap_{j=1}^{n(s)} P\{Y_{i(j)} = y_{i(j)}\} = p(s) \bigcap_{j=1}^{n(s)} \xi_{i(j)}(y_{i(j)}),$$

when we use the stochastic independence between S and the collection $\{Y_i: i = 1, \ldots, N\}$ to reduce the big conditional probability to $P\{S = s\}$ in the next to final line above. Thus we have just proved that

$$P\{Y_i = y_i \text{ for all } i \in S \mid S = s\} = \prod_{j=1}^{n(s)} \xi_{i(j)}\,(y_{i(j)}), \tag{20}$$

i.e., given the sample S, the distribution of the sample observations $\{Y_i(j)\}$ is the same as it would be in an imagined exhaustive census whose data had the same stochastic properties as those in the sample survey. The *likelihood* of the sample data D is

$$p(S)\prod_{i \in S} \xi_i(Y_i). \tag{21}$$

Since $p(.)$ must be independent of the unknown parameters and other unknown characteristics of the $\{\xi_i\}$, likelihood maximization does not involve $p(.)$ in any way, and in particular it does not involve any reciprocal inclusion probabilities $1/\pi_i$. The sample S is an ancillary statistic, and inference may be based on formula (20) alone if you follow the ancillarity principle.

The essential role of the sampling plan is to provide a randomizing vehicle to determine which life histories to include in the sample in a way that induces cost-effective analysis and helps make sure that S actually is stochastically independent of the life courses Y_i after time 0. Lack of such a randomizing mechanism entails the risk that S becomes informative, as well as the usual problems of generalizability of results. (See Smith, 1983, and Royall, 1985, for critical assessments of the need for randomization, and Smith, 1984, for a discussion of its meaning.)

Any likelihood analysis depends of course on the specification of the $\{\xi_i\}$. If the model ξ_i of individual behavior is incorrect or unrealistic, then the outcome of the analysis must be affected unfavorably. Indeed, as everybody professes to realize, *any* model is incorrect or unrealistic in many respects. This is inherent in all analyses that use statistical models, and the analysis of sample survey data is no exception. One must not let this fact stifle one's ability to use models productively for the analysis of sample data any more than for other kinds of data. The value of a model lies in its ability to pick up important aspects of behavior and to serve as a guide to our inference about reality. Some sampling practitioners display an evident ambivalence (or even aversion) towards behavioral modeling, but an investigator interested in analyzing a particular aspect of behavior by means of sample data should not let this dictate his own choice of method. The preparation of the information in a major data set for publication in a book of official statistics, say, is quite a different operation than the penetration of a subarea for an analytical purpose. The concerns of data producing agencies are certainly real and important enough, but there is no need to let them dominate the picture the way they have done so far. There is little reason why others should feel restricted by the same considerations.

In response to the need for "infinite population" modeling concepts in the analysis of survey samples, some statisticians have begun to provide a kind of half-way house where new finite population statistics are *defined* in terms of model parameters but the properties of their estimators are *studied* as if there were no model but only the finite population. For instance, Chambless and Boyle (1985) have suggested that a parametric likelihood

$$\Lambda(\beta) = \prod_{i=1}^{N} \Lambda_i(\beta)$$

that would apply to the entire finite population (of size N) under a given behavioral model, be estimated by its sample counterpart

$$\hat{\Lambda}(\beta) = \prod_{i \in S} [\Lambda_i(\beta)]^{1/\pi(i)}$$

for individual inclusion probabilities $\{\pi(i)\}$. If $\Lambda(\beta)$ is maximized when $\beta = B$, they suggest that B be regarded as the finite population quantity of interest, and that the value $\hat{B}$ of β that maximizes $\hat{\Lambda}(\beta)$ be regarded as an estimator of B. This is in line with previous suggestions for generalized linear models by Binder (1983) and his predecessors. It has been followed up by Folsom et al. (this volume), who have also extended the theory to the (nonparametric) Kaplan-Meier product limit estimator.

We discussed an application of these ideas to the estimation of Markov chain transition probabilities in Section 2.3. In that setting, B is a set of (unobserved) transition proportions in the finite population, which it may certainly make sense to estimate. Similarly, the (unobserved) Kaplan-Meier product limit function for the finite population describes the distribution of a positive variable over the members of that population, and again it may make sense to estimate it from the sample data. In a case-by-case consideration of other situations, we are bound to find more models where a finite population estimator B is a meaningful statistic in its own right for which the sample counterpart $\hat{B}$ is a sensible estimator. For such situations, the statistical theory developed will be useful. This does not mean, however, that a finite population estimator B *always* is a meaningful descriptive statistic, irrespective of any appeal to an underlying model, or that one should restrict one's analysis of survey sample data to situations where B is meaningful in this manner, or indeed that statistical inference from survey samples must be only to the finite population level.

It is important to maintain that the specification of a model of individual behavior is an issue separate from the role of sampling weights. Whether the investigator has got ξ_i right or not, the likelihood has the form in formula (21) so long as sampling and analysis are not outcome-dependent. Of course, there is no compulsion to rely on the likelihood approach. One is free to use any inference procedure available, subject only to the assessment of the statistical properties of the procedure. Some such procedures may involve sampling weights. In fact, properties of the sampling plan will generally enter into likelihood analysis if the sampling is outcome-based, for then the likelihood has a form like

$$\Lambda = [\prod_{i \in S} \xi_i(Y_i)] \sum_{i \in S} \sum_{y_i} \left[\{\prod_{i \in S} \xi_i(y_i)\} \, p(S \mid Y_i : i \in S; \, y_i ; i \notin S) \right],$$

where the double sum is taken over all $i \notin S$ and over all values y_i that the sample path of target population member i outside the sample can attain, and where $p(s \mid y_1, \ldots, y_N) = P\{S = s \mid Y_i = y_i$ for $i = 1, \ldots, N\}$ is the conditional probability of drawing the sample s when the sample path outcomes are as specified.

In certain cases, the sampling mechanism turns out to enter the likelihood only via the sampling fractions of outcome-dependent strata. The example of our Section 4.3 is a case in point. However, weighting is no panacea that can solve most problems of survey analysis, including model misspecification, nor can it replace modeling and make behavioral models superfluous.

REFERENCES

Aalen, O. O., Borgan, Ø., Keiding N., & Thormann. J. (1980), "Interaction between Life History Events: Nonparametric Analysis for Prospective and Retrospective Data in the Presence of Censoring," *Scandinavian Journal of Statistics*, 7, 161–171.

Allison, P. D. (1982), "Discrete-Time Methods for the Analysis of Event Histories," in *Sociological Methodology 1982* (S. Leinhardt, ed.), Jossey-Bass, San Francisco, pp. 61–98.

Anderson, T. W., & Goodman L. A. (1957), "Statistical Inference about Markov Chains," *The Annals of Mathematical Statistics*, 28, 89–110.

Berk, R. A. (1983), "An Introduction to Sample Selection Bias," *American Sociological Review*, 48, 386–398.

Berk, R. A., & Subhash C. R. (1982), "Selection Biases in Sociological Data," *Social Science Research*, 11, 352–398.

Binder, D. A. (1983), "On the Variances of Asymptotically Normal Estimators from Complex Surveys," *International Statistical Review*, 51, 279–292.

Chambless, L. E., & Boyle, K. E. (1985), "Maximum Likelihood Methods for Complex Sample Data: Logistic Regression and Discrete Proportional Hazards Models," *Communications in Statistics—Theory and Methods*, 14, 1377–1392.

Cohen, R., & Cohen, J. (1984), "The Clinician's Illusion," *Archives of General Psychiatry*, 41, 1178–1182.

Colding-Jørgensen, H., & Simonsen, W. (1940), "Statistical Appendix," in *Graviditet og Gallestensdannelse* (J. M. Wollesen), Nyt Nordisk Forlag, Copenhagen, pp. 81–101.

Cosslett, S. R. (1981), "Maximum Likelihood Estimator for Choice-Based Samples," *Econometrica*, 49, 1289–1316.

Duncan, G. J. (1982), "The Implications of Changing Family Composition for the Dynamic Analysis of Family Economic Well-Being," in *Panel Data on Incomes: A Selection of the Papers Presented at a Conference on the Analysis of Panel Data on Incomes 24/25 June 1982 at the International Centre for Economics and Related Disciplines*, pp. 203–239.

Duncan, G. J., & Kalton, G. (1985), "Issues of Design and Analysis of Surveys across Time," *Bulletin of the International Statistical Institute*, 51, 14.1. Also in *International Statistical Review*, 1987, 55, 97–117.

Fay, R. (1982), "Contingency Table Analysis for Complex Sample Designs: CPLX," *Proceedings of the Section on Survey Research Methods*, American Statistical Association, 44–53.

Fienberg, S. E. (1980), "The Measurement of Crime Victimization: Prospects for Panel Analysis of a Panel Survey," *The Statistician*, 29, 313–350.

Funck Jensen, U. (1982), *The Feller-Kolmogorov Differential Equation and the State Hierarchy Present in Models in Demography and Related Fields* (Stockholm Research Reports in Demography No. 9), Section of Demography, University of Stockholm.

Hansen, M. H., Madow, W. G., & Tepping, B.J.(1983), "An Evaluation of Model-Dependent and Probability-Sampling Inferences in Sample Surveys" (with discussion), *Journal of the American Statistical Association*, 78, 776–807.

Hoem, J. M. (1969), "Purged and Partial Markov Chains," *Skandinavisk Aktuarietidskrift*, 52, 147–155.

Hoem, J. M. (1985), "Weighting, Misclassification, and Other Issues in the Analysis of Survey Samples of Life Histories," in *Longitudinal Analysis of Labor Market Data* (J. J. Heckman & B. Singer, eds.), Cambridge University Press, pp. 249–293. (List of misprints available from the author.)

Hoem, J. M., & Funck Jensen, U. (1982), "Multistate Life Table Methodology: A

Probabilist Critique," in *Multidimensional Mathematical Demography* (K. C. Land & A. Rogers, eds.), Academic Press, New York, pp. 155–264.

Hoem, J. M., Rennermalm, B., & Selmer, R. (1986), *Restriction Biases in the Analysis of Births and Marriages to Cohabiting Women from Data on the Most Recent Conjugal Union Only* (Stockholm Research Reports in Demography No. 18, rev.), Section of Demography, University of Stockholm.

Holt, D., Smith T. M. F., & Winter P. D. (1980), "Regression Analysis of Data from Complex Surveys," *Journal of the Royal Statistical Society*, A, **143**, 474–487.

Jewell, N. P. (1985), "Least Squares Regression with Data Arising from Stratified Samples of the Dependent Variable," *Biometrika*, **72**, 11–21.

Kalbfleisch, J. D., & Lawless, J. F. (1985), "The Analysis of Panel Data under a Markov Assumption," *Journal of the American Statistical Association*, **80**, 863–871.

Kalton, G. (1981), "Models in the Practice of Survey Sampling," *Bulletin of the International Statistical Institute*, **49**, 514–531. Also in *International Statistical Review*, 1983, **51**, 175–188.

Keiding, N., & Gill, R. D. (1987), *Random Truncation Models and Markov Processes* (Research Report No. 87/3), Statistical Research Unit, University of Copenhagen.

Kish, L. (1981), "Contribution to Discussion on Conceptual and Theoretical Framework for Survey Sampling," *Bulletin of the International Statistical Institute*, **49**, 536–539.

Klevmarken, A. (1986), "On the Estimation of Recursive and Interdependent Models from Sample Survey Data," paper contributed to the International Symposium on Panel Surveys, Washington, DC, November.

Little, R. J. A. (1982), "Models for Nonresponse in Sample Surveys," *Journal of the American Statistical Association*, **77**, 237–250.

Manski, C. F., & McFadden, D. (1981), "Alternative Estimators and Sample Designs for Discrete Choice Analysis," in *Structural Analyses of Discrete Data* (C. F. Manski & D. McFadden, eds.), MIT Press, Cambridge, MA, pp. 1–50.

Marini, M. M., Olsen, A. R., & Rubin, D. B. (1979), "Maximum-Likelihood Estimation in Panel Studies with Missing Data," in *Sociological Methodology 1980* (K. F. Schuessler, ed.), Jossey-Bass, San Francisco, pp. 314–357.

Nathan, G., & Holt, D. (1980), "The Effect of Survey Design on Regression Analysis," *Journal of the Royal Statistical Society*, B, **42**, 377–386.

O'Muircheartaigh, C., & Wong, S. T. (1981), "The Impact of Sampling Theory on Survey Sampling Practice: A Review," *Bulletin of the International Statistical Institute*, **49**, 465–493.

Rao, C. R. (1985), "Weighted Distributions Arising out of Methods of Ascertainment: What Population Does a Sample Represent?," in *A Celebration of Statistics: The ISI Centenary Volume* (A. C. Atkinson & S. E. Fienberg, eds.), Springer-Verlag, New York, pp. 543–569.

Rao, J. N. K., & Scott, A. J. (1984), "On Chi-Squared Tests for Multiway Contingency Tables with Cell Proportions Estimated from Survey Data," *Annals of Statistics*, **12**, 46–60.

Rindfuss, R. R., Bumpass L. L., & Palmore, J. A. (1985), "Analyzing Selected Fertility Histories: Do Restrictions Bias Results?," *Demography*, **24**, 113–122.

Rindfuss, R. R., Swicegood, C. G., & Rosenfeld, R. A. (1986), "Disorder in the Life Course: How Common and Does It Matter?" (Unpublished manuscript).

Royall, R. (1985), "Current Advances in Sampling Theory: Implications for Human Observational Studies," *American Journal of Epidemiology*, **104**, 463–474.

Schirm, A., Trussell, J., Menken, J., & Grady, W. R. (1982), "Contraceptive Failure in the United States: The Impact of Social, Economic, and Demographic Factors," *Family Planning Perspectives*, **14**, 68–75.

Schweder, T. (1970), "Composable Markov Processes," *Journal of Applied Probability*, **7**, 400–410.

Singer, B. (1981), "Estimation of Nonstationary Markov Chains from Panel Data," in *Sociological Methodology 1981* (S. Leinhardt, ed.), Jossey-Bass, San Francisco, pp. 319–337.

Singer, B., & Spilerman, S. (1977), "Fitting Stochastic Models to Longitudinal Survey Data: Some Examples in the Social Sciences," *Bulletin of the International Statistical Institute*, **47**, 283–300.

Smith, T. M. F. (1983), "On the Validity of Inferences from Nonrandom Samples," *Journal of the Royal Statistical Society*, A, **146**, 394–403.

Smith, T. M. F. (1984), "Present Position and Potential Developments: Some Personal Views: Sample Surveys," *Journal of the Royal Statistical Society*, A, **147**, 208–221.

Stasny, E. A. (1986a), "Estimating Gross Flows Using Panel Data with Nonresponse: An Example from the Canadian Labor Force Survey," *Journal of the American Statistical Association*, **81**, 42–47.

Stasny, E. A. (1986b), "Some Markov Chain Models for Nonresponse in Estimating Gross Labor Force Flows," paper contributed to the International Symposium on Panel Surveys, Washington, DC, November.

Sugden, R. A., & Smith, T. M. F. (1984), "Ignorable and Informative Designs in Survey Sampling Inference," *Biometrika*, **71**, 495–506.

Survey Research Center (1983), *User Guide for the Panel Study of Income Dynamics* (Preliminary draft), ICPSR, The University of Michigan, Ann Arbor. (Final version issued in 1984.)

Vardi, Y. (1985), "Empirical Distributions in Selection Bias Models" (with discussion), *Annals of Statistics*, **50**, 178–205.

Wellek, S. (1986), *A Nonparametric Model for Product-Limit Estimation under Right Censoring and Left Truncation* (Technical report), Institut für Medizinische Statistik und Dokumentation der Universität Mainz.

NOTE

*The author has benefited from discussion with Ralph Folsom, D. Holt, Graham Kalton, Jan Lanke, T. M. F. Smith, T. N. Srinivasan, and Rick Williams on topics in this paper, as well as from editorial advice from Anders Klevmarken.

Modeling Considerations: Discussion from a Modeling Perspective*

Stephen E. Fienberg

---

1. INTRODUCTION

It is a great pleasure to serve as a discussant for these informative and thought-provoking papers, not only because of their high quality and detailed content, but also because their presentation signals a fundamental shift in the attitudes of survey researchers that has taken place in recent years. About ten years ago, I discussed the use of stochastic models for victimization in the design and analysis of the National Crime Survey (Fienberg, 1978). The models were intended to utilize the rotating panel structure of the NCS and I suggested that both the design and analysis features of such panel surveys were inherently different from the more traditional surveys focusing on cross-sectional descriptive information. Those commenting on my paper roundly condemned my views and suggested that I simply did not understand anything about sample surveys. Shortly thereafter, I was involved in related discussions about the value of panel data on employment and unemployment in connection with the work of the National Commission on Employment and Unemployment Statistics. My views had a more receptive audience, consisting mainly of economists who wished to study the dynamics of the employment process, but I was once again rebuffed by the survey samplers at least in part because my views involved the development and estimation of statistical models. The current and widespread acceptance of modeling and its implications for panel research, as evidenced by the present papers and others in this volume, stands in stark contrast to the attitudes expressed just a decade ago.

My discussion is organized around a series of recurring themes that tie the papers together. These themes take the form of questions:

1. To model or not to model?
2. (a) What models require panel data?
   (b) What models can be profitably examined with panel data?
3. If we are to conduct a panel survey, what data should we collect?
4. What population does the sample "represent"?
5. What is the role of sampling weights and adjustments for the

probabilities of selection?
6. What is the role of random selection or randomization?
7. What is the role of quality data collection?

The first of these questions is obviously rhetorical since all four papers are about the use of models of one sort or another. While these authors acknowledge that panel data can be and often are used for purely descriptive purposes, for them as for me, the primary value of panel data comes from the ways they can be used to estimate and understand the adequacy of statistical models of behavioral phenomena. Thus the real issue is what models should we use rather than whether we should use any.

In subsequent sections, I discuss the remaining recurring themes and indicate how my views on them relate to those expressed in the four papers at this session.

2. WHAT MODELS REQUIRE PANEL DATA?

There are several types of data, which may or may not be of value in the investigation of behavioral models, that could be measured from a single survey but whose quality can be substantially enhanced by the use of cumulative data from several waves of a panel survey (see Fienberg & Tanur, 1986). On the other hand, there is considerable agreement among the papers at this session (e.g., see the Solon and the Hoem papers) that certain types of statistical models require some form of panel or other longitudinal data because they relate to phenomena extending over time—namely, dynamic, individual-level models focusing on events, durations, or transitions—i.e., changes of status. Hoem goes on to note that these phenomena of interest do not necessarily come neatly organized to coincide with the traditional view of panel surveys as involving repeated measurements at regularly spaced intervals.

Solon and Heckman and Robb both point out that the estimation of parameters in other models, especially those with specific forms of time-dependent correlation structure, can profit from the availability of panel data, but that the importance of the panel data (for identification purposes or for efficient estimation) depends on the assumptions one is willing to make. The use of repeated cross sections to estimate parameters in seemingly longitudinal models has, of course, been noted in a variety of other settings—e.g., for the estimation of age, period, and cohort effects (the volume edited by Mason & Fienberg, 1985, summarizes this literature). Nonetheless, the focus on the roles that repeated cross sections and panel data can play in the estimation of a specific class of regression-like models by Solon and by Heckman and Robb is especially welcome.

Solon introduces a class of regression-like models, commonly used in economics, of the form

$$y_{it} = a + bx_{it} + u_i + e_{it}, \tag{1}$$

where t indexes time periods. The quantity u_i reflects the effects of time-invariant unobserved individual characteristics that are correlated with the explanatory variable but not with the error term. Note that there is nothing inherently dynamic about this model. Nonetheless, the presence of the unobserved variable prevents the use of cross-sectional data and ordinary least squares for regression analysis to estimate the coefficient b. Solon then goes on to show how panel data can be used to address this problem, but he points out the need for other assumptions about measurement errors and the correlation structure of the errors over time that are required for the panel setting.

Heckman and Robb pursue this theme in the related but somewhat different context of an attempt to measure the impact of training on earnings, where the selection of participants for the training program depends on an unobserved mechanism. Their paper is a technical tour de force, in which we are systematically taken through a series of approaches to data collection and provided with the minimal assumptions required to specify the model (including the selection mechanism). Some of these assumptions are major advances over those in the literature, and others such as the assumption of zero third and fifth moments for the error term are not really a practical advance over the usual assumptions of symmetry or normality. His primary conclusion is that many models thought to be basically longitudinal can be fit using repeated cross-sectional data.

This conclusion is technically correct but it needs to be qualified. First, the details of the model specification Heckman and Robb work with are behaviorally questionable. I note such memorable phrases from the paper as "assume that trainees use a common discount factor" and "to simplify expressions, assume that people live forever." Hartigan (1986), in commenting on an earlier version of this material, noted: "Of course, you'll have to agree that this is a weird thing you're proposing. You wouldn't want to go on public TV saying this." Second, the assumptions invoked to allow for identification of the key parameters, while seemingly simple in form, are far stronger than Heckman and Robb's paper suggests. Moreover, given the absence of data from the paper it is hard to tell how realistic these assumptions are even if one accepts the basic model as "behaviorally defensible." Third, the basic model that Heckman and Robb use does not appear to have a full longitudinal structure:

$$y_{it} = b_t + d_i a + e_{it} \quad t > k \tag{2}$$

$$= b_t + e_{it} \quad t \leq k,$$

where $d_i = 1$ if the ith person trains in period k, $d_i = 0$ otherwise, and the selection mechanism reflected in d_i is correlated with the error term.

As with the approach described by Solon, the longitudinal feature of the model comes from the intervention of training and, depending on the assumptions invoked, a model for the correlation structure for the error terms over time. Had Heckman and Robb chosen to examine dynamically structured models, e.g., for the duration of subsequent spells of employment or for job search given unemployment (e.g., see Heckman & Singer, 1984), the value of panel data would have not been in doubt.

3. WHAT DATA SHOULD WE COLLECT?

This question is really a series of questions. From whom should we collect data? For what periods of time should we collect data? How frequently should we interview respondents? What use should we make of retrospective data? In a sense we cannot really address these questions without talking about the models of interest. Since the major ongoing panel surveys in the United States are intended to have multiple analytical uses (this is, in part, how their cost has been justified to funding agencies), it is still worthwhile to discuss some of the data collection issues for omnibus panel surveys.

Lillard's paper helps us think about a subset of these issues in the context of the Panel Study of Income Dynamics (PSID). PSID's focus is on families and these have a dynamic structure over time that includes the entry and exit of individuals not in the study at the beginning, the splitting and merging of units, and so on. The data collection rule adopted by PSID includes in the survey at any given wave all individuals who are part of the families of original sample members. If those individuals were not in the sample at the outset but had joined a sample household at some point, they are included "in the sample" until such time as they leave the sample household and then they are dropped from subsequent waves. Lillard describes the impact of such individuals on the sample size at a given wave. Given the omnibus nature of the PSID, I have a feeling that the investigators actually should have been collecting more data—retrospectively, to learn about relevant past information for entering individuals, and prospectively, for all individuals who were at some time part of the families of original sample members. In fact, I believe that a case could even be made for following the members of all subsequent families created or joined by individuals who enter data collection at any stage. The primary issue here would seem to be one of cost and not one involving the values of sample weights for individuals so included.

Having mentioned a preference for collecting retrospective data for entering individuals, I also note that retrospective data may also be an effective way to deal with individuals who are absent from a household for an interview but are still members of that household. This suggestion should not be taken as a blanket endorsement of retrospective data in place of concurrent data, but rather as the view that the collection of some data, even if they are subject to substantial measurement errors, may be better than the collection of no data at all. The reliability and validity of retrospective data remains a virtually unexplored topic. A recent review by Dawes and Pearson (1986), however, suggests that one should be very cautious in the use of retrospective data (see also Markus, 1986).

4. WHAT POPULATION DOES THE SAMPLE REPRESENT?

I have adapted the heading for this section from Rao (1985), who notes that there is not much discussion in the literature on how a statistician decides on an appropriate specification given a sample, although any statistical inference presupposes some kind of specification. The heading invokes the notoriously vague concept of "representativeness" (Kruskal & Mosteller, 1979) and reverses the more traditional approach, which first specifies a behavioral model (and thus the population for which it is appropriate) and then asks what data are appropriate for judging the adequacy of that model. To me the question as posed here gets matters backwards.

One of the difficulties faced by the analyst of data from omnibus panel surveys such as PSID is that, despite the inclusion of "official sample weights" on all data files, different data are relevant for different substantive questions of interest. As Lillard notes, the dynamics of family composition exacerbate this problem. But we should not let the panoply of data available drive the choice of substantive questions.

The message of this section is brief but clear. Instead of asking "What population does the sample represent?" we should ask, first, "What population (or model) is of interest?" and then "What data associated with a particular panel survey are relevant for analysis of the model?"

5. WHAT IS THE ROLE OF SAMPLING WEIGHTS AND ADJUSTMENTS FOR THE PROBABILITIES OF SELECTION?

In addressing this question, both Hoem and Lillard make a clear distinction between the use of panel data for cross-sectional descriptive purposes and their use for the study of dynamic population models. While the use of weights can be justified for drawing inferences about static finite population aggregates, there is an extensive but confusing debate regarding the relevance of weights for the study of formal statistical models, especially if the analyst wishes to adopt a standard approach to inference such as one based on the likelihood function. At the heart of this debate is the traditional sample-survey approach that uses the randomization in the sample selection procedure as providing a probabilistic basis for inference. Lillard's discussion of weights for data from the

PSID and the quote taken by Hoem from the PSID User Guide capture the nature of the confusion resulting from attempts to adapt this approach to the model-based inference setting. A notable exception to this confusion is the paper by Folsom et al. in this volume, which presents a clear but rather narrow approach to design-based inference for panel survey data.

Hoem's response to the question of whether or not to use sampling weights is clear and unequivocal (and consistent with my own; e.g., see Fienberg & Tanur, 1986, 1987). Sampling weights, as they are usually constructed, are at best irrelevant to a likelihood-based approach to statistical inference. Hoem gives a clear and convincing argument using a simple dynamic model to support this position and, in the process, he notes that weighting and model misspecification are separate issues. In principle, Hoem's paper should settle the issue about weighting once and for all, but confusion in practice is unfortunately likely to continue.

The one exception in which the use of weights may be appropriate is outcome-based sampling, where the sampling plan may be informative for the model of interest. Here a weighted adjustment, in the spirit of that described by Rao (1985), is the appropriate likelihood-based approach. Such an adjustment sometimes coincides with the use of weights equal to the reciprocals of the sampling fractions in outcome-based strata. Most practitioners are not fully aware of the dangers of outcome-based sampling and its relevance to prospective panel surveys such as PSID. Consider, for example, an individual who joins or leaves a sample family, after the beginning of a panel survey, in order to qualify for non-cash benefits from some program. If program participation is the substantive focus of analysis, then the creation of new family units is clearly a form of outcome-based sampling and the analysis of any model using these data may need to incorporate the selection mechanism into the model under study. (Heckman and Robb also have a section dealing with outcome-based sampling and selection models.)

I have only one note of disagreement with Hoem, in connection with his statement that there is a fairly complete sampling theory approach to the use of loglinear models in contingency tables. The work of Rao and Scott (1984), to which he refers, and that of Bedrick (1983) are not quite as complete and useful as they might appear on the surface. First, their results produce consistent estimates for loglinear model problems but not necessarily optimal estimates, e.g., in the sense of maximizing the likelihood resulting from the complex sampling models considered. Second, the useful approximations derived by these authors are applicable only to decomposable loglinear models, i.e., for ones that have direct estimates under simple random sampling. Thus, even for survey-based categorical data problems, there is not really an easily implemented and rigorously justifiable alternative to a standard model-based approach to inference.

6. WHAT IS THE ROLE OF RANDOM SELECTION OR RANDOMIZATION?

To help focus the discussion of the interpretation of their model, Heckman and Robb ask two questions regarding training programs:

Q1. What would be the mean impact of training on earnings if people were randomly assigned to training?

Q2. How do the postprogram mean earnings of the trained compare to what they would have been in the absence of training?

In his paper, he focuses on Q2 while I think that Q1 is the more interesting question. The random assignment of persons to training does provide information relevant to policy options, especially in changing environments, and it takes us back to the Fisherian notion of randomization in experiments. The role of randomization in experimentation can, in turn, be formally linked to the role of random selection in survey settings (Fienberg & Tanur, 1987).

Alternative approaches to the generic panel survey involving quasi-experimental designs exist, but they usually do not formally incorporate the selection bias focus suggested by Heckman and Robb. Such designs may be useful for exploratory purposes, but the survey analyst must resist the temptation to make causal inferences such as those embedded in Heckman and Robb's Q1. These inferences flow "naturally" from quasi-experimental studies, even when there is no conceptual experiment that can be done (see Holland, 1986). If the focus is one of the evaluation of policy options, good panel surveys, even those in which subsets of respondents experience interventions, are not adequate substitutes for proper randomized experiments. I make this statement recognizing the difficulties involved in implementing a randomized socioeconomic experiment (e.g., see the discussion in Fienberg et al., 1985).

Random selection and randomization play several different roles that are relevant to the present discussion:

1. Scientific objectivity.
2. The uncoupling of the selection mechanism from the model of interest.
3. (The relevance of second-order statistical theory based on the assumption of unit-treatment additivity.)
4. The possibility of randomization inferences conditional on the design as an alternative to likelihood and other analyses conditional on the data.

These roles are listed in what I take to be their descending order of importance, and I have put the third one in parentheses because it is the one role in the list that is linked to the experimental setting rather than both the experimental and sampling settings. (Note that Heckman and Robb also use the unit-treatment additivity formulation in expression (2) above.) Given that we are primarily concerned with publicly funded panel studies, the importance of maintaining the perception of scientific objectivity is critical. The second role of random selection and randomization provides the justification for the statements by Hoem on weighting, and the absence of this uncoupling is the reason why Heckman and Robb must spend so much effort focusing on selection bias. It is of course true that attrition and missing data compromise this crucial role of random selection and randomization and this is why we must pay serious attention to minimizing their impact. I also note that the second role is even important from the Bayesian perspective—this is relevant, since Heckman and Robb's approach does have a Bayesian flavor to it.

7. WHAT IS THE ROLE OF QUALITY DATA COLLECTION?

Although this question comes at the end of my list, one could easily make the argument that it should come at the beginning. This is because the quality of panel data is all-important to the modeling enterprise. A critical aspect of the data collection process is the design itself. We are surprisingly ignorant about the implications of dynamic behavioral models for the design of panel or longitudinal surveys. Examples of the research issues here include:

1. The optimal timing of interviews in repeat waves of a panel. See the discussions of this aspect of design in Becker and Kersting (1983) and in Titterington (1980).
2. The need for multiple (staggered) panels to make identifiable certain panel effects or model parameters of interest (e.g., effects similar to the age, period, and cohort effects described in Mason & Fienberg, 1985). See also the discussion in Fienberg and Tanur (1986).
3. The development of alternative rules for inclusion and exclusion of individuals from families and other longitudinally dynamic units of observation.
4. The collection of auxiliary data on the determinants of dynamic self-selection and the implications of such data collection on the identification of alternative models. See the discussion of this issue by Solon.

Several of the papers have mentioned measurement problems and both Solon and Heckman and Robb focus on models with formal components for measurement error. The use of such models is not sufficient, however. We must also reexamine the data collection instruments themselves in order to gain a deeper understanding of the cognitive processes that respondents are called upon to use when answering survey questions (e.g., see Jabine et al., 1984). There are distinctive cognitive aspects of survey design that are relevant for panel surveys.

Finally, in the actual process of data collection we must work hard to minimize the impact of attrition and other forms of nonresponse.

REFERENCES

Becker, N. G., & Kersting, G. (1983), "Design Problems for the Pure Birth Process," *Advances in Applied Probability*, **15**, 255–273.

Bedrick, E. (1983), "Adjusted Chi-Squared Tests for Cross-Classified Tables of Survey Data," *Biometrika*, **70**, 591–595.

Dawes, R., & Pearson, R. (1986), "The Effect of Theory-Based Schemas on Retrospective Data" (Unpublished manuscript).

Fienberg, S. E. (1978), "Victimization and the National Crime Survey: Problems of Design and Analysis," in *Survey Sampling and Measurement* (K. Namboodiri, ed.), Academic Press, New York, pp. 89–106.

Fienberg, S. E., Singer, B., & Tanur, J. M. (1985), "Large-Scale Social Experimentation in the United States," in *A Celebration of Statistics: The ISI Centenary Volume* (A. C. Atkinson & S. E. Fienberg, eds.), Springer-Verlag, New York, pp. 287–326.

Fienberg, S. E., & Tanur, J. M. (1986), "The Design and Analysis of Longitudinal Surveys: Controversies and Issues of Cost and Continuity," in *Survey Research Designs: Towards a Better Understanding of the Costs and Benefits* (R. W. Pearson & R. F. Boruch, eds.), Lecture Notes in *Statistics*, **38**, Springer-Verlag, New York, 60–93.

Fienberg, S. E., & Tanur, J. M. (1987), "Experimental and Sampling Structures: Parallels Diverging and Meeting," *International Statistical Review*, **55**, 75–96.

Hartigan, J. (1986), "Discussion 3: Alternative Methods for Evaluating the Impact of Interventions," in *Drawing Inferences from Self-Selected Samples* (H. Wainer, ed.), Springer-Verlag, New York, p. 58.

Heckman, J., & Singer, B. (1984), "Econometric Duration Analysis," *Journal of Econometrics*, **24**, 63–132.

Holland, P. W. (1986), "Statistics and Causal Inference" (with discussion), *Journal of the American Statistical Association*, **81**, 945–970.

Jabine, T. B., Straf, M. L., Tanur, J. M., & Tourangeau, R., eds. (1984), *Cognitive Aspects of Survey Methodology: Building a Bridge between Disciplines*, National Academy Press, Washington, DC.

Kruskal, W. H., & Mosteller, F. (1979), "Representative Sampling III: The Current Statistical Literature," *International Statistical Review*, **47**, 245–265.

Markus, G. B. (1986), "Stability and Change in Political Attitudes: Observed, Recalled, and 'Explained,'" *Political Behavior*, **8**, 945–970.

Mason, W. M., & Fienberg, S. E., eds. (1985), *Cohort Analysis in Social Research: Beyond the Identification Problem*, Springer-Verlag, New York.

Rao, C. R. (1985), "Weighted Distributions Arising out of Methods of Ascertainment: What Population Does a Sample Represent?," in *A Celebration of Statistics: The ISI Centenary Volume* (A. C. Atkinson & S. E. Fienberg, eds.), Springer-Verlag, New York, pp. 543–569.

Rao, J. N. K., & Scott, A. J. (1984), "On Chi-Squared Tests for Multiway Contingency Tables with Cell Proportions Estimated from Survey Data," *Annals of Statistics*, **12**, 46–60.

Titterington, D. M. (1980), "Aspects of Optimal Design in Dynamic Systems," *Technometrics*, **22**, 289–299.

NOTE

*The preparation of this discussion was supported in part by the National Science Foundation under Grant No. SES-84–06952 to Carnegie Mellon University.

Modeling Considerations: Discussion from a Survey Sampling Perspective

Graham Kalton

Populations are subject to temporal change because of changes in both composition and the characteristics of population elements. Changes in composition occur through "births" and "deaths," that is, new entrants to the population and leavers from it. Examples of changes in characteristics are: a person who is employed one month and unemployed the next; an elector who votes Democratic at one election and Republican at the next; a girl who is 85 cm tall at age 2, 94 cm tall at age 3, and 102 cm tall at age 4.

The dynamic nature of populations gives rise to the need for the collection of survey data over time. If the survey aims to estimate only the population parameters at different time points (e.g., unemployment rates on a monthly basis), or to estimate net change (e.g., the rise or fall in the unemployment rate from one month to the next), then the requisite data may be obtained from a series of repeated cross-sectional surveys with different samples conducted at the different time points. The limitation of repeated cross-sectional surveys is that since they do not provide longitudinal data for individuals, they preclude some important forms of longitudinal analysis. Sometimes individual longitudinal data can be obtained from records or through retrospective reports from respondents. However, the records of interest are often unavailable and respondents' memories are too unreliable and subject to distortion to provide data of acceptable accuracy. Then panel surveys are needed to collect data while they are still relatively fresh in respondents' minds. Other benefits of panel surveys are discussed by Duncan and Kalton (1987).

The papers in this session consider some important issues relating to the analysis of longitudinal data. I will focus my discussion around three main themes: measuring gross change, whether longitudinal data are necessary, and the issue of weighting. These themes are considered in turn in the following sections.

1. MEASURING GROSS CHANGE

The key advantage of longitudinal data over repeated cross-sectional data is the ability to measure gross change, that is, change over time at the individual level. The early work of Lazarsfeld and his colleagues (e.g., Lazarsfeld & Fiske, 1938; Lazarsfeld, 1948) drew attention to the value of studying gross change. With a main concern with categorical variables, Lazarsfeld promoted the "turnover table" as an analytic tool, that is, the cross-tabulation of the individual responses at the two time points. The joint distribution in the body of the table summarizes the gross changes, while the difference between the two marginal distributions to the table indicates the net change. The marginal distributions could be obtained from repeated cross-sectional data but the joint distribution requires longitudinal data.

In the section of his paper on dynamic analysis, Solon provides a number of good examples of the value of analyzing changes at the individual level in economic research. Take, for instance, the example he gives from Hill's research on the persistence of poverty. Repeated cross-sectional surveys can produce measures of the proportions of families in poverty at each year, but only a panel survey can provide the data needed to study the extent to which the *same* families remain in poverty over time.

Although the potential to measure gross change is the core analytic advantage of a panel survey over repeated cross-sectional surveys, the actual estimation of gross change is bedeviled by measurement error. As Solon observes, the problem is not that panel surveys are more prone to measurement errors than cross-sectional surveys (the contrary may in fact be the case), but rather that measures of gross change and associated analyses are particularly sensitive to the effects of measurement error. (See also the paper by Kalton et al. in this volume.) As Solon demonstrates, the serial correlation of measurement error is an important factor to be considered, yet there is almost no empirical evidence on its magnitude. The critical issue of measurement error in panel surveys is one that needs more research in order that panel surveys may yield their full analytic potential. This research should be directed at reducing the amount of measurement error, estimating the magnitude of measurement error, and developing additional techniques. to compensate for measurement error in the dynamic analysis of panel survey data.

2. ARE LONGITUDINAL DATA NECESSARY?

A major form of gross change analysis examines whether a change in variable x causes a change in variable y. As Solon notes, relating changes in y to changes in x, rather than relating y to x, serves to control for time-invariant characteristics. As a special case, x may be a dichotomous variable, either receiving a treatment (e.g., attending a training program) or not. One wave of data collection takes place before some sample members receive the treatment, and then a second wave takes place afterwards. If the variable y is measured at both waves, a panel survey of this nature is equivalent to the widely used quasi-experimental design variously known as "the before–after design with control group," "the untreated control group design with pretest and posttest," and "the pre/post control group design." The sizable literature on quasi-experimental designs is thus relevant in this context (see, for instance, Cook & Campbell, 1979; Cochran, 1983; and Kish, 1987).

The paper by Heckman and Robb addresses the issue of estimating the effect of a treatment in a quasi-experimental design setting. The authors provide an excellent systematic analysis of the minimal assumptions needed to identify the treatment effect with three alternative types of data: a single posttreatment survey, repeated cross-sectional surveys, and a panel survey. (In the quasi-experimental design literature, the single posttreatment survey is equivalent to a control group comparison design or to a posttest-only design with nonequivalent groups; repeated cross-sectional surveys are equivalent to an untreated control group design with separate pretest and posttest samples.)

Heckman and Robb's results provide valuable insights into the identification of a treatment effect based on data from these different designs. From their analysis of identifiability, they conclude that "the relative benefits of longitudinal data have been overstated." I am not fully convinced by their arguments for this conclusion. In the first place, I would like to see further justification of the applicability of some of the assumptions made in forming the estimators of treatment effects with the different designs, and an assessment of the sensitivity of the estimators to departures from the assumptions. Extreme caution is needed in estimating treatment effects from quasi-experiments, and due attention must be given to all the potential threats to validity (see Kish, 1987; Cook & Campbell, 1979). Second, I believe that the issue of sampling error, even if not relevant in the discussion of identification, is a factor that cannot be ignored in practice. For example, Heckman and Robb show that under certain assumptions it is possible to identify the treatment effect α from repeated cross-sectional data even when the participant status of individuals is unknown. However, the estimator of α, $(\bar{Y}_t - \bar{Y}_{t'})/\hat{p}$, is clearly likely to be subject to extremely large sampling error (especially when p is small, as will often be the case), unless the sample sizes are very large. This estimator may, therefore, be of little practical utility. Third, I think the comparison between panel surveys and cross-sectional surveys should be made on the basis of the sample sizes achieved for a given budget, not on the relative sizes of available cross-sectional and panel samples. As Duncan et al. (1984) show, the costs of two rounds of a panel survey may be lower than the costs of two comparable cross-sectional surveys. The full assessment of the relative merits of panel surveys, repeated cross-sectional surveys, and single posttreatment surveys for estimating treatment effects requires careful consideration to be given to the realism of the assumptions made in forming the estimators, to the sensitivity of the estimators to departures from the assumptions, to the multitude of threats to the validity of the conclusions, and to the practicability of the alternative survey designs.

An important factor to be considered in measuring the effect of many types of treatment is what Kish (1987) calls the time curve of response. The effects of treatments are not always immediate or permanent. Thus, for instance, the effect of a training program (an increase in pay) may not occur until six to nine months after the training has been completed, and the effect may disappear within two or three years. In medical studies, drug treatments are often not one-time affairs, but rather are ongoing. A patient may initially react positively to the drug, but after prolonged use it may become ineffective. Time-series, but not necessarily longitudinal, data are needed to study such effects. Thus both panel surveys and repeated cross-sectional surveys can provide the necessary data. However, a design that includes only a single posttreatment survey cannot produce the data needed to examine the time curve of response.

Heckman and Robb discuss both the case where the treatment effect α is assumed to be a constant for all individuals and the case where it is taken to be a random effect, α_i, specific to the ith individual (with α_i even possibly being negative for some individuals and positive for others). With the fixed-effects model, the parameter α is population-free, and it does not matter how the sample is chosen to estimate α. In general this model is implausible. As a rule, one might expect the treatment to affect individuals differently. It may be possible to identify subgroups of the population that are affected differently by the treatment (e.g., the pay of young and old, male and female, white and nonwhite workers may be affected differently by the training program), in which case an analysis can be conducted to estimate the treatment effect within subgroups. Even when this is done, however, it is unrealistic to assume that all the variability in the α_i is accounted for by the subgroup differences. Rather, there will remain some variation in the α_i within subgroups; that is, the random-effects model is the appropriate one. A summary measure of the effect of a treatment with a random-effects model is the mean effect, $\bar{\alpha} = E(\alpha_i)$. As Heckman and Robb point out, this measure is not population-free: In their terms, it makes a difference whether you are measuring the average effect of the treatment as if persons were randomly assigned to treatment or measuring the average effect on those receiving the treatment. With the random-effects model, it also matters how the sample is drawn. This relates to the issue of weighting that Hoem addresses in his paper.

3. THE USE OF WEIGHTS

In standard survey practice, weights are used to compensate for unequal selection probabilities, for nonresponse and noncoverage, and for poststratification purposes. Unequal selection probabilities may arise because analytic benefits accrue from oversampling certain subgroups of the population, because it is cost-effective to sample different subgroups at different rates, or because of features of the sampling frame (e.g., duplicate listings, clusters of elements). The final composition of the weights is often complex. The use of weights in forming estimates of descriptive parameters for the population surveyed (e.g., the unemployment rate, the proportion of persons in poverty) is widely accepted. There is, however, a long-standing debate on the appropriateness of using weights in the construction of analytic models that seek to describe causal systems (e.g., DuMouchel & Duncan, 1983; Hansen et al., 1983; Kalton, 1983; Smith, 1976; and Smith, 1984). This issue becomes especially pertinent with panel surveys, since panel data are widely used for causal modeling. I see, however, no special considerations that affect panel surveys differently from cross-section surveys with regard to the arguments for and against the use of weights in analytic modeling.

I do not believe that there is any disagreement among the protagonists in the debate on the use of weights in modeling about the value of modeling in scientific research, or about the need for the utmost care in model construction and testing. I believe that all would also agree that no model will ever hold exactly. As I see it, the disagreements center on whether to rely completely on the assumptions of a carefully developed and tested model or whether to seek some protection against model misspecification.

Hoem and others who oppose the use of weights argue that one should rely on the model. If one accepts all the model assumptions, weights become irrelevant and their use gives rise to a loss in precision in the estimates of the model parameters. Another consequence of reliance on the model is that the standard errors of the parameter estimates are computed according to the model. They do not therefore need to take the survey sample design into account. My own view is that in most—but not all—circumstances it is preferable to conduct a weighted analysis and to compute standard errors appropriate for the sample design employed (Kalton, 1983). I believe that this provides some limited protection against model misspecification at a cost that is generally acceptable.

I will try to explain my position by an example that is vastly over-simplified for ease of presentation. This example is helpful for convey-ing the essence of my argument, but it risks the rebuttal that the misspecifications I am postulating can easily be addressed by modifying the model. While that is indeed true in this case, I ask the reader to bear in mind that in large scale surveys weighting is a far more compli-cated matter that cannot be readily handled by a simple elaboration of the model. For the example, following Hoem I will consider the estima-tion of the transition probabilities in a simple Markov chain model with two waves of data collection. Suppose that the population is divided into two strata by income (high and low) and that a larger sampling fraction is employed in the high income stratum.

If the simple Markov chain model holds, every individual has the same transition probabilities, and those probabilities can be estimated by the simple unweighted sample proportions. Suppose, however, that the model is false, with older persons having different transition probabilities from younger persons, and that older persons tend to have higher incomes than younger persons. The unweighted sample propor-tions will in this case be meaningless quantities; they are arbitrary averages of the sample proportions in the two strata. They depend on the choice of sampling fractions adopted, and they would change if the sampling fractions were changed.

Consider now the weighted estimates of the transition probabilities, with weights inversely proportional to selection probabilities. The use of weights does not correct the false model, but it does produce mean-ingful estimates. The weighted sample proportions estimate the cor-responding population proportions. The correct model would contain dif-ferent transition probabilities for older and younger persons. The weighted sample proportions are estimates of averages of these transi-tion probabilities for younger and older people combined, taking account of the mix of younger and older persons in the population surveyed. As such, with the age differences having not been recognized, they provide the best predictions of transitions for the total population.

As this example illustrates, the use of weights provides only a limited protection against model misspecification. The use of weights leads to consistent estimates of the finite population parameters, and these es-timates are applicable only for that population (or for the superpopula-tion of which the finite population is a random realization). In the ex-ample, the weighted sample estimates are good estimates of average transition probabilities for older and younger persons combined in the proportions that they appear in this survey population at this particular time. They would not provide good estimates for another population with a different age composition. Thus the protection offered by the use of weights applies only with a restrictive inference. It treats the model more as a predictive equation than as a causal model. Whether this limited protection is useful depends on the proposed uses of the model. I believe that in many applications it is beneficial, but there are certainly some applications where the protection is irrelevant.

A pertinent consideration in the choice between weighted and unweighted analyses is whether they give different results. If the model is correct, they will not. The use of weights with the correct model does not cause a bias in the estimates, but only leads to some loss of precision. An appreciable difference between the weighted and unweighted results indicates that the model is false. A modification to the model should therefore be sought. DuMouchel and Duncan (1983) present an interesting example where a regression model was modified to eliminate the original discrepancy between the weighted and unweighted results.

An issue closely related to weighting is that of standard error estimation. If a model is correct, the standard errors of the parameter estimates can be computed according to the model specifications without reference to the sample design. If the standard errors based on the sample design were computed in this case, they should produce the same results in expectation; however, they are likely to be less reliable and more difficult to compute. In practice, when standard errors of estimates of model parameters are computed to take account of complex multistage designs, they generally tend to be larger than the model-based standard errors. This indicates that the model is misspecified to some extent. Failure to take account of the sample design in computing standard errors can tend to overstate the precision of the parameter estimates.

The issue of weighting also arises in Lillard's paper on the sample dynamics in the Panel Study of Income Dynamics (PSID). With panels such as the PSID, which aim to provide data for both longitudinal and cross-sectional analyses, two types of dynamics need to be considered: population dynamics and sample dynamics.

Population dynamics refer to changes in the population over time. With panel surveys such as the PSID, new entrants to the population occur because of births, immigration, returning from abroad, and returning from the institutional population. Leavers from the population occur because of deaths, emigration, and entering an institution. Population changes add markedly to the complexity of the analysis of panel data. For cross-sectional analysis, the population of inference changes from one period to the next. For longitudinal analysis, careful consideration needs to be given to the definition of the longitudinal population of inference (see, for instance, Goldstein, 1979; Judkins et al., 1984; Kasprzyk & Kalton, 1983; and Kalton & Lepkowski, 1985). For many purposes, this longitudinal population may be taken as those in the population throughout the period of interest; however, this definition is not always appropriate.

Sample dynamics refer to additions to and losses from the original sample over time. Changes in the sample may come about as a result of population changes, for instance the addition of births and the losses from deaths, or as a result of other factors, such as the addition of the nonsample people in the PSID and the losses from nonresponse. These two types of changes need to be clearly distinguished since they have different implications for the analysis.

The PSID started with a sample of households in 1968, and the individuals in those households became sample persons to be followed in the panel. Over the course of time, the size of the original panel has declined markedly because of nonresponse, deaths, and other losses. If the nonresponse attrition were random, it would reduce the sample size but not otherwise affect the analysis of the survey data. However, as Lillard's analyses indicate, attrition is not random; nonwhite, mobile, older, and low-income individuals are more likely to drop out of the panel than their counterparts. In an attempt to counteract the biasing effects of nonrandom nonresponse, some form of nonresponse compensation procedure may be applied. In the PSID, weighting adjustments are made for attrition. This general-purpose form of adjustment seems appropriate to me for most analyses. However, a tailor-made adjustment may be required for certain types of analysis. Since Lepkowski provides a full treatment of adjustments for panel nonresponse elsewhere in this volume, I will not discuss the matter further.

Besides following original sample persons, the PSID also collects data for nonsample persons, that is persons not in the original sample who subsequently live with sample persons. The purpose of collecting data on nonsample persons is to provide background information on the sample persons. Consequently, nonsample persons are included in the PSID only for the period that they live with sample persons. They are given weights of zero, and hence are excluded from person-level analyses. It may be noted that the addition of nonsample persons is not peculiar to the PSID, but also occurs with other surveys such as the Survey of Income and Program Participation (SIPP) and the National Medical Care Utilization and Expenditure Survey (NMCUES). It is, however, a more significant concern with the PSID because of the length of time the panel has been running. In recent years of the PSID, nonsample persons have constituted approximately 20 percent of persons for whom data are collected. The exclusion of nonsample people from PSID analyses, especially cross-sectional analyses, thus results in a considerable wastage of data.

As Lillard indicates, one way to include the nonsample individuals in an analysis is to run the analysis unweighted (that is, give all respondents a weight of 1). However, this approach ignores the complex variation in sampling rates in the original PSID sample, and in particular the oversampling of low-income households. It also ignores the fact, noted by Lillard, that nonsample persons enter the panel because of events related to migration, living arrangements, and possibly marriage, and hence are atypical of the general population at least in terms of these characteristics. While the unweighted sample is larger than the weighted one, it is a collection of individuals that represents no meaningful population. Unweighted estimates of descriptive population parameters, such as the percentage of persons in poverty, are clearly likely to be seriously biased. The consequences of unweighted analysis for the estimation of model parameters are not obvious, but there is certainly no protection against model misspecification of the type discussed above. For some models there may be clear risks of choice-based sampling biases.

The reason that nonsample persons are excluded from individual-level weighted analyses is that their selection probabilities and those of sample persons cannot be precisely determined if they are included. There are, however, ways in which approximate estimates of probabilities of being included in the panel at a particular wave can be made, and these can then be used to create weights for cross-sectional analyses. Consider, for instance, a family that comprises one sample person and one nonsample person at the current wave. The probability of that family being included in the panel is the sum of the probability that the sample person's original family was selected in 1968 and the corresponding probability for the nonsample person's original family. The first of these probabilities is known, but the second is not. Given the complex pattern of selection probabilities and nonresponse adjustments for the original PSID sample and the length of time the PSID has been running, it is not possible to determine the selection probabilities for the original families of nonsample persons. In addition, nonresponse adjustments for panel attrition make the problem more complex. Nevertheless, some approximations to the probabilities of inclusion in the 1968 sample might be developed for the original families of nonsample persons. If this is done, the current family's selection probability can be estimated, and this is equal to the selection probability of each of the current family members. This method for determining selection probabilities for sample and nonsample persons is a multiplicity approach. The multiplicities can be applied at either the family or person level (see Huang, 1984, for further details). The approach is used in the SIPP and was also used in the NMCUES. Although its application with the PSID is more difficult, I believe that it merits investigation.

The preceding discussion of multiplicity weighting relates to cross-sectional estimation. Weighting nonsample persons for inclusion in longitudinal analysis is more difficult. If the longitudinal population of inference is taken to be those present throughout a specified period, nonsample persons will be excluded if they enter after the start or leave before the end of the period. This combination of reasons for exclusion makes the development of a straightforward weighting scheme difficult. The difficulty would be reduced if nonsample persons were retained in the panel when they stopped living with sample persons.

As is clear from the vast literature on the subject, longitudinal analysis of survey data is an important subject. There is a great deal to be learned from studies of how individuals (or other units of enquiry) change over time. The temporal patterns of individuals' characteristics and the changes in characteristics over time provide an important extra dimension for analysis. The papers in this session make a valuable contribution to this literature, bringing out both the potentialities and the limitations of longitudinal analysis. They also serve an important function in drawing attention to the complexities of longitudinal analyses, in particular the serious problem of measurement error in analyses that involve individual change and the difficulties arising from changes in the population and changes in the sample over time.

REFERENCES

Cochran, W. G. (1983), *Planning and Analysis of Observational Studies*, Wiley, New York.

Cook, T. D., & Campbell, D. T. (1979), *Quasi-Experimentation. Design and Analysis for Field Settings*, Houghton Mifflin, Boston.

DuMouchel, W. H. & Duncan, G. (1983), "Using Sample Survey Weights in Multiple Regression Analysis of Stratified Samples," *Journal of the American Statistical Association*, 78, 535–543.

Duncan, G. J., Juster, F. T., & Morgan, J. N. (1984), "The Role of Panel Data in a World of Scarce Resources," in *The Collection and Analysis of Economic and Consumer Behavior Data* (S. Sudman & M. A. Spaeth, eds.), Bureau of Economic and Business Research, Champaign, IL, pp. 301–328.

Duncan, G. J., & Kalton, G. (1987), "Issues of Design and Analysis of Surveys across Time," *International Statistical Review*, 55, 97–117.

Goldstein, H. (1979), *The Design and Analysis of Longitudinal Studies*, Academic Press, London.

Hansen, M. H., Madow, W. G., & Tepping, B. J. (1983), "An Evaluation of Model-Dependent and Probability-Sampling Inferences in Sample Surveys" (with discussion), *Journal of the American Statistical Association*, 78, 776–807.

Huang, H. (1984), "Obtaining Cross-Sectional Estimates from a Longitudinal Survey: Experiences of the Income Survey Development Program," *Proceedings of the Section on Survey Research Methods*, American Statistical Association, 670–675.

Judkins, D., Hubble, D., Dorsch, J., McMillen D., & Ernst, L. (1984), "Weighting of Persons for SIPP Longitudinal Tabulations," *Proceedings of the Section on Survey Research Methods*, American Statistical Association, 676–681.

Kalton, G. (1983), "Models in the Practice of Survey Sampling," *International Statistical Review*, 51, 175–188.

Kalton, G., & Lepkowski, J. (1985), "Following Rules in SIPP," *Journal of Social and Economic Measurement*, 13, 319–329.

Kasprzyk, D., & Kalton, G. (1983), "Longitudinal Weighting in the Income Survey Development Program" in *Technical, Conceptual, and Administrative Lessons of the Income Survey Development Program* (M. H. David, ed.), Social Science Research Council, Washington, D.C., pp. 155–170.

Kish, L. (1987), *Statistical Design for Research*, Wiley, New York.

Lazarsfeld P. F., (1948), "The Use of Panels in Social Research," *Proceedings of the American Philosophical Society*, 42, 405–410.

Lazarsfeld, P. F., & Fiske, M. (1938), "The 'Panel' as a New Tool for Measuring Opinion," *Public Opinion Quarterly*, **2**, 596–612.

Smith, T. M. F. (1976), "The Foundations of Survey Sampling: A Review" (with discussion), *Journal of the Royal Statistical Society, A*, **139**, 183–204.

Smith, T. M. F. (1984), "Present Position and Potential Developments: Some Personal Views. Sample Surveys" (with discussion), *Journal of the Royal Statistical Society, A*, **147**, 208–221.

Index

Adjustment cell imputation method, 408
Age effect, population estimate, 342
Arrays, *see* Relational Data Base
   Management System
Attribute-based studies, 163
Attrition, 498, 506
   pattern weighting, 370
   population estimate, 342
Available-case analysis, 415

Behavior, weighting, 539
Behavioral issues, 497
Birth, handling in economic survey, 91
Bounding, 11, 14, 35
British Social Attitudes Panel, 319. *See also*
   Social Attitudes Survey
BSRP, *see* Business Survey Redesign Project
Business Register, 83-85
Business Survey Redesign Project, 104

Carry-over wave imputation, 370
Causal modeling, 375
   nonresponse example, 377
Causal models, 432
   evaluating assumptions, 394
   implementation:
      through imputation, 396
      through weighting, 395
CDF, *see* Cumulative distribution functions
Classification maintenance, 80
Closed-form estimators, causal models and,
   388
Cohort studies, 163
Collection, 101
Complete-case analysis, 415
Complete wave respondent weighting, 370
Computer algorithm, reporting overlap, 41
Conditional imputation, 407
Conditioning, 254, 276, 340. *See also*
   Consumer Expenditure Interview
   Survey; Repeated interviewing
   definition, 254, 319

desirability, 320
effects, 324, 344
effects on subgroups, 326
literature review, 323
measurement, 320
real change *vs.* reporting change, 320
rotation, 255
Conditioning effect, population estimate, 341
Confidentiality, 55–56, 77
Consumer Expenditure Interview Survey,
   289
   analysis method, 294
   data adjustment, 292
   data collection, 291
   design features, 290
   recall effects, 300
   repeated interview effects, 289
   time-in-sample effects, 296
Contigency table, 132
Control-function estimators, 533
Corporate Tax Data Base, 86
Covariance stationarity, 528
Coverage, 3, 4–10, 80
   checks, 8
   error, 5
   error reduction, 9
CPS, *see* Current Population Survey
Cross-sectional estimation, 583
Cross-sectional imputation, 410
Cross-sectional procedures, 518
Cross-sectional survey, nonresponse and, 5
Cross-wave hot-deck imputation, 370
Cumulative distribution functions (CDF), 117
Current Population Survey (CPS), 377

Data base, designing, 245
Data base management, 242. *See also*
   Relational Data Base Management
   Systems
   aggregate-time groups, 199
   data matrices, 194
   general-purpose, 201

Data base management (*Continued*)
  group relationships, 194
  hierarchical structures, 202
  household panel studies, 190. *See also*
      Household panels
  identification variables, 195
  implementation of data structures, 199
  logical structure, 195
  match-merges, 199
  problems, 194
  PSID directly structured file, 217, 221
  PSID examples, 206
  PSID hierarchical file, 209, 213
  PSID identification variables, 210, 217
  rectangular structure, 205
  sorts, 199
  specialized, 201
Data Base Management Systems (DBMS), 184
  acquiring, 185
  limits, 185
  statistical functions, 185
Data base organization, *see also* Data base
      strategies
  individual-based studies, 178
  longitudinal applications, 178
  multi-level files, 172
  single-level files, 173
Data base strategies, *see also* Data
      organization
  attribute-based studies, 163
  cohort studies, 163
  complexity issues, 164
  conflicting uses, 166
  data access options, 183
  hypothetical example, 169
  longitudinal aggregate studies, 163
  overview, 163
  survey design, 165
  terminology, 171
  unit of analysis, 167
Data collection, 75, 569. *See also* Panel data
  quality, 573
Data collection mode, 42, 281
  changes, 45
  cold contact, 43
  effectiveness, 44
  follow-up cooperation, 42
  initial, 42
  nonsampling errors, 260
  response quality, 44
  response rates, 43
  tracking, 43
Data matrices, 194
Data organization, 172. *See also* Data base
      organization

cross-sectional applications, 172
Data structure, *see* Data base management
DBMS, *see* Data Base Management Systems
Death, handling in economic survey, 91
Descriptive analysis, 109
Design-consistent estimation, 129
Discrete proportional hazards model, 136
Dynamic analysis, 486

Economic research, 486
Economic survey, 80
  characteristic features, 82
  classification maintenance procedures, 91
    small units, 99–101
  collection, 101
  coverage:
    large units, 93–95
    procedures, 91
    small units, 96–99
  environment, 81
  estimation, 102
  future direction, 104
  general themes, 104
  large unit treatment, 105
  processing systems, 90
  response burden, 101, 105
  rotation, 101
  sampling, 102
  small unit treatment, 105
  sources, 88
  standardization, 104
  Statistics Canada, 81
Estimation, 160

Fixed-effects method, 524
Forward tracing, 53

GCM, *see* Growth curve model
General incomplete model (GIM), 119
General linear multivariate model (GLMM),
      118
GIM, *see* General incomplete model
GLMM, *see* General linear multivariate
      model
Gross change, measuring, 264, 284, 576
Growth curve model (GCM), 125

Hazard function, 131
Health Interview Survey (HIS), comparison
      with NMCUES, 304
HIS, *see* Health Interview Survey
Hot deck, 396
Household panels, *see also* Longitudinal
      household
  analytic needs, 191

composition changes, 193
design features, 190

Imputation, 358, 396, 406
  adjustment cell method, 408
  combined with weighting, 367
  comparison with weighting, 364
  compensating for wave nonresponse, 358
  conditional, 407
  cross-sectional, 410
  cross-wave methods, 360
  general methods, 358
  inferences, 415
  longitudinal, 410
  marginal, 407
  matching method, 409
  mean, 408
  multivariate, 407
  properties for government surveys, 375
  regression, 409
  row and column fits, 411
  stochastic, 408
Income Survey Developmental Program
    (ISDP), 13
"Infinite population" modeling concept, 561
Information needs, 1
Instrumental-variable estimator, 520
  choice-based sampling, 530
  contamination bias, 532
Interval between waves, 278
Interview spacing:
  behavior effects, 29
  bounding status, 35
  migration, 35
  omissions, 28
  panel effects, 29
  primary considerations, 26–27
  respondent burden, 31, 32
  respondent conditioning, 30
  respondent learning, 32
  telescoping, 28
  tracking mobile respondents, 33
Item nonresponse, 348, 400
  causal modeling of, 392

Kalman filter:
  measurement equation, 460
  transition equation, 461
Kaplan-Meier product limit estimator, 134

Likelihood analysis, 415, 545, 554
Longitudinal aggregate studies, 163
Longitudinal descriptive estimation, 115
Longitudinal estimates:
  family, 139

household, 139
  weighting issues, 139
Longitudinal household:
  common estimators, 145
  definition, 110, 141
  missing subject-matter data, 148
  period of existence, 149
  problems, 146
  unbiased estimates, 143
  unknown weight, 147
  weighting, 112
    adjustments, 150–158
    procedures, 145
Longitudinal imputation, 410
Longitudinal population definitions, 109
Longitudinal procedures, 524
Longitudinal survey, 2
  coverage checks, 8
  data collection, 75
  data collection mode, 42
  design features, 25
  geographical restrictions, 34
  interview spacing, 26
  respondent selection, 36
Lost to follow-up (LTF), 33
LTF, see Lost to follow-up

MAR (missing at random), 404
Marginal imputation, 407
Markov chain model, 543
Maximum likelihood (ML) method, 415
MCAR (missing completely at random), 404,
    417
Mean imputation, 408
Measurement bias, population estimate,
    341
Measurement error, 1, 3, 265, 438, 480, 481,
    484, 490
  change scores estimate, 442, 444, 447
  consequences of covariances, 449
  cross-sectional scores estimate, 441, 443
  level scores estimate, 444, 447
  net change estimates, 439
  stability estimates, 440
Migration, 35
Missing data, 348, 400. See also Nonresponse
  mechanisms, 402
Modeling, 342, 566, 575
Multiplicity estimation, 112
Multivariate analysis, model-based, 118
Multivariate imputation, 407

National Center for Health Statistics
    (NCHS), 305
National Crime Survey (NCS), 11, 15, 25, 35

National Medical Care Expenditure Survey
    (NMCES), 108, 139
National Medical Care Utilization and
    Expenditure Survey (NMCUES), 108,
    139, 251, 304
    comparisons with HIS, 304–317
NCS, *see* National Crime Survey
Nonattrition pattern weighting, 370
Nonrandom sampling plans, 530
Nonrespondents, treatment of, 75
Nonresponse, 3, 4–10, 250, 275. *See also*
    Wave nonresponse
    adjustments, 426
    attrition, 7
    causal models, 386
    error reduction, 9
    estimating causal models, 388
    ignorable, 382
    implementing causal models, 394
    item, 348, 400
    longitudinal survey, 5–6
    nonignorable, 375
    solutions, 7
    survey variance, 7
    unit, 348
Nonsample persons, 582
Nonsampling error, 16, 249. *See also*
    Conditioning; Nonresponse
    design effects, 256
    interval between waves, 256, 279
    investigating, 271
    measuring gross change, 264
    mode of data collection, 260, 281
    quality control, 273
    respondent rules, 258, 280
    rules for following sample persons, 261
    telescoping, 257

Outcome-dependent sampling, 551

Panel conditioning, 16. *See also* Conditioning
Panel data:
    analytical benefits, 486
    collection guidelines, 226
    complexities, 227
    controlling for unobserved variables, 489
    definition, 432
    dynamic self-selection, 493
    managing, 231
    measurement error, 438, 490
    modeling, 567
    special management requirements, 242
Panel data analysis, 108, 433
    design-consistent estimation, 129
    estimation, 121

growth curve models, 125
longitudinal descriptive estimation, 115
longitudinal household definitions, 110
longitudinal household weighting, 112
longitudinal population definitions, 109
model-based multivariate analysis, 118
model definitions, 118
reduced models, 122
repeated measurements, 118, 123
robust interference, 129
survival analysis, 130
variance component estimation, 127
Panel data management, see Relational Data
    Base Management Systems
Panel Study of Income Dynamics (PSID),
    222, 251, 487, 497, 509, 569
    data structures, 206
    directly structured file, 217, 221
    hierarchical file, 209, 213
    identification variables, 210, 217
Period effect, population estimate, 341
Polynomial Reparameterization, 125
Population dynamics, 109, 575
Population estimate:
    age effect, 342
    attrition–item nonresponse, 342
    attrition–unit nonresponse, 342
    conditioning effect, 341
    measurement bias, 341
    period effect, 341
Population mean, time-adjusted, 116
Population median, 117
Preliminary estimate, 457
    alternate method, 469
    change estimates, 477
    choice of alpha, 458
    Kalman filter, 460
    level estimates, 473, 476
    parameters estimation, 465
    prediction problem, 457
    stochastic least squares, 468
    sampling approach, 457, 458
Probabilities of selection, 570
Probability sampling, 108
Procedural languages, 186
Product limit estimator, *see* Kaplan–Meier
    product limit estimator
PSID, *see* Panel Study of Income Dynamics

Quality control, 273

Random coefficients, 516
Randomization, 572
Random selection, 572

RDBMS, *see* Relational Data Base
    Management Systems
Recall, 10
    bounding, 14–15
    model, 28
    period length, 13
    reference period, 11
Recall effects, Consumer Expenditure
    Interview Survey, 300
Recall error, 3
Reduced models, fitting, 126
Reference period, 11
Refusal rates, 9
Regression imputation, 370, 409
Relational Data Base Management Systems
    (RDBMS), 184
    characteristics, 228
    documentation of data organization, 232
    documenting metadata, 238
    enhanced storage, 231
    error correction, 237
    examples, 229
    extracting longitudinal data structure, 233
    housekeeping, 235
    legibility, 235
    managing panel data, 231
    portability, 235
    program documentation, 236
Repeated interviewing, *see also* Conditioning
    attitude changes, 327
    attitude freezing, 328
    honest reporting, 330
    reporting improvement, 334
    respondent burden, 335
    understanding interview process, 334
Repeated surveys, 2
Respondent, tracing, 52. *See also* Tracing
Respondent burden, 31, 32
Respondent characteristics, 282
Respondent conditioning, 30, 42
Respondent rules, 258, 280
Respondent selection, 36
    panel waves, 39
    reporting overlap, 39
    respondent conditioning, 42
    response quality, 37
    rules, 37
    status, 37
Respondent status, types, 37
Response burden, 101, 105
Response propensity weighting, 395
Response rates, 43
Reverse Record Check (RRC), 65–72
Reverse tracing, 54
Robust inference, 129

Rotation, 101, 255
RRC, *see* Reverse Record Check

Sample dynamics, 497, 582
    attrition, 498
    entry of nonsample individuals, 499
Sampling approach, 480
Sampling weights, 570
Scientific analysis, 226
"Seam", 14
Selection bias, 512
Single-time surveys, 1
Social Attitudes Survey, 321. *See also*
    Conditioning; Repeated interviewing
    attrition, 322
    estimating conditioning effects, 325
    investigating conditioning effects, 324
    literature review, 323
    research method, 323
    response, 322
Standard error estimation, 581
Standardization, in BSRP, 104
Statistical design, 160
Statistical matching procedure, 396
Statistical packages, 186. *See also* Data Base
    Management Systems
Statistics Canada, 65, 80
    policies and guidelines, 82
Stochastic coefficient model, 125
Stochastic imputation, 408
Stochastic least squares, 468
SURREGR, 121
Survey of Income and Program Pariciration
    (SIPP), 6, 14, 27, 139, 251, 259, 263, 390,
    497
Survey regression package, *see* SURREGR
Survey of Residential Alterations and
    Repairs (SORAR), 11
Survival analysis, 130, 178
    contigency table approach, 132
    discrete proportional hazards model,
        136
    hazard function, 131
    Kaplan–Meier product limit estimator, 134
    survival function, 131
Survival function, 131

Target populations, 4
Tax Record Access Program, 86
Telescoping, 14, 28, 300
    definition, 257
Time-in-sample bias:
    definition, 16
    differential response probabilities, 20
    potential reasons, 18–20

Time-in-sample effects, Consumer
    Expenditure Interview Survey, 296
Time-in-sample error, 4
Time series, 460, 480
Tracing, 52
    case study, 63
    census-related studies, 66
    choosing, 57
    confidentiality, 55–56, 77
    environment, 53
    forward, 53
    information, 55, 59–60
    intersurvey period, 54
    nature of survey, 57
    not-traced cases, 63
    planning costs, 61
    retrospective, 53
    reverse, 54
    Reverse Record Check, 67
    techniques, 59
    verification, 62
Tracking, 43. *See also* Tracing
Transition probabiltiies, 580
Treatment, 577
    assignment rules, 515
    definition, 512
    impact on outcomes, 512
    nonrandom samples, 530
    outcome function, 514
    time curve of response, 578

Unit nonresponse, 348, 549

causal modeling of, 389
Univariate imputation, 407

Variance component estimation, 127
Verification, 62

Wave nonresponse, 348
    combined weighting and imputation, 367
    compensation strategy selection, 369-373
    flexibility, 371
    forms of analysis, 349
    frequency distribution, 350
    imputation, see Imputation
    patterns, 350
    practicality, 371
    precision, 372
    preservation of relationships, 372
    quality, 372
    types of data, 349
    weighting, 352
    weighting *vs.* imputation, 364
Weighting:
    adjustments, 9-10, 395
    combined with imputation, 367
    comparison with imputation, 364
    compensating for wave nonresponse, 352
    individual behavior and, 539
    multiplicity, 583
    response propensity, 395
Weights, 579
Weights, model misspecification, 580